SELECTED LOGIC PAPERS

World Scientific Series in 20th Century Mathematics

Published

Vol. 1 The Neumann Compendium
edited by F. Bródy and T. Vámos

Vol. 2 40 Years in Mathematical Physics
by L. D. Faddeev

Vol. 3 Selected Papers of Yu I Manin
by Y. Manin

Vol. 4 A Mathematician and His Mathematical Work
— Selected Papers of S. S. Chern
edited by S. Y. Cheng, P. Li and G. Tian

Vol. 5 Fields Medallists' Lectures
edited by Sir Michael Atiyah and Daniel Iagolnitzer

Vol. 6 Selected Logic Papers
by G. E. Sacks

World Scientific Series in 20th Century Mathematics – Vol. 6

SELECTED LOGIC PAPERS

Gerald E. Sacks
Department of Mathematics
Harvard University
and
Massachusetts Institute of Technology
USA

World Scientific
Singapore • New Jersey • London • Hong Kong

Published by

World Scientific Publishing Co. Pte. Ltd.

P O Box 128, Farrer Road, Singapore 912805

USA office: Suite 1B, 1060 Main Street, River Edge, NJ 07661

UK office: 57 Shelton Street, Covent Garden, London WC2H 9HE

Library of Congress Cataloging-in-Publication Data

Sacks, Gerald E.

Selected logic papers / Gerald E. Sacks.

p. cm. -- (World Scientific series in 20th century mathematics; vol. 6)

Includes bibliographical references.

ISBN 9810232675 (alk. paper)

1. Logic, symbolic and mathematical. I. Title. II. Series.

QA9.2.S23 1999

511.3--dc21 99-22812

CIP

British Library Cataloguing-in-Publication Data

A catalogue record for this book is available from the British Library.

The author and publisher would like to thank the following publishers of the various journals and books for their assistance and permission to include the selected reprints found in this volume:

Academic Press
American Mathematical Society
Association for Symbolic Logic
Elsevier Science Publishers
Gauthier-Villars
Johns Hopkins University Press
Polish Academy of Sciences
Princeton University Press

While every effort has been made to contact the publishers of reprinted papers prior to publication, we have not been successful in some cases. Where we could not contact the publishers, we have acknowledged the source of the material. Proper credit will be accorded to these publishers in future editions of this work after permission is granted.

Printed in Singapore by Uto-Print.

In Memory Of My Mother

Ethel Sacks

Contents

General Introduction

Nothing is forgotten, fortunately or not.

-H. M.

Looking backward I see a problem solver more than a system builder and still do. I always felt more comfortable working on someone else's problem. Avoidance of some sort I suppose. Two exceptions come to mind immediately, splitting an re set into incomparable re sets [1] and solving Post's problem invariantly. The first was my initial foray into re sets and I am still working on the second. The problem with problem solving is being unable to let go of an intractable problem. This did not happen to me, fortunately, until fairly late in the game, about fifteen years ago I think.

The papers I selected for reprinting in this volume are more interesting to me than the ones I left out. I do not know if there is a common theme. For me logic is about definability, but that could be a recursion theorist talking. On the other hand the principal methods of recursion theory, priority arguments and forcing, do apply in other areas of mathematical logic.[a]

[a]T. Slaman put it as follows. "There is a particular perspective obtained by consistently including an understanding of 'definability' in the solution to a mathematical problem. It is the assumption of that vantage point which I identify with recursion theory. Of course not only 'recursion theorists', but also computer scientists, descriptive set theorists and others have claim to it; many of us ascend but we do not all look in the same direction. We are all trying to understand the interaction between mathematical objects and the means needed to speak about them."

On the degrees less than $0'$ [1] appeared in 1963. The paper presents two results from my Ph.D. thesis (Cornell 1961) directed by J. B. Rosser. The first is a splitting theorem for re sets with several corollaries. My favorite then and now is: each degree strictly between 0 and $0'$ is incomparable with some re degree. The second is the existence of a minimal degree below $0'$. I was intrigued by the idea of using the priority method to construct a non-re degree, an approach taught to me by Shoenfield [r1]. The paper is devoid of intuitive remarks save for the final paragraph. It is tempting to blame the lack on Kleene's influence, but the truth is that in 1961 I greatly distrusted intuitive remarks and regarded them as intellectually corrupting. These days formal reasoning seems too weak to stand alone, like a skeleton without a body.

Recursive enumerability and the jump operator [2] was written in the fall of 1961. The main result, a determination of the range of the jump restricted to re degrees, was inspired by Friedberg's determination of the range of the jump [r2], and by Shoenfield [r1] on the jump restricted to degrees below $0'$. This paper in harmony with its predecessor was free of intuitive remarks. The search for the proof led me to a simple instance of the infinite injury method. Phrases such as "finite injury" or "infinite injury" did not occur.

The recursively enumerable degrees are dense [3] needed something more than infinite injury, an insight that was painfully long in coming. Density says that if B and C are re sets such that $B <_T C$, then there is an re D such that $B <_T D <_T C$. Infinite injury readily leads to a D such that $B \leq_T D <_T C$. The real stumbling block was how to arrange $B <_T D$. It was overcome by a planting strategy. Potentially infinite initial segments of C are planted in D to insure $D \not\leq_T B$, a strategy that succeeds because $C \not\leq_T B$. The paper was submitted without any intuitive remarks. The referee objected so violently I added several. Far from enough but at least the phrase, "planting", was added.

A simple set which is not effectively simple [4] owes its existence to my desire to use Kleene's fixed point theorem in a proof, preferably a priority argument. The fundamental, yet nonintuitive aspect of the fixed point theorem appealed to me. In this paper there are less equations and more words. I was beginning to come under the beneficial influence of Post and Friedberg.

Metarecursive sets [5] (with G. Kreisel) initiated a big change in my thinking about re sets. I have written elsewhere about Kreisel's powerful and relentless influence. One point worth repeating is: I fought him off as long as I could, but have never regretted my final surrender. Together we found a way to make sense out of infinitely long convergent computations in a setting that allowed priority arguments. Infinitely long computations were tractable as long as they retained some of the combinatoric properties of finite computations essential to ordinary recursion theory. We delighted in the term, "metafinite". I began to say good-bye to finite computations.

Post's problem, admissible ordinals, and regularity [6] continued the study of metarecursion and pushed it upward to Kripke–Platek admissible ordinals. Most of the questions in this paper were raised by Kreisel. Some of the answers given are as follows. Let α be a Σ_1 admissible ordinal. There exists a non-α-recursive, α-re, subgeneric set. (A subgeneric means $\langle L(\alpha), A\rangle$ is Σ_1 admissible.) There exists a countable α such that no α-re set is maximal. (A complete understanding of maximal α-re sets was found by Lerman [r3] and Simpson.) For every α there is an intermediate α-re degree. (A better solution to Post's problem was offered in [12].) In retrospect the result I am most fond of has to do with regularity. A subset A of α is *regular* if $(A \cap \beta) \in L(\alpha)$ for every $\beta < \alpha$. Regularity is the same as Jensen's notion of amenability. In other words the intersection of A with any α-finite set is α-finite. Each α-re set has the same α-degree as some regular α-re set by an argument that goes back to Dekker's deficiency set construction in classical recursion theory.

On a theorem of Lachlan and Martin [7] reproves their result that there exists an re degree $\mathbf{d}$ such that

$$\forall\, n[\mathbf{0}^{(n)} < \mathbf{d}^{(n)} < \mathbf{0}^{(n+1)}].$$

Lachlan's proof combined a priority argument with Kleene's fixed point theorem. My reproof was an application of the fixed point theorem to a uniformity implicit in the proof of the jump operator result in [2]. There is also a twist in the use of the fixed point theorem. Let t be a total recursive function, A a set, and $\{R_e^A \mid e \geq 0\}$ a standard enumeration of all sets re in A. According to Kleene there is a fixed point c such that

$$R_{t(c)}^A = R_c^A.$$

The twist, which enlivened the paper, is that c is independent of A.

A minimal hyperdegree [8] (with R. O. Gandy) was an application of perfect forcing to hyperdegrees. Gandy and I worked independently on the problem of finding a minimal hyperdegree, and after conferring with him at his cottage in Wales, I wrote it up in a style more ponderous than I would choose now. This was my first excursion into class forcing over a countable structure with severely restricted closure properties, a setting where considerations of effectiveness matter more than they do in models of set theory. The experience gained here was helpful in the 1-*section* [13] and *k-section* [16] papers.

Measure-theoretic uniformity in recursion theory and set theory [9] started with an interest in the question: what is the measure of $\{X \mid \omega_1^X = \omega_1^{CK}\}$? The answer turned out to be 1, although every authority I consulted beforehand thought it should be 0. Along the way I noticed that a Π_1^1 set of positive measure had to have a hyperarithmetic member. The argument looked at definable aspects of adjoining an arbitrary real T to the hyperarithmetic hierarchy. Soon after I could not resist looking further at $L(\alpha, T)$ when $L(\alpha)$ is a countable model of ZF & $V = L$. Let $\mathcal{F}$ be a sentence of in the language of set theory with parameters from $L(\alpha, T)$. Define $p(\mathcal{F})$, the probability that $\mathcal{F}$ is true in $L(\alpha, T)$, to be the Lebesgue measure of

$$\{T \mid L(\alpha, T) \models \mathcal{F}\}.$$

The $L(\alpha)$-definability of $p(\mathcal{F})$ readily shows that $L(\alpha, T)$ is a model of ZFC & $V \neq L$ with probability 1. Thus Cohen's historic results were obtainable by familiar facts about the constructive nature of the measurability of Borel sets and prenex normal form. I later learned that the above naive measure-theoretic point of view was equivalent to Solovay's forcing with closed sets of positive measure. I spoke on $p(\mathcal{F})$ at a logic seminar at Stanford in 1965. The small audience had the highest density of distinguished logicians I ever faced. There sat Cohen, Kreisel, Solovay, Barwise, Feferman and Platek.

Recursion in objects of finite type [10] begins with an elaborate joke that went unnoticed when the paper was delivered in Nice (ICM) in 1970. My mistake was thinking that France is a Catholic country. The paper announced some results about 1-*sections* building on Kleene's work. A type n object F is *normal* according to Kleene if ${}^nE$, the equality predicate for objects of type less than n, is recursive in F. (The objects of type 0 are non-negative integers.) S_1F, the 1-section of F, is the set of all reals, i.e. objects of type 1, recursive in F. The main theorem says (i) is equivalent to (ii). Let $A_1(S_1F)$ be the transitive set whose elements are coded by the reals in S_1F.

(i) A is a countable Σ_1 admissible set that satisfies Σ_1 dependent choice.

(ii) There exists a normal type 2 F such that $A = A_1(S_1F)$.

The proof that (i) implies (ii) is a class forcing argument over a Σ_1 admissible set, a technique I learned from [8] and extended to E-closed, inadmissible structures in [21]. The corollary of most interest to me was: if F is normal of type > 2, then there is a normal G of type 2 such that $S_1F = S_1G$.

Forcing with perfect closed sets [11] looks at applications of perfect forcing to set theory, hyperarithmetic theory, and Turing degrees. My inspiration was Spector's construction of a minimal Turing degree via recursive

Cantor sets. I added the sequential (or fusion) lemma: let $\mathcal{F}_i$ $(i < \omega)$ be a sequence of sentences; then

$$\begin{aligned}\forall\, P(\exists\, Q \subseteq P)(\forall\, i < \omega)\,\exists\, R_0\,\exists\, R_1 \\ Q = R_0 \cup R_1, \\ R_1 \neq Q \longrightarrow R_0 \Vdash \mathcal{F}_i, \\ R_0 \neq Q \longrightarrow R_1 \Vdash (\neg \mathcal{F}_i).\end{aligned}$$

In set theory minimal sidewise extensions of countable initial segments of L were obtained. In the hyperarithmetic case there was a wrinkle. Forcing was defined to be identical with truth for all ranked sentences in order to make it immediate that the forcing relation, restricted to ranked sentences, was Π^1_1 in V, hence Σ_1 over $L(\omega_1^{CK})$.

In recursion theory a real X was found such that (a) $X^{(2)} \equiv_T 0^{(\omega)}$ and (b) $0^{(n)} \leq_T X$ for all n. (Putnam had shown earlier that (b) implies $0^{(\omega)} \leq_T X^{(2)}$.) The construction of X drew on the notion of *pointedness.* A perfect condition P is *recursively pointed* if

$$\forall\, X[X \in P \longrightarrow P \leq_T X].$$

(P ambiguously denotes the perfect condition P and its canonical real code.) Hyperarithmetic pointedness proved useful in the study of hyperdegrees as in [15] below.

The α-finite injury method [12] (with S. Simpson) lifts the Friedberg–Muchnik (FM) solution to Post's problem from ω to every Σ_1 admissible α by means of condensation (i.e. downward Skolem–Löwenheim) arguments in admissible $L(\alpha)$'s. Simpson and I focused on this problem only after Shore demolished an earlier approach based on ordinal arithmetic suggested by Stillwell. The FM solution works straightforwardly if $L(\alpha)$ satisfies Σ_2 replacement. As in [5] the FM solution is readily adapted when α^*, the Σ_1^α projectum of α, is less than α. The most interesting case is when $\alpha^* = \alpha$ and there is a largest α-cardinal in the sense of $L(\alpha)$, call it $\aleph$. Then $L(\alpha)$ is the limit of Σ_1 substructures (hence of α-stable ordinals), requirements can be arranged in blocks of length $\aleph$, and injuries can be bounded by α-stable ordinals. Later Lerman's notion of *tame* and Shore's notion of *block* clarified the dynamics of the construction.

A simple combinatoric lemma turned out to be useful. If β is an infinite regular α-cardinal, then β is regular with respect to Π_1^α functions. (A regular α-cardinal is by definition regular with respect to α-finite functions.)

The 1-section of a type n object [13] gave the details of the proof of the main result of [10]. I tried for a long time with a notable lack of success to stretch the argument to show the k-section $(k > 1)$ of a normal type

n object ($n > k$) was the k-section of a normal type $(k+1)$ object. The missing ingredient is described below [16].

In *Remarks against foundational activity* [14] I was out of place, an amateur among professionals, but I enjoyed speaking against foundational activity, dismissing as groundless the fears of contradiction in mathematics, and pointing out the dangers of excessive precision and clarity.

Countable admissible ordinals and hyperdegrees [15] began as the write-up of "each countable admissible ordinal greater than ω is of the form ω_1^X for some $X \subseteq \omega$", a result announced in 1967, but gradually evolved into an obsession with class forcing over a countable admissible set. I doubt if it was worth the effort expended to show: if α is a countable admissible $> \omega$, then there is an $X \subseteq \omega$ such that $\omega_1^X = \alpha$ and $\omega_1^Y < \alpha$ for all Y of lower hyperdegree than X.

The main result of the paper is: if A is a countable Σ_1 admissible set that satisfies Σ_1 dependent choice, the set of hyperdegrees of reals in A has a minimal upper bound. The notion of pointedness [11] was extended to the hyperarithmetic case. P, a perfect condition, is pointed if

$$\forall\, T[T \in P \longrightarrow P \leq_h T].$$

(P ambiguously denotes a condition and its canonical real code.) To cope with the reality that A may fall far short of resembling an $L(\alpha)$, a triple (so-called) forcing relation, $\Vdash\!\!\vdash^A$, was introduced. $\Vdash\!\!\vdash^A$combined with Σ_1 DC made it possible to carry out in A the sort of fusion argument with hyperarithmetically pointed P's more readily done in an admissible $L(\alpha)$.

I spent a lot of time trying to eliminate the use of Σ_1 DC above until F. Abramson showed it was impossible to do so. He constructed a countable admissible A such that no G generic over A is a minimal upper bound for the hyperdegrees contained in A. He assumed that the forcing relation underlying G, restricted to Σ_1 sentences, was Σ_1 over A.

The k-section of a type n object [16] completed the proof of the Plus-One theorem: if $n > k > 1$ and F is a normal type n object, then there is a normal type $(k+1)$ object G whose *k-section* equals that of F. The ingredient absent from [13] was reflection. I noticed one day, I forget why, that Grilliot selection for sets of reals *re* in ${}^3E$ led to a simple reflection principle in the setting of recursion in normal objects that stretched the forcing argument of [13] from $k = 1$ to $k > 1$. I asked Harrington to extend and apply reflection further in the above setting in his thesis. His success inspired much of my work on E-closed initial segments of L [20, 21, 22].

Effective bounds on Morley rank [17] attempted to pin down the absolute nature of Morley rank with the help of hyperarithmetic theory and model theory on admissible sets. I had long been intrigued by the thought that the central notions of model theory were absolute. A typical result was as follows. Let $\mathcal{A}$ be a countable substructure of a model of a complete, substructure complete, countable theory T, and $p \in S_1\mathcal{A}$. If p has a Morley rank, then p is hyperarithmetic in $\mathcal{A}$ and its Morley rank is an ordinal recursive in $\mathcal{A}$. (Earlier Lachlan had shown that Morley rank is always countable.) It follows that the Morley rank of a totally transcendental theory T is an ordinal recursive in T. The impetus for this paper goes back to Kreisel's effective analysis of the Cantor–Bendixson theorem [r5]. Deeper absoluteness results on the Morley derivative were obtained by Calmer [r6].

Post's problem, absoluteness and recursion in finite types [18] worked directly with Kleene's theory of recursion in finite types and tried to reach some definite conclusions about Post's problem. In retrospect it served as a warm-up for [20] and [21]. The main result in the case of ${}^3E$ was: the class

$$\{R \mid R \subseteq 2^\omega \wedge \exists b \in 2^\omega (R \textit{ recursive in } {}^3E, b)\}$$

is not *re* in ${}^3E$. It was proved by a forcing argument that relied on a regularity assumption introduced by Normann: there exists a recursively regular well-ordering $\prec$ of 2^ω such that $\prec$ is recursive in ${}^3E$; *recursively regular* means there is no function f recursive in ${}^3E, a$ (for some $a \subseteq \omega$) such that $dom(f)$ is a bounded initial segment of $\prec$ and $range(f)$ is unbounded in $\prec$.

Two projecta related to ${}^3E$, η and ρ, were defined and proved equal with the aid of Normann's regularity assumption. In a footnote a claim was made that a positive solution of an absolute version of Post's problem for ${}^3E$ followed from $\eta = \rho$. The proof of the claim is extended in [20] to E-closed initial segments of L, where η and ρ reappear, and without any regularity assumption lead to a positive solution to Post's problem.

On the number of countable models [19] was my first foray into the wilderness of Vaught's conjecture. I have been planning to rewrite, clarify and extend this paper for some time. The main result slightly updated is: if $r(M) \leq \omega_1^M$ for every M a model of T, then $r(T) < \omega_2^T$. A brief explanation may help.

T is a countable theory. M is a countable model of T. $r(M)$ is the Morley rank of M, the ordinal height of $\mathcal{L}_M$, a canonical fragment of $\mathcal{L}_{\omega_1,\omega}$. M is the atomic model of $Th(M)$, the complete theory of M in $\mathcal{L}_M$. $r(M)$ is at most $\omega_1^M + 1$. The differences between

$$(1)\ r(M) = \omega_1^M \quad \text{and} \quad (2)\ r(M) = \omega_1^M + 1$$

are noteworthy. (1) implies M is the atomic model of its complete theory T_M in $\mathcal{L}_{\omega_1,\omega}$ restricted to $L(\omega_1^M, M)$, but does **not** imply T_M is ω-categorical. (2) means that M realizes a non-principal type of T_M.

$r(T)$ is the least ordinal where a witness to the truth of Vaught's conjecture is constructible from T. The witness has two forms: a perfect kernel of models of T, or a countable family of Scott sentences that characterize all countable models of T.

ω_2^T is the least γ such that $L(\gamma, T)$ satisfies Σ_2 replacement.

A similar result holds for $PC_{\omega_1,\omega}$ classes.

Post's problem in E-recursion [20] completes the attempt on Post's problem begun in [18]. Suppose $L(\kappa)$ is E-closed: if f is a partial E-recursive function, $b \in L(\kappa)$, and $f(b)\downarrow$, then $f(b) \in L(\kappa)$. The partial E-recursive functions, discovered by Normann [r7], and independently and subsequently by Moschovakis, are a set-theoretic version of Kleene recursion in normal objects of finite type. The main result of the paper is a positive solution to Post's problem: there exist two subsets of $L(\kappa)$, each E-recursively enumerable on $L(\kappa)$, such that neither is E-reducible to the other. The proof has some curious features. None of the Friedberg–Muchnik requirements is injured. Each requirement is attempted at most once. A form of reflection is key to the argument, in particular a Harringtonesque connection between reflection and Moschovakis divergence witnesses. The presence of divergence witnesses in $L(\kappa)$ makes it possible to avoid injuries.

The construction is guided by two projecta, η and ρ, seen to be equal by a fine structure argument. One proposition, 3.2, is stated and proved in a way that to this day I do not understand. The sole purpose of 3.2 is to assist in the proof of 3.3, a theorem whose statement seems intelligible.

The limits of E-recursive enumerability [21] continues a theme of [18]: certain naturally enumerable sets are not "recursively enumerable". [21] assumes $L(\kappa)$ is E-closed but not Σ_1 admissible and answers the following question. Let $gc(\kappa)$ be the greatest cardinal in the sense of $L(\kappa)$.

$$\text{Is } 2^{gc(\kappa)} \cap L(\kappa) \text{ } E\text{-recursively enumerable in some } a \in L(\kappa)?$$

The answer breaks into 3 cases.

(1) if $L(\kappa) \Vdash [gc(\kappa) \text{ is regular}]$, then **no**.

(2) if $L(\kappa) \Vdash [\omega < cofinality(gc(\kappa)) < gc(\kappa)]$, then **yes**.

(3) if $L(\kappa) \Vdash [\omega = cofinality(gc(\kappa))]$, then **no** if and only if there exists an amenable subset of $gc(\kappa)$ not in $L(\kappa)$.

Note that κ is not assumed to be countable.

Case (1) extends the forcing argument of [18] with the help of a Jensenesque fine structure fact: if $V = L$ and λ is a singular limit *ordinal,* then there exists an amenable singular limit sequence through λ.

Case (2) is an application of combinatoric and fine structural ideas of Silver [r8] and S. Friedman [r9]. Friedman showed that the ω_{ω_1}-degrees above $0'$ are well-ordered if $V = L$. His approach shows there is a $d \in L(\kappa)$ such that $O(x)$ is E-recursive in x, d (uniformly) for all $x \in 2^{gc(\kappa)} \cap L(\kappa)$. ($O(x)$ is the least ordinal by which x is constructible.) The computability of $O(x)$ yields the E-recursive enumerability of $2^{gc(x)} \cap L(\kappa)$.

Case (3) is an application of infinitary model theory. The principal inspiration is Green's compactness theorem [r10]: $R(\omega+\omega)^+$, the least Σ_1 admissible set with $R(\omega+\omega)$ as a member, is Σ_1 compact.

Inadmissible forcing [22] (with T. Slaman) applies pointed perfect forcing [15] to solve a problem in E-recursion. The result is: suppose $L(\kappa)$ is countable, E-closed, and not Σ_1 admissible; assume

$$L(\kappa) \models [cofinality(gc(\kappa)) > \omega];$$

($gc(\kappa)$ is the greatest cardinal in the sense of $L(\kappa)$); then there exists a $T \subseteq gc(\kappa)$ such that $L(\kappa, T) = E(T)$, the least E-closed set with T as a member.

A forcing condition p is a suitable uncountable tree in $L(\kappa)$. p is E-*pointed* in the sense that the standard code for p is E-recursive in any path through p. Fusion and reflection combine to show $L(\kappa, T)$ is E-closed for generic T.

Effective versus proper forcing [23] tries to characterize those notions of set forcing that preserve E-closure. Recall: if A is a Σ_1 admissible set, $\mathcal{P} \in A$ is a forcing relation, and G is $\mathcal{P}$-generic over A, then $A(G)$ is Σ_1 admissible. The situation is more complex for an E-closed $L(\kappa)$. Consider $E(\omega_1)$, the E-closure of ω_1. $E(\omega_1) = L(\kappa)$ for some κ. If G collapses ω_1 to ω, then $L(\kappa, G)$ is not E-closed.

The main result says that $\mathcal{P} \in E(\omega_1)$ preserves E-closure iff $\mathcal{P}$ is locally proper, a variant of Shelah's notion of proper. Let $L(\kappa)$ be E-closed. $I \in L(\kappa)$ is a *co-re substructure* of $L(\kappa)$ if I is E-closed and for all $m, a \in I$ and $e < \omega$,

$$\exists x[x \in m \wedge \{e\}(x,a)\uparrow] \longrightarrow \exists x[x \in (m \cap I) \wedge \{e\}(x,a)\uparrow].$$

I is *pointed* if $\exists b[b \in I \wedge \kappa_r^b \notin I]$. κ_r^b is the greatest reflecting ordinal relative to b.

$\mathcal{P}$ is *locally proper* if for all $p \in \mathcal{P}$ and $a \in L(\kappa)$, there is a countable pointed co-re substructure I such that $p, a \in I$ and

$$\exists q[q \in \mathcal{P} \wedge p \geq q \wedge q \text{ generic over } I].$$

The last means: if D is a dense set of conditions, suitably defined over I, then $q \subseteq \cup(D \cap I)$. The genericity of q implies p cannot force $\{e\}(a)$ to converge via a computation not in I.

[23] ends with a sentence that fits here:

Farewell to higher recursion theory, but not to recursion theory; there is no way to say good-bye to recursion theory.

Cambridge, Massachusetts
February 1999

References

[1] to [23] refer to the list of Selected Logic Papers.

[r1] Shoenfield J. R., *On degrees of unsolvability*, *Ann. Math.* **69** (1959) 644–655.

[r2] Friedberg R., *A criterion for completeness of degrees of unsolvability*, *J. Symb. Logic* **22** (1957) 159–160.

[r3] Lerman M., *Maximal α – re sets*, *Trans. Amer. Math. Soc.* **260** (1980) 363–378.

[r4] Abramson F., *Sacks forcing does not always produce a minimal upper bound*, *Adv. Math.* **31** (1979) 110–130.

[r5] Kreisel G., *Set theoretic methods suggested by the notion of potential totality*, in *Infinitistic Methods* (Pergamon, 1961), pp. 325–369.

[r6] Calmer C., Ph.D. Thesis, Stanford University (1978).

[r7] Normann D., *Set recursion*, in J. E. Fenstad *et al.* eds., *Generalized Recursion Theory II* (North-Holland, 1978), pp. 303–320.

[r8] Silver J., *On the singular cardinals problem*, in *Proc. Int. Cong. Math.* Vancouver, 1974.

[r9] Friedman S., *Uncountable admissibles II: compactness*, *Israel J. Math.* **40** (1984) 129–149.

[r10] Green J., Σ_1 *compactness for next admissible sets*, *J. Symb. Logic* **39** (1974) 105–116.

SELECTED LOGIC PAPERS

ANNALS OF MATHEMATICS
Vol. 77, No. 2, March, 1963

ON THE DEGREES LESS THAN 0′

BY GERALD E. SACKS*

(Received October 2, 1961)

By *degree*, we mean degree of recursive unsolvability as defined by Kleene and Post in [6]. Two of the principal results of this paper are:

(1) there exists a degree $\mathbf{h}$ such that $\mathbf{0} < \mathbf{h} < \mathbf{0}'$, and such that $\mathbf{0} < \mathbf{d} < \mathbf{h}$ for no degree $\mathbf{d}$;

(2) if $\mathbf{c}$ is a non-zero degree less than $\mathbf{0}'$, then there exists a degree $\mathbf{b}$ which is the degree of a recursively enumerable set, and which is incomparable with $\mathbf{c}$.

The first result answers in the affirmative a question raised by Shoenfield in [10]: "Does there exist a minimal degree $\mathbf{a}$ such that $\mathbf{a} \leq \mathbf{0}'$?" The second result has as a consequence: a set is recursive if and only if it is recursive in every non-recursive, recursively enumerable set.

Our notation, unless otherwise specified, is that of [4], [6] and [11]; we assume a considerable amount of familiarity with [4] and [6]. If e is a natural number and $h(n)$ is a function, then we employ $\{e\}^h(n)$ to denote the partial function

$$U(\mu z T_1^1(\tilde{h}(z), e, n, z)) .$$

This use of $\{e\}^h(n)$ is consistent with the definition of it given by Kleene in [4]. Recall that a non-empty set D is recursively enumerable in a set B if there is a function recursive in B whose range is D.

THEOREM 1. *Let B, C and D be sets such that C is recursive in D, D is recursively enumerable in B, and C is not recursive in B. Then there exist disjoint sets D_0 and D_1 such that $D = D_0 \cup D_1$, and such that for each $i < 2$, D_i is recursively enumerable in B, and C is not recursive in B, D_i.*

PROOF. Let $B^* = \{2n \mid n \in B\}$ and $D^* = \{2n + 1 \mid n \in D\}$. Then C is recursive in D^*, D^* is recursively enumerable in B^*, and C is not recursive in B^*. Suppose there are disjoint sets D_0^* and D_1^* such that $D_0^* \cup D_1^* = D^*$, and such that for each $i < 2$, D_i^* is recursively enumerable in B^*, and C is not recursive in B^*, D_i^*. Then the disjoint sets $D_0 = \{n \mid 2n + 1 \in D_0^*\}$ and $D_1 = \{n \mid 2n + 1 \in D_1^*\}$ are such that $D_0 \cup D_1 = D$, and such that for each $i < 2$, D_i is recursively enumerable in B, and C is not recursive in B, D_i. Thus we may assume without any loss of generality that every member of B is even, and every member of D is odd.

* The author is a National Science Foundation Postdoctoral Fellow.

Let $b(n)$ be the representing function of B: $n \in B \rightarrow b(n) = 0$ and $n \notin B \rightarrow b(n) = 1$. Let $c(n)$ and $d(n)$ be the representing functions of C and D respectively. Then $b(n) \cdot d(n)$ is the representing functions of $B \cup D$. Since there exists a set, namely B, in which D is not recursive, it must be that D is infinite. Let $f(n)$ be a function recursive in B which enumerates D without repetitions:

$$D = \{f(n) \mid n \geqq 0\} \quad \text{and} \quad (n, m)(n \neq m \rightarrow f(n) \neq f(m)) .$$

Let $d(s, n) = 0$ if there is a $k \leqq s$ such that $f(k) = n$, and let $d(s, n) = 1$ otherwise. It is clear that the function $d(s, n)$ is recursive in B, and that for each n, $\lim_s d(s, n)$ exists and is equal to $d(n)$. Since C is recursive in D, there exists a Gödel number g with the property that $c(n) = \{g\}^a(n)$ for all n. We make use of g to define a function $c(s, n)$ which is recursive in B:

$$c(s, n) = \begin{cases} U(\mu y T_1^1(\prod_{j<y} p_j^{d(s,j)}, g, n, y)) & \text{if } (Ey)_{y \leq s} T_1^1(\prod_{j<y} p_j^{d(s,j)}, g, n, y) \\ s + 1 & \text{otherwise.} \end{cases}$$

Since $\lim_s d(s, n) = d(n)$ for all n, and since $c(n) = \{g\}^a(n)$ for all n, it follows that $\lim_s c(s, n)$ exists and is equal to $c(n)$ for all n.

A very useful property of the Gödel numbering devised by Kleene in [4] to arithmetize his formalism for recursive functions is: the Gödel number of a deduction d is greater than the integer denoted by any formal numeral occurring in d.[1] This property was utilized by Friedberg in [1]. It follows from GND that $c(s, n) = s + 1$ whenever $n \geqq s$.

It is our intention to define twelve functions simultaneously by induction: $d_i(s, n)$, $t_i(s)$, $y_i(s, n, e)$, $P_i(s, n, e)$, $m_i(s, e)$ and $K_i(s, e)$ $(i = 0, 1)$. Each of these functions will be recursive in B. At stage s $(s \geqq 0)$ of the induction, we will define $d_i(s, n)$, $t_i(s)$, $y_i(s, n, e)$, $P_i(s, n, e)$, $m_i(s, e)$ and $K_i(s, e)$ for all $i < 2$, all n and all e. The sole purpose of stage s is to put the natural number $f(s)$ (the sth member of D) in D_0 or in D_1, but not in both. Thus at stage s, we set either $d_0(s, f(s))$ or $d_1(s, f(s))$ equal to 0, but not both.

Stage $s = 0$. We set $d_1(0, n) = 1$ for all $n \neq f(0)$, and we set $d_1(0, f(0)) = 0$. We set $P_i(0, n, e) = 2$, $d_0(0, n) = t_i(0) = y_i(0, n, e) = 1$ and $m_i(0, e) = K_i(0, e) = 0$ for all $i < 2$, all e and all n.

Stage $s > 0$. For each $i < 2$, let

$$t_i(s) = \begin{cases} \mu e(f(s) < K_i(s-1, e)) & \text{if } (Ee)_{e \leq s}(f(s) < K_i(s-1, e)) \\ s + 1 & \text{otherwise.} \end{cases}$$

[1] We will denote this fact by GND.

Let $z(s) = 1$ if $t_1(s) \geqq t_0(s)$, and let $z(s) = 0$ otherwise. Thus $t_{z(s)}(s) \geqq t_{1-z(s)}(s)$. We set $d_{z(s)}(s, n) = d_{z(s)}(s-1, n)$ for all $n \neq f(s)$, and we set $d_{z(s)}(s, f(s)) = 0$. We set $d_{1-z(s)}(s, n) = d_{1-z(s)}(s-1, n)$ for all n.

We now define $y_i(s, n, e)$ and $P_i(s, n, e)$ for all $i < 2$, all n and all e:

$$y_i(s,n,e) = \begin{cases} \mu y T_1^1\left(\prod_{j<y} p_j^{b(j)\cdot d_i(s,j)}, e, n, y\right) & \text{if } (Ey)_{y \leqq s} T_1^1\left(\prod_{j<y} p_j^{b(j)\cdot d_i(s,j)}, e, n, y\right) \\ s+1 & \text{otherwise;} \end{cases}$$

$$P_i(s,n,e) = \begin{cases} U\big(y_i(s, n, e)\big) & \text{if } y_i(s, n, e) \leqq s \\ s+2 & \text{otherwise.} \end{cases}$$

Before we define $m_i(s, e)$, we note that

$$(i)_{i<2}(e)(Et)_{t \leqq s}\big(c(s, t) \neq P_i(s, t, e)\big) .$$

This last is clear, since it is immediate from GND that $c(s, s) = s + 1$, and that $P_0(s, s, e) = P_1(s, s, e) = s + 2$ for all e. We complete stage $s > 0$ by defining $m_i(s, e)$ and $K_i(s, e)$ for all $i < 2$ and all e:

$$m_i(s, e) = \mu t\big(c(s, t) \neq P_i(s, t, e)\big) ;$$

$$K_i(s, e) = \begin{cases} K_i(s-1, e) & \text{if } (n)_{n<m_i(s,e)}\big(y_i(s, n, e) \leqq K_i(s-1, e)\big) \\ \mu t (n)_{n<m_i(s,e)}\big(y_i(s, n, e) \leqq t\big) & \text{otherwise.} \end{cases}$$

We list two remarks which will be needed in Lemmas 1, 3 and 5:

(R1) $(i)_{i<2}(s)(e)\big(K_i(s+1, e) \geqq K_i(s, e)\big)$;

(R2) $(i)_{i<2}(s)(e)(n)_{n<m_i(s,e)}\big(y_i(s, n, e) \leqq K_i(s, e)\big)$.

It can easily be verified by the methods of [6] that each of the twelve functions we have defined is recursive in B. Such a verification is possible because of two facts: each of the functions $f(s)$, $b(s)$ and $c(s, n)$ is recursive in B; only bounded quantifiers occur in the case hypotheses of the definitions of the functions $t_i(s)$, $y_i(s, n, e)$ and $K_i(s, e)$ $(i = 0, 1)$.

It is clear that

$$(i)_{i<2}(s)(n)\big(0 \leqq d_i(s+1, n) \leqq d_i(s, n) \leqq 1\big) .$$

It follows that for each $i < 2$ and each n, $\lim_s d_i(s, n)$ exists and is less than 2. For each $i < 2$, let $d_i(n) = \lim_s d_i(s, n)$, and let D_i be the set whose representing function is $d_i(n)$. Then $D_0 \cap D_1 = \varnothing$, and $D = D_0 \cup D_1$, since at stage s, $f(s)$ was put in D_0 or in D_1, but not in both.

We now show by means of a *reductio ad absurdum* argument that, for each $i < 2$, C is not recursive in B, D_i. Suppose then that C is recursive in B, $D_{i'}$, where $i' < 2$. It is our intention to prove that C is recursive in B, contrary to the hypothesis of our theorem. Recall that B contains only even numbers, and D only odd. This means that B and $D_{i'}$

are each recursive in $B \cup D_{i'}$, since

$$n \in B \equiv (n \text{ is even } \& \ n \in B \cup D_{i'})$$
$$n \in D_{i'} \equiv (n \text{ is odd } \ \& \ n \in B \cup D_{i'}) .$$

The representing function of $B \cup D_{i'}$ is $b(n)\cdot d_{i'}(n)$. It follows that there is a Gödel number e' such that $c(n) = \{e'\}^{b\cdot d_{i'}}(n)$ for all n.

LEMMA 1. *The set* $\{K_{i'}(s, e') \mid s \geqq 0\}$ *is infinite.*

PROOF. Suppose, for the sake of a *reductio ad absurdum*, that $\{K_{i'}(s, e') \mid s \geqq 0\}$ is finite; let m be its largest member. Let s' be such that $c(s, n) = c(n)$ for all s and n such that $s \geqq s'$ and $n \leqq m$. Let $y(n)$ denote the partial function

$$\mu y T_1^1(\textstyle\prod_{j<y} p_j^{b(j)\cdot d_{i'}(j)}, e', n, y) .$$

Since $c(n) = \{e'\}^{b\cdot d_{i'}}(n)$ for all n, $y(n)$ is defined and $c(n) = U(y(n))$ for all n. Let y' be the largest member of $\{y(n) \mid n \leqq m\}$. Let s'' be such that $s'' \geqq s'$, $s'' \geqq y'$ and $d_{i'}(s, n) = d_{i'}(n)$ for all n and s such that $n \leqq y'$ and $s \geqq s''$. It follows that $y(n) = y_{i'}(s'', n, e')$ for all $n \leqq m$. Then $P_{i'}(s'', n, e') = U(y(n)) = c(n) = c(s'', n)$ for all $n \leqq m$, since $y(n) \leqq y' \leqq s''$ for all $n \leqq m$, and since $s'' \geqq s'$. This means that $m_{i'}(s'', e') > m$. But then by remark (R2), $y_{i'}(s'', m, e') \leqq K_{i'}(s'', e') \leqq m$. This last is absurd, since by GND,[1] we have $m < y(m) = y_{i'}(s'', m, e')$.

Let e^* be the least e such that $\{K_i(s, e) \mid i < 2 \ \& \ s \geqq 0\}$ is infinite; by Lemma 1, $e^* \leqq e'$. Let i^* be the least i such that $\{K_i(s, e^*) \mid s \geqq 0\}$ is infinite.

LEMMA 2. *There is an* s' *such that for all* $s \geqq s'$, $z(s) = 1 - i^*$ *or* $e^* < t_{i^*}(s)$.

PROOF. Suppose, for the sake of a *reductio ad absurdum*, that there are infinitely many s with the property that $z(s) = i^*$ and $e^* \geqq t_{i^*}(s)$. Let S be an infinite set such that for each $s \in S$, $s \geqq e^*$, $z(s) = i^*$ and $e^* \geqq t_{i^*}(s)$. Recall that $t_{z(s)}(s) \geqq t_{1-z(s)}(s)$. Then for each $s \in S$, $e^* \geqq t_{i^*}(s) \geqq t_{1-i^*}(s)$. As S is infinite, there must be an infinite subset of S, call it T, such that for each r and $s \in T$, $t_{1-i^*}(r) = t_{1-i^*}(s)$.

Thus there is an e^{**} such that for each $s \in T$, we have $s \geqq e^* \geqq e^{**} = t_{1-i^*}(s)$. It follows from the definition of the function $t_{1-i^*}(s)$ that, for each $s \in T$, $f(s) < K_{1-i^*}(s-1, e^{**})$. Since T is infinite, and since the function $f(s)$ enumerates D without repetitions, it must be that $\{K_{1-i^*}(s, e^{**}) \mid s \geqq 0\}$ is infinite. It follows from the definition of e^* that $e^* = e^{**}$; thus $\{K_{1-i^*}(s, e^*) \mid s \geqq 0\}$ is infinite. This means that $1 - i^* > i^* = 0$. But then for any $s \in T$, we have $e^* = t_{i^*}(s) = t_{1-i^*}(s) = e^{**}$ and $z(s) = i^* = 0$. This last is impossible because $z(s) = 1$ if $t_1(s) = t_0(s)$ and $s > 0$.

Let s^* be the least s' such that $s' \geqq e^*$, and such that for all $s \geqq s'$, $z(s) = 1 - i^*$ or $e^* < t_{i^*}(s)$.

LEMMA 3. *If* $s \geqq s^*$ *and if* $n < m_{i^*}(s, e^*)$, *then* $d_{i^*}(s', j) = d_{i^*}(s, j)$ *for all* j *and* s' *such that* $j < y_{i^*}(s, n, e^*)$ *and* $s' \geqq s$.

PROOF. Let s and n be such that $s \geqq s^*$ and $n < m_{i^*}(s, e^*)$. We prove the lemma by induction on $s' \geqq s$. Suppose $s' \geqq s$ and $d_{i^*}(s', j) = d_{i^*}(s, j)$ for all $j < y_{i^*}(s, n, e^*)$. Since $s' \geqq s^*$, we know that

(a) $z(s' + 1) = 1 - i^*$, or

(b) $e^* < t_{i^*}(s' + 1)$

is the case. If (a) holds, then $d_{i^*}(s' + 1, j) = d_{i^*}(s', j)$ for all j. Suppose now that (b) is true. Note that $e^* \leqq s^*$. Then $f(s' + 1) \geqq K_{i^*}(s', e^*)$, since otherwise (b) is false. By remark (R1), $K_{i^*}(s', e^*) \geqq K_{i^*}(s, e^*)$, since $s' \geqq s$. By remark (R2), $K_{i^*}(s, e^*) \geqq y_{i^*}(s, n, e^*)$, since $n < m_{i^*}(s, e^*)$. But then $d_{i^*}(s' + 1, j) = d_{i^*}(s', j)$ for all $j < y_{i^*}(s, n, e^*) \leqq f(s' + 1)$.

LEMMA 4. *If* $s \geqq s^*$ *and if* $n < m_{i^*}(s, e^*)$, *then*

$$y_{i^*}(s, n, e^*) = y_{i^*}(s', n, e^*) \leqq s \qquad \textit{for all } s' \geqq s .$$

PROOF. Let s and n be such that $s \geqq s^*$ and $n < m_{i^*}(s, e^*)$. Let $s' \geqq s$. Since $n < m_{i^*}(s, e^*)$, it must be that $c(s, n) = P_{i^*}(s, n, e^*)$. Since (by GND) $c(s, n) \leqq s + 1$, we have $P_{i^*}(s, n, e^*) = U(y_{i^*}(s, n, e^*))$ and $y_{i^*}(s, n, e^*) \leqq s$. But then

$$y_{i^*}(s, n, e^*) = \mu y T_1^1(\textstyle\prod_{j<y} p_j^{b(j)\cdot d_{i^*}(s,j)}, e^*, n, y) .$$

By Lemma 3, $d_{i^*}(s', j) = d_{i^*}(s, j)$ for all $j < y_{i^*}(s, n, e^*)$. This means that

$$(y)_{y \leqq y_{i^*}(s,n,e^*)}\big(T_1^1(\textstyle\prod_{j<y} p_j^{b(j)\cdot d_{i^*}(s,j)}, e^*, n, y) \equiv T_1^1(\prod_{j<y} p_j^{b(j)\cdot d_{i^*}(s',j)}, e^*, n, y)\big) .$$

It follows that $y_{i^*}(s, n, e^*) = y_{i^*}(s', n, e^*)$.

LEMMA 5. *The set* $\{m_{i^*}(s, e^*) \mid s \geqq 0\}$ *is infinite.*

PROOF. Suppose, for the purpose of a *reductio ad absurdum*, that the set $\{m_{i^*}(s, e^*) \mid s \geqq 0\}$ is finite; let m' be its greatest member. Let s' be such that $s' \geqq s^*$, and such that $c(s, n) = \lim_s c(s, n) = c(n)$ for all s and n such that $s \geqq s'$ and $n \leqq m'$. Let m'' be the greatest member of $\{m_{i^*}(s, e^*) \mid s \geqq s'\}$. Let s'' be such that $s'' \geqq s'$ and $m_{i^*}(s'', e^*) = m'' \leqq m'$.

We now show by induction on $s \geqq s''$ that $m_{i^*}(s, e^*) = m''$ and $K_{i^*}(s, e^*) = K_{i^*}(s'', e^*)$ for all $s \geqq s''$. Suppose $s \geqq s''$, $m_{i^*}(s, e^*) = m''$, and $K_{i^*}(s, e^*) = K_{i^*}(s'', e^*)$. By Lemma 4, $y_{i^*}(s + 1, n, e^*) = y_{i^*}(s, n, e^*) \leqq s$ for all $n < m_{i^*}(s, e^*)$, since $s \geqq s'' \geqq s' \geqq s^*$. This means that $P_{i^*}(s + 1, n, e^*) = P_{i^*}(s, n, e^*) = c(s, n)$ for all $n < m_{i^*}(s, e^*) = m''$. But $c(s + 1, n) = c(s, n) = c(n)$ for all $n \leqq m'' \leqq m'$, since $s \geqq s'$. Then $P_{i^*}(s + 1, n, e^*) =$

$c(s+1, n)$ for all $n < m_{i^*}(s, e^*) = m''$; consequently, $m_{i^*}(s+1, e^*) \geqq m''$. It follows from the definition of m'' that $m_{i^*}(s+1, e^*) = m''$. By remark (R2) and Lemma 4, $y_{i^*}(s+1, n, e^*) = y_{i^*}(s, n, e^*) \leqq K_{i^*}(s, e^*)$ for all $n < m_{i^*}(s, e^*) = m'' = m_{i^*}(s+1, e^*)$. From this last, it is clear that $K_{i^*}(s, e^*) = K_{i^*}(s+1, e^*)$.

Thus we have shown that $K_{i^*}(s, e^*) = K_{i^*}(s'', e^*)$ for all $s \geqq s''$; but this is absurd, since the set $\{K_{i^*}(s, e^*) \mid s \geqq 0\}$ is infinite.

LEMMA 6. *If $s \geqq s^*$ and if $n < m_{i^*}(s, e^*)$, then $c(n) = P_{i^*}(s, n, e^*)$.*

PROOF. Let $n < m_{i^*}(s, e^*)$ and $s \geqq s^*$. By Lemma 4, $y_{i^*}(s, n, e^*) = y_{i^*}(s', n, e^*) \leqq s$ for all $s' \geqq s$. By Lemma 5, there is an $s'' \geqq s$ such that $m_{i^*}(s'', e^*) \geqq m_{i^*}(s, e^*) > n$ and such that $c(s'', n) = \lim_s c(s, n) = c(n)$. But then $c(s'', n) = P_{i^*}(s'', n, e^*) = U(y_{i^*}(s'', n, e^*))$, since $n < m_{i^*}(s'', e^*)$ and $s'' \geqq s$. It follows that $c(n) = c(s'', n) = U(y_{i^*}(s, n, e^*)) = P_{i^*}(s, n, e^*)$, since $y_{i^*}(s'', n, e^*) = y_{i^*}(s, n, e^*) \leqq s$.

Let

$$t(n) \cong \mu s(s \geqq s^* \;\&\; n < m_{i^*}(s, e^*))\ ;$$

by Lemma 5, $t(n)$ is defined for all n. By Lemma 6, $c(n) = P_{i^*}(t(n), n, e^*)$ for all n. Since each of the functions $m_{i^*}(s, e)$ and $P_{i^*}(s, n, e)$ is recursive in B, it must be that C is recursive in B. But this last is contrary to the hypothesis of our theorem.

Thus, by a *reductio ad absurdum* argument, we have shown that C is not recursive in B, D_i for any $i < 2$. It follows that $D_0 \neq \varnothing$, because otherwise $D_1 = D$, and C is recursive in B, D_1. We know from stage 0 that $D_1 \neq \varnothing$. For each $i < 2$, we define a function $g_i(n)$ which is recursive in B and which enumerates D_i:

$$g_i(n) = \begin{cases} (n)_1 & \text{if } d_i((n)_0, (n)_1) = 0 \\ \mu t(t \in D_i) & \text{otherwise}\ . \end{cases}$$

We will use the following elementary lemma to obtain some corollaries to Theorem 1.

LEMMA 7. *If A_0 and A_1 are disjoint sets, and if for each $i < 2$, A_i is recursively enumerable in $A_0 \cup A_1$, then the degree of $A_0 \cup A_1$ is equal to the least upper bound of the degrees of A_0 and A_1.*

PROOF. Let $\boldsymbol{a}_0$, $\boldsymbol{a}_1$ and $\boldsymbol{a}$ be the degrees of A_0, A_1 and $A_0 \cup A_1$ respectively. Clearly $\boldsymbol{a} \leqq \boldsymbol{a}_0 \cup \boldsymbol{a}_1$. We show $\boldsymbol{a}_0 \leqq \boldsymbol{a}$. Suppose we wish to know if $n \in A_0$. We first ask if $n \in A_0 \cup A_1$. If the answer is no, then $n \notin A_0$. If the answer is yes, we simultaneously enumerate A_0 and A_1; eventually n will turn up in A_0 or in A_1, but not in both, and our original question will be answered. Each of the enumeration functions is recursive in $A_0 \cup A_1$. Thus $\boldsymbol{a}_0 \leqq \boldsymbol{a}$, and similarly, $\boldsymbol{a}_1 \leqq \boldsymbol{a}$.

Following Shoenfield [10], we say a degree $\boldsymbol{b}$ is recursively enumerable in a degree $\boldsymbol{c}$ if there is a set of degree $\boldsymbol{b}$ which is recursively enumerable in a set of degree $\boldsymbol{c}$, and we say a degree is recursively enumerable if it is the degree of a recursively enumerable set. Thus $\boldsymbol{b}'$ is recursively enumerable in $\boldsymbol{b}$, and a recursively enumerable degree is a degree which is recursively enumerable in $\boldsymbol{0}$.

COROLLARY 1. *Let* $\boldsymbol{b}$ *and* $\boldsymbol{c}$ *be degrees such that* $\boldsymbol{b} < \boldsymbol{c}$ *and* $\boldsymbol{c}$ *is recursively enumerable in* $\boldsymbol{b}$. *Then there exist degrees* $\boldsymbol{c}_0$ *and* $\boldsymbol{c}_1$ *such that* $\boldsymbol{c}_0 \bigcup \boldsymbol{c}_1 = \boldsymbol{c}$, $\boldsymbol{c}_0 \mid \boldsymbol{c}_1$, *and such that for each* $i < 2$, $\boldsymbol{b} < \boldsymbol{c}_i < \boldsymbol{c}$ *and* $\boldsymbol{c}_i$ *is recursively enumerable in* $\boldsymbol{b}$.

PROOF. Let B and C be sets of degrees $\boldsymbol{b}$ and $\boldsymbol{c}$ respectively, and let C be recursively enumerable in B. We set D equal to C and apply Theorem 1. Thus we obtain disjoint sets D_0 and D_1 such that $C = D_0 \cup D_1$, and such that for each $i < 2$, C is not recursive in B, D_i, and D_i is recursively enumerable in B. For each $i < 2$, let $\boldsymbol{d}_i$ be the degree of D_i and let $\boldsymbol{c}_i = \boldsymbol{b} \bigcup \boldsymbol{d}_i$. Note that for each $i < 2$, D_i is recursively enumerable in C, since B is recursive in C. It follows from Lemma 7 that $\boldsymbol{c} = \boldsymbol{d}_0 \bigcup \boldsymbol{d}_1$. Thus $\boldsymbol{c}_0 \bigcup \boldsymbol{c}_1 = \boldsymbol{b} \bigcup \boldsymbol{d}_0 \bigcup \boldsymbol{d}_1 = \boldsymbol{c}$. Since C is not recursive in B, D_i for any $i < 2$, it must be that $\boldsymbol{c}_i = \boldsymbol{b} \bigcup \boldsymbol{d}_i < \boldsymbol{c}$ for all $i < 2$. If $\boldsymbol{c}_0$ and $\boldsymbol{c}_1$ were comparable, one of them would equal $\boldsymbol{c}$. Since $\boldsymbol{c}_0 \mid \boldsymbol{c}_1$, it follows that $\boldsymbol{b} < \boldsymbol{c}_i$ for all $i < 2$.

The first corollary strengthens a result of Muchnik [7]. He proved that if $\boldsymbol{c}$ is a non-zero, recursively enumerable degree, then there exist recursive enumerable degrees $\boldsymbol{c}_0$ and $\boldsymbol{c}_1$ such that $\boldsymbol{c}_0 \bigcup \boldsymbol{c}_1 \leqq \boldsymbol{c}$ and $\boldsymbol{c}_0 \mid \boldsymbol{c}_1$.

COROLLARY 2. *Each non-recursive, recursively enumerable set is the union of two disjoint, recursively enumerable sets whose degrees are incomparable.*

PROOF. Let $B = \varnothing$ and $D = C$, where C is a non-recursive, recursively enumerable set, and then apply Theorem 1. Then $C = D = D_0 \cup D_1$, $D_0 \cap D_1 = \varnothing$, and for each $i < 2$, C is not recursive in D_i and D_i is recursively enumerable. Then by Lemma 7, $\boldsymbol{c} = \boldsymbol{d}_0 \bigcup \boldsymbol{d}_1$, where $\boldsymbol{c}, \boldsymbol{d}_0$ and $\boldsymbol{d}_1$ are the degrees of C, D_0 and D_1 respectively. If $\boldsymbol{d}_0$ and $\boldsymbol{d}_1$ were comparable, then C would be recursive in D_i for some $i < 2$.

Corollary 2 improves Theorem 1 of Friedberg [3]. His theorem states that each non-recursive, recursively enumerable set is the union of two disjoint, non-recursive, recursively enumerable sets.

Corollary 3. *The set of all degrees less than* $\mathbf{0}'$ *has no maximal member.*

Proof. Let $\boldsymbol{b}$ be a degree less than $\mathbf{0}'$. Then $\mathbf{0}'$ is recursively enumerable in $\boldsymbol{b}$, and we apply Corollary 1 with $\boldsymbol{c} = \mathbf{0}'$ to obtain a degree $\boldsymbol{c}_0$ such that $\boldsymbol{b} < \boldsymbol{c}_0 < \mathbf{0}'$.

Shoenfield [10] proved that the set of all degrees less than $\mathbf{0}'$ has no greatest member.

Corollary 4. *If* $\boldsymbol{b}$, $\boldsymbol{c}$ *and* $\boldsymbol{d}$ *are degrees such that* $\boldsymbol{b} < \boldsymbol{c} < \boldsymbol{d}$ *and* $\boldsymbol{d}$ *is recursively enumerable in* $\boldsymbol{b}$, *then there exists a degree* $\boldsymbol{h}$ *such that* $\boldsymbol{b} < \boldsymbol{h} < \boldsymbol{d}$, $\boldsymbol{c} \mid \boldsymbol{h}$ *and* $\boldsymbol{h}$ *is recursively enumerable in* $\boldsymbol{b}$.

Proof. Let B, C and D be sets of degrees $\boldsymbol{b}$, $\boldsymbol{c}$ and $\boldsymbol{d}$ respectively, and let D be recursively enumerable in B. We apply Theorem 1 to obtain sets D_0 and D_1. For each $i < 2$, let $\boldsymbol{d}_i$ be the degree of D_i and let $\boldsymbol{h}_i = \boldsymbol{b} \cup \boldsymbol{d}_i$. By Lemma 7, $\boldsymbol{d} = \boldsymbol{d}_0 \bigcup \boldsymbol{d}_1$, and consequently $\boldsymbol{d} = \boldsymbol{h}_0 \bigcup \boldsymbol{h}_1$. Thus for each $i < 2$, $\boldsymbol{b} \leqq \boldsymbol{h}_i \leqq \boldsymbol{d}$. Since C is not recursive in B, D_i for any $i < 2$, it must be that $\boldsymbol{c} \nleqq \boldsymbol{h}_i$ for any $i < 2$. If $\boldsymbol{h}_i \leqq \boldsymbol{c}$ for all $i < 2$, then $\boldsymbol{d} = \boldsymbol{h}_0 \bigcup \boldsymbol{h}_1 \leqq \boldsymbol{c} < \boldsymbol{d}$; consequently, either $\boldsymbol{h}_0$ or $\boldsymbol{h}_1$ is incomparable with $\boldsymbol{c}$.

Shoenfield [9] proved that if $\boldsymbol{b}$ and $\boldsymbol{c}$ are degrees such that $\boldsymbol{b} < \boldsymbol{c} < \boldsymbol{b}'$, then there exists a degree $\boldsymbol{h}$ such that $\boldsymbol{b} < \boldsymbol{h} < \boldsymbol{b}'$ and $\boldsymbol{c} \mid \boldsymbol{h}$. The next two corollaries are immediate consequences of Corollary 4.

Corollary 5. *If* $\boldsymbol{c}$ *is a degree such that* $\mathbf{0} < \boldsymbol{c} < \mathbf{0}'$, *then there exists a recursively enumerable degree* $\boldsymbol{h}$ *such that* $\boldsymbol{c} \mid \boldsymbol{h}$.

Corollary 6. *A set is recursive if and only if it is recursive in every non-recursive, recursively enumerable set.*

Corollary 7. *If* $\boldsymbol{c}$ *is a degree which is recursively enumerable in some degree less than itself, then* $\boldsymbol{c}$ *is the least upper bound of the set of all degrees less than* $\boldsymbol{c}$.

Proof. Let $\boldsymbol{c}$ be recursively enumerable in $\boldsymbol{b}$, where $\boldsymbol{b} < \boldsymbol{c}$, and apply Corollary 1. Then $\boldsymbol{c} = \boldsymbol{c}_0 \bigcup \boldsymbol{c}_1$, and for each $i < 2$, $\boldsymbol{c}_i < \boldsymbol{c}$. If $\boldsymbol{g}$ is greater than or equal to every degree less than $\boldsymbol{c}$, then $\boldsymbol{g} \geqq \boldsymbol{c}_0 \bigcup \boldsymbol{c}_1 = \boldsymbol{c}$.

We know from [1] that there exists a recursively enumerable degree $\boldsymbol{d}$ such that $\mathbf{0} < \boldsymbol{d} < \mathbf{0}'$; clearly $\boldsymbol{d}$ is not complete. By Corollary 7, $\boldsymbol{d}$ is the least upper bound of the set of all degrees less than $\boldsymbol{d}$. Since each complete degree is recursively enumerable in some degree less than itself, it follows that Corollary 7 strengthens a related result of Spector [11]; he showed that each complete degree is the least upper bound of the set of all degrees less than itself. We do not know if there exists a degree $\boldsymbol{b}$ such that $\boldsymbol{b}$ is the least upper bound of the set of all degrees less than $\boldsymbol{b}$ and such that $\boldsymbol{b}$ is not recursively enumerable in some degree less than itself; however, it seems likely that such a $\boldsymbol{b}$ exists. (*Note added in proof*: We now can show there exist degrees $\boldsymbol{b}$, $\boldsymbol{c}$, and $\boldsymbol{d}$, each greater than $\mathbf{0}$ and less than $\mathbf{0}'$, such that the only degrees less than $\boldsymbol{b}$ are $\boldsymbol{c}$, $\boldsymbol{d}$, and $\mathbf{0}$, and such that $\boldsymbol{c}$ and $\boldsymbol{d}$ are incomparable. Then $\boldsymbol{b}$ is the least upper bound of all degrees less than $\boldsymbol{b}$, but by Corollaries 1 and 4, $\boldsymbol{b}$ is not recursively enumerable in any degree less than itself. This result will appear in [9].)

COROLLARY 8. *If* $\boldsymbol{c}$ *is a degree which is recursively enumerable in some degree less than itself, then there exists an infinite, ascending sequence of degrees* $\boldsymbol{b}_0 < \boldsymbol{b}_1 < \boldsymbol{b}_2 < \cdots$ *such that* $\boldsymbol{b}_0 < \boldsymbol{b}_1 < \boldsymbol{b}_2 < \cdots < \boldsymbol{c}$ *and such that* $\boldsymbol{b}_0 < \boldsymbol{b}_1 < \boldsymbol{b}_2 < \cdots < \boldsymbol{g} < \boldsymbol{c}$ *for no degree* $\boldsymbol{g}$.

PROOF. Let $\boldsymbol{c}$ be recursively enumerable in $\boldsymbol{b}$, where $\boldsymbol{b} < \boldsymbol{c}$. Let $B = \{\boldsymbol{d} \mid \boldsymbol{b} \leqq \boldsymbol{d} < \boldsymbol{c}\}$. We regard B as being partially ordered according to the partial ordering of degrees. It follows from Corollary 1 that B has no maximal member. By Zorn's lemma, B contains a maximal, linearly ordered subset T. Since B has no maximal member, it must be that T has no greatest member. Let $T = \{\boldsymbol{t}_0, \boldsymbol{t}_1, \boldsymbol{t}_2, \cdots\}$. Let $r(0) = 0$, and for each $n > 0$, let $r(n)$ be the least k such that $\boldsymbol{t}_k > \boldsymbol{t}_{r(n-1)}$. Let $\boldsymbol{b}_n = \boldsymbol{t}_{r(n)}$ for all n. Then $\boldsymbol{b}_0 < \boldsymbol{b}_1 < \boldsymbol{b}_2 < \cdots < \boldsymbol{c}$. If there were a $\boldsymbol{g}$ such that $\boldsymbol{b}_0 < \boldsymbol{b}_1 < \boldsymbol{b}_2 < \cdots < \boldsymbol{g} < \boldsymbol{c}$, then $\boldsymbol{g}$ would be greater than every member of T, and T would not be maximal.

In [11] Spector left open the following question: does there exist an infinite, ascending sequence of degrees with a minimal upper bound? Corollary 8 answers this question in the affirmative. In particular, there is an infinite, ascending sequence of degrees which has $\mathbf{0}'$ as a minimal upper bound. From Corollary 1 of Theorem 3 of [6], we know that the infinite, ascending sequence $\mathbf{0} < \mathbf{0}' < \mathbf{0}'' < \cdots$ does not have $\mathbf{0}^{(\omega)}$ as a minimal upper bound; however, by [2] $\mathbf{0}^{(\omega)}$ is complete, and so by Corollary 8, there exists an infinite, ascending sequence of degrees which has $\mathbf{0}^{(\omega)}$ as a minimal upper bound. (*Note added in proof*: We now can show that every countable set of degrees has a minimal upper bound. We can also show that the set of all arithmetical degrees has a minimal upper bound less than $\mathbf{0}^{(\omega)}$. These results will appear in [9].)

Friedberg [1] and Muchnik [7] independently solved Post's problem by discovering an ingenious method of defining recursively enumerable sets to meet certain stringent requirements. We call their procedure, or any variation of it, the priority method. Thus our proof of Theorem 1 made use of the priority method; Shoenfield employed it to prove Theorem 4 of [9]. In Theorem 2 below, we make further use of the priority method to solve a problem outside the theory of recursively enumerable degrees.

In general it appears that a problem P requires the priority method for its solution S if and only if the nature of P is such that it severely restricts the number of quantifiers that may appear in S. We discuss this matter further in [8], where, insofar as is possible, we isolate the priority method from recursion theory in order to expose its combinatorial nature.

Following Kleene [5], we represent the sequence of natural numbers $f(0), f(1), \cdots, f(n-1)$ by the sequence number

$$\bar{f}(n) = \prod_{i<n} p_i^{f(i)+1} ,$$

and we write Seq(a) when a is a sequence number, i.e.,

$$\text{Seq}(a) \equiv a > 0 \;\&\; (i)_{i<\text{lh}(a)}((a)_i > 0) .$$

We will need two recursive predicates defined in [11]:

$$\text{Cpt}(a, b) \equiv (i)_{i<\text{lh}(a)}((a)_i = 0 \vee (b)_i = 0 \vee (a)_i = (b)_i) ,$$
$$\text{Ext}(b, a) \equiv \text{Seq}(a) \;\&\; \text{Seq}(b) \;\&\; \text{Cpt}(a, b) \;\&\; b \geqq a .$$

We will need an additional recursive predicate and a recursive function:

$$\text{In}(b, a) \equiv \sim\text{Ext}(a, b) \;\&\; \sim\text{Ext}(b, a) ,$$
$$\text{Cv}(a) = \prod_{i<\text{lh}(a)} p_i^{(a)_i-1} .$$

Note that if $f(n)$ is a function, then $\text{Cv}(\bar{f}(n)) = \tilde{f}(n)$ for all n.

Let $\{a_0, a_1, a_2, \cdots\}$ be a sequence of sequence numbers such that for each i, $\text{Ext}(a_{i+1}, a_i) \;\&\; a_{i+1} > a_i$. Then there is a unique function $f(n)$ with the property that for each i there is an n such that $\bar{f}(n) = a_i$. If A is a set of sequence numbers such that $\{a_0, a_1, a_2, \cdots\} \subseteqq A$, then we say the function $f(n)$ is a member of the closure of A, or in symbols, $f(n) \in \bar{A}$.

In [11] Spector defined a contracting sequence $F_0 \supset F_1 \supset F_2 \supset \cdots$ of sets of sequence numbers and then obtained a function of minimal degree as the unique member of $\bigcap_{i=0}^{\infty} \bar{F}_i$. He defined his sets one at a time; that is, he defined a function G such that for all n, $F_{n+1} = G(F_n)$. Our construction of a minimal degree has his as an ancestor, but it differs from his in that we define our sets of sequence numbers simultaneously. They nonetheless manage to have properties similar to those of his sets because of our use of the priority method. Possibly the most apparent difference between the two proofs is that each of his sets of sequence numbers is recursive, whereas each of ours is merely recursively enumerable.

We define a recursive predicate which is central to our construction of a minimal degree less than $\mathbf{0}'$:

$$H(c, t, e, x, m, b, d) \equiv (i)_{i<2}(\mathrm{Ext}((x)_i, t) \,\&\, T_1(c, (m)_{i,0}, (m)_{i,1}) \\ \&\, (x)_i = U((m)_{i,1}) \,\&\, T_1^1(\mathrm{Cv}((x)_i), e, b, (d)_i) \\ \&\, (d)_i \leqq \mathrm{lh}((x)_i)) \,\&\, U((d)_0) \neq U((d)_1)\,.$$

We define a partial recursive function:

$$Y(c, t, e) \cong \mu x H(c, t, e, (x)_0, (x)_1, (x)_2, (x)_3)\,.$$

It is important to observe that if $Y(c, t, e)$ is defined, then

$$\mathrm{In}((Y(c, t, e))_{0,0}, (Y(c, t, e))_{0,1})\,.$$

We prove this observation with the help of GND.[1] Let c, t and e be such that $Y(c, t, e)$ is defined, and let $y = Y(c, t, e)$. Then for each $i < 2$, we have

$$T_1^1(\mathrm{Cv}((y)_{0,i}), e, (y)_2, (y)_{3,i}) \,\&\, (y)_{3,i} \leqq \mathrm{lh}((y)_{0,i})\,.$$

Suppose $\mathrm{Ext}((y)_{0,0}, (y)_{0,1})$. It follows from GND that

$$T_1^1(\mathrm{Cv}((y)_{0,0}), e, (y)_2, (y)_{3,1})\,,$$

since $(y)_{3,1} \leqq \mathrm{lh}((y)_{0,1}) \leqq \mathrm{lh}((y)_{0,0})$; this means that $(y)_{3,0} = (y)_{3,1}$, since $T_1^1(\mathrm{Cv}((y)_{0,0}), e, (y)_2, z)$ is true for at most one z. But $U((y)_{3,0}) \neq U((y)_{3,1})$. Thus $\sim\mathrm{Ext}((y)_{0,0}, (y)_{0,1})$, and similarly, $\sim\mathrm{Ext}((y)_{0,1}, (y)_{0,0})$.

We define a recursively enumerable set of sequence numbers denoted by $W(c, t, e)$: if $\mathrm{Seq}(t)$, then $t \in W(c, t, e)$; if $u \in W(c, t, e)$ and if $Y(c, u, e)$ is defined, then $(Y(c, u, e))_{0,i} \in W(c, t, e)$ for all $i < 2$; the only members of $W(c, t, e)$ are those given by the last two clauses. If W is a recursively enumerable set, we say e is a Gödel number of W if W is the range of the partial recursive function $U(\mu z T_1(e, n, z))$. Note that if $Y(c, t, e)$ is defined, then both $(Y(c, t, e))_{0,0}$ and $(Y(c, t, e))_{0,1}$ are members of the unique recursively enumerable set which has c as a Gödel number. It is clear that there exists an effective procedure P for obtaining a Gödel number of $W(c, t, e)$ from the triple (c, t, e). We define the recursive function $V(c, t, e)$ to be that function whose value for the triple (c, t, e) is equal to the result of applying procedure P to (c, t, e).

THEOREM 2. *There exists a degree* $\mathbf{h}$ *such that* $\mathbf{0} < \mathbf{h} < \mathbf{0}'$ *and such that* $\mathbf{0} < \mathbf{d} < \mathbf{h}$ *for no degree* $\mathbf{d}$.

PROOF. We will define three functions, $u(s)$, $v(s)$ and $t(s)$, and two partial functions, v_i^s and $Q(i, s)$, simultaneously by induction; neither of the partial functions will be defined when $i > s$. For each s, we will have $\mathrm{Ext}(u(s+1), u(s)) \,\&\, u(s+1) > u(s)$; our function of minimal degree will be the unique member of the closure of $\{u(s) \mid s \geqq 0\}$.

Let $u(0) = 2$, and let q be a Gödel number of the recursively enumerable set of sequence numbers $\{t \mid \mathrm{Ext}(t, u(0))\}$. We define a recursive function $w(t)$:

$$w(t) = \begin{cases} 0 & \text{if } \sim\mathrm{Seq}(t) \\ \mu x(\mathrm{In}((x)_0, (x)_1) \ \& \ (i)_{i<2} \mathrm{Ext}((x)_i, t)) & \text{otherwise} . \end{cases}$$

We set $v(0) = w(u(0))$, $t(0) = 0$, $v_0^0 = q$ and $Q(0, 0) = 0$. That completes stage 0 of our definition by induction.

Before we proceed with stage $s > 0$, we make several assumptions concerning what took place at stage $s - 1$ of the definition:

(A1) $u(s-1)$, $v(s-1)$ and $t(s-1)$ were defined;

(A2) if $k \leqq t(s-1)$, then v_k^{s-1} was defined;

(A3) $Q(i, s-1)$ was defined if and only if $i < s$;

(A4) $\mathrm{In}((v(s-1))_0, (v(s-1))_1)$;

(A5) $(i)_{i<2} \mathrm{Ext}((v(s-1))_i, u(s-1))$;

(A6) $t(s-1)$ is less than or equal to the cardinality of the set $\{i \mid Q(i, s-1) = 1\}$.

Clearly these six assumptions are true if $s = 1$.

Let $0(n)$ be the function whose value is 0 for all n. From (A4) and (A5) we know that $(v(s-1))_0$ and $(v(s-1))_1$ are sequence numbers such that neither one extends the other. If $\{s\}^0(i)$ is defined for all $i < v(s-1)$, then the sequence number

$$\prod\nolimits_{i<v(s-1)} p_i^{1+\{s\}^0(i)}$$

must fail to extend at least one of the sequence numbers $(v(s-1))_0$ and $(v(s-1))_1$. With this last remark in mind, we define $u(s)$:

$$m(s) = \begin{cases} 0 & \text{if } (Ei)_{i<v(s-1)}(\{s\}^0(i) \text{ is not defined}) \\ \mu t(\sim\mathrm{Ext}(\prod_{i<v(s-1)} p_i^{1+\{s\}^0(i)}, (v(s-1))_t)) & \text{otherwise} , \end{cases}$$

$$u(s) = (v(s-1))_{m(s)} .$$

It follows from (A4) and (A5) that $\mathrm{Ext}(u(s), u(s-1)) \ \& \ u(s) > u(s-1)$.

Let $\{i \mid Q(i, s-1) = 1\} \cup \{s\} = \{i_1^s, i_2^s, \cdots, i_{r(s)}^s\}$, where $i_1^s < i_2^s < \cdots < i_{r(s)}^s = s$. By (A6), $t(s-1) < r(s)$. We define v_k^s for all $k \leqq r(s)$:

$$v_k^s = \begin{cases} v_k^{s-1} & \text{if } k \leqq t(s-1) \\ V(v_{k-1}^s, u(s), i_k^s) & \text{if } t(s-1) < k \leqq r(s) . \end{cases}$$

It follows from (A2) that v_k^s is well-defined for all $k \leqq r(s)$. Note that $v_{r(s)}^s = V(v_{r(s)-1}^s, u(s), s)$.

Let $t(s)$ be the greatest k such that for all $j < k$, $Y(v_j^s, u(s), i_{j+1}^s)$ is defined; clearly $0 \leqq t(s) \leqq r(s)$. We define $v(s)$ and $Q(i, s)$ for all $i \leqq s$:

$$v(s) = \begin{cases} w(u(s)) & \text{if } t(s) = 0 \\ \left(Y(v^s_{t(s)-1}, u(s), i^s_{t(s)})\right)_0 & \text{if } t(s) > 0\,; \end{cases}$$

$$Q(i, s) = \begin{cases} Q(i, s-1) & \text{if } i < s \\ 1 & \text{if } i = s \end{cases} \text{ if } t(s) = r(s)$$
$$\begin{cases} Q(i, s-1) & \text{if } i < i^s_{t(s)+1} \\ 0 & \text{if } i = i^s_{t(s)+1} \\ 1 & \text{if } i^s_{t(s)+1} < i \leqq s \end{cases} \text{ if } t(s) < r(s)\,.$$

That completes stage s. It is readily seen that (A1) – (A6) remain true when $s+1$ is replaced by s throughout. For example, we have $\mathrm{In}((v(s))_0, (v(s))_1)$ as a consequence either of the definition of $w(t)$ or of the observation made immediately after the definition of $Y(c, t, e)$. Note that $Q(0, s) = 0$ and $v^s_0 = q$ for all s.

LEMMA 8. *For each i there is an s^* such that for all $s > s^*$, $Q(i, s-1) = Q(i, s)$.*

PROOF. By induction on i. Let $i \geqq 0$; suppose s^{**} has the property that $Q(j, s-1) = Q(j, s)$ for all $j < i$ and $s > s^{**}$. Suppose, for the sake of a *reductio ad absurdum*, that there is no s^* such that $Q(i, s-1) = Q(i, s)$ for all $s > s^*$. Since $Q(i, s)$ equals either 0 or 1 for all $s \geqq i$, there must be an s' such that $Q(i, s'-1) = 0$, $Q(i, s') = 1$ and $s' > s^{**}$. It follows that $i^{s'}_{t(s')+1} < i < s'$. But then $Q(i^{s'}_{t(s')+1}, s'-1) = 1$ and $Q(i^{s'}_{t(s')+1}, s') = 0$; however, this is impossible since $i^{s'}_{t(s')+1} < i$ and $s' > s^{**}$.

For each i, let $s(i)$ be the least s^* such that $Q(i, s-1) = Q(i, s)$ for all $s > s^*$. Thus $Q(i, s) = Q(i, s(i)) = \lim_s Q(i, s)$ for all $s \geqq s(i)$. In addition, if $Q(i, s(i)-1)$ is defined, then $Q(i, s(i)-1) \neq Q(i, s(i))$. Let $s^*(i)$ be the largest member of $\{s(j) \mid j \leqq i\}$.

LEMMA 9. *If $\lim_s Q(i, s) = 0$, then $s(j) < s(i)$ for all $j < i$.*

PROOF. Suppose $\lim_s Q(i, s) = 0$, $s(j) \geqq s(i)$ and $j < i$. Since $Q(i, s(i))$ is defined, we have $i \leqq s(i)$. Then $Q(j, s(j)-1)$ is defined, since $j \leqq i-1 \leqq s(i)-1 \leqq s(j)-1$. It follows that $i^{s(j)}_{t(s(j))+1} \leqq j < i$, since

$$Q(j, s(j)-1) \neq Q(j, s(j))\,.$$

But then $Q(i, s(j)) = 1$, which is impossible, since $s(j) \geqq s(i)$ and $\lim_s Q(i, s) = 0$.

It is an immediate consequence of Lemma 9 that if $\lim_s Q(i, s) = 0$, then $s^*(i) = s(i) > s^*(j)$ for all $j < i$.

Let $S = \{i \mid \lim_s Q(i, s) = 1\}$. We define a partial function i_k by induction:

$$i_1 \cong \mu i(i \in S)$$
$$i_{k+1} \cong \mu i(i \in S \,\&\, i > i_k) \qquad (k > 0)\,.$$

Thus if S is infinite, $S = \{i_1, i_2, i_3, \cdots\}$, where $i_1 < i_2 < i_3 < \cdots$. At this point in our argument we do not know anything about the cardinality of S; however, we will eventually see that i_k is defined for all $k > 0$.

LEMMA 10. *If i_k is defined and if $s > s^*(i_k)$, then $i_k^s = i_k$.*

PROOF. Let i_k be defined, and let $s > s^*(i_k)$. Then $Q(j, s-1)$ is defined for all $j \leqq i_k$, since $s - 1 \geqq s^*(i_k) \geqq i_k$. But then $Q(j, s-1) = \lim_s Q(j, s)$ for all $j \leqq i_k$, since $s > s^*(i_k) \geqq s(j)$ for all $j \leqq i_k$. It follows that $i_j^s = i_j$ for all j such that $1 \leqq j \leqq k$.

LEMMA 11. *If i_k is defined and if $s > s^*(i_k)$, then $k \leqq t(s)$.*

PROOF. Let i_k be defined and let $s > s^*(i_k)$. Suppose $t(s) < k$. Then by Lemma 10, $i^s_{t(s)+1} = i_{t(s)+1}$, since $s > s^*(i_k) \geqq s^*(i_{t(s)+1})$. This means that $Q(i_{t(s)+1}, s) = Q(i^s_{t(s)+1}, s) = 0$. But $Q(i_{t(s)+1}, s) = 1$, since $s > s(i_{t(s)+1})$.

For each k such that i_k is defined, let $s'(i_k)$ be the least s^* such that $k \leqq t(s)$ for all $s \geqq s^*$; by Lemma 11, $s'(i_k) \leqq s^*(i_k) + 1$. It is immediate from the definition of v_k^s that $v_k^{s+1} = v_k^s$ for all s such that $k \leqq t(s)$. Thus if i_k is defined, we have $v_k^s = v_k^{s'(i_k)}$ for all $s \geqq s'(i_k)$. For each k such that i_k is defined, let $v_k = v_k^{s'(i_k)}$. Recall that $v_0^s = v_0^0 = q$ for all s; let $v_0 = q$. Observe that if i_k is defined, then $v_k = V(v_{k-1}^{s'(i_k)}, u(s'(i_k)), i_k^{s'(i_k)})$, since $t(s'(i_k) - 1) < k$.

LEMMA 12. *If i_k is defined and if $s \geqq s'(i_k)$, then $i_k^s = i_k$.*

PROOF. Let i_k be defined. First we show by induction on s that $i_k^s = i_k^{s'(i_k)}$ for all $s \geqq s'(i_k)$. Let s be such that $i_k^s = i_k^{s'(i_k)}$ and $s \geqq s'(i_k)$. Then by definition of $s'(i_k)$, $k \leqq t(s)$. Recall that $Q(j, s-1) = Q(j, s)$ for all $j < i^s_{t(s)}$. This means that

(a) $Q(j, s-1) = Q(j, s)$ for all $j < i_k^s = i_k^{s'(i_k)} \leqq i^s_{t(s)}$.

Since $k \leqq t(s)$, we have

(b) $Q(i_k^s, s) = 1$.

It follows from (a) and (b) that $i_j^{s+1} = i_j^s$ for all j such that $1 \leqq j \leqq k$.

It is clear that $s'(i_k) \geqq s'(i_j)$ for all j such that $1 \leqq j \leqq k$. Thus we have that $i_j^s = i_j^{s'(i_k)}$ for all j and s such that $s \geqq s'(i_k)$ and $1 \leqq j \leqq k$. It follows that $Q(j, s'(i_k)) = \lim_s Q(j, s)$ for all $j \leqq i_k^{s'(i_k)}$; consequently, $i_j^{s'(i_k)} = i_j$ for all j such that $1 \leqq j \leqq k$.

LEMMA 13. *If i_k is defined, then $v_k = V(v_{k-1}, u(s'(i_k)), i_k)$.*

PROOF. Let i_k be defined. If $k = 1$, then $v_{k-1}^{s'(i_k)} = v_0^{s'(i_k)} = q = v_0 = v_{k-1}$, since $v_0^s = q$ for all s. If $k > 1$, then $v_{k-1}^{s'(i_k)} = v_{k-1}^{s'(i_{k-1})} = v_{k-1}$, since

$s'(i_k) \geqq s'(i_{k-1})$. Thus we have established that $v_{k-1}^{s'(i_k)} = v_{k-1}$. By Lemma 12, $i_k^{s'(i_k)} = i_k$. It follows that $v_k = v_k^{s'(i_k)} = V(v_{k-1}^{s'(i_k)}, u(s'(i_k)), i_k^{s'(i_k)}) = V(v_{k-1}, u(s'(i_k)), i_k)$.

Let $(c)^*$ be the unique recursively enumerable set which has c as a Gödel number; thus $(c)^*$ is the range of the partial recursive function $U(\mu z T_1(c, n, z))$. Note that $(V(c, t, e))^* = W(c, t, e)$. We list four *notes* which will be needed in Lemmas 14, 15, 17 and 18:

(N1) $(V(c, t, e))^* - \{t\} \subseteqq (c)^*$;

(N2) if $\sim \text{Ext}(m, t)$, then $m \notin (V(c, t, e))^*$;

(N3) if $t_1 \in (V(c, t, e))^*$ and if $t_2 = (Y(c, t_1, e))_{0,i}$ for some $i < 2$, then $t_2 \in (V(c, t, e))^*$;

(N4) if $Y(c, t, e)$ is defined, then $(Y(c, t, e))_{0,i} > t$ and $\text{Ext}((Y(c, t, e))_{0,i}, t)$ for all $i < 2$.

Note (N4) is a consequence of the observation made immediately after the definition of $Y(c, t, e)$.

Suppose k and s are such that v_k^s is defined and $k > 0$. Then there is a unique $s^* \leqq s$ such that $v_k^{s'} = V(v_{k-1}^{s^*}, u(s^*), i_k^{s^*})$ for all s' such that $s^* \leqq s' \leqq s$; the existence of s^* is evident from the definition of v_k^s. For each k and s such that v_k^s is defined and $k > 0$, we define $y(k, s)$ to be the unique s^* described above. We define $y(0, s) = 0$ for all s. If v_k^s is defined and if $k > 0$, then $y(k-1, s)$ is defined and $y(k-1, s) \leqq y(k, s) \leqq s$. This last follows from the definition of v_k^s.

LEMMA 14. *If v_k^s is defined and if $k \geqq 1$, then $(v_k^s)^* - \{u(y(k, s))\} \subseteqq (v_{k-1}^s)^* - \{u(y(k-1, s))\}$.*

PROOF. Let v_k^s be defined and $k \geqq 1$. We know that $v_k^s = V(v_{k-1}^{y(k,s)}, u(y(k, s)), i_k^{y(k,s)})$. It follows from *note* (N1) that $(v_k^s)^* - \{u(y(k, s))\} \subseteqq (v_{k-1}^{y(k,s)})^*$. But $v_{k-1}^{y(k,s)} = v_{k-1}^{y(k-1,s)} = v_{k-1}^s$, since $y(k-1, s) \leqq y(k, s) \leqq s$. Thus we have that $(v_k^s)^* - \{u(y(k, s))\} \subseteqq (v_{k-1}^s)^*$. We need only show that $u(y(k-1, s)) \notin (v_k^s)^* - \{u(y(k, s))\}$; if $y(k-1, s) = y(k, s)$, there is nothing to prove. Suppose $y(k-1, s) < y(k, s)$. Recall that $\text{Ext}(u(s), u(s-1))$ and $u(s) > u(s-1)$ for all $s > 0$. This means that $\sim \text{Ext}(u(y(k-1, s)), u(y(k, s)))$. It follows from *note* (N2) that $u(y(k-1, s)) \notin (v_k^s)^*$.

LEMMA 15. *If i_k is defined and if $s \geqq s'(i_k)$, then $u(s) \in (v_k)^*$.*

PROOF. Let i_k be defined; then $k > 0$. We prove the lemma by induction on s. By Lemma 13, $v_k = V(v_{k-1}, u(s'(i_k)), i_k)$; consequently, $u(s'(i_k)) \in (v_k)^*$. Suppose $u(s) \in (v_k)^*$, where $s \geqq s'(i_k)$. Then from the definition of $s'(i_k)$, we have $0 < k \leqq t(s)$. It must then be that

$$u(s+1) = (v(s))_i = (Y(v_{t(s)-1}^s, u(s), i_{t(s)}^s))_{0,i}$$

for some $i < 2$; furthermore, $u(s+1) \in (v_{t(s)-1}^s)^*$. In order to show

$u(s+1) \in (v_k)^*$, we consider two cases:

(a) $k = t(s)$;

(b) $k < t(s)$.

First suppose (a) holds. Then by Lemma 12, $i^s_{t(s)} = i^s_k = i_k$, since $s \geqq s'(i_k)$. In addition $v^s_{t(s)-1} = v^s_{k-1} = v_{k-1}$, since $v^s_{k-1} = q$ if $k = 1$, and since $s \geqq s'(i_k) \geqq s'(i_{k-1})$ if $k > 1$. Thus we have

$$u(s+1) = \big(Y(v_{k-1}, u(s), i_k)\big)_{0,i}$$

for some $i < 2$. It follows from *note* (N3) that $u(s+1) \in (v_k)^*$, since $u(s) \in (v_k)^*$.

Now suppose (b) is the case. We have that $u(s+1) \in (v^s_{t(s)-1})^*$. We know $u(s+1) > u(y(t(s)-1, s))$, because $s \geqq y(t(s)-1, s)$. It follows that $u(s+1) \in (v^s_{t(s)-1})^* - \{u(y(t(s)-1, s))\}$. Let $r = t(s) - 1 - k$; then $r \geqq 0$. We apply Lemma 14 r times in succession in order to obtain

$$(v^s_{t(s)-1})^* - \{u(y(t(s)-1, s))\} \subseteqq (v^s_k)^* - \{u(y(k, s))\} .$$

Then $u(s+1) \in (v^s_k)^* = (v_k)^*$, since $s \geqq s'(i_k)$.

Let $h(n)$ be the unique function which is a member of the closure of $\{u(0), u(1), u(2), \cdots\}$. Thus for all s, $u(s) = \bar{h}(\mathrm{lh}(u(s)))$. By Lemma 15, if i_k is defined, then $h(n)$ is a member of the closure of $(v_k)^*$. We will show $h(n)$ has minimal degree less than $\mathbf{0}'$. To see that $h(n)$ is not recursive, it is merely necessary to recollect the definition of $u(s)$ at stage $s > 0$. If $s > 0$, there is an $i < v(s-1)$ such that either $\{s\}^0(i)$ is not defined or $\{s\}^0(i)$ is defined but not equal to $h(i)$. Thus $h(n)$ is not recursive with Gödel number s in $0(n)$ for any $s > 0$, and consequently, $h(n)$ is not recursive.

To see that the function $u(s)$, and thereby the function $h(n)$, has degree less than or equal to $\mathbf{0}'$, we study stage $s > 0$ (the induction step) of our definition by induction and apply the method of Kleene and Post [6]. We first define

$$F_1(s) = \prod_{i \leq s} p_i^{Q(i,s)} ,$$

$$F_2(s) = \begin{cases} \prod_{i \leq r(s)} p_i^{v^s_i + 1} & \text{if } s > 0 \\ 2^{q+1} & \text{if } s = 0 , \end{cases}$$

$$F(s) = 2^{u(s)} \cdot 3^{v(s)} \cdot 5^{t(s)} \cdot 7^{F_1(s)} \cdot 11^{F_2(s)} .$$

Our simultaneous definition by induction of the functions $u(s)$, $v(s)$ and $t(s)$ and the partial functions v^s_i and $Q(i, s)$ can be regarded as a definition by induction of the function $F(s)$. There exists a function $L(n)$ such that

$$F(0) = 2^2 \cdot 3^{w(2)} \cdot 5^0 \cdot 7^{2^0} \cdot 11^{2^{q+1}}$$

$$F(s) = L(F(s-1)) \qquad (s > 0) .$$

If $L(n)$ has degree less than or equal to $\mathbf{0}'$, then so do $F(s)$ and $u(s)$. It is not necessary for us to exhibit the function $L(n)$ in order to indicate why its degree is less than or equal to $\mathbf{0}'$. It is sufficient to examine stage $s > 0$.

Let g be a Gödel number of the partial recursive function $Y(c, t, e)$; i.e., $Y(c, t, e) \cong U(\mu y T_3(g, c, t, e, y))$. We claim that the function $L(n)$ is recursive in the predicates $(Ey)T_1^1(\tilde{0}(y), s, i, y)$, $(Ey)T_3(g, c, t, e, y)$. The first predicate is needed to determine which case holds in the definition of $m(s)$ $(s > 0)$. Thus $m(s) = 0$ if (but not only if)

$$(Ei)_{i<v(s-1)} \sim(Ey)T_1^1(\tilde{0}(y), s, i, y) .$$

The second predicate is needed in the definition of $t(s)$ $(s > 0)$. Thus $t(s) = 0$ if and only if

$$\sim(Ey)T_3(g, v_0^s, u(s), i_1^s, y) .$$

Since both the first and the second predicates have degree less than or equal to $\mathbf{0}'$, it follows that $L(n)$, $F(s)$ and $h(n)$ also have degree less than or equal to $\mathbf{0}'$.

LEMMA 16. *If* $e > 0$, *if* $\lim_s Q(e, s) = 0$, *and if* $\{e\}^h(b)$ *is defined for all* b, *then the function* $\{e\}^h(b)$ *is recursive.*

PROOF. By Lemma 9, $s(e) = s^*(e) > s(j)$ for all $j < e$. This means that $Q(j, s(e) - 1) = Q(j, s(e)) = \lim_s Q(j, s)$ for all $j < e$. Since $Q(e, s(e) - 1)$ is either undefined or equal to 1, there must be a k such that $i_k^{s(e)} = e$. Then $Q(i_k^{s(e)}, s(e) - 1)$ is either undefined or equal to 1 and $Q(i_k^{s(e)}, s(e)) = 0$. It follows that $t(s(e)) + 1 = k$ and that $Y(v_{k-1}^{s(e)}, u(s(e)), i_k^{s(e)})$ is undefined. We show that $v_{k-1}^{s(e)} = v_{k-1}$. If $k = 1$, then $v_{k-1}^{s(e)} = q = v_0 = v_{k-1}$. Suppose $k > 1$; then i_j is defined and is equal to $i_j^{s(e)}$ for all j such that $i \leqq j < k$, since $Q(j, s(e) - 1) = Q(j, s(e)) = \lim_s Q(j, s)$ for all $j < e = i_k^{s(e)}$; in particular, $i_{k-1}^{s(e)} = i_{k-1}$. Recall that $s'(i_{k-1}) \leqq s^*(i_{k-1}) + 1$. Then

$$s(e) = s^*(e) = s^*(i_k^{s(e)}) > s^*(i_{k-1}^{s(e)}) = s^*(i_{k-1}) \geqq s'(i_{k-1}) - 1 .$$

Since $s(e) \geqq s'(i_{k-1})$, we have $v_{k-1}^{s(e)} = v_{k-1}$.

Thus we have shown that $Y(v_{k-1}, u(s(e)), e)$ is undefined. It is clear that $u(s) \in (v_{k-1})^*$ for all $s \geqq s(e)$, since if $k = 1$, then $v_{k-1} = q$, and since if $k > 1$, then i_{k-1} is defined, $s(e) \geqq s'(i_{k-1})$, and Lemma 15 applies.

We define a recursive predicate $L(x, b)$:

$$\begin{aligned} L(x, b) \equiv \operatorname{Ext}\big((x)_0, u(s(e))\big) \ \& \ T_1\big(v_{k-1}, (x)_{1,0}, (x)_{1,1}\big) \\ \& \ (x)_0 = U\big((x)_{1,1}\big) \ \& \ T_1^1\big(\operatorname{Cv}((x)_0), e, b, (x)_3\big) \\ \& \ (x)_3 \leqq \operatorname{lh}\big((x)_0\big) . \end{aligned}$$

We claim that for each b, $U((\mu x L(x, b))_3)$ is defined and is equal to $\{e\}^h(b)$. Since $u(s) \in (v_{k-1})^*$ for all $s \geqq s(e)$, and since $\{e\}^h(b)$ is defined for all b, it follows that

$$(b)(Ex)\big(L(x, b) \ \& \ \{e\}^h(b) = U((x)_3)\big) .$$

Thus the partial recursive function $U((\mu x L(x, b))_3)$ is defined for all b and hence is recursive. It remains only to see that $U((\mu x L(x, b))_3) = \{e\}^h(b)$ for all b. If this last were not the case, then there would exist y, z and b such that

$$L(y, b) \ \& \ L(z, b) \ \& \ U\big((y)_3\big) \neq U\big((z)_3\big) .$$

But then we would have

$$H\big(v_{k-1}, u(s(e)), e, 2^{(y)_0}\cdot 3^{(z)_0}, 2^{(y)_1}\cdot 3^{(z)_1}, b, 2^{(y)_3}\cdot 3^{(z)_3}\big) ,$$

and $Y(v_{k-1}, u(s(e)), e)$ would be defined, which it is not.

LEMMA 17. *If* $u \in W(c, t, e)$ & $w \in W(c, t, e)$ & $w > u$ & $\mathrm{Ext}(w, u)$, *then* $Y(c, u, e)$ *is defined and* $\mathrm{Ext}(w, (Y(c, u, e))_{0,i})$ *for some* $i < 2$.

PROOF. Since $u \in W(c, t, e)$, there must exist $u_0, u_1, \cdots, u_n$ and $y_1, y_2, \cdots, y_n$ such that

$$\begin{aligned} u_0 &= t \\ u_k &= \big(Y(c, u_{k-1}, e)\big)_{0,y_k} \qquad (0 < k \leqq n) \\ u_n &= u . \end{aligned}$$

It is of course possible that $u = t$, in which case we take $n = 0$. Similarly, there exist $w_0, w_1, \cdots, w_m$ and $z_1, z_2, \cdots, z_m$ such that

$$\begin{aligned} w_0 &= t \\ w_k &= \big(Y(c, w_{k-1}, e)\big)_{0,z_k} \qquad (0 < k \leqq m) \\ w_m &= w . \end{aligned}$$

Let d be the largest k such that $u_j = w_j$ for all $j \leqq k$. Clearly $d \leqq n$. In addition, $d < m$ by *note* (N4), since $w > u$. For the sake of a *reductio ad absurdum*, suppose $d < n$. Then $u_{d+1} = (Y(c, w_d, e))_{0,y_{d+1}}$ and $w_{d+1} = (Y(c, w_d, e))_{0,z_{d+1}}$. But $y_{d+1} \neq z_{d+1}$, since $u_{d+1} \neq w_{d+1}$. It follows from the observation made immediately after the definition of $Y(c, t, e)$ that $\mathrm{In}(u_{d+1}, w_{d+1})$. This means that $\sim\mathrm{Ext}(w, u_{d+1})$, since $\mathrm{Ext}(w, w_{d+1})$. But then $\sim\mathrm{Ext}(w, u)$, since $\mathrm{Ext}(u, u_{d+1})$. This last contradicts the hypothesis of the lemma, so it must be that $d = n$. Thus $u = u_n = u_d = w_d < w$. It follows with the help of *note* (N4) that $Y(c, u, e)$ is defined and that $\mathrm{Ext}(w, (Y(c, u, e))_{0,z_{d+1}})$.

LEMMA 18. *If* $\lim_s Q(e, s) = 1$ *and if* $\{e\}^h(b)$ *is defined for all* b, *then the function* $h(b)$ *is recursive in the function* $\{e\}^h(b)$.

PROOF. Let k be such that $i_k = e$. By Lemma 15, $u(s) \in (v_k)^*$ for all $s \geqq s'(e)$. This last fact enables us to define a function $p(r)$:

$$p(0) = u(s'(e))$$
$$p(r+1) = \mu x(x \in (v_k)^* \mathbin{\&} x > p(r) \mathbin{\&} \mathrm{Ext}(x, p(r)) \mathbin{\&} (Es)\,\mathrm{Ext}(u(s), x)) .$$

It is clear that for each r there is an n such that $\bar{h}(n) = p(r)$. Thus it is sufficient for us to show that the function $p(r)$ is recursive in the function $\{e\}^h(b)$.

By Lemma 13, $v_k = V(v_{k-1}, u(s'(e)), e)$. Thus for each r, $p(r) \in W(v_{k-1}, u(s'(e)), e) \mathbin{\&} p(r+1) \in W(v_{k-1}, u(s'(e)), e) \mathbin{\&} p(r+1) > p(r) \mathbin{\&} \mathrm{Ext}(p(r+1), p(r))$. It follows from Lemma 17 that for each r, $Y(v_{k-1}, p(r), e)$ is defined and $\mathrm{Ext}(p(r+1), (Y(v_{k-1}, p(r), e))_{0,i})$ for some $i < 2$. For each r, let

$$z(r) = Y(v_{k-1}, p(r), e) ,$$
$$t(r) = \mu i \mathrm{Ext}(p(r+1), (z(r))_{0,i}) .$$

We now show for all r that

(a) $p(r+1) = (z(r))_{0,t(r)}$;

(b) $t(r) = \mu i(U((z(r))_{3,i}) = \{e\}^h((z(r))_2))$.

Let r be fixed. First we prove (a). By *note* (N3), $(z(r))_{0,t(r)} \in (v_k)^*$, since $p(r) \in (v_k)^*$. By *note* (N4), $(z(r))_{0,t(r)} > p(r)$ and $\mathrm{Ext}((z(r))_{0,t(r)}, p(r))$. There is an s such that $\mathrm{Ext}(u(s), (z(r))_{0,t(r)})$, since $\mathrm{Ext}(p(r+1), (z(r))_{0,t(r)})$, and since there is an s such that $\mathrm{Ext}(u(s), p(r+1))$. Clearly $p(r+1) \geqq (z(r))_{0,t(r)}$; (a) readily follows. From the definition of the partial recursive function $Y(c, t, e)$ it is evident that

$$T_1^1(\mathrm{Cv}((z(r))_{0,t(r)}), e, (z(r))_2, (z(r))_{3,t(r)}) \mathbin{\&} (z(r))_{3,t(r)} \leqq \mathrm{lh}((z(r))_{0,t(r)}) .$$

This last means that $U((z(r))_{3,t(r)}) = \{e\}^h((z(r))_2)$, since $(z(r))_{0,t(r)} = p(r+1) = \bar{h}(n)$ for some n. But then (b) follows, since $U((z(r))_{3,0}) \neq U((z(r))_{3,1})$.

It is immediate from (a) and (b) that the following system of equations defines the function $p(r)$ recursively from the function $\{e\}^h(b)$:

$$p(0) = u(s'(e)) ,$$
$$z(r) = Y(v_{k-1}, p(r), e) ,$$
$$t(r) = \mu i(U((z(r))_{3,i}) = \{e\}^h((z(r))_2)) ,$$
$$p(r+1) = (z(r))_{0,t(r)} .$$

We conclude from Lemmas 16 and 18 that for each $e > 0$, either $\{e\}^h(b)$

is not defined for some b, or the function $\{e\}^h(b)$ is recursive, or the function $h(b)$ is recursive in the function $\{e\}^h(b)$. (It is now clear that i_k is defined for all $k > 0$, since the function $h(b)$ is recursive in the function $\{e\}^h(b)$ for infinitely many e.) Thus $h(b)$ has minimal degree less than $\mathbf{0}'$, since $\mathbf{0}'$ was shown to non-minimal by Kleene and Post in [6].

THEOREM 3. *For each degree $\mathbf{b}$, there is a degree $\mathbf{c}$ such that $\mathbf{b} < \mathbf{c} < \mathbf{b}'$, and such that $\mathbf{b} < \mathbf{d} < \mathbf{c}$ for no degree $\mathbf{d}$.*

PROOF. We simply retrace the proof of Theorem 2 and relativize all functions and predicates, when appropriate, with respect to $b(n)$, a function of degree $\mathbf{b}$. For example, we define

$$\begin{aligned} H_b(c, t, e, x, m, b, d) \equiv (i)_{i<2}(&\mathrm{Ext}((x)_i, t) \;\&\; T_1^1(\tilde{b}((m)_{i,1}), c, (m)_{i,0}, (m)_{i,1}) \\ &\& \; (x)_i = U((m)_{i,1}) \;\&\; T_1^{1,1}(\tilde{b}((d)_i), \mathrm{Cv}((x)_i), e, b, (d)_i) \\ &\& \; (d)_i \leqq \mathrm{lh}((x)_i)) \;\&\; U((d)_0) \neq U((d)_1) \,. \end{aligned}$$

In this manner we obtain a function $h(n)$ which is not recursive in the function $b(n)$ and whose degree is less than or equal to $\mathbf{b}'$. It has the property that for each $e > 0$, either $\{e\}^{b,h}(n)$ is undefined for some n or the function $\{e\}^{b,h}(n)$ is recursive in the function $b(n)$ or the function $h(n)$ is recursive in the functions $\{e\}^{b,h}(n)$, $b(n)$. The desired degree $\mathbf{c}$ is the least upper bound of the degrees of $b(n)$ and $h(n)$.

We end with a word on the role of the priority method in the proof of Theorem 2. Let $v(k, s) = v_k^s$ if v_k^s is defined, and let $v(k, s) = 0$ otherwise. Then for each k, $\lim_s v(k, s) = \lim_s v_k^s = v_k$, since for each k, v_k^s is defined for all sufficiently large s. It is readily seen that the degree of the function $v(k, s)$ is less than or equal to $\mathbf{0}'$. It follows that the degree of the function v_k is less than or equal to $\mathbf{0}''$. Spector [11] worked directly with a function resembling v_k, and that is why the best upper bound he could supply for his minimal degree was $\mathbf{0}''$. By working indirectly with the function $v(k, s)$, we were able to obtain $\mathbf{0}'$ as an upper bound. It is generally true of the priority method that one works with a convergent sequence of functions rather than with their limit in order to avoid an unwanted quantifier.

INSTITUTE FOR ADVANCED STUDY

REFERENCES

1. Richard M. FRIEDBERG, *Two recursively enumerable sets of incomparable degrees of unsolvability*, Proc. Nat. Acad. Sci U.S.A., 43 (1957), 236–238.
2. ———, *A criterion for completeness of degrees of unsolvability*, J. Symb. Logic, 22 (1957), 159–160.

3. ———, *Three theorems on recursive enumeration*, J. Symb. Logic, 23 (1958), 309-316.
4. S. C. KLEENE, Introduction to Metamathematics, New York, Toronto, Amsterdam and Gröningen, 1952.
5. ———, *On the forms of the predicates in the theory of constructive ordinals* (second paper), Amer. J. Math., 77 (1955), 405-428.
6. ——— and Emil L. POST, *The upper semi-lattice of degrees of recursive unsolvability*, Ann. of Math., 59 (1954), 379-407.
7. A. A. MUCHNIK, *Negative answer to the problem of reducibility of the theory of algorithms* (in Russian), Dokl. Akad. Nauk SSSR, 108 (1956), 194-197.
8. Gerald E. SACKS, *On the priority method of Friedberg and Muchnik*, in preparation.
9. ———, Degrees of Unsolvability, Annals of Mathematics Studies, Princeton, forthcoming.
10. J. R. SHOENFIELD, *On degrees of unsolvability*, Ann. of Math., 69 (1959), 644-653.
11. Clifford SPECTOR, *On degrees of recursive unsolvability*, Ann. of Math., 64 (1956), 581-592.

RECURSIVE ENUMERABILITY AND THE JUMP OPERATOR

BY

GERALD E. SACKS(1)

By degree we mean degree of recursive unsolvability as defined by Kleene and Post in [4]. Following Shoenfield [7], we say a degree $\mathbf{c}$ is recursively enumerable in a degree $\mathbf{b}$ if there is a set of degree $\mathbf{c}$ which is the range of a function of degree less than or equal to $\mathbf{b}$, and we call a degree recursively enumerable if it is recursively enumerable in $\mathbf{0}$ (i.e., if it is the degree of a recursively enumerable set). The jump operator, which takes the degree $\mathbf{d}$ to the degree $\mathbf{d}'$ (the completion of $\mathbf{d}$), was defined in [4] and has the following properties: if $\mathbf{h}$ is recursively enumerable in $\mathbf{d}$, then $\mathbf{h} \leqq \mathbf{d}'$; $\mathbf{d}' > \mathbf{d}$; and $\mathbf{d}'$ is recursively enumerable in $\mathbf{d}$. In [4] a degree $\mathbf{c}$ is said to be complete if there exists a degree $\mathbf{d}$ such that $\mathbf{d}' = \mathbf{c}$. Friedberg [1] showed that a degree $\mathbf{c}$ is complete if and only if $\mathbf{c} \geqq \mathbf{0}'$.

For any degree $\mathbf{b}$, if $\mathbf{b} \leqq \mathbf{d} \leqq \mathbf{b}'$, then $\mathbf{b}' \leqq \mathbf{d}' \leqq \mathbf{b}$ and $\mathbf{d}'$ is recursively enumerable in $\mathbf{b}'$. Shoenfield [7] proved that if $\mathbf{b}' \leqq \mathbf{c} \leqq \mathbf{b}''$ and $\mathbf{c}$ is recursively enumerable in $\mathbf{b}'$, then there is a degree $\mathbf{d}$ such that $\mathbf{b} \leqq \mathbf{d} \leqq \mathbf{b}'$ and $\mathbf{d}' = \mathbf{c}$. Thus the degrees which lie between $\mathbf{b}'$ and $\mathbf{b}''$ and are recursively enumerable in $\mathbf{b}'$ can be viewed as the completions of the degrees which lie between $\mathbf{b}$ and $\mathbf{b}'$. He also showed there is a degree greater than $\mathbf{b}$ and less than $\mathbf{b}'$ which is not recursively enumerable in $\mathbf{b}$.

Our main result below is that the degrees which lie between $\mathbf{b}'$ and $\mathbf{b}''$ and are recursively enumerable in $\mathbf{b}'$ can be viewed as the completions of the degrees which lie between $\mathbf{b}$ and $\mathbf{b}'$ and are recursively enumerable in $\mathbf{b}$. Our notation is that of [3].

THEOREM 1. *Let* $\mathbf{a}$, $\mathbf{b}$ *and* $\mathbf{c}$ *be degrees such that* $\mathbf{a} \nleqq \mathbf{b}$, $\mathbf{a} \leqq \mathbf{b}' \leqq \mathbf{c}$ *and* $\mathbf{c}$ *is recursively enumerable in* $\mathbf{b}'$. *Then there exists a degree* $\mathbf{d}$ *such that* $\mathbf{a} \nleqq \mathbf{d}$, $\mathbf{b} \leqq \mathbf{d}$, $\mathbf{d}' = \mathbf{c}$ *and* $\mathbf{d}$ *is recursively enumerable in* $\mathbf{b}$.

Proof. We first prove the theorem when $\mathbf{b} = \mathbf{0}$, and then indicate the changes needed when $\mathbf{b} > \mathbf{0}$. Thus we have degrees $\mathbf{a}$ and $\mathbf{c}$ such that $\mathbf{a} > \mathbf{0}$, $\mathbf{a} \leqq \mathbf{0}' \leqq \mathbf{c}$ and $\mathbf{c}$ is recursively enumerable in $\mathbf{0}'$, and we wish find a recursively enumerable degree $\mathbf{d}$ such that $\mathbf{a} \nleqq \mathbf{d}$ and $\mathbf{d}' = \mathbf{c}$.

Let f be a function of degree less than or equal to $\mathbf{0}'$ whose range is a set C of

Presented to the Society, April 19, 1962; received by the editors May 4, 1962.

(1) The author is a National Science Foundation Postdoctoral Fellow.

degree **c**. Let c be the representing function of C. Let g be a recursive function whose range is a set J of degree $\mathbf{0}'$. Let j be the representing function of J. We define

$$j(s,n) = \begin{cases} 0 & \text{if } (Ek)_{k<s}(g(k)=n), \\ 1 & \text{otherwise.} \end{cases}$$

It is clear that $j(s, n)$ is a recursive function, and that for each n, $\lim_s j(s,n)$ exists and is equal to $j(n)$. Since f is recursive in j, there is a Gödel number z_1 such that

$$f(n) = \{z_1\}^j(n) = U(\mu y T_1^1(\bar{j}(y), z_1, n, y))$$

for all n. We define a recursive function $f(s,n)$ of supreme importance to our argument;

$$f(s,n) = \begin{cases} U\left(\mu y T_1^1\left(\prod_{i<y} p_i^{j(s,n)}, z_1, n, y\right)\right) \\ \qquad \text{if} (Ey)_{y \leqq s} T_1^1\left(\prod_{i<y} p_i^{j(s,n)}, z_1, n, y\right), \\ s+1 \qquad \text{otherwise.} \end{cases}$$

We claim that $\lim_s f(s,n)$ exists and is equal to $f(n)$ for all n. Our claim is a consequence of the fact that $f(n) = \{z_1\}^j(n)$ and $\lim_s j(s,n) = j(n)$ for all n.

Let a be an everywhere positive function of degree **a**, and let z_2 be a Gödel number such that $\{z_2\}^j(n) = a(n)$ for all n. We define

$$a(s,n) = \begin{cases} U\left(\mu y T_1^1\left(\prod_{i<y} p_i^{j(s,n)}, z_2, n, y\right)\right) \\ \qquad \text{if } (Ey)_{y \leqq s}\left[T_1^1\left(\prod_{i<y} p_i^{j(s,n)}, z_2, n, y\right) \& \, U(y) \geqq 1\right], \\ 1 \qquad \text{otherwise.} \end{cases}$$

The function $a(s,n)$ is recursive; for each n, $\lim_s a(s,n)$ exists and is equal to $a(n)$.

A useful property of the Gödel numbering devised by Kleene in [3] to arithmetize his formalism for recursive functions is: the Gödel number of a deduction is greater than the intuitive counterpart of any formal numeral occurring in the deduction. We will denote this fact by GND. It follows from GND that $a(s,n) = 1$ whenever $n \geqq s$.

We define two recursive functions, $t(s,n)$ and $h(s,n)$, by means of an induction on s:

$$\begin{aligned} t(s,n) &= \mu m_{m<s}(f(s,m) = n); \\ h(0,n) &= 0; \\ h(s+1,n) &= h(s,n) + \mathrm{sg}(|t(s+1,n) - t(s,n)|). \end{aligned}$$

Recall that the bounded least number operator is defined in such a way that $t(s,n) = s$ if and only if there is no $m < s$ such that $f(s,m) = n$.

We now proceed to define four recursive functions, $y(s,n,e)$, $m(s,e)$, $r(s,n,e)$ and $d(s,n)$, simultaneously by induction on s. The function $d(s,n)$ will be such that

$$0 \leqq d(s+1,n) \leqq d(s,n) \leqq 1$$

for all s and n. Thus for each n, $\lim_s d(s,n)$ will exist; furthermore, $\lim_s d(s,n)$ will be the representing function of a recursively enumerable set D. The degree of D will be the desired degree $\mathbf{d}$. At stage s of the construction we put finitely or infinitely many natural numbers in D; our main objective is to see that $\mathbf{c} \leqq \mathbf{d}'$; however, with the aid of a system of priorities, we exercise restraint when we add members to D in order to insure that $\mathbf{a} \nleqq \mathbf{d}$ and $\mathbf{d}' \leqq \mathbf{c}$.

Stage $s = 0$. We set $y(0,n,e) = r(0,n,e) = 0$, $m(0,e) = e+1$ and $d(0,n) = 1$ for all n and e.

Stage $s > 0$. We define $y(s,n,e)$ for all n and e:

$$y(s,n,e) = \begin{cases} \mu y T_1^1\left(\prod_{i<y} p_i^{d(s-1,i)}, e, n, y\right) \\ \qquad \text{if } n \geqq e \,\&\, (Ey)_{y \leqq s} T_1^1\left(\prod_{i<y} p_i^{d(s-1,i)}, e, n, y\right), \\ 0 \quad \text{otherwise.} \end{cases}$$

It follows from GND that $y(s,n,e) = 0$ whenever $n \geqq s$.

We define $m(s,e)$ for all e; there are three mutually exclusive cases.

Case 1. $y(s,e,e) = 0$. We set $m(s,e) = e+1$.

Case 2. $y(s,e,e) > 0$ and there is an n such that

$$e < n < m(s-1,e) \,\&\, y(s,n,e) \neq y(s-1,n,e) \,\&\, a(s,n) \neq U(y(s,n,e)).$$

We set

$$m(s,e) = \mu n_{e<n}[y(s,n,e) \neq y(s-1,n,e) \,\&\, a(s,n) \neq U(y(s,n,e))].$$

Case 3. Otherwise.

We set

$$m(s,e) = \mu n[m(s-1,e) \leqq n < 2m(s-1,e) + s \\ \&(Et)\ (e < t \leqq n \,\&\, a(s,t) \neq U(y(s,t,e))].$$

Note that Case 3 of the definition of $m(s,e)$, the least number operator is bounded.

We define $r(s,n,e)$ and $d(s,p_e^n)$ for all n and e by means of a simultaneous induction on e. Let $e \geqq 0$ and suppose $r(s,n,i)$ and $d(s,p_i^n)$ have been defined for all $i < e$ and all n; we define $r(s,n,e)$ and $d(s,p_e^n)$ for all n as follows:

$$r(s,n,e) = \begin{cases} 0 & \text{if } (Ei)(Em)(Et)[i < e \leqq t \leqq n \,\&\, p_i^m < y(s,t,e) \\ & \qquad \&\, d(s,p_i^m) \neq d(s-1,p_i^m)], \\ 1 & \text{otherwise;} \end{cases}$$

$$d(s, p_e^n) = \begin{cases} d(s-1, p_e^n) & \text{if } n \geqq h(s,e), \\ d(s-1, p_e^n) & \text{if } (Ei)(Em)[i \leqq e \,\&\, i \leqq m < m(s,i) \\ & \quad \&\, r(s,m,i) = 1 \,\&\, p_e^n < y(s,m,i)], \\ 0 & \text{otherwise.} \end{cases}$$

We conclude the construction by setting $d(s,n) = d(s-1,n)$ for all n not a power of a prime. It is readily verified by the method of [4] that each of the four functions just defined is recursive. Such a verification is possible for two reasons: each of the functions $a(s,n)$ and $h(s,n)$ is recursive; at stage $s > 0$, all quantifiers, as well as all applications of the least number operator, are bounded. For each n, let

$$d(n) = \lim_s d(s,n);$$

it is clear that $d(n) = 0$ if and only if there is an s such that $d(s,n) = 0$. Thus d is the representing function of a recursively enumerable set. Let **d** be the degree of d.

We list some remarks which will be needed in vital parts of the body of our argument:

(R1) $(s)(e)[m(s,e) > e]$;

(R2) $(s)(n)(e)[r(s,n,e) = 0 \rightarrow r(s,n+1,e) = 0]$;

(R3) $(s)(n)(e)[(y(s,n,e) = 0 \,\&\, n > e) \rightarrow m(s,e) \leqq n]$.

Remark (R1) is easily proved by induction on s if the definition of the bounded least number operator is kept in mind.

We prove remark (R3) by induction on s. We have

$$(n)(e)[(y(0,n,e) = 0 \,\&\, n > e) \rightarrow m(0,e) \leqq n].$$

Let s be such that $s > 0$ and

$$(n)(e)[(y(s-1,n,e) = 0 \,\&\, n > e) \rightarrow m(s-1,e) \leqq n].$$

Let e and n be such that

$$y(s,n,e) = 0 \qquad \text{and } n > e.$$

Then $a(s,n) \neq U(y(s,n,e))$, since $a(s,n) \geqq 1$ and $U(0) = 0$. First we suppose $n < m(s-1,e)$. Then $y(s-1,n,e) > 0$ as a consequence of the induction hypothesis. But then either Case 1 or Case 2 of the definition of $m(s,e)$ holds, and so $m(s,e) \leqq n$. Now we suppose $m(s-1,e) \leqq n$. If either Case 1 or Case 2 of the definition of $m(s,e)$ holds, then $m(s,e) \leqq m(s-1,e) \leqq n$ by remark (R1). If Case 3 holds and $n < 2m(s-1,e) + s$, then $m(s,e) \leqq n$. If Case 3 holds and $n \geqq 2m(s-1,e) + s$, then $m(s,e) \leqq 2m(s-1,e) + s \leqq n$. (Note that if Case 3 holds and

$$(t)[e < t < 2m(s-1,e) + s \rightarrow a(s,t) = U(y(s,t,e))],$$

then $m(s,e) = m(s-1,e) + s$; this last is a consequence of the definition of the bounded least number operator.)

We introduce two predicates:

$A(e)$: if the set $\{m(s,e) \mid s \geqq 0\}$ is infinite, then there is an $n \geqq e$ such that $\lim_s y(s,n,e)$ either does not exist or is equal to 0.

$B(e)$: $\lim_n d(p_e^n)$ exists and is equal to $1 - c(e)$.

We will prove $(e)A(e)$ and $(e)B(e)$ by means of a simultaneous induction on e. From $(e)A(e)$ it will follow that $\mathbf{a} \not\leqq \mathbf{d}$. From $(e)B(e)$ it will follow that $\mathbf{c} \leqq \mathbf{d}'$. Fix $e^* \geqq 0$ and suppose $A(e)$ and $B(e)$ are true for all $e < e^*$. We proceed to prove $A(e^*)$ and $B(e^*)$.

LEMMA 1. *Let $y(s,n,e^*) > 0$ and $m(s,e^*) > n \geqq e^*$. Let $d(s,p_i^m) = d(s-1,p_i^m)$ for all i, m and t such that $i < e^* \leqq t \leqq n$ and $p_i^m < y(s,t,e^*)$. Then $y(s,n,e^*) = y(s+1,n,e^*)$.*

Proof. Since $y(s,n,e^*) > 0$, we have

$$y(s,n,e^*) = \mu y T_1^1\left(\prod_{i<y} p_i^{d(s-1,i)}, e^*, n, y\right).$$

We suppose $y(s+1,n,e^*) \neq y(s,n,e^*)$ and then show there is an i, an m and a t such that

$$i < e^* \leqq t \leqq n \;\&\; p_i^m < y(s,t,e^*) \;\&\; d(s,p_i^m) \neq d(s-1,p_i^m).$$

Since $y(s+1,n,e^*) \neq y(s,n,e^*)$, there must be a $j < y(s,n,e^*)$ such that $d(s,j) \neq d(s-1,j)$. Recall that $d(s,w) = d(s-1,w)$ for all w not a power of a prime. Thus there is an i' and an m' such that

$$d(s,p_{i'}^{m'}) \neq d(s-1,p_{i'}^{m'})$$

and $p_{i'}^{m'} < y(s,n,e^*)$. But then by the hypothesis of the lemma, $e^* \leqq i'$. Thus we have

$$e^* \leqq i' \;\&\; e^* \leqq n < m(s,e^*) \;\&\; p_{i'}^{m'} < y(s,n,e^*) \;\&\; d(s,p_{i'}^{m'}) \neq d(s-1,p_{i'}^{m'}).$$

It follows from the definition of $d(s,p_{i'}^{m'})$ that $r(s,n,e^*) = 0$. But this last means the desired i, m and t exist.

LEMMA 2. *Let $y(s,n,e^*) > 0$ and $m(s,e^*) > n > e^*$. Let $d(s,p_i^m) = d(s-1,p_i^m)$ for all i, m and t such that $i < e^* \leqq t \leqq n$ and $p_i^m < y(s,t,e^*)$. Then $m(s+1,e^*) > n$.*

Proof. Since $m(s,e^*) > n > e^*$, it follows from remark (R3) and Case 1 of the definition of $m(s,e)$ that

$$y(s,t,e^*) > 0$$

for all t such that $e^* \leqq t \leqq n$. But then by Lemma 1,

$$y(s,t,e^*) = y(s+1,t,e^*)$$

for all t such that $e^* \leqq t \leqq n$. Suppose $m(s+1,e^*) \leqq n$. Then $m(s+1,e^*) < m(s,e^*)$, and consequently, Case 2 of the definition of $m(s+1, e^*)$ holds. This means there is a t (namely, $m(s+1, e^*)$) such that

$$e^* < t \leqq n \;\&\; y(s, t, e^*) \neq y(s, + 1, t, e^*).$$

LEMMA 3. $A(e^*)$.

Proof. By the hypothesis of our theorem the function a is nonrecursive. We suppose $A(e^*)$ is false and show a is recursive. Thus the set $\{m(s, e^*) \mid s \geqq 0\}$ is infinite, and for each $n \geqq e^*$, $\lim_s y(s, n, e^*)$ exists and is positive. Let $R(n, s)$ denote the predicate

$$m(s, e^*) > n \;\&\; (e)(m)(t)[(p_e^m < y(s, t, e^*) \;\&\; e < e^* \leqq t \leqq n) \rightarrow d(s-1, p_e^m) = d(p_e^m)].$$

We know $B(e)$ is true for all $e < e^*$. This means $\lim_m d(p_e^m)$ exists for all $e < e^*$. For each $e < e^*$, let $g(e)$ be such that

$$(m)[m \geqq g(e) \rightarrow d(p_e^m) = d(p_e^{g(e)})].$$

We define a recursive function $z(n)$ as follows: first we require that $z(n) = 1$ for all n not a power of a prime; then we specify

$$z(p_e^m) = \begin{cases} d(p_e^{g(e)}) & \text{if } e > e^* \;\&\; m \geqq g(e), \\ d(p_e^m) & \text{if } e > e^* \;\&\; m < g(e), \\ 1 & \text{otherwise.} \end{cases}$$

The predicate $R(n, s)$ can now be rewritten as

$$m(s, e^*) > n \;\&\; (e)(m)(t)[(p_e^m < y(s, t, e^*) \;\&\; e < e^* \leqq t \leqq n) \rightarrow d(s-1, p_e^m) = z(p_e^m)].$$

It is clear that $R(n, s)$ is recursive, since the functions m, y and z are recursive.

Now we show $(n)(Es)R(n, s)$. Fix n. Since $\lim_s y(s, n, e^*)$ exists for all $n \geqq e^*$ there is a y such that

$$y \geqq y(s, t, e^*)$$

for all t and s such that $e^* \leqq t \leqq n$. Let s' be so large that

$$d(s-1, w) = d(w)$$

for all s and w such that $s \geqq s'$ and $w < y$. Since the set $\{m(s, e^*) \mid s \geqq 0\}$ is infinite, there is an $s \geqq s'$ such that $m(s, e^*) > n$. But then $R(n, s)$.

Let $w(n)$ denote the recursive function $\mu s R(n, s)$. Note that $w(n+1) \geqq w(n)$ for all n.

Next we prove $y(w(n), n, e^*) = \lim_s y(s, n, e^*)$ for all $n > e^*$. Fix $n > e^*$. We show by induction on s that $y(w(n), n, e^*) = y(s, n, e^*)$ for all $s \geqq w(n)$. Let s be such that $s \geqq w(n)$ and

$$y(w(n), e^*) = y(s, n, e^*) \,\&\, R(n, s).$$

Since $m(s, e^*) > n > e^*$, it follows from remark (R3) and Case 1 of the definition of $m(s, e^*)$ that $y(s, t, e^*) > 0$ for all t such that $e^* \leqq t \leqq n$. By the definition of $R(n, s)$, we have

$$d(s-1, p_e^m) = d(p_e^m)$$

for all e, m and t such that $e < e^* \leqq t \leqq n$ and $p_e^m < y(s, t, e^*)$. Recall that if $d(s-1, w) = d(w)$ then $d(s', w) = d(w)$ for all $s' \geqq s$. It follows from Lemma 1 that

$$y(s, t, e^*) = y(s+1, t, e^*)$$

for all t such that $e^* \leqq t \leqq n$. It follows from Lemma 2 that

$$m(s+1, e^*) > n.$$

But then

$$y(w(n), n, e^*) = y(s+1, n, e^*) \,\&\, R(n, s+1).$$

Thus $y(w(n), n, e^*) = y(s, n, e^*)$ for all $s \geqq w(n)$, and $\lim_s y(s, n, e^*) = y(w(n), n, e^*)$.

Finally, we show by means of a reductio ad absurdum that

$$a(n) = U(y(w(n), n, e^*))$$

for all $n > e^*$. It will then follow that a is recursive, since w is recursive. Fix $n > e^*$ and suppose $a(n) \neq U(y(w(n), n, e^*))$. Since $y(w(n), n, e^*) = \lim_s y(s, n, e^*)$, and since $a(n) = \lim_s a(s, n)$, there is an s^* such that for all $s \geqq s^*$,

$$a(n) = a(s, n) \neq U(y(s, n, e^*)) = U(y(w(n), n, e^*)).$$

Let $s > s^*$ and suppose $m(s-1, e^*) \leqq m(s^*, e^*) + n + e^* + 1$. If either Case 1 or Case 2 of the of the definition of $m(s, e^*)$ holds, then

$$m(s, e^*) \leqq \max(e^* + 1, m(s-1, e^*)) \leqq m(s^*, e^*) + n + e^* + 1.$$

If Case 3 holds and $n < 2m(s-1, e^*) + s$, then $m(s, e^*) \leqq n$. If Case 3 holds and $2m(s-1, e^*) + s \leqq n$, then $m(s, e^*) \leqq n$. Thus we have shown by induction on s that

$$m(s, e^*) \leqq m(s^*, e^*) + n + e^* + 1$$

for all $s \geqq s^*$. But this last is absurd, since the set $\{m(s, e^*) \mid s \geqq 0\}$ is infinite.

For each $e \geqq 0$, we say e is stable if for all $n \geqq e$, $\lim_s y(s, n, e)$ exists and is positive. Note that if e is not the Gödel number of a system of equations, then $y(s, n, e) = 0$ for all s and n, and consequently, e is not stable. It follows that there are infinitely many e which are not stable, since there are infinitely many e which are not Gödel numbers of systems of equations. We define

$$e_0 = \mu e \text{ (e is not stable)};$$

$$e_{j+1} = \mu e \text{ ($e > e_j$ and e is not stable)}.$$

Thus $e_0 < e_1 < e_2 < \cdots$ is a listing of all the e which are not stable. For each $j \geqq 0$, let n_j be the least $n \geqq e_j$ such that $\lim_s y(s, n, e_j)$ either does not exist or is equal to 0.

The most important part of our argument is contained in Lemma 4. If the proof of our theorem is a heavy meal, then the proof of Lemma 4 is the main course; furthermore, it is there that the combinatorial flavor of our reasoning is strongest.

LEMMA 4. *For each k and v, there is an $s \geqq v$ such that*

$$(j)_{j<k}[m(s, e_j) \leqq n_j \vee r(s, n_j, e_j) = 0 \vee y(s, n_j, e_j) = 0].$$

Proof. Fix k and v. We suppose there is no s with the properties required by the lemma, and then show it is possible to define an infinite, descending sequence of natural numbers.

We propose the following system of equations as a means of defining two functions, $S(t)$ and $M(t)$, simultaneously by induction:

$$S(0) = \mu s \ (s \geqq v);$$

$$M(t) = \mu j[j < k \,\&\, n_j < m(S(t), e_j) \,\&\, r(S(t), n_j, e_j) = 1 \,\&\, y(S(t), n_j, e_j) > 0];$$

$$S(t+1) = \mu s(Em)[s \geqq S(t) \,\&\, m < y(S(t), n_{M(t)}, e_{M(t)}) \,\&\, d(s, m) \neq d(S(t) - 1, m)].$$

Clearly $S(0)$ is well defined and greater than or equal to v. Suppose $t \geqq 0$ and $S(t)$ is well defined and greater than or equal to v. Then $M(t) < k$, since we have supposed the lemma to be false. Thus

$$y(S(t), n_{M(t)}, e_{M(t)}) > 0$$

and $\lim_s y(s, n_{M(t)}, e_{M(t)})$ does not exist or is equal to 0. Then there must be an $s > S(t)$ such that

$$y(s, n_{M(t)}, e_{M(t)}) \neq y(S(t), n_{M(t)}, e_{M(t)});$$

note that $S(t) > 0$, since $y(0, n, e) = 0$ for all n and e; this means there is an $s > S(t)$ and an m such that

$$m < y(S(t), n_{M(t)}, e_{M(t)}) \,\&\, d(s-1, m) \neq d(S(t) - 1, m).$$

Then $S(t+1)$ is well defined and greater than or equal to v.

For each $t \geqq 0$, let

$$u(t) = \mu m[d(S(t+1), m) \neq d(S(t) - 1, m)].$$

Now we show $u(t) < u(t-1)$ for all $t > 0$. Fix $t > 0$. Since we have

$$u(t) < y(S(t), n_{M(t)}, e_{M(t)})$$

by definition of u, it will be sufficient to show

$$y(S(t), n_{M(t)}, e_{M(t)}) \leqq u(t-1).$$

Since $d(w) = 1$ for all w not a power of a prime, there must exist i and m such that $u(t-1) = p_i^m$. Note that $d(S(t),\ u(t-1)) \neq d(S(t)-1,\ u(t-1))$; this last follows from the definitions of $S(t)$ and $u(t-1)$. Let

$$e = e_{M(t)},\ s = S(t) \text{ and } n = n_{M(t)}.$$

First we suppose $i < e$. This means

$$i < e \leqq n \;\&\; d(s, p_i^m) \neq d(s-1, p_i^m) \;\&\; r(s,n,e) = 1,$$

since $M(t) < k$. But then it follows from the definition of $r(s,n,e)$ that $y(s,n,e) \leqq p_i^m$. Now we suppose $i \geqq e$. This means

$$e \leqq i \;\&\; e \leqq n < m(s,e) \;\&\; r(s,n,e) = 1 \;\&\; d(s, p_i^m) \neq d(s-1, p_i^m),$$

since $M(t) < k$. But then it follows from the definition of $d(s, p_i^m)$ that $y(s,n,e) \leqq p_i^m = u(t-1)$.

LEMMA 5. *If* $c(e^*) = 0$, *then* $\lim_n d(p_{e^*}^n)$ *exists and is equal to* 1.

Proof. Let t be the least m such that $f(m) = e^*$. Let s' be so large that $s' > t$ and $f(s,m) = f(m)$ for all s and m such that $s \geqq s'$ and $m \leqq t$. Then $t(s,e^*) = t$ for all $s \geqq s'$, and consequently, $h(s,e^*) = h(s',e^*)$ for all $s \geqq s'$. But then

$$d(s, p_{e^*}^n) = d(s-1, p_{e^*}^n)$$

for all s and n such that $s > 0$ and $n \geqq h(s',e^*)$, since $h(s,e^*) \leqq h(s',e^*)$ for all $s \leqq s'$. It follows that $\lim_s d(s, p_{e^*}^n) = 1$ for all $n \geqq h(s',e^*)$, since $d(0,w) = 1$ for all w. Then $d(p_{e^*}^n) = 1$ for all $n \geqq h(s',e^*)$, and $\lim_n d(p_{e^*}^n) = 1$.

LEMMA 6. *If* $c(e^*) = 1$, *then* $\lim_n d(p_{e^*}^n)$ *exists and is equal to* 0.

Proof. First we show that the set $\{t(s,e^*) \mid s \geqq 0\}$ is infinite. Suppose $t(s,e^*) \leqq t$ for all s. Let s' be so large that $s' > t$ and $f(s,m) = f(m)$ for all s and m such that $s \geqq s'$ and $m \leqq t$. Then $f(s', t(s',e^*)) = e^*$, since $t(s',e^*) < s'$. But

$$f(s', t(s',e^*)) = f(t(s',e^*)),$$

since $t(s',e^*) \leqq t$. But then $f(t(s',e^*)) = e^*$; this last is impossible because C is the range of f and $c(e^*) = 1$.

Since the set $\{t(s,e^*) \mid s \geqq 0\}$ is infinite, it is clear that the set $\{h(s,e^*) \mid s \geqq 0\}$ is infinite.

By Lemma 3, we know $A(e)$ holds for all $e \leqq e^*$. This means that if $e \leqq e^*$ and e

is stable, then the set $\{m(s,e^*)\,|\,s \geqq 0\}$ is finite. If $e \leqq e^*$ and e is stable, let $m(e)$ be the greatest member of $\{m(s,e^*)\,|\,s \geqq 0\}$; if $e \leqq e^*$ and e is not stable, let $m(e) = n_j$, where j is such that $e = e_j$. If $e \leqq e^*$ and $e \leqq m < m(e)$, then $\lim_s y(s,m,e)$ exists. Let y be so large that

$$y \geqq y(s,m,e)$$

for all s, m and e such that $e \leqq e^*$ and $e \leqq m < m(e)$.

Fix $n > y$. We show $d(p^n_{e^*}) = 0$. It will suffice to find an s such that $d(p^n_{e^*}, s) = 0$. Let v be such that $h(v,e^*) > n$. Let k be such that if $e \leqq e^*$ and e is not stable, then $e = e_j$ for some $j < k$. By Lemma 4 there is an $s \geqq v$ such that

$$(j)_{j<k}[m(s,e_j) \leqq n_j \vee r(s,n_j,e_j) = 0 \vee y(s,n_j,e_j) = 0].$$

We will show:

$$h(s,e^*) > n;$$

$$(e)(m)[(e \leqq e^* \ \&\, e \leqq m < m(s,e)) \rightarrow (r(s,m,e) = 0 \vee p^n_{e^*} \geqq y(s,m,e))].$$

It will then follow from the definition of $d(s,p^n_{e^*})$ that $d(s,p^n_{e^*}) = 0$. We have $h(s,e^*) > n$, since $s \geqq v$ and $h(s,e^*)$ is a nondecreasing function of s. Fix e and m so that $e \leqq e^*$ and $e \leqq m < m(s,e)$. Suppose e is stable. Then $m < m(e)$, since $m(s,e) \leqq m(e)$. But then $y \geqq y(s,m,e)$, and consequently $p^n_{e^*} \geqq y(s,m,e)$, since $n > y$.

Now suppose e is not stable. Then $e = e_j$, where $j < k$, and $m(e) = n_j$. If $m < n_j$, then $m < m(e)$ and $p^n_{e^*} \geqq y(s,m,e)$. Suppose $m \geqq n_j$. Then $m(s,e_j) > n_j$. This last means that either $r(s,n_j,e) = 0$ or $y(s,n_j,e) = 0$. If $r(s,n_j,e) = 0$, then by remark (R2), $r(s,m,e) = 0$, since $m \geqq n_j$. Suppose $y(s,n_j,e) = 0$. Since $n_j \leqq m < m(s,e)$, it follows from remark (R3) that $n_j = e$. But then $y(s,e,e) = 0$, and Case 1 of the definition of $m(s,e)$ holds. It follows that $m(s,e) = e+1, m = e$ and $y(s,m,e) = 0$.

Thus $d(p^n_{e^*}) = 0$ for all $n > y$, and $\lim_n d(p^n_{e^*})$ exists and is equal to 0.

Lemmas 5 and 6 constitute a proof of $B(e^*)$. That concludes our proof by induction of $(e)A(e)$ and $(e)B(e)$. It is now easily seen that $\mathbf{c} \leqq \mathbf{d}'$. Observe that

$$(e)(Et)[(m)_{m \geqq t}(d(p^m_e) = 1) \vee (m)_{m \geqq t}(d(p^m_e) = 0)]$$

is an immediate consequence of $(e)B(e)$. We define

$$k(e) = \mu t[(m)_{m \geqq t}(d(p^m_e) = 1) \vee (m)_{m \geqq t}(d(p^m_e) = 0)].$$

The function k has degree less than or equal to $\mathbf{d}'$, and by $(e)\mathrm{B}(e)$,

$$c(e) = 1 - d(p^{k(e)}_e)$$

for all e.

LEMMA 7. $\mathbf{a} \nleqq \mathbf{d}$.

Proof. We suppose there is a Gödel number e such that

$$a(n) = \{e\}\ (n)$$

for all n, and then show $A(e)$ is false. First we show that $\lim_s y(s,n,e)$ exists and is positive for all $n \geqq e$. Fix $n \geqq e$; let

$$w = \mu y T_1^1(\tilde{d}(y), e, n, y).$$

Let s' be so large that $d(s,m) = d(m)$ whenever $s \geqq s'$ and $m < w$. Then

$$y(s,n,e) = w \ \&\ U(w) = a(n)$$

for all $s \geqq s' + w$; $w > 0$, since 0 is not the Gödel number of a deduction.

Now we show the set $\{m(s,e) \mid s \geqq 0\}$ is infinite. We fix $m > e$ and obtain an s' such that $m(s',e) > m$. Let s be so large that $s > m$ and

$$a(s,t) = a(t) = \{e\}^d(t) = U(y(s,t,e))$$

for all t such that $e \leqq t \leqq m$. If $m(s-1,e) > m$, then $s-1$ is the desired s'. Suppose $m(s-1,e) \leqq m$. This means

$$a(s,t) = U(y(s,t,e))$$

for all t such that $e \leqq t \leqq m(s-1,e)$; in addition, $y(s,e,e) > 0$, since $a(s,e) \geqq 1$ and $U(0) = 0$. But then Case 3 of the definition of $m(s,e)$ holds. It follows that $m(s,e) > m$, since $s > m$.

LEMMA 8. $\mathbf{d}' \leqq \mathbf{c}$.

Proof. We will define two functions, $E(e,n)$ and $L(e)$, simultaneously by induction on e so that each is recursive in the function $c(n)$. We will combine the definition of E and L with a proof by induction on e of the following:

$$(e)(n)\{E(e,n) = 0 \leftrightarrow [n \geqq e \,\&\, (w)(Es)(s > w \,\&\, m(s,e) > n)$$
$$\&\, (m)(n \geqq m \geqq e \rightarrow (Ey)\, T_1^1(\tilde{d}(y), e, m, y))]\};$$

$$(e)\,(m)[m \geqq L(e) \rightarrow d(p_e^m) = d(p_e^{L(e)})].$$

It follows immediately from the above and remark (R1) that

$$(e)[E(e,e) = 0 \leftrightarrow (Ey) T_1^1(\tilde{d}(y), e, e, y)];$$

but then if E is recursive in c, we have $\mathbf{d}' \leqq \mathbf{c}$.

Fix $e \geqq 0$. Our induction hypothesis has two parts:

(1, e) for each $i < e$ and each n, $E(i,n)$ has been defined and

$$E(i,n) = 0 \leftrightarrow [n \geqq i \,\&\, (w)(Es)(s > w \,\&\, m(s,i) > n)$$
$$\&\, (m)(n \geqq m \geqq i \rightarrow (Ey) T_1^1(\tilde{d}(y), i, m, y))];$$

(2, e) for each $i < e$, $L(i)$ has been defined and

$$(m)[m \geqq L(i) \rightarrow d(p_i^m) = d(p_i^{L(i)})].$$

We proceed to define $E(e,n)$ for all n, verify $(1, e+1)$, define $L(e)$ and verify $(2, e+1)$.

Let $(1, e+1, n)$ denote the following predicate: for each $t < n$, $E(e,t)$ has been defined and

$$E(e,t) = 0 \leftrightarrow [t \geqq e \,\&\, (w)(Es)(s > w \,\&\, m(s,e) > t)$$
$$\&\, (m)(t \geqq m \geqq e \rightarrow (Ey)T_1^1(\tilde{d}(y), e, m, y))].$$

To verify $(1, e+1)$, it suffices to prove $(1, e+1, n)$ for all n. We define $E(e,n)$ and prove $(1, e+1, n)$ for all n by means of an induction on n. First we set $E(e,t) = 1$ for all $t < e$. Then it is clear that $(1, e+1, t)$ holds for all $t \leqq e$. Now we fix $n \geqq e$ and suppose $(1, e+1, n)$ holds. We define $E(e,n)$ and then prove $(1, e+1, n+1)$. The definition of $E(e,n)$ has two cases.

Case 1. $(Et)(e \leqq t < n \,\&\, E(e,t) \neq 0)$. We set $E(e,n) = 1$.

Case 2. Otherwise. It follows from $(1, e+1, n)$ that

$$(m)(n > m \geqq e \rightarrow (Ey)T_1^1(\tilde{d}(y), e, m, y)).$$

For each m such that $n > m \geqq e$, let

$$y(m) = \mu y T_1^1(\tilde{d}(y), e, m, y).$$

Let y^* be the largest member of $\{y(m) \mid n > m \geqq e\} \cup \{0\}$. Let

$$s^* = \mu s[(i)(i < y^* \rightarrow d(s-1, i) = d(i)) \,\&\, s > y^*].$$

Recall that for each $s > 0$ and i, if $d(s-1, i) = d(i)$, then $d(s', i) = d(i)$ for all $s' \geqq s$. It follows from the definition of $y(s,m,e)$ and the fact that 0 is not the Gödel number of a deduction that

$$(s)(m)[(s \geqq s^* \,\&\, n > m \geqq e) \rightarrow y(s,m,e) = y(m) > 0].$$

We define

$$E(e,n) = \begin{cases} 0 & \text{if } (Es)\{y(s,n,e) > 0 \,\&\, n < m(s,e) \,\&\, s > s^* \,\&\, \\ & \quad (i)_{i<e}[(m)(m < L(i) \rightarrow d(s-1, p_i^m) = d(p_i^m)) \\ & \quad \&\, (m)(y(s,n,e) > m \geqq L(i) \rightarrow d(s-1, p_i^m) \\ & \qquad = d(p_i^{L(i)}))]\}, \\ 1 & \text{otherwise.} \end{cases}$$

To verify $(1, e+1, n+1)$, it suffices to prove

$$E(e,n) = 0 \leftrightarrow [n \geqq e \,\&\, (w)(Es)(s > w \,\&\, m(s,e) > n)$$
$$\&\, (m)(n \geqq m \geqq e \rightarrow (Ey)T_1^1(\tilde{d}(y), e, m, y))].$$

Suppose $E(e,n) = 0$. Then Case 2 of the definition of $E(e,n)$ must hold. Let s be the natural number whose existence is required by the fact $E(e,n) = 0$; thus

$$y(s,n,e) > 0 \mathbin{\&} n < m(s,e) \mathbin{\&} s > s^*.$$

It is our aim now to prove

$$y(s,n,e) = y(s',n,e) \mathbin{\&} n < m(s',e)$$

for all $s' \geqq s$ by means of an induction on s'. Fix $s' > s$ and suppose

$$y(s,n,e) = y(s'-1,n,e) \mathbin{\&} n < m(s'-1,e).$$

Suppose for the sake of a reductio ad absurdum that there is a w such that $d(s'-1,w) \neq d(s'-2,w)$ and $w < y(s'-1,n,e)$. Then there must be an i and an m such that (0) and (1) are true:

(0) $\qquad d(s'-1,p_i^m) \neq d(s'-2,p_i^m) \mathbin{\&} p_i^m < y(s'-1,n,e) = y(s,n,e);$

(1) $\qquad (i')_{i'<i}(m')[d(s'-1,p_{i'}^{m'}) = d(s'-2,p_{i'}^{m'}) \vee p_{i'}^{m'} \geqq y(s'-1,n,e)].$

If $i < e$, then it follows from the second half of (0), the definition of s and $(2, e)$ that $d(s-1,p_i^m) = d(p_i^m)$; but this last contradicts the first half of (0), since $s' > s$. Thus $i \geqq e$. Since $e \leqq n < m(s'-1,e)$, it follows from (0) and the definition of $d(s'-1,p_i^m)$ that $r(s'-1,n,e) = 0$. This last means there is an i', an m' and a t such that

$$i' < e \leqq t \leqq n \mathbin{\&} p_{i'}^{m'} < y(s'-1,t,e) \mathbin{\&} d(s'-1,p_{i'}^{m'}) \neq d(s'-2,p_{i'}^{m'}).$$

If $t = n$, this last contradicts (1), since $i' < e \leqq i$. Thus $t < n$, and

$$p_{i'}^{m'} < y(s'-1,t,e) = y(s^*,t,e) \leqq y^*,$$

since $s' > s > s^*$ and $e \leqq t < n$. But this is absurd, since

$$d(s'-1,p_{i'}^{m'}) = d(s'-2,p_{i'}^{m'}) = d(s-1,p_{i'}^{m'}) = d(p_{i'}^{m'})$$

is a consequence of the fact that $s' > s > s^*$ and $p_{i'}^{m'} < y^*$.

Since $d(s'-1,w) = d(s'-2,w)$ for all $w < y(s'-1,n,e)$, and since $y(s'-1,n,e) = y(s,n,e) > 0$, it must be that

$$y(s',n,e) = y(s'-1,n,e) = y(s,n,e).$$

Then we have

$$(m)[n \geqq m \geqq e \rightarrow y(s',n,e) = y(s'-1,n,e) > 0],$$

since $s' > s > s^*$. It follows that either Case 2 or Case 3 of the definition of $m(s',e)$ holds. If Case 2 of the definition of $m(s',e)$ holds, then it is clear $n < m(s',e)$. If Case 3 holds, then $n < m(s',e)$ because $n < m(s'-1,e)$.

Thus we have shown that

$$y(s',n,e) = y(s,n,e) > 0 \mathbin{\&} n < m(s',e)$$

for all $s' \geqq s$. It follows immediately that $(Ey)T_1^1(\tilde{d}(y),e,n,y)$ and

$$(w)(Es)(s > w \,\&\, m(s,e) > n).$$

Note that

$$(m)(n > m \geqq e \rightarrow (Ey)T_1^1(\tilde{d}(y), e, m, y))$$

is a consequence of the fact that Case 2 of the definition of $E(e,n)$ holds. That completes the first half of the verification of $(1, e+1, n+1)$; in order to verify the second half, we suppose

$$(w)(Es)(s > w \,\&\, m(s,e) > n) \,\&\, (m)(n \geqq m \geqq e \rightarrow (Ey)T_1^1(\tilde{d}(y), e, m, y)),$$

and then show $E(e,n) = 0$. It follows from $(1, e+1, n)$ that Case 2 of the definition of $E(e,n)$ holds. Let v be so large that $v > L(i)$ for all $i < e$. Let $z = \mu y T_1^1(\tilde{d}(y), e, n, y)$. Let w be so large that $w > z$ and

$$d(w-1, t) = d(t)$$

for all $t < p_e^{z+v}$. Let s be such that $s > w + s^*$ and $m(s,e) > n$. It follows easily from $(2, e)$ that s has the properties required to conclude $E(e,n) = 0$; note that $y(s,n,e) = z > 0$.

The definition of $L(e)$ has two cases:

Case 1. $c(e) = 0$. Then by $B(e)$, $\lim_m d(p_e^m) = 1$. We set

$$L(e) = \mu t(s)(m)[m \geqq t \rightarrow d(s, p_e^m) = 1].$$

Case 2. $c(e) = 1$. It is a consequence of $(1, e+1)$ and of the definition of $y(s,n,i)$ that for each $i \leqq e$ and each n,

$$E(i,n) = 0 \rightarrow [(Es)(m(s,i) > n) \,\&\, \lim_s y(s,n,i) \text{ exists and is positive}].$$

For each $i \leqq e$, it follows from $A(i)$ that there is a t such that $t \geqq i$ and $E(i,t) = 1$. For each $i \leqq e$, let

$$t_i = \mu t(E(i,t) = 1 \,\&\, t \geqq i).$$

It follows from $(1, e+1)$ that for each $i \leqq e$, t_i satisfies either (2) or (3):

(2) $(Ew)(s)(s > w \rightarrow m(s,i) \leqq t_i)$;

(3) $\lim_s y(s, t_i, i)$ does not exist or is equal to 0.

Note that since $E(i,t) = 0$ whenever $t_i > t \geqq i$, it follows from $(1, e+1)$ that

$$(m)(t_i > m \geqq i \rightarrow \lim_s y(s,m,i) \text{ exists and is positive}).$$

But then

$$(Ey)(i)(m)(s)[(i \leqq e \,\&\, t_i > m \geqq i \,\&\, s \geqq 0) \rightarrow y(s,m,i) \leqq y];$$

let $L(e)$ be the least such y. We now verify $(2, e+1)$. What follows is similar to the proof of Lemma 6. Fix $n \geqq L(e)$. We must show $d(p_e^n) = \lim_m d(p_e^m)$. If Case 1 of the definition of $L(e)$ holds, there is nothing to prove. Suppose Case 2 of the definition of $L(e)$ holds. Then $c(e) = 1$, and by $B(e)$, $\lim_m d(p_e^m) = 0$. In order to show $d(p_e^n) = 0$, it suffices to find an s such that $d(s, p_e^n) = 0$. Let k be such that if $i \leqq e$ and i is not stable, then $i = e_j$ for some $j < k$. Let w be so large that for all $i \leqq e$, if (2) holds, then

$$(s)(s > w \rightarrow m(s, i) \leqq t_i).$$

By the same argument as in Lemma 6, there is a $v > w$ such that

$$(s)(s \geqq v \rightarrow h(s, e) > n).$$

By Lemma 4, there is an $s \geqq v$ such that

$$(j)_{j<k}[m(s, e_j) \leqq n_j \vee r(s, n_j, e_j) = 0 \vee y(s, n_j, e_j) = 0].$$

If we can show for each i and m that

$$(i \leqq e \,\&\, i \leqq m < m(s, i)) \rightarrow (r(s, m, i) = 0 \vee p_e^n \geqq y(s, m, i)),$$

then it will be clear that $d(s, p_e^n) = 0$. Fix i and m so that $i \leqq e$ and $i \leqq m < m(s, i)$. Suppose (2) holds. Then $m(s, i) \leqq t_i$, since $s \geqq v > w$. But then $t_i > m \geqq i$, and consequently,

$$p_e^n > n \geqq L(e) \geqq y(s, m, i).$$

Now suppose (3) holds. Then i is not stable, and there is a $j < k$ such that $e_j = i$; in addition, $n_j = t_i$. If $m < n_j = t_i$, then $p_e^n \geqq y(s, m, i)$, since $n \geqq L(e)$. Suppose $m \geqq n_j$. Then $m(s, i) > n_j$. This last means either

$$r(s, n_j, e_j) = 0 \qquad \text{or} \qquad y(s, n_j, e_j) = 0.$$

If $r(s, n_j, i) = 0$, then by remark (R2), $r(s, m, i) = 0$, since $m \geqq n_j$. Suppose $y(s, n_j, i) = 0$. Since $n_j \leqq m < m(s, i)$, it follows from remark (R3) that $n_j = i$. But then $y(s, i, i) = 0$, $m(s, i) = i + 1$, $m = i$ and $y(s, m, i) = 0 < p_e^n$.

Inspection of the definitions of E and L readily reveals they are recursive in c. We make some informal remarks to indicate how to write equations defining E and L recursively in c. Fix $e \geqq 0$ and $n \geqq e$, and consider the definition of $E(e, n)$. The choice between Case 1 and Case 2 can be made effectively once the values of $E(e, t)(t < n)$ are known. Suppose Case 2 holds. The values of y^* and s^* are defined by means of a predicate recursive in d. Then the value of $E(e, n)$ is found from the values of $d(p_i^m)$ ($m \leqq L(i)$ and $i < e$) by means of a predicate of degree less than or equal to $\mathbf{0}'$. Thus the value of $E(e, n)$ can be expressed in terms of the values of $E(e, t)(t < n)$ and $L(i)$ ($i < e$) with the aid of a predicate of degree less than or equal to $\mathbf{d} \cup \mathbf{0}'$. Similarly, the value of $L(e)$ can be expressed in terms of the values of $E(i, n)$ ($i \leqq e$ and $n \geqq 0$) with the aid of a predicate of degree $\mathbf{c} \cup \mathbf{0}'$. Then $\mathbf{d}' \leqq \mathbf{c}$, since $\mathbf{d}'$ is the degree of the function $E(e, e)$, and since $\mathbf{d} \cup \mathbf{0}' \cup \mathbf{c} = \mathbf{c}$.

When $\mathbf{b} > \mathbf{0}$, the changes needed in the above argument are largely notational. The notion of recursiveness is replaced throughout by the notion of recursiveness in a function of degree $\mathbf{b}$. The arguments contained in Lemmas 1–8 are retained unaltered save for relativization to a function of degree **b**. The functions $a(s,n)$, $f(s,n)$ and $d(s,n)$ are now recursive in a function of degree **b**. We set $d(s,2\cdot 3^n) = b(n)$ for all s and all $n > 0$, where b has degree **b**, in order to insure $\mathbf{b} \leqq \mathbf{d}$. The value of $E(e,n)$ is now obtained from the values of $E(e,t)$ $(t < n)$ and $L(i)\,(i < e)$ with the aid of a predicate of degree less than or equal to $\mathbf{b}' \cup \mathbf{d}$.

COROLLARY 1. *If* **b** *and* **c** *are degrees, then the following conditions are equivalent*:

(i) $\mathbf{b}' \leqq \mathbf{c} \leqq \mathbf{b}''$ *and* **c** *is recursively enumerable in* $\mathbf{b}'$;

(ii) *there is a* **d** *such that* $\mathbf{b} \leqq \mathbf{d} \leqq \mathbf{b}'$ *and* $\mathbf{d}' = \mathbf{c}$;

(iii) *there is a* **d** *such that* $\mathbf{b} \leqq \mathbf{d} \leqq \mathbf{b}'$, **d** *is recursively enumerable in* **b** *and* $\mathbf{d}' = \mathbf{c}$.

For each degree **b**, let $R_{\mathbf{b}}$ denote the set of all degrees greater than or equal to **b**, recursively enumerable in **b** and less than or equal to $\mathbf{b}'$. Let j denote the jump operator. Then Corollary 1 tells us that the order-preserving map

$$j: R_{\mathbf{b}} \rightarrow R_{\mathbf{b}'}$$

is onto. It also follows from Theorem 1 that any element of $R_{\mathbf{b}'}$ greater than $\mathbf{b}'$ is the image of more than one element of $R_{\mathbf{b}}$; Friedberg (result unpublished) has shown that $\mathbf{b}'$ does not have a unique pre-image in $R_{\mathbf{b}}$. We do not know if $R_{\mathbf{b}}$ and $R_{\mathbf{b}'}$ are order-isomorphic, but we conjecture that they are. We can show (announced in [6] for $\mathbf{b} = \mathbf{0}$): for any degree **b**, $R_{\mathbf{b}}$ is a universal, countable partial ordering.

COROLLARY 2. *There exists a recursively enumerable degree* **d** *such that* $\mathbf{d} < \mathbf{0}' < \mathbf{0}'' = \mathbf{d}'$.

Proof. Let $\mathbf{b} = \mathbf{0}$, $\mathbf{c} = \mathbf{0}''$ and $\mathbf{a} = \mathbf{0}'$, and apply Theorem 1 to obtain **d**. Then **d** is recursively enumerable, $\mathbf{0}' \nleqq \mathbf{d}$ and $\mathbf{d}' = \mathbf{0}''$.

Note that Corollary 2 provides still another solution to Post's problem.

COROLLARY 3. *For each degree* **b** *and each natural number n, there is a degree* **d** *recursively enumerable in* **b** *such that*

$$\mathbf{b} < \mathbf{d} < \mathbf{b}' < \mathbf{d}' < \mathbf{b}'' < \cdots < \mathbf{b}^{(n)} < \mathbf{d}^{(n)} < \mathbf{b}^{(n+1)}.$$

Proof. We know from [2] that there exists a degree **g** such that **g** is recursively enumerable in $\mathbf{b}^{(n)}$ and $\mathbf{b}^{(n)} < \mathbf{g} < \mathbf{b}^{(n+1)}$. By Theorem 1, there is a degree $\mathbf{h}_1$ such that $\mathbf{h}_1$ is recursively enumerable in $\mathbf{b}^{(n-1)}$, $\mathbf{b}^{(n-1)} < \mathbf{h}_1 < \mathbf{b}^{(n)}$ and $\mathbf{h}_1' = \mathbf{g}$. By making $n-1$ further applications of Theorem 1, we obtain degrees $\mathbf{h}_2, \mathbf{h}_3, \cdots, \mathbf{h}_n$ such that for $2 \leqq i \leqq n$, $\mathbf{h}_i$ is recursively enumerable in $\mathbf{b}^{(n-i)}$,

$$\mathbf{b}^{(n-i)} < \mathbf{h}_i < \mathbf{b}^{(n-i+1)} \quad \text{and} \quad \mathbf{h}_i' = \mathbf{h}_{i-1}.$$

Let $\mathbf{d} = \mathbf{h}_n$. Then **d** is recursively enumerable in **b**, and for all $i \leqq n$, $\mathbf{b}^{(i)} < \mathbf{d}^{(i)} < \mathbf{b}^{(i+1)}$.

Corollary 3 improves a result of Shoenfield [7]; he showed that for each degree $\mathbf{b}$, there is a degree $\mathbf{d}$ such that $\mathbf{b} < \mathbf{d} < \mathbf{b}' < \mathbf{d} < \mathbf{b}''$. We do not know if for any degree $\mathbf{b}$ there exists a degree $\mathbf{d}$ such that for all $n \geqq 0$,

$$\mathbf{b}^{(n)} < \mathbf{d}^{(n)} < \mathbf{b}^{(n+1)};$$

if for some $\mathbf{b}$, such a $\mathbf{d}$ exists, then by Theorem 1, $\mathbf{d}$ can be given the additional property of recursive enumerability in $\mathbf{b}$.

Theorem 1 can be extended without any radical alteration of its proof. For example, we can show: if $\mathbf{g}$ is a recursively enumerable degree such that $\mathbf{g}' < \mathbf{0}''$, then there is a recursively enumerable degree $\mathbf{d}$ such that

$$\mathbf{g} < \mathbf{d} < \mathbf{0}' < \mathbf{0}'' = \mathbf{d}'.$$

We say a sequence $\mathbf{a}_0, \mathbf{a}_1, \mathbf{a}_2, \cdots$ of degrees is simultaneously recursively enumerable if there is a sequence $A_0, A_1, A_2, \cdots$ of simultaneously recursively enumerable sets such that $\mathbf{a}_i$ is the degree of A_i for all i. Using the method underlying the proof of Theorem 1, we can show: if $\mathbf{a}_0 < \mathbf{a}_1 < \mathbf{a}_2 < \cdots$ is an infinite, ascending sequence of simultaneously recursively enumerable degrees, then there exists a recursively enumerable degree $\mathbf{d}$ such that

$$\mathbf{a}_0 < \mathbf{a}_1 < \mathbf{a}_2 < \cdots < \mathbf{d} < \mathbf{0}'.$$

We end with a conjecture: the upper semi-lattice of recursively enumerable degrees is dense (i.e., if $\mathbf{b}$ and $\mathbf{c}$ are recursively enumerable degrees such that $\mathbf{b} < \mathbf{c}$, then there exists a recursively enumerable degree $\mathbf{d}$ such that $\mathbf{b} < \mathbf{d} < \mathbf{c}$). The only evidence we have to offer in favor of this conjecture is contained in the results we announced above and the result of Muchnik [5] that there is no minimal, nonzero, recursively enumerable degree([2]).

References

1. R. M. Friedberg, *A criterion for completeness of degrees of unsolvability*, J. Symbolic Logic **22** (1957), 159–160.

2. ———, *Two recursively enumerable sets of incomparable degrees of unsolvability*, Proc. Nat. Acad. Sci. U.S.A. **43** (1957), 236–238.

3. S. C. Kleene, *Introduction to metamathematics*, Van Nostrand, New York, 1952.

4. S. C. Kleene and E. L. Post, *The upper semi-lattice of degrees of recursive unsolvability*, Ann. of Math. (2) **59** (1954), 379–407.

5. A. A. Muchnik, *Negative answer to the problem of reducibility of algorithms*, Dokl. Akad. Nauk SSSR **108** (1956), 194–197. (Russian)

6. Gerald E. Sacks, *The universality of the recursively enumerable degrees*, Abstract No. 60T-14, Notices Amer. Math. Soc. **7** (1960), 996.

7. J. R. Shoenfield, *On degrees of unsolvability*, Ann. of Math. (2) **69** (1959), 644–653.

8. Clifford Spector, *On degrees of recursive unsolvability*, Ann. of Math. (2) **64** (1956), 581–592.

Institute for Advanced Study,
Princeton, New Jersey

---

(2) *Added in proof*. Our conjucture is true; we will give a proof elsewhere.

The Recursively Enumerable Degrees are Dense*

By GERALD E. SACKS

Our main result is: if **b** and **c** are recursively enumerable degrees such that $\mathbf{b} < \mathbf{c}$, then there exists a recursively enumerable degree **d** such that $\mathbf{b} < \mathbf{d} < \mathbf{c}$. By degree we mean degree of recursive unsolvability as defined by Kleene and Post [2]. A degree is called recursively enumerable if it is the degree of a recursively enumerable set. The upper semi-lattice of recursively enumerable degrees has a least member, **0**, the degree of all recursive sets, and a greatest member, $\mathbf{0}'$, the degree of all complete sets. Post [4] asked: does there exist a recursively enumerable degree **d** such that $\mathbf{0} < \mathbf{d} < \mathbf{0}'$? Friedberg [1] and Muchnik [3] answered Post's question in the affirmative. Sacks [5] showed that every countable, partially ordered set is imbeddable in the upper semi-lattice of recursively enumerable degrees. Muchnik [3] announced that if **c** is a non-zero, recursively enumerable degree, then there exists a recursively enumerable degree **d** such that $\mathbf{0} < \mathbf{d} < \mathbf{c}$.

The arguments of Friedberg [1], Muchnik [3] and Sacks [5] have a great deal in common. All three authors make use of a method which may be roughly described as follows (a precise abstract description is given in Sacks [5]): One or more sets are to be recursively enumerated. Certain requirements are to be met. A typical requirement is: the set B is not recursive in the set A with Gödel number e. Unfortunately, the requirements tend to conflict. Thus we may wish to add n to B to insure that $B \neq \{e\}^A$, but the addition of n to B may ruin what we did earlier in the enumeration to insure $A \neq \{f\}^B$. We resolve all conflicts by appealing to a system of priorities assigned to the requirements before the enumeration begins. Then it can be shown that each requirement is "ruined" only finitely often, and that, consequently, each requirement is eventually met.

In the argument below we assign priorities to requirements, but we are forced to permit each requirement to be "ruined" infinitely often. Nonetheless, we are still able to meet each requirement. Thus there is an important combinatorial difference between the method of Friedberg [1] and Muchnik [3] and the method we use below. The difference is a consequence of the fact that Friedberg and Muchnik deal only with finite, initial segments of functions while we are forced to deal indirectly with entire functions. Almost all

* This research was supported in part by National Science Foundation Grant GP-124.

of the complications we encounter below arise from the non-recursiveness of B. We are given a non-recursive, recursively enumerable set B, and we wish to find a recursively enumerable set D such that B is recursive in D but D is not recursive in B. Since B is to be recursive in D, we are forced to add many members to D in complete disregard of all priorities. Thus even the requirement of highest priority may be "ruined" infinitely often. If B were recursive, we could put all of B in D in one step, and then the arguments of [1] and [3] would suffice.

The basic combinatorial principle of this paper appears in primitive form in Sacks [6]. Our basic source of notation is [2]. We conclude with intuitive remarks and an open problem.

THEOREM. *If* **b** *and* **c** *are recursively enumerable degrees such that* $\mathbf{b} < \mathbf{c}$, *then there exists a recursively enumerable degree* **d** *such that* $\mathbf{b} < \mathbf{d} < \mathbf{c}$.

PROOF. Let f, g respectively, be a one-one recursive function whose range is a set B, C respectively, of degree **b**, **c** respectively. We define:

$$b(s, n) = \begin{cases} 0 & \text{if } (Et)_{t<s}(f(t) = n) \\ 1 & \text{otherwise ;} \end{cases}$$

$$c(s, n) = \begin{cases} 2 & \text{if } (Et)_{t<s}(g(t) = n) \\ 1 & \text{otherwise .} \end{cases}$$

Then $\lim_s b(s, n)$ exists for each n; in addition, $\lim_s b(s, n)$ is the representing function of B. Similarly, $2 - \lim_s c(s, n)$ is the representing function of C. Let $k(s)$ be a recursive function such that, for each s, $s < k(s) < k(s+1)$. We define three recursive functions:

$$y_b(s, n, e) = \begin{cases} \mu y T_1^1(\prod_{j<y} p_j^{b(s,j)}, e, n, y) & \text{if } (Ey)_{y<s} T_1^1(\prod_{j<y} p_j^{b(s,j)}, e, n, y) \\ k(s) & \text{otherwise ;} \end{cases}$$

$$h(0, n, e) = 0 \ ;$$

$$h(s+1, n, e) = h(s, n, e) + sg(| y_b(s+1, n, e) - y_b(s, n, e) |) \ ;$$

$$t(s, i, e) = \sum_{j \leq i} h(s, j, e) \ .$$

We define four recursive functions, $y(s, n, e)$, $m(s, e)$, $r(s, n, e)$ and $d(s, n)$, simultaneously by induction on s. The desired degree **d** will be the degree of the predicate $(Es)(d(s, n) = 0)$.

Stage $s = 0$. We set $y(0, n, e) = m(0, e) = 0$ and $r(0, n, e) = d(0, n) = 1$ for all n and e, except that $d(0, p(f(0), 1)) = 0$.

Stage $s > 0$. We set

$$y(s, n, e) = \begin{cases} \mu y T_1^1(\prod_{j<y} p_j^{d(s-1,j)}, e, n, y) & \text{if } (Ey)_{y<s} T_1^1(\prod_{j<y} p_j^{d(s-1,j)}, e, n, y) \\ 0 & \text{otherwise .} \end{cases}$$

The definition of $m(s, e)$ has two cases:

Case m1. There exists an $n < m(s-1, e)$ such that

$$c(s, n) \neq \mathrm{U}(y(s, n, e)) \ \& \ y(s, n, e) \neq y(s-1, n, e) .$$

We set $m(s, e)$ equal to the least such n.

Case m2. Otherwise.

We set $m(s, e)$ equal to the least n such that

$$\begin{aligned} & m(s-1, e) \leqq n < 2m(s-1, e) + s \\ & \& \ (Et)_{t \leqq n}(c(s, t)) \neq \mathrm{U}(y(s, t, e)) . \end{aligned}$$

If no such n exists, then $m(s, e) = m(s-1, e) + s$; this last is in accord with the definition of the bounded, least number operator.

Let $p(i, m) = p_i^{2+m}$ and $p(e, i, m) = p(e, p(i, m))$. We define $r(s, n, e)$ and $d(s, p(e, i, m))$ for all e, n, i and m by a simultaneous induction on e. First we set $d(s, p(f(s), 1)) = 0$.

$$r(s, n, e) = \begin{cases} 0 & \text{if } (Ei)_{i<e}(Et)_{t \leqq n}(Em)[p(i, m) < y(s, t, e) \\ & \quad \& \ d(s, p(i, m)) \neq d(s-1, p(i, m))] , \\ 0 & \text{if } (Et)_{t \leqq n}[p(f(s), 1) < y(s, t, e)] , \\ 1 & \text{otherwise} . \end{cases}$$

The definition of $d(s, p(e, i, m))$ has three cases.

Case d1. $t(s, i, e) > m$. We set

$$d(s, p(e, i, m)) = \begin{cases} d(s-1, p(e, i, m)) & \text{if } (Eu)_{u \leqq e}(Ev)_{v < m(s, u)}[r(s, v, u) = 1 \\ & \quad \& \ p(e, i, m) < y(s, v, u)] , \\ 0 & \text{otherwise} . \end{cases}$$

Case d2. $t(s, i, e) = m$. We set

$$d(s, p(e, i, m)) = \begin{cases} 0 & \text{if } c(s, i) = 2 \\ & \quad \& \ (j)_{j<i}[d(s-1, j) = \mathrm{U}(y_b(s, j, e))] \\ & \quad \& \sim (Eu)_{u \leqq e}(Ev)_{v < m(s, u)}[r(s, v, u) = 1 \\ & \quad \quad \& \ p(e, i, m) < y(s, v, u)] , \\ d(s-1, p(e, i, m)) & \text{otherwise} . \end{cases}$$

Case d3. $t(s, i, e) < m$. We set $d(s, p(e, i, m)) = d(s-1, p(e, i, m))$.

The construction is concluded by setting $d(s, w) = d(s-1, w)$ for all w not equal to $p(f(s), 1)$ or $p(e, i, m)$ for some e, i and m. Let $d(n) = \lim_s d(s, n)$ for each n. Let D be the set whose representing function is d; D is recursively enumerable since $n \in D$ if and only if $(Es)(d(s, n) = 0)$. We list two notes which will be needed later:

(N1) $(s)(n)(e)(r(s, n, e) = 0 \rightarrow r(s, n+1, e) = 0)$;

(N2) $(s)(n)(e)(y(s, n, e) = 0 \to n \geqq m(s, e))$.

We prove (N2) by induction on s. Certainly,

$$(n)(e)(y(0, n, e) = 0 \to n \geqq m(0, e)) .$$

Fix $s > 0$, n and e. Suppose $y(s, n, e) = 0$, and

$$(n)(e)(y(s - 1, n, e) = 0 \to n \geqq m(s - 1, e)) .$$

Suppose *Case* m1 of the definition of $m(s, e)$ holds. If $y(s - 1, n, e) = 0$, then $n \geqq m(s - 1, e) > m(s, e)$. Suppose $y(s - 1, n, e) > 0$. Then

$$y(s, n, e) \neq y(s - 1, n, e)$$
$$\& \ 0 = \mathrm{U}(y(s, n, e)) \neq c(s, n) > 0 ;$$

Consequently, $n \geqq m(s, e)$. Now suppose *Case* m2 of the definition of $m(s, e)$ holds. Once again, $\mathrm{U}(y(s, n, e)) \neq c(s, n)$; it follows $n \geqq m(s, e)$.

LEMMA 1. *Let $y(s, n, e) > 0$, $m(s, e) > n$ and $p(f(s), 1) \geqq y(s, t, e)$ for all $t \leqq n$. Let $d(s, p(i, m)) = d(s - 1, p(i, m))$ for all i, t and m such that $i < e$, $t \leqq n$ and $p(i, m) < y(s, t, e)$. Then $y(s, n, e) = y(s+1, n, e)$ and $m(s+1, e) > n$.*

PROOF. Since $y(s, n, e) > 0$, we have

$$y(s, n, e) = \mu y T_1^1(\textstyle\prod_{j<y} p_j^{d(s-1,j)}, e, n, y) .$$

If $d(s, j) = d(s - 1, j)$ for all $j < y(s, n, e)$, then $y(s + 1, n, e) = y(s, n, e)$. The hypothesis of our lemma tells us that $r(s, n, e) = 1$. But then it follows from the definition of $d(s, j)$ that $d(s, j) = d(s - 1, j)$ for all $j < y(s, n, e)$, since $m(s, e) > n$.

Note (N1) makes clear that the above argument also works for any $t < n$. Thus we have

$$(t)_{t \leqq n}(y(s, t, e) = y(s + 1, t, e)) .$$

Suppose $m(s + 1, e) \leqq n$. Then $m(s + 1, e) < m(s, e)$, and *Case* m1 of the definition of $m(s + 1, e)$ holds. But then we have a t (namely, $m(s + 1, e)$) such that $t \leqq n$, and $y(s, t, e) \neq y(s + 1, t, e)$.

For each e, we say e is stable if $\lim_s y(s, n, e)$ exists and is positive for all n. Since there are infinitely many e which are not Gödel numbers of systems of equations, there are infinitely many non-stable e. Let $\{e_0 < e_1 < e_2 < \cdots\}$ be the set of all non-stable e. For each j, let n_j be the least n such that $\lim_s y(s, n, e_j)$ either does not exist or is zero. Lemma 2 is our basic combinatorial principle (cf. [6, Lemma 4]).

LEMMA 2. *For each k and v, there is an $s \geqq v$ such that*

$$(j)_{j<k}[m(s, e_j) \leqq n_j \vee r(s, n_j, e_j) = 0 \vee y(s, n_j, e_j) = 0] .$$

PROOF. Fix k and v. We suppose there is no s with the desired properties and then define an infinite descending sequence of natural numbers.

We propose the following system of equations as a means of defining two functions, $S(t)$ and $M(t)$, simultaneously by induction:

$$\begin{aligned} S(0) &= v\,; \\ M(t) &= \mu j_{j<k}[n_j < m(S(t), e_j) \\ &\qquad \& \ r(S(t), n_j, e_j) = 1 \\ &\qquad \& \ y(S(t), n_j, e_j) > 0]\,; \\ S(t+1) &= \mu s(Em)[s \geqq S(t) \ \& \ m < y(S(t), n_{M(t)}, e_{M(t)}) \\ &\qquad \& \ d(s, m) \neq d(S(t)-1, m)]\,. \end{aligned}$$

Clearly $S(0) \geqq v$. Suppose $t \geqq 0$, $S(t)$ is well-defined and $S(t) \geqq v$. Then $M(t) < k$, since we have supposed the lemma to be false. Thus

$$y(S(t), n_{M(t)}, e_{M(t)}) > 0\,,$$

and $\lim_s y(s, n_{M(t)}, e_{M(t)})$ does not exist or does equal 0. Then there must be an $s > S(t)$ such that

$$\begin{aligned} &m < y(S(t), n_{M(t)}, e_{M(t)}) \\ &\& \ d(s-1, m) \neq d(S(t)-1, m)\,. \end{aligned}$$

But then $S(t+1)$ is well-defined and $S(t+1) \geqq v$.

For each $t \geqq 0$, let

$$u(t+1) = \mu m[d(S(t+1), m) \neq d(S(t)-1, m)]\,.$$

Fix $t > 0$. We show $u(t+1) < u(t)$. Since we know

$$u(t+1) < y(S(t), n_{M(t)}, e_{M(t)})\,,$$

it suffices to show

$$y(S(t), n_{M(t)}, e_{M(t)}) \leqq u(t)\,.$$

It follows from the definition of S that

$$d(w, m) = d(S(t-1)-1, m)$$

whenever $S(t) > w \geqq S(t-1)$ and $m < y(S(t-1), n_{M(t-1)}, e_{M(t-1)})$. Consequently,

$$d(S(t), u(t)) \neq d(S(t)-1, u(t))\,.$$

First suppose $u(t) = p(f(S(t)), 1)$. Then $y(S(t), n_{M(t)}, e_{M(t)}) \leqq u(t)$, since $r(S(t), n_{M(t)}, e_{M(t)}) = 1$. Now suppose $u(t) = p(i, f, m)$ for some i, f and m. Let $s = S(t)$, $n = n_{M(t)}$ and $e = e_{M(t)}$. If $i < e$, then the definition of r tells us that $y(s, n, e) \leqq u(t)$, since $r(s, n, e) = 1$. If $i \geqq e$, then the definition of d tells us that $y(s, n, e) \leqq u(t)$, since $n < m(s, e)$, $r(s, n, e) = 1$ and $d(s, u(t)) \neq d(s-1, u(t))$.

We introduce two predicates:

$A(e)$: if e is stable, then the set $\{m(s, e) \mid s \geqq 0\}$ is finite.

$B(e)$: the set $\{m \mid d(p_e^m) = 0\}$ is recursive in B.

We will prove $(e)A(e)$ and $(e)B(e)$ by means of a simultaneous induction on e. From $(e)A(e)$ it will follow that $\mathbf{c} \not\leqq \mathbf{d}$. From $(e)B(e)$ it will follow that $\mathbf{d} \not\leqq \mathbf{b}$.

LEMMA 3. $(i)_{i<e}B(i) \rightarrow A(e)$.

PROOF. We know $\mathbf{c} \not\leqq \mathbf{b}$. We suppose that $B(i)$ is true for each $i < e$ and that $A(e)$ is false, and we show $\mathbf{c} \leqq \mathbf{b}$. Thus the set $\{m(s, e) \mid s \geqq 0\}$ is infinite, and for each n, $\lim_s y(s, n, e)$ exists and is positive. Let $R(n, s)$ denote the predicate

$$\begin{aligned} & m(s, e) > n \\ & \& \ (i)(m)(t)[p(i, m) < y(s, t, e) \ \& \ i < e \ \& \ t \leqq n \\ & \qquad\qquad \rightarrow d(s-1, p(i, m)) = d(p(i, m))] \\ & \& \ (z)(t)[z \in B \ \& \ p(z, 1) < y(s, t, e) \ \& \ t \leqq n \\ & \qquad\qquad \rightarrow d(s-1, p(z, 1)) = 0] \,. \end{aligned}$$

(Recall that $z \in B$ if and only if $d(p(z, 1)) = 0$.) Since $B(i)$ is true for all $i < e$, it follows that R is recursive in B. Since the set $\{m(s, e) \mid s \geqq 0\}$ is infinite, and since $\lim_s y(s, n, e)$ exists and is positive for each n, it follows that $(n)(Es)R(n, s)$. Let

$$w(n) = \mu s R(n, s) \,.$$

Then w is recursive in B, and for each n, $w(n+1) \geqq w(n)$.

Fix n. We show $\lim_s y(s, n, e) = y(w(n), n, e)$. Let s be such that $s \geqq w(n)$ and

$$y(w(n), n, e) = y(s, n, e) \ \& \ R(n, s) \,.$$

Since $m(s, e) > n$, it follows from *note* (N2) that $y(s, t, e) > 0$ for all $t \leqq n$. We know f enumerates B without repetitions. It follows from the definition of R that $p(f(s), 1) \geqq y(s, t, e)$ for all $t \leqq n$. But then Lemma 1 tells us that

$$y(s+1, t, e) = y(s, t, e)$$

for all $t \leqq n$, and that $m(s+1, e) > n$. Thus

$$y(w(n), n, e) = y(s+1, n, e) \quad \text{and} \quad R(n, s+1) \,.$$

It follows that $\lim_s y(s, n, e) = y(w(n), n, e)$.

Finally, we show

$$\lim_s c(s, n) = \mathrm{U}(y(w(n), n, e))$$

for all n. If this last is true, then C is recursive in B, since $2 - \lim_s c(s, n)$ is the representing function of C; and since w is recursive in B. Fix n and suppose $\lim_s c(s, n) \neq \text{U}(y(w(n), n, e))$. There must be an s^* such that for all $s \geqq s^*$,

$$c(s, n) = \lim_s c(s, n) \neq \text{U}\big(y(w(n), n, e)\big) = \text{U}\big(y(s, n, e)\big) .$$

Fix $s > s^*$ and suppose $m(s-1, e) \leqq m(s^*, e) + n$. If *Case* m1 of the definition of $m(s, e)$ holds, then $m(s, e) \leqq m(s^*, e) + n$. If *Case* m2 holds and $n < 2m(s-1, e) + s$, then $m(s, e) \leqq m(s^*, e) + n$, since $c(s, n) \neq \text{U}(y(s, n, e))$. If $n \geqq 2m(s-1, e) + s$, then $m(s, e) \leqq n$. Thus

$$m(s, e) \leqq m(s^*, e) + n$$

for all $s \geqq s^*$. This last is impossible, since the set $\{m(s, e) \mid s \geqq 0\}$ is infinite.

Suppose $\{e\}^b(n)$ is defined for all n. Then $\lim_s y_b(s, n, e)$ exists for each n, and $\lim_s t(s, i, e)$ exists for each i. In addition, $\lim_s t(s, i, e)$ (regarded as a function of i) is recursive in B. All this is clear from the definition of y_b and t.

LEMMA 4. *If* $\{e\}^b(n)$ *is defined for all* n *and* $(u)_{u \leqq e} A(u)$, *then* $d(p(e, i, \lim_s t(s, i, e))) = 0$ *for only finitely many* i.

PROOF. We suppose the lemma is false and show C is recursive in B. Our first claim is:

$$d = \{e\}^b .$$

Fix j. Let s be so large that

$$d(j) = d(w-1, j) \;\&\; \text{U}\big(y_b(w, j, e)\big) = \{e\}^b(j)$$

for all $w \geqq s$. Let $t(i) = \lim_s t(s, i, e)$ for all i. Let w and i be such that $j < i$, $w \geqq s$, $c(w, i) = 2$,

$$0 = d\big(w, p\big(e, i, t(i)\big)\big) \neq d\big(w-1, p\big(e, i, t(i)\big)\big) ,$$

and $t(i) = t(w, i, e)$. It follows from *Case* d2 of the definition of d that

$$d(w-1, j) = \text{U}\big(y_b(s, j, e)\big) ,$$

since $j < i$. But then $d(j) = \{e\}^b(j)$, since $w \geqq s$, and our first claim is proved.

Our second claim is: for all sufficiently large i,

$$d\big(p\big(e, i, t(i)\big)\big) = \lim_s c(s, i) .$$

Our first and second claims, together with the fact that t is recursive in B, imply that C is recursive in B. Our second claim is a consequence of Lemma 2 and $(u)_{u \leqq e} A(u)$. If $u \leqq e$ and u is stable, then $A(u)$ tells us that the set $\{m(s, u) \mid s \geqq 0\}$ is finite. If $u \leqq e$ and u is stable, let $m(u)$ be the greatest member of $\{m(s, u) \mid s \geqq 0\}$. If $u \leqq e$ and u is non-stable, then $u = e_i$ for some i; let $m(u) = n_i$. Recall that n_i is the least witness to the fact that e_i is non-stable. Thus $\lim_s y(s, v, u)$ exists if $u \leqq e$ and $v < m(u)$. Let y be so large that $y(s, v, u) \leqq y$ if $s \geqq 0$, $u \leqq e$ and $v < m(u)$. Fix $i \geqq y$. We show

$d(p(e, i, t(i))) = \lim_s c(s, i)$. If $\lim_s c(s, i) = 1$, then $c(s, i) = 1$ for all s, and $d(s, p(e, i, t(i))) = 1$ for all s. Suppose $\lim_s c(s, i) = 2$. Let w be so large that $c(s, i) = 2$, $t(s, i, e) = t(i)$, and

$$(j)_{j<i}[d(s-1, j) = \mathrm{U}(y_b(s, j, e))]$$

for all $s \geqq w$. The existence of w follows from our first claim. By Lemma 2, there is an $s \geqq w$ such that for all $u \leqq e$, if u is non-stable (hence equal to e_i for some i), then

$$m(s, u) \leqq n_i \vee r(s, n_i, u) = 0 \vee y(s, n_i, u) = 0 .$$

We show $d(s, p(e, i, t(i))) = 0$. *Case* d2 is such that we need only show

$$\sim(Eu)_{u \leqq e}(Ev)_{v<m(s,u)}[r(s, v, u) = 1 \;\&\; p(e, i, t(i)) < y(s, v, u)] .$$

Fix $u \leqq e$ and $v < m(s, u)$. Suppose u is stable. Then $y(s, v, u) \leqq y \leqq i \leqq p(e, i, t(i))$. Suppose u is non-stable. Let $u = e_i$. If $v < n_i$, then $v < m(u)$ and $y(s, v, u) \leqq y \leqq p(e, i, t(i))$. Suppose $v \geqq n_i$. Then $m(s, u) > n_i$, and consequently,

$$r(s, n_i, u) = 0 \vee y(s, n_i, u) = 0 .$$

If $r(s, n_i, u) = 0$, then $r(s, v, u) = 0$ by *Note* (N1), since $n_i \leqq v$. Suppose $y(s, n_i, u) = 0$. Then $n_i \geqq m(s, u)$ by *Note* (N2). But $m(s, u) > n_i$ since we have supposed $v \geqq n_i$.

LEMMA 5. $(u)_{u \leqq e}A(u) \rightarrow B(e)$.

PROOF. Our first claim is (L).

(L) $$(Ey)(m)(x)(i)[m < t(x, i, e) \;\&\; p(e, i, m) \geqq y \rightarrow d(p(e, i, m)) = 0] .$$

The proof of our first claim is virtually identical with the proof of the second claim of Lemma 4. For each $u \leqq e$, we define $m(u)$ as in Lemma 4. Then $\lim_s y(s, v, e)$ exists if $u \leqq e$ and $v < m(u)$. Define y as in Lemma 4. Fix m, x and i so that $m < t(x, i, e)$ and $p(e, i, m) \geqq y$. We show $d(p(e, i, m)) = 0$. By Lemma 2 there is an $s \geqq x$ such that for all $u \leqq e$, if u is non-stable (hence equal to e_i for some i), then

$$m(s, e_i) \leqq n_i \vee r(s, n_i, e_i) = 0 \vee y(s, n_i, e_i) = 0 .$$

We show $d(s, p(e, i, m)) = 0$. Since $s \geqq x$, we have

$$m < t(x, i, e) \leqq t(s, i, e) .$$

Thus the definition of $d(s, p(e, i, m))$ is given by *Case* d1. We need only show

$$\sim(Eu)_{u \leqq e}(Ev)_{v<m(s,u)}[r(s, v, u) = 1 \;\&\; p(e, i, m) < y(s, v, u)] .$$

Fix $u \leqq e$ and $v < m(s, u)$. If u is stable, then $y(s, v, u) \leqq y \leqq p(e, i, m)$. Suppose u is non-stable. Then we repeat the argument at the end of Lemma 4 to show $r(s, v, u) = 0$ or $p(e, i, m) \geqq y(s, v, u)$.

We are ready to prove $B(e)$. We will use only Lemma 4 and (L). First we suppose that $\{e\}^b(n)$ is undefined for some n; let N be the least such n. It follows from the definition of y_b that $\lim_s t(s, i, e) = \infty$ for all $i \geqq N$. Let y have the property described by (L). Then

$$(i)(m)[i \geqq N \ \& \ p(e, i, m) \geqq y \rightarrow d(p(e, i, m)) = 0] .$$

If $i < N$, then $\lim_s t(s, i, e)$ is finite; consequently, the set

$$\{p(e, i, m) \mid d(p(e, i, m)) = 0 \ \& \ i < N\}$$

is finite by *Case* d3 of the definition of d. Now $d(p_e^n) = 0$ only if $p_e^n = p(e, i, m)$ for some i and m or $p_e^n = p(f(s), 1)$. It follows that the set $\{n \mid d(p_e^n) = 0\}$ is recursive, hence recursive in B.

Now suppose $\{e\}^b(n)$ is defined for all n. Then $\lim_s t(s, i, e)$ is finite for all i. Let $\lim_s t(s, i, e) = t(i)$. Recall that $t(s + 1, i, e) \geqq t(s, i, e)$ for all s and i. It follows from (L) that

$$(i)(m)[m < t(i) \ \& \ p(e, i, m) \geqq y \rightarrow d(p(e, i, m)) = 0] .$$

By Lemma 4, the set

$$\{p(e, i, t(i)) \mid d(p(e, i, t(i))) = 0\}$$

is finite. By *Case* d3, $d(p(e, i, m)) = 1$ if $m > t(i)$. Since $t(i)$ is recursive in B, it follows that the set $\{n \mid d(p_e^n) = 0\}$ is recursive in B.

Lemmas 3 and 5 constitute a proof of $(e)A(e)$ and $(e)B(e)$.

LEMMA 6. *C is not recursive in D.*

PROOF. Suppose $\lim_s c(s, n) = \{e\}^a(n)$ for all n. Then e is stable, and by $A(e)$, the set $\{m(s, e) \mid s \geqq 0\}$ has a greatest member; call it M. Let w be so large that

$$\lim_s c(s, n) = c(s, n) = \mathrm{U}(y(s, n, e))$$

when $s \geqq w$ and $n \leqq M$. If $s \geqq w$, then either

(a) $c(s, n) \neq \mathrm{U}(y(s, n, e))$ for some $n \leqq m(s, e)$, or

(b) $m(s, e) = m(s - 1, e) + s$.

If (a) holds, then $M < m(s, e)$. Thus (b) must hold for infinitely many s. But then $M < m(s, e)$ for some s.

LEMMA 7. *D is not recursive in B.*

PROOF. Suppose $d(n) = \{e\}^b(n)$ for all n, where $b(n) = \lim_s b(s, n)$ for all n. Then $\lim_s t(s, i, e)$ is finite for each i. Let $t(i) = \lim_s t(s, i, e)$. By Lemma 4, there is an n such that $d(p(e, i, t(i))) = 1$ for each $i \geqq n$. We shall find an i such that $i \geqq n$ and $d(p(e, i, t(i))) = 0$. We proceed as in Lemma 4. Define y

as in Lemma 4. Fix $i \geqq y + n$. Define w and s as in Lemma 4. Then the final argument of Lemma 4 tells us that $d(s, p(e, i, t(i))) = 0$.

Let $\mathbf{d}$ be the degree of D. It follows from Lemmas 6 and 7 that $\mathbf{c} \nleqq \mathbf{d}$ and $\mathbf{d} \nleqq \mathbf{b}$. We have $\mathbf{b} \leqq \mathbf{d}$, since $d(p(m, 1)) = 0$ if and only if $m \in B$. It remains only to show that D is recursive in C. We give an intuitive description of a procedure E for computing D from C. Our description is such that the translation of E into a system of equations which define D recursively in C is not difficult. We need a recursive predicate $Q(u, v, s, e, i, m)$:

$$u \leqq e \ \& \ v < m(s, u) \ \& \ r(s, v, u) = 1 \ \& \ p(e, i, m) < y(s, v, u) .$$

We also need a predicate $R(s, e, n)$:

$$\begin{aligned} &(u)(i)(m)(t)[u < e \ \& \ t \leqq n \ \& \ p(u, i, m) < y(s, t, e) \\ &\qquad \rightarrow d(s-1, p(u, i, m)) = d(p(u, i, m))] \\ &\& \ (i)(t)[i \in B \ \& \ t \leqq n \ \& \ p(i, 1) < y(s, t, e) \\ &\qquad \rightarrow d(s-1, p(i, 1)) = d(p(i, 1))] . \end{aligned}$$

We need the next two lemmas to describe E.

LEMMA 8. $Q(u, v, s, e, i, m) \ \& \ R(s, u, v) \rightarrow (w)_{w \geqq s} Q(u, v, w, e, i, m)$.

PROOF. We proceed with an induction on $w \geqq s$. Fix $w \geqq s$ and suppose $Q(u, v, w, e, i, m)$ and $R(w, u, v)$ hold. Recall that f, the recursive function whose range is B, is one-one. It follows from Lemma 1 that $v < m(w+1, u)$ and that

$$y(w, t, u) = y(w+1, t, u)$$

for all $t \leqq v$. But then $R(w+1, u, v)$ holds and $r(w+1, v, u) = 1$, and consequently, $Q(u, v, w+1, e, i, m)$.

LEMMA 9. $(w)(Es)_{s \geqq w}(u)(v)[Q(u, v, s, e, i, m) \rightarrow R(s, u, v)]$.

PROOF. Fix w, e, i and m. For each $u \leqq e$, define $m(u)$ as in the second half of the proof of Lemma 4. Define y as in Lemma 4. Thus $y(s, v, u) \leqq y$ if $s \geqq 0$, $u \leqq e$ and $v < m(u)$. It is safe to suppose that w is so large that

$$(s)(v)(s > w \ \& \ v \leqq y \rightarrow d(s-1, v) = d(v)) .$$

By Lemma 1 there is an $s > w$ such that for any non-stable $u \leqq e$, we have

$$m(s, u) \leqq n_j \ \bigvee \ r(s, n_j, u) = 0 \ \bigvee \ y(s, n_j, u) = 0 ,$$

where $u = e_j$; recall that if u is non-stable, then $m(u) = n_j$.

Fix u and v and suppose $Q(u, v, s, e, i, m)$ holds. We show $R(s, u, v)$. First we suppose u is stable. Then $v < m(s, u) \leqq m(u)$, since $Q(u, v, s, e, i, m)$ holds; consequently,

$$(n)(t)[t \leqq v \ \& \ n < y(s, t, u) \rightarrow d(s-1, n) = d(n)] ,$$

since $s > w$. But then $R(s, u, v)$ holds. Now we suppose u is non-stable. Let $u = e_j$. Then $m(u) = n(j)$. If $m(s, u) \leqq m(u)$, then $v < m(u)$, and as above, $R(s, u, v)$ holds. Thus

$$r(s, m(u), u) = 0 \vee y(s, m(u), u) = 0 .$$

If $r(s, m(u), u) = 0$, then $v < m(u)$, since $r(s, v, u) = 1$; this last follows from (N1). Suppose $y(s, m(u), u) = 0$. Then it follows from (N2) that $m(u) \geqq m(s, u)$. But then $v < m(u)$.

We return to the description of procedure E. Fix z. We compute $d(z)$ with the help of C. If z is not of the form $p(b, 1)$ or $p(e, i, m)$, then $d(z) = 1$. If $z = p(b, 1)$, then $d(z) = 0$ if and only if $b \in B$. Suppose $z = p(e, i, m)$. The predicate $(Es)(t(s, i, e) = j)$ is recursive in B, since

$$t(s, i, e) \leqq t(s + 1, i, e)$$

for all s. We consider three cases:

(a) $(s)(t(s, i, e) < m)$;

(b) $(Es)(t(s, i, e) > m)$;

(c) $\lim_s t(s, i, e) = m$.

With the help of B, we can decide which one of (a), (b) and (c) holds; note that the monotonicity of $t(s, i, e)$, regarded as a function of s, is vital. If (a) holds, then *Case* d3 tells us that $d(z) = 1$. Suppose (b) holds. Let w^* be such that $t(s, i, e) > m$ for all $s \geqq w^*$. We can determine w^* with the help of B. Suppose $d(s, z) = 1$ for all $s < w^*$. Then $d(z) = 0$ if and only if

$$(Es)(u)(v)[s \geqq w^* \ \& \ \sim Q(u, v, s, e, i, m)] ,$$

since *Case* d1 applies when $s \geqq w^*$. By Lemma 9, there is an $s \geqq w^*$ such that

$$(u)(v)[Q(u, v, s, e, i, m) \rightarrow R(s, u, v)] .$$

We can find s if we know the value of $d(x)$ for some of the following x: $x = p(u, j, k)$, where $u < e$; $x = p(j, 1)$, where $j \in B$. (Recall that $d(p(j, 1)) = 0$ if and only if $j \in B$.) If $\sim Q(u, v, s, e, i, m)$ holds for all u and v, then $d(z) = 0$; we only have to check $u \leqq e$ and $v < m(s, u)$. Suppose for some u and v, $Q(u, v, s, e, i, m)$ holds. Then we have $R(s, u, v)$ by definition of s. By Lemma 8, it follows that

$$(w)_{w \geqq s} Q(u, v, w, e, i, m) .$$

But then $d(z) = 0$ if and only if $d(w, z) = 0$ for some $w < s$.

Finally, we suppose (c) holds. Let w' be such that $t(s, i, e) = m$ for all $s \geqq w'$. We can determine w' with the help of B. Suppose $d(s, z) = 1$ for all $s < w'$. Then $d(z) = 0$ if and only if

$$(Es)[s \geqq w' \ \& \ c(s, i) = 2 \\ \& \ (u)(v) \sim Q(u, v, s, e, i, m) \\ \& \ (j)_{j<i}(d(s-1, j) = U(y_b(s, j, e)))] ,$$

since *Case* d2 applies when $s \geqq w'$. If $j < i$, then

$$y_b(s, j, e) = \lim_s y_b(s, j, e)$$

for all $s \geqq w'$, since $t(w', i, e) = \lim_s t(s, i, e) = m$. Let $v^* \geqq w'$ be such that for all $s \geqq v^*$,

$$c(s, i) = \lim_s c(s, i) \ \& \ (j)_{j<i}[d(s-1, j) = d(j)] .$$

We can determine v^* with the help of C and the values of $d(x)$ for $x < z$, since $j < i$ implies that $j < p(e, i, m)$. If $\lim_s c(s, i) = 1$ or $d(j) \neq \text{U}(\lim_s y_b(s, j, e))$ for some $j < i$, then $d(z) = 1$. Suppose this last hypothesis is false. Suppose also that $d(s, z) = 1$ for all $s < v^*$. Then $d(z) = 0$ if and only if

$$(Es)(u)(v)[s \geqq v^* \ \& \ \sim Q(u, v, s, e, i, m)] ,$$

since *Case* d2 applies when $s \geqq v^*$. We now continue as in *Case* (*b*). With the help of Lemmas 8 and 9, $d(z)$ is easily determined.

That completes our computation of $d(z)$ from C. To find $d(z)$, we had to know the value of $d(x)$ for finitely many $x : x < z$; $x = p(u, j, k)$, where $u < e$; $x = p(j, 1)$, where $j \in B$. We had to know that $d(p(j, 1)) = 0$ if and only if $j \in B$. We used very heavily the fact that B is recursive in C.

The basic combinatorial principle of this paper is contained in Lemma 2; this same principle appeared in simpler form in [6]. We combined this principle with a further combinatorial principle expressed by Lemma 4 in order to show the recursively enumerable degrees are dense. Lemma 4 was needed to show D is not recursive in B. The prethinking which inspired Lemma 4 may be described as follows. C is not recursive in B. So let us keep planting members of C in D until D looks enough like C to guarantee that D is not recursive in B. But let us not plant members of C in D with utter abandon, because we wish to have C not recursive in D. At the same time let us plant B in D with utter abandon so that B will be recursive in D. We plant a member of C in D when we set $d(s, p(e, i, t(s, i, e))) = 0$; this happens only if $c(s, i) = 2$. (Recall $c(s, i) = 2$ only if $i \in C$.) In order to prevent us from planting too much of C in D, we must have a method of unplanting members of C already planted in D. Suppose we have planted $i \in C$ in D; that is, we set $d(s, p(e, i, t(s, i, e))) = 0$. If for some $w > s$, $t(w, i, e) > t(s, i, e)$, then *Case* d1 or d2 may give us a chance of setting $d(w, p(e, i, 1 + t(s, i, e))) = 0$; if this last happens, we have unplanted i. We have the opportunity to unplant i if and only if $\lim_s t(s, i, e) > t(s, i, e)$. This happens if and only if $\{e\}^b(i)$ is undefined or unequal to $y_b(s, i, e)$. If for some i, $\{e\}^b(i)$ is undefined, there is no need to plant any member of C in D. If $\{e\}^b(i)$ is defined for all i, then Lemma 4 tells us that we do not permanently plant infinitely much of C in D. Of course the proof of Lemma 4 turns on the fact that C is not recursive in B.

There are still many open questions concerning the recursively enumerable degrees; however, it may be possible to say something about the elementary theory of recursively enumerable degrees if the following question is answered: if $\mathbf{b}$ and $\mathbf{c}$ are non-zero, recursively enumerable degrees such that $\mathbf{b} < \mathbf{c}$, does there exist a recursively enumerable degree $\mathbf{d}$ such that $\mathbf{d} < \mathbf{c}$ and $\mathbf{d} \cup \mathbf{b} = \mathbf{c}$? In a talk at the Berkeley Model Theory Symposium (1963), J. R. Shoenfield conjectured that the answer is yes. We agree with him, but we are unable to prove it.

INSTITUTE FOR DEFENSE ANALYSES, PRINCETON
CORNELL UNIVERSITY

REFERENCES

1. RICHARD M. FRIEDBERG, *Two recursively enumerable sets of incomparable degrees of unsolvability*, Proc. Nat. Acad. Sci., U.S.A., 43 (1957), 236-238.
2. S. C. KLEENE and EMIL L. POST, *The upper semi-lattice of degrees of recursive unsolvability*, Ann. of Math., 59 (1954), 379-407.
3. A. A. MUCHNIK, *Negative answer to the problem of reducibility of the theory of algorithms* (in Russian), Dokl. Akad. Nauk SSSR, 108 (1956), 194-197.
4. EMIL L. POST, *Recursively enumerable sets of positive integers and their decision problems*, Bull. Amer. Math. Soc., 50 (1944), 284-316.
5. GERALD E. SACKS, Degrees of Unsolvability, Annals of Mathematics Studies Number 55, Princeton, 1963.
6. ———, *Recursive enumerability and the jump operator*, Trans. Amer. Math. Soc., in press.

(Received August 6, 1963)

A SIMPLE SET WHICH IS NOT EFFECTIVELY SIMPLE

GERALD E. SACKS[1]

For each e, let f_e be the partial recursive function

$$U(\mu y T_1(e, n, y)),$$

and let W_e be the range of f_e. Then $W_0, W_1, W_2, \cdots$ is the Kleene enumeration of the recursively enumerable sets. Post [5] calls a recursively enumerable set simple if its complement is infinite but does not contain any infinite, recursively enumerable set. Raymond Smullyan calls a recursively enumerable set W effectively simple if its complement is infinite, and if there is a partial recursive function f such that for each e, if W_e is contained in the complement of W, then $f(e)$ is defined and is greater than the cardinality of W_e.[2] Clearly, an effectively simple set is simple. The simple set S constructed by Post in [5] is effectively simple. This latter is no accident. In fact it is not unreasonable to claim that any direct attack on the problem of constructing a simple set must result in an effectively simple set. Our purpose here is to obtain a simple set which is not effectively simple. We will make strong use of the recursion theorem of Kleene [2]; however, we will use it in the informal manner of Myhill [4]. Our notation is that of [2].

We introduce a recursive function E:

$$E(0) = \mu x T_1((x)_0, (x)_1, (x)_2);$$
$$E(s+1) = \mu x[x > E(s) \;\&\; T_1((x)_0, (x)_1, (x)_2)].$$

We will need E to simultaneously enumerate all the recursively enumerable sets in a fashion suitable for the proving of our theorem. It is a peculiarity of our proof that we cannot rely merely on the usual properties associated with any standard enumeration of the recursively enumerable sets; instead, we are forced to specify a particular enumeration. For each e and s we define a finite set W_e^s: for each m, $m \in W_e^s$ if and only if for some $i \leqq s$,

$$m = U((E(i))_2) \;\&\; e = (E(i))_0.$$

Then for each e, $W_e^0 \subseteq W_e^1 \subseteq W_e^2 \subseteq \cdots$, and $W_e = \bigcup \{W_e^s \mid s \geqq 0\}$. We

Presented to the Society January 24, 1963; received by the editors October 25, 1962.

[1] The preparation of this paper was partially supported by N.S.F. Grant GP-124.

[2] Smullyan actually requires that f be recursive, but it is easy to show the two definitions equivalent.

say s defines $f_e(n)$ if $e=(E(s))_0$ and $n=(E(s))_1$. If s defines $f_e(n)$, then $f_e(n)=U((E(s))_2)$. For each e and n, let

$$S(e, n) \simeq \mu s \quad (s \text{ defines } f_e(n)).$$

S is partial recursive, and $S(e, n)$ is defined if and only if $f_e(n)$ is defined.

Let 0 denote the empty set. It is clear there exists a recursive function g such that for each e, i and z, we have

$$W_{g(e,i,z)} = \begin{cases} \{2^i \cdot 3^t \mid E(S(e,z)) < t \leqq E(S(e,z)) + f_e(z)\} & \text{if } f_e(z) \text{ is defined,} \\ 0 & \text{otherwise.} \end{cases}$$

The recursion theorem tells us that there exists a recursive function z such that for each e and i, we have

$$W_{z(e,i)} = W_{g(e,i,z(e,i))} = \begin{cases} \{2^i \cdot 3^t \mid E(S(e,z(e,i))) < t \leqq E(S(e,z(e,i))) + f_e(z(e,i))\} \\ \qquad \text{if } f_e(z(e,i)) \text{ is defined,} \\ 0 \quad \text{otherwise.} \end{cases}$$

We note some properties of z:

(1) if $f_e(z(e, i))$ is defined, then $f_e(z(e, i))$ is equal to the cardinality of $W_{z(e,i)}$;

(2) if $i \neq j$, then $W_{z(e,i)} \cap W_{z(e,j)} = 0$;

(3) if $f_e(z(e, i))$ is defined, then for all n, $W_n^{S(e,z(e,i))} \cap W_{z(e,i)} = 0$;

(4) if $i \neq j$ and both $f_e(z(e, i))$ and $f_e(z(e, j))$ are defined, then $z(e, i) \neq z(e, j)$.

To prove (3), let $s=S(e, z(e, i))$ and let $m \in W_n^s \cap W_{z(e,i)}$. Then $m \leqq E(s)$, since $m=U((E(i))_2)$ for some $i \leqq s$, and since E is an increasing function. (Recall that $U(x) \leqq x$ for all x.) But $m > E(s)$, since $m=2^i \cdot 3^t$ for some $t > E(s)$.

THEOREM 1. *There exists a simple set which is not effectively simple.*

PROOF. We will define a sequence A, B, Q_0, Q_1, Q_2, $\cdots$ of simultaneously recursively enumerable sets. A will be simple, but not effectively simple. B will be such that if $e \in B$, then $W_e \cap A \neq 0$. We will see to it that if W_e is infinite, then $e \in B$. Each Q_e will be finite and will contain a set that will serve as a witness to the fact that f_e does not effectively bound the cardinalities of the finite subsets of the complement of A.

Stage $s=0$. We set $A^0=B^0=Q_i^0=0$ for all i.

Stage $s>0$. Let $e=(E(s))_0$ and $n=(E(s))_1$. Thus s defines $f_e(n)$. We perform the following two operations in the indicated order:

(a) We set $Q_j^s=Q_j^{s-1}$ for all $j\neq e$. If there is no i such that $i\leqq e$ and $n=z(e, i)$, we set $Q_e^s=Q_e^{s-1}$. If there is such an i, then by (4) it is unique. In addition, $S(e, z(e, i))$ is defined and

$$W_{z(e,i)} = \{2^i\cdot 3^t \mid E(S(e, z(e, i))) < t \leqq E(S(e, z(e, i))) + f_e(z(e, i))\}.$$

We set $Q_e^s=Q_e^{s-1}\cup W_{z(e,i)}$.

(b) If $e\in B^{s-1}$ or if there is no m such that

$$m \in W_e^s \;\&\; (j)_{j\leqq e}\,(m \notin Q_j^s),$$

then we set $B^s=B^{s-1}$ and $A^s=A^{s-1}$. If $e\notin B^{s-1}$ and there is an m with the above property, let i be the least one. We set $B^s=B^{s-1}\cup\{e\}$ and $A^s=A^{s-1}\cup\{i\}$.

Let $A=\bigcup\{A^s \mid s\geqq 0\}$ and $B=\bigcup\{B^s \mid s\geqq 0\}$. Since E and z are recursive, it follows A is recursively enumerable. For each e, let $Q_e=\bigcup\{Q_e^s \mid s\geqq 0\}$. Q_e is finite; in fact,

$$Q_e = \bigcup\,\{W_{z(e,i)} \mid i \leqq e\},$$

since $Q_e^{s-1}\neq Q_e^s=Q_e^{s-1}\cup W_{z(e,i)}$ if and only if $i\leqq e$ and $s=S(e, z(e, i))$.

LEMMA 1. *If W_e is infinite, then $A\cap W_e\neq 0$.*

PROOF. We know Q_j is finite for every j. Let m be a member of W^e which is greater than every member of Q_j for all $j\leqq e$. Let s be such that $m\in W_e^s$. First we suppose $e\in B^{s-1}$. Then there must be a $t<s$ such that $e\notin B^{t-1}$ and $e\in B^t$. At stage t we must have performed operation (b) in such a manner that $B^t=B^{t-1}\cup\{e\}$ and $A^t=A^{t-1}\cup\{i\}$, where $i\in W_e^s$. Now we suppose $e\notin B^{s-1}$. We have

$$m \in W_e^s \;\&\; (j)_{j\leqq e}\,(m \notin Q_j^s).$$

But then operation (b) at stage s forces us to put a member of W_e^s in A^s.

LEMMA 2. *If $m\in W_e^s-Q_j^s$, then $m\notin Q_j$.*

PROOF. Suppose for the sake of a reductio ad absurdum that $m\in W_e^s-Q_e^s$ and $m\in Q_j$. Since $Q_j=\bigcup\{W_{z(j,i)} \mid i\leqq j\}$, there must be an $i\leqq j$ such that $m\in W_{z(j,i)}$. Since $W_{z(j,i)}$ is nonempty, $f_j(z(j, i))$ is defined. Let $t=S(j, z(j, i))$. Then t defines $f_j(z(j, i))$, and consequently,

$$Q_j^t = Q_j^{t-1} \cup W_{z(j,i)},$$

since $i \leqq j$. Since $m \notin Q_j^s$, we must have $s < t$. Since $m \in W_e^s$, we have

$$m \in W_e^t \cap W_{z(j,i)} \neq 0.$$

But this last contradicts (3).

Lemma 3. *If $m \in Q_i \cap A$, then there exists an s and an e such that $(E(s))_0 = e < i$ and $\{e\} = B^s - B^{s-1}$ and $\{m\} = A^s - A^{s-1}$.*

Proof. Since $m \in A$, there is an s such that $\{m\} = A^s - A^{s-1}$. Let $e = (E(s))_0$. Since $A^s \neq A^{s-1}$, we must have $\{e\} = B^s - B^{s-1}$. In addition,

$$m \in W_e^s \;\&\; (j)_{j \leqq e}\, (m \notin Q_j^s).$$

It follows from Lemma 2 that $(j)_{j \leqq e}\, (m \notin Q_j)$. But then $e < i$, since $m \in Q_i$.

Lemma 4. *The set $Q_i \cap A$ has at most i members.*

Proof. Suppose m and n are distinct members of $Q_i \cap A$. Lemma 3 guarantees the existence of $s(m)$, $e(m)$, $s(n)$ and $e(n)$ with properties as stated in the conclusion of Lemma 3. Thus

$$\{m\} = A^{s(m)} - A^{s(m)-1} \;\&\; \{n\} = A^{s(n)} - A^{s(n)-1}.$$

Since $m \neq n$, it follows $s(m) \neq s(n)$. But then $e(m) \neq e(n)$, since

$$\{e(m)\} = B^{s(m)} - B^{s(m)-1} \;\&\; \{e(n)\} = B^{s(n)} - B^{s(n)-1}.$$

We also know from Lemma 3 that $e(m) < i$ and $e(n) < i$. Thus we can map the set $Q_i \cap A$ in a one-to-one fashion into the set $\{e \mid e < i\}$.

Lemma 5. *For each e, there is a z such that W_z is contained in the complement of A and such that either $f_e(z)$ is undefined or $f_e(z)$ is not greater than the cardinality of W_z.*

Proof. Fix e. We show that some member of the sequence, $z(e, 0)$, $z(e, 1), \cdots, z(e, e)$ serves as the desired z. Suppose there is an $i \leqq e$ such that $f_e(z(e, i))$ is undefined. Then $W_{z(e,i)} = 0$, and the lemma is proved. Suppose then that $f_e(z(e, i))$ is defined for all $i \leqq e$. By (1), $f_e(z(e, i))$ is not greater than the cardinality of $W_{z(e,i)}$ for any $i \leqq e$. Thus it suffices to find an $i \leqq e$ such that $W_{z(e,i)} \cap A = 0$. The sets,

$$W_{z(e,0)}, W_{z(e,1)}, \cdots, W_{z(e,e)},$$

are nonempty and disjoint. If each of them has a member in A, then their union has at least $e+1$ members in A. But their union is Q_e, and Lemma 4 tells us that Q_e has at most e members in A.

It follows from Lemma 5 that A is not effectively simple. It also follows from Lemma 5 that the complement of A is infinite, since otherwise, the constant function

$$f(n) = 1 + \text{cardinality of the complement of } A$$

would constitute a counterexample to Lemma 5. Finally, by Lemma 1, A is simple.

Post [5] calls a recursively enumerable set W hyper-simple if its complement is infinite, and if there does not exist a recursively enumerable sequence of disjoint, finite sets, each one of which contains a member of the complement of W. It can be shown with the help of Lemma 4 that A is not hyper-simple.

The proof of Theorem 1 above is, as far as we know, the first proof in recursion theory to make simultaneous use of the recursion theorem and the priority method of Friedberg [1] and Muchnik [3]. The priority method was needed to resolve the inevitable conflict between putting elements in A as required by Lemma 1 and keeping them out of A as required by Lemma 4. Thus in operation (b), we are not allowed to take m from W_e^s and add it to A^s if for some $j \leqq e$, $m \in Q_j^s$. The recursion theorem was needed to prove that our system of priorities does eventually resolve all conflicts happily; in particular, the recursion theorem made possible the proof of Lemma 2.

References

1. R. M. Friedberg, *Two recursively enumerable sets of incomparable degrees of unsolvability*, Proc. Nat. Acad. Sci. U. S. A. **43** (1957), 236–238.

2. S. C. Kleene, *Introduction to metamathematics*, Van Nostrand, New York, 1952.

3. A. A. Muchnik, *Negative answer to the problem of reducibility of the theory of algorithms*, Dokl. Akad. Nauk SSSR **108** (1956), 194–197. (Russian)

4. John Myhill, *Creative sets*, Z. Math. Logik Grundlagen Math. **1** (1955), 97–108.

5. E. Post, *Recursively enumerable sets of positive integers and their decision problems*, Bull. Amer. Math. Soc. 50 (1944), 284–316.

Cornell University

The Journal of Symbolic Logic
Volume 30, Number 3, Sept. 1965

METARECURSIVE SETS

G. KREISEL AND GERALD E. SACKS*

1. Introduction. Our ultimate purpose is to give an axiomatic treatment of recursion theory sufficient to develop the priority method. The direct or abstract approach is to keep in mind as clearly as possible the methods actually used in recursion theory, and then to formulate them explicitly. The indirect or experimental approach is to look first for other mathematical theories which seem similar to recursion theory, to formulate the analogies precisely, and then to search for an axiomatic treatment which covers not only recursion theory but also the analogous theories as particular cases.

The first approach is more general because it does not depend on the existence of (familiar) analogues. A concrete mathematical theory, it seems, need have no such analogues and still be important, as e.g. classical number theory. In such a case, an axiomatic treatment may still be useful for exhibiting the mathematical structure of the theory considered and the assumptions on which it rests. However, it will lack one of the most heavily advertised advantages of the axiomatic method, namely, the "economy of thought" which results from an uniform theory for several different and interesting cases: we cannot hope for this if, by hypothesis, we know of only one particular case. In contrast, the second approach, if successful at all, is bound to achieve such "economy" because we start out with several interesting particular cases. Another possible virtue of the second approach is that of field work over insight: the abstract pattern that we are looking for and hoping to formalize in axioms, may not be evident in any one mathematical theory, but may spring to the eye if one happens to look simultaneously at several theories which happen to realize the pattern.

The present paper is intended as a contribution to the second approach. But, for the time being, we play safe and stress its independent interest as a *recursion theory on the recursive ordinals*, or, at most, as an important special case of: recursion theory on initial segments of the ordinals.

Since almost any generalization of recursion theory that comes to mind, will preserve pre-Post [14] recursion theory, such as Kleene's enumeration theorem, we treat the elementary parts briefly, and concentrate on post-Friedberg recursion theory. We generalize and discuss Friedberg's solution [2] to Post's problem and the maximal set constructions of Friedberg [3]

Received October 21, 1964.
* This research was partially supported by NSF Grant GP-124.
The authors wish to thank Saul Kripke for many helpful conversations.

and Yates [20]. We also present some applications to the theory of Π_1^1 sets of natural numbers.

By use of unique notations for recursive ordinals, e.g. Gandy [4], the concepts of our *metarecursion theory* may be formulated in familiar terms: Let Q be a Π_1^1 set of natural numbers (the set of unique notations), ordered by the restriction of a recursive relation $<$ to Q, and let $|n|$ be the ordinal of the predecessors of n, $n \in Q$. Let the order type of Q be Kleene's ω_1. The definitions below are invariant for all such $(Q, <)$.

Metafinite sets: $\{|n| \mid n \in X\}$ for any hyperarithmetic $X \subseteq Q$.

Metarecursive sets: $\{|n| \mid n \in X\}$, where X is Δ_1 *on* Q, i.e. there are two disjoint Π_1^1 sets, Y_1 and Y_2, such that $X = Y_1$ and $Q - X = Y_2$.

Metarecursive functions: $\{\langle |n|, |m| \rangle \mid \langle n, m \rangle \in X\}$ (graph of the function), where X is a Π_1^1 subset of Q^2, and $(\forall n \in Q)(E_1 m \in Q) X(n, m)$.

Metarecursively enumerable sets: If non-empty, the range of a metarecursive function; it turns out: $\{|n| \mid n \in X\}$, where X is any Π_1^1 subset of Q.

Bounded sets: X is bounded if $X \subset \{|n| \mid |n| < \alpha\}$ for some $\alpha < \omega_1$; all metafinite sets are bounded; all bounded, metarecursive sets are metafinite, but some bounded, metarecursively enumerable sets are not metafinite.

We discuss the notion of relative metarecursiveness in section 3.

The definitions above in terms of notations are given only for comparison with the literature. One of the advantages of our treatment is that the theory is notation-free, instead of first being developed in terms of notations and then proved independent of the notation system used. (The use of notations, particularly non-unique ones, is a thoughtless hangover from a constructive theory of ordinals, where one wanted the usual ordinal functions to be represented by constructive functions. For this one needs non-unique notations. But once the jump operator is introduced, constructivity is gone by the board, and the use of such notations becomes pointless. But once one has unique notations, it is natural to work with abstract ordinals instead.)

The remainder of the paper can be read independently of the next section, which sets out the considerations which led to the correct definitions.

2. Analysis of the problem. (*a*) *Language of recursion theory*. The first step in any generalization of an informal theory is to pick out the primitive notions and the formal language. A simple-minded choice is this:

A two-sorted language, with variables $x, y, \ldots$ for individuals, $f, g, h, \ldots$ for monadic functions; $=$ between individuals and applications of functions to individuals; two distinct individuals, say 0, 1.

In ordinary recursion theory, the interpretation is this: $x, y, \ldots$ range over ω, $f, g, \ldots$ over the recursive number-theoretic functions; 0, 1 have their usual meaning.

The theory of non-empty, recursively enumerable sets $\subseteq \omega$ can be interpreted, by defining the relation $f \subseteq g$ (read: the range of f is contained in the range of g) explicitly by $(x)(\mathrm{E}y)(f(x) = g(y))$. And f enumerates a recursive set if g enumerates its complement, i.e.,
$R(f) \leftrightarrow (\mathrm{E}g)(x)(y)(\mathrm{E}z)[f(x) \neq g(y) \ \&\ (f(z) = x \vee g(z) = x)]$,
or if it has a characteristic function g, i.e.,
$R_{+}(f) \leftrightarrow (\mathrm{E}g)(x)[g(x) = 0 \leftrightarrow (\mathrm{E}y)(f(y) = x)]$. $Fin(f) \leftrightarrow (g)[g \subseteq f \rightarrow R(g)]$ is a definition of finite set suggested by Lacombe, i.e., every r.e. set contained in the set (enumerated by) f is recursive, or if the range of every recursive function restricted to f is recursive, i.e.,
$F(f) \leftrightarrow (g)(\mathrm{E}h)(x)[R(h) \ \&\ (h(x) = 0 \leftrightarrow (\mathrm{E}y)(g(f(y)) = x))]$.

The theory of recursive functions of finite sequences can be interpreted in the above theory by use of inverse pairing functions in the usual way, by assuming function constants j_1, j_2, such that
$(x, y)(\mathrm{E}_1 z)[j_1(z) = x \ \&\ j_2(z) = y \ \&\ j_1(0) = 0 \ \&\ j_2(0) = 0]$.

Evidently, relative recursiveness (degree theory) of r.e. sets does not need an extension of the language.

(*b*) *Transfer principles or axiomatic theory*? Keeping to the language (a), there are two kinds of generalizations. First, we may look for other interpretations (i.e. different ranges for the variables, but requiring the second sort to range over $Y \subseteq X^X$ if X is the range of the first) such that for a suitable class of formulae, the interpretations are equivalent, as for the language of real algebra ($+$ and $\cdot$) when one permits all real closed fields as interpretations. Thus, if in our case the class of formulae involved includes formally different definitions like R and R_{+}, or Fin and F, these pairs will be equivalent in all interpretations. Second, we pick on several interpretations first and then look for an axiomatic theory which is valid for them, but may leave some statements undecided; and, of course, even an axiomatically decided statement may have proofs in the special case which are not valid (e.g., from the axioms for groups the right and left inverses are equal; but a simpler "proof" is possible for commutative groups!). This is the common situation in mathematics, where different concepts, i.e., formally different definitions, may have the same extension in one interpretation, but diverge in another.

We are convinced that non-trivial recursion theory is not primarily concerned with constructivity. But it remains a fact that certain parts of it have a close relation to constructive (intuitionistic) concepts, e.g., the distinction between a finite set given by: its sequence number or by: $\{x \mid x \leqq n \ \&\ (\mathrm{E}y)(f(y) = x)\}$, where f is recursive, cf. Post [14]. This suggests that correct generalizations may use an axiomatic theory based on the formal rules of intuitionistic logic. Since these formal rules have other interpretations than the intended constructive ones, the suggestion is quite consistent with our conviction. We do not pursue this matter here. The reader may note that the source of our definitions (described immediately below) is the theory of model-theoretic invariants of [9]. But for an axiomatic theory it is not convenient to have merely these invariants; one wants to develop their logic, i.e., give an appropriate interpretation of logical constants. There is a precise analysis in the literature (for so-called extensionally definite formulae of [8]), and this logic is indeed non-classical.

(*c*) *Model-theoretic foundations*. The very first appearance of the class of number-theoretic recursive relations and functions was in Gödel [5], not in connection with Church's problematic thesis, but under the name: *Entscheidungsdefinite Relationen*. In the spirit of the times this was formulated proof-theoretically, but the model-theoretic meaning was clear; one has a sequence of terms, the *numerals* e.g., 0, $S0$, ..., $S^n 0$ in the usual systems Φ of arithmetic with 0 for zero and S for the successor function; granted that in each model $\mathscr{A}$ of Φ, any given numeral has the same denotation, say the *number* n, then the formula $\alpha(x)$ is entscheidungsdefinit if the set $\omega \cap \{x \mid \models_{\mathscr{A}} \alpha(x)\}$ is independent of $\mathscr{A}$; we call it invariant on ω. In Mostowski's exposition of Gödel's theory, much of the work is independent of the assumption that the set (of Gödel numbers of) Φ is recursively enumerable, this being the case for which sets invariant on ω are recursive. Note in passing that Mostowski speaks of definability or strong representability; we add "on ω" because true invariance means that $\{x \mid \models_{\mathscr{A}} \alpha(x)\}$ is independent of $\mathscr{A}$. Since the definition of invariance makes sense for any class of models, and not only general models, and in fact for any language, and not only the usual languages, it is natural to investigate what happens: Mostowski and his pupils studied ω-models, and gave a polished exposition in Mostowski [13]. But several results people expected to hold had shown to be false, e.g., in Kreisel [8], p. 120.

In terms of the language (a), the trouble was this: Mostowski took for granted that the range of the individuals was to remain ω, and that the functions used were to be those invariant on ω in all ω-models of the usual systems of analysis. Recursively enumerable sets were not reduced to recursive functions, but were regarded as "semi-invariantly" defined: $\bigcap_{\mathscr{A}}(\omega \cap \{x \mid \models_{\mathscr{A}} \alpha(x)\})$, and thus turned into Π_1^1 sets; but of course not into the range of any of the functions considered because the range of a hyperarithmetic function on ω is hyperarithmetic.

In more general terms, the defect was conceptual: just what is special about the numerals in general models of arithmetic? Granted that invariant definitions in our sense above, must define subsets of ω, a simple application of the finiteness theorem shows that they are finite. But in ω-models, ω itself is invariantly definable. Moreover, if we really only considered first order systems of arithmetic, we should have just one ω-

model. But if we consider, as was done in Mostowski [13], second order systems, and all hyperarithmetic subsets of ω are invariantly definable, we should regard these definitions of sets as "just as good as the numerals". So, the modification suggested in Kreisel [9], was to use invariants of ω-models for recursion theory on hyperarithmetic sets of natural numbers. If, instead of second order arithmetic, one considers a theory of ordinals, one gets precisely the definitions in the introduction for Π_1^1-systems Φ by the following

Fundamental principle of the generalization: One takes results of ordinary recursion theory, and replaces the word "recursive" by "invariantly definable on ω," some (or all) occurrences of "finite" by "invariantly definable;" all this for general models. To get the analogue of recursion theory on ω_1, one replaces ω by ω_1, and "general models" by "ω-models". Other occurrences of "finite" may be replaced by "bounded" or retained, in accordance with the second approach of (b).

	General Models	ω-models
Invariantly definable	Finite subsets of ω (all Φ)	Subsets of ω_1 which are Δ_1, where Φ is Π_1^1.
Invariantly definable *on*	ω: recursive subsets of ω, where Φ is r.e.	ω_1: subsets of ω_1 which are Δ_1 *on* ω_1, where Φ is Π_1^1.

To get all subsets of ω, resp. ω_1 mentioned in the table, Φ must of course contain certain minimal axioms. In the case of general models, Robinson's system Q is a well known sufficient set; in the case of ω-models one needs the supremum axiom and (certain restricted forms of) the principle of proof and definition by generalized induction. For details, cf. 1.2 (c) of [9], where also a more abstract (and very much restrained) exposition of the general notions above is given in 1.1.

A completely parallel treatment can be given based on (transfinite) concatenation theory where an ω-model contains the sequence $|^*|^*\ldots$ of length precisely $= \omega$, and instead of the numerals $|$, $|^*|$, $|^*|^*|$, ... one has explicit definitions of sequences of $|$ of length $< \alpha$, $\alpha < \omega_1$. In view of the elegance of Smullyan [19], it may be worth pursuing a recursion theory based on this.

Besides generalizations, the model-theoretic foundation leads also to an analysis of familiar concrete results. For, different general notions may have the same extension in a special application; thus given a concrete result one has the problem of analyzing just which general notion is involved. In the case of general models (of arithmetic) or ordinary recursion theory, the collection N^ω of finite sets of natural number N has the following properties: its cardinal is N, moreover $N^\omega \simeq N$ by functions invariantly definable on N, N^{ω} respectively; each finite set has cardinal $< N$, is invariantly definable (i.e. included in an invariantly definable set) ***and*** conversely. Different proofs of the same theorem involving finiteness, may use different properties of N^{ω}, and will generalize to proofs of different theorems in generalizations. Similarly, the distinction between regular and non-regular sets breaks down, a set being regular if its intersection with any invariantly definable set is also invariantly definable. All sets invariantly definable on the hard core are automatically regular. In the case of general models, every set is regular, but not for other classes of models. (Intuitionistically, not every r.e. set X is regular in the sense that $(n)(\mathrm{E}a)(a = \{x \mid x \epsilon X \ \&\ x \leqq n\})$ where a ranges over finite sequences presented explicitly; in fact, on Church's thesis, regular r.e. sets are recursive because a would be a recursive function of n. If intuitionistic logic should turn out to be exactly correct for the axiomatic theory, a little sophistication will be needed to interpret non-prenex formulas. Thus $(n){\sim}{\sim}(\mathrm{E}a)(a = \{x \mid x \epsilon X \ \&\ x \leqq n)$ at first sight, conflicts with the existence of non-regular, metarecursively enumerable sets: this overlooks the fact that intuitionistic higher order systems are not subsystems of the "corresponding" classical systems, intuitionistic negation being weak.)

N.B. Our point of view differs from Mostowski [12] who regards the model-theoretic foundation as *the* approach to recursion theory, while we regard it as a prop for an abstract theory. Consequently, he formulates explicitly conditions on consequence relations, satisfied by those for general and for ω-models, from which the relevant properties of definability-*on* can be derived. The proofs below make the parallel quite clear, but are set out informally; their principal function is to resolve the ambiguities described above in the analysis of familiar results.

(*d*) *An equation calculus*. For technical convenience we use an equation calculus got from one of Kleene's by taking metafinite sets of premises instead of finite ones. Kripke [10] has developed an equation calculus suitable for recursion theory on arbitrary, initial segments of the ordinals. We are indebted to Kripke [10] for the equation calculus we describe in section 3 below, which is suitable for recursion theory on the recursive ordinals. Now, in the case of ordinary recursion theory we find most expositions based on an equation calculus repulsive because a set of rules is put down, without any clear reason for the choice. The justification lies in the theorems proved, e.g., closure under diagonalization; but one seldom says in so many words with respect to what notion of definability the rules are supposed to be complete. For the model-theoretic foundation above the situation is much more satisfactory since, once one is concerned with general models, the only arbitrary element is the use of finite formulae. Here we would argue that at least the definitions should be given unambiguously; the restriction seems satisfactory since for arbitrary sets Φ the only invariantly definable (syntactic) objects are finite. But once one considers ω-models, the matter is more delicate, and one has to study just what infinite formulae are appropriate (see [9] § 2.2).

(*e*) *Further generalization.* The notions of (c) extend to other classes of models, e.g., β-models; of a theory of ordinals in which $<$ must be well-founded, i.e., all models considered are abstractly simply segments of the ordinals; or of a set theory in which ϵ must be well-founded. In the latter case we have a recursion theory on the elements of the Shepherdson-Cohen minimal models. But for a systematic treatment, the classes of models considered must not be chosen by ad hoc considerations ("absoluteness" of some historically or intrinsically important notion like finite or well-founded) but by a more flexible method, e.g., the use of infinite formulae. In fact, an alternative definition of ω-model is this: we consider all general models satisfying the infinite formula $(x < \omega)(x = 0 \vee x = S^1 0 \vee x = S^2 0 \vee \ldots)$. Now, the syntax of infinitary languages presupposes a recursion theory on appropriate ordinals, and these languages are then used for a model theory which, by (c), is to provide the recursion theory on larger ordinals, and so on. (It is understood that due respect must be paid to the satisfaction relation, and not only to purely external (syntactic) properties.)

In such transfinite progressions there are usually catching points where new difficulties are met. It is therefore particularly encouraging that a direct development of recursion theory based on arbitrary ordinals is in progress, independently by Kripke [10], based on equation calculi, and by Platek, based on schemata.

It is not yet known just how much of what we call non-trivial recursion theory generalizes to (say) uncountable ordinals. But there is reason to hope that these other developments will reveal useful information for abstract recursion theory not provided by (c) above.

3. The equation calculus. Following Kripke [10], we make use of a formalism similar to that of Kleene [6]. For each recursive ordinal, we have a corresponding numeral. (It is convenient to think of the set of numerals as a Π_1^1 set of unique notations.) We have a sequence, $f_0, f_1, f_2, \ldots$, of function letters, and a sequence, $x_0, x_1, x_2, \ldots$, of variables. Additional primitive symbols are: $'$(successor), E (there exists), and $<$ (less than). The terms are the numerals, the variables, and expressions of the form t' where t is a term, or $f(t_1, \ldots, t_n)$ where f is a function letter and $t_1, \ldots, t_n$ are terms ($n \geqq 0$), or $(\mathrm{E}x < t_1)(t_2)$ where x is a variable and t_1 and t_2 are terms. An equation is $t_1 = t_2$, where t_1 and t_2 are terms.

The rules are Kleene's rules R1 and R2, and Kripke's rule R3: (a) if γ and β are numerals for recursive ordinals γ and β such that $\gamma < \beta$, then $((\mathrm{E}x < \beta)(t(x))) = 0$ is an immediate consequence of $t(\gamma) = 0$; (b)

$((\mathrm{E}x < \beta)(t(x))) = 1$ is an immediate consequence of the set of premises $\{t(\gamma) = 1 \mid \gamma < \beta\}$. The role of R3 is easily understood if one considers a slight extension of Kleene's formalism for recursion theory. Suppose we add propositional connectives, number-quantifiers bounded by terms, and appropriate rules to Kleene's formalism. The resulting formalism can be used to define the recursive functions by finite systems of equations in a manner similar to the use of Kleene's formalism. The two formalisms are equivalent in the sense that both define the same class of functions, the recursive functions; however, their generalizations to metarecursion theory differ in that the extended formalism permits deductions from infinite sets of premises. Rule R3 is nothing more than the familiar rule concerning the use of the bounded quantifier.

For each finite set E of equations, we define $\mathrm{S^E}$, the set of all equations deducible from E, by transfinite induction on the *recursive* ordinals. This elegant definition is due to Kripke [10]. Let $\mathrm{S_0^E} = \mathrm{E}$, and $\mathrm{S^E_{\alpha+1}} =$ the set of all conclusions obtained by applying rules R1–R3 to $\mathrm{S^E_\alpha}$; when we apply R3 to S_α, we require that $\beta < \alpha$. If α is a limit ordinal, let $\mathrm{S^E_\alpha} = \bigcup\{\mathrm{S^E_\beta} \mid \beta < \alpha\}$. Finally, let $\mathrm{S^E} = \bigcup\{\mathrm{S^E_\alpha} \mid \alpha < \omega_1\}$.

Let Q be a Π_1^1 subset of unique notations for recursive ordinals; it is convenient, but not necessary, to assume Q is a subset of Kleene's O linearly ordered by $<_O$. Each equation of our formalism can be viewed as a finite sequence of members of Q, and consequently, a member of Q can be assigned effectively to it to serve as its Gödel number. For each $m \epsilon Q$, let S^{E}_m be the set of Gödel numbers of members of $\mathrm{S^E_{|m|}}$, where $|m|$ is the recursive ordinal represented by m. Then the predicate

$$m \epsilon Q \;\&\; n \epsilon S^{\mathrm{E}}_m$$

is Π_1^1; and there exists a recursive function f such that for each $m \epsilon Q$, S^{E}_m is a hyperarithmetic set and $f(m)$ is an index for it. The proofs of these last two facts are along the same lines as Kleene's argument [7] that O is Π_1^1 and Spector's argument [18] that for each $b \epsilon O$, O_b is hyperarithmetic.

Now let Q serve as the set of unique notations needed in section 1 to define metafinite, metarecursive, and metarecursively enumerable, and think of the Gödel numbers as designating recursive ordinals. Then the set of Gödel numbers of all equations deducible from E becomes a metarecursively enumerable set of recursive ordinals, and for each α, $\mathrm{S^E_\alpha}$ becomes a metafinite set. These developments are in accord with the transfer principle of section 2; the finite deductions of Kleene's formalism for recursion theory have evolved into the metafinite "deductions" of the formalism for metarecursion theory.

A partial function ψ from the recursive ordinals into the recursive ordinals is partial metarecursive if and only if there is a finite system of equations E with principal function letter f such that $\psi(\alpha)$ is defined and

$= \beta$ if and only if $f(\boldsymbol{\alpha}) = \boldsymbol{\beta}$ is a member of S^E. A function is metarecursive if it is partial metarecursive and total. (Note that we have required E to be finite rather than metafinite; if E is permitted to be metafinite (or even metarecursive), the class of metarecursive functions is not enlarged. The transfer principle of section 2 allowed for the possibility that certain occurrences of "finite" in recursion theory would not become occurrences of "metafinite" in metarecursion theory. Kripke [10] shows that finite systems of equations suffice even for recursion theory on uncountable initial segments of the ordinals. The definition of maximal set given in section 4 is another example of an occurrence of "finite" in recursion theory which is retained in metarecursion theory.) A non-empty set is metarecursively enumerable if it is the range of a metarecursive function.

We show that the above definitions are equivalent to those given in section 1. (Kripke [11] greatly clarifies the connection between recursion theory on arbitrary initial segments of ordinals and the analytic hierarchy.) It is enough to show that a function metarecursive according to section 3 is metarecursive according to section 1, and that a set metarecursively enumerable according to section 1 is metarecursively enumerable according to section 3. Let ψ be a metarecursive function defined by a finite system E. For each recursive ordinal α, let $\boldsymbol{\alpha}$ be the unique member of Q such that $|\boldsymbol{\alpha}| = \alpha$. For each equation e, let $[e]$ be its Gödel number. Then

$$\psi(\alpha) = \beta \leftrightarrow (Em)(m \epsilon Q \ \& \ [f(\boldsymbol{\alpha}) = \boldsymbol{\beta}] \ \epsilon \ S^E_m),$$

and consequently, $G_\psi = \{(\boldsymbol{\alpha}, \boldsymbol{\beta}) \mid \psi(\alpha) = \beta\}$ is a Π^1_1 subset of Q^2. Thus a function metarecursive in the sense of the present section is metarecursive in the sense of section 1. Now let A be a set of recursive ordinals which is metarecursively enumerable in the sense of section 1. Then $A = \{|m| \mid m \epsilon B\}$, where B is a Π^1_1 subset of Q. There exists a 1–1 recursive function h such that $m \epsilon B \leftrightarrow h(m) \epsilon O$. The usual inductive definition of O can be translated into a finite set of equations in our formalism which define a metarecursive function whose range is O. By combining these equations with the equations for h, we can define a metarecursive function whose range is B. Finally, we can find a metarecursive function whose range is A, since there exists a finite set of equations defining a partial metarecursive function whose domain is Q and which takes each member of Q to the ordinal it represents.

Note that in the above, we are thinking of the recursive ordinals as an extension of the natural numbers, and that we have shown that each Π^1_1 set of natural numbers is a metarecursively enumerable set of recursive ordinals. Similarly, each metafinite (or metarecursive) set of natural numbers is hyperarithmetic.

Our first theorem is an enumeration theorem for partial metarecursive functions. A proof of it could be given by developing the formalism of metarecursion theory along the same lines that Kleene [6] develops the formalism of recursion theory. We regard it, however, as a direct consequence of the enumeration theory for Π_1^1 predicates; i.e., each Π_1 predicate $P(x_1, \ldots, x_n)$ can be put in the form

$$(f)(\mathrm{E}y)T_1^1(\bar{f}(y), e, x_1, \ldots, x_n, y).$$

THEOREM 1. *For each $n > 0$: there exists a partial metarecursive function $\varphi_n(z, x_1, \ldots, x_n)$ such that for each partial metarecursive function $\psi(x_1, \ldots, x_n)$ there exists a finite recursive ordinal e such that*

$$(x_1)\ldots(x_n)[\psi(x_1, \ldots, x_n) \simeq \varphi_n(e, x_1, \ldots, x_n)].$$

In Theorem 1 the x_i's of course range over the recursive ordinals, and $\simeq$, the symbol for strong equality, has the same meaning it has in Kleene [6].

We need a metarecursive indexing of the metafinite sets. Let j and k be metarecursive functions such that for each metafinite set K there exists a unique α such that

$$(x)[x \epsilon K \leftrightarrow (x < j(\alpha) \;\&\; k(\alpha, x) = 0)].$$

For each α, we define $K_\alpha = \{x | x < j(\alpha) \;\&\; k(\alpha, x) = 0\}$. For each α, K_α is metafinite, since it is bounded and metarecursive. If $K = K_\alpha$, we call α the index of K. The existence of j and k follows from the fact that the hyperarithmetic sets are defined by transfinite recursion on O. The indexing provided by j and k enables us to speak of partial metarecursive functions which take metafinite sets as arguments. Thus when we say $\varphi(K)$ is a partial metarecursive function, we mean there exists a partial metarecursive function $\psi(\alpha)$ such that $\varphi(K) \simeq \psi(\alpha)$, where $K = K_\alpha$.

Let A and B be arbitrary sets of recursive ordinals, and let $a(\alpha)$ be the characteristic function of A. We say *A is weakly metarecursive in B* if there exists a partial metarecursive function φ such that

$$(i)_{i<2}(\alpha)[a(\alpha) = i \leftrightarrow (\mathrm{E}M)(\mathrm{E}N)(\varphi(M, N, \alpha) = i \;\&\; M \subseteq B \;\&\; N \subseteq cB)],$$

where M and N range over the metafinite sets. The definition of "weakly metarecursive in" is analogous to the definition of "recursive in". In recursion theory, a question about the membership of A is decided after asking a finite set of questions about the membership of B. In metarecursion theory, a question about the membership of A is decided after asking a metafinite set of questions about B. Let $s(H, A)$ be the representing function of the predicate $H \subseteq A$. We say *A is metarecursive in B* if there exist partial metarecursive functions φ_0 and φ_1 such that

$$(H)(i)_{i<2}[s(H, A) = i \leftrightarrow (\mathrm{E}M)(\mathrm{E}N)(\varphi_0(M, N, H) = i \;\&\; M \subseteq B \;\&\; N \subseteq cB)]$$

and

$$(H)(i)_{i<2}[s(H, cA) = i \leftrightarrow (\mathrm{E}M)(\mathrm{E}N)(\varphi_1(M, N, H) = i \;\&\; M \subseteq B \;\&\; N \subseteq cB)],$$

where H, M and N range over the metafinite sets. The definition of "metarecursive in" is also analogous to the definition of "recursive in", since in recursion theory, a finite set of questions about A is decided after asking a finite set of questions about B. Clearly, if A is metarecursive in B, then A is weakly metarecursive in B. But the converse is not true! This is a far from abvious result recently obtained by G. Driscoll. In fact, Driscoll shows that the relation "weakly metarecursive" is not transitive, whereas it is an exercise to show that the relation "metarecursive in" is transitive.

Let $\psi(M, N, \alpha)$ be a partial metarecursive function, where the variables M and N range over the metafinite sets. By Theorem 1, it has a normal form, $\varphi(e, M, N, \alpha)$. Let $\langle e, M, N, \alpha, \gamma\rangle$ denote the Gödel number of the ordered 5-tuple, $(e, M, N, \alpha, \gamma)$; we assume 0 is not a Gödel number. Since φ is partial metarecursive, there exists a metarecursive function $t(e, \sigma)$ such that: if $\varphi(e, M, N, \alpha)$ is defined and equal to γ, then $t(e, \sigma) = \langle e, M, N, \alpha, \gamma\rangle$ for some σ; if $t(e, \sigma) \neq 0$, then $t(e, \sigma) = \langle e, M, N, \alpha, \gamma\rangle$ for some M, N, α and γ, and $\varphi(e, M, N, \alpha)$ is defined and equal to γ. We define a T-predicate analogous to that of Kleene [6]; let B be an arbitrary set of recursive ordinals

$$T_1^1(B, e, \alpha, \sigma) \leftrightarrow t(e, \sigma) = \langle e, M, N, \alpha, \gamma\rangle \;\&\; M \subseteq B \;\&\; N \subseteq cB.$$

Let $U(e, \sigma)$ be a metarecursive function such that if $t(e, \sigma) = \langle e, M, N, \alpha, \gamma\rangle$, then $U(e, \sigma) = \gamma$. We define two partial functions:

$$\{e\}_\sigma^B(\alpha) = \gamma \leftrightarrow T_1^1(B, e, \alpha, \sigma) \;\&\; U(e, \sigma) = \gamma;$$
$$\{e\}^B(\alpha) = \gamma \leftrightarrow (\mathrm{E}\sigma)[T_1^1(B, e, \alpha, \sigma)] \;\&\; (\sigma)[T_1^1(B, e, \alpha, \sigma) \to U(e, \sigma) = \gamma].$$

We have a normal form theorem as a consequence of Theorem 1.

THEOREM 2. *If A is weakly metarecursive in B, then there exists a finite ordinal e such that*

$$(\alpha)[a(\alpha) = \{e\}^B(\alpha)],$$

where $a(\alpha)$ is the characteristic function of A.

Note that if $\{e\}^B(\alpha)$ is defined and < 2 for all α, then $\{e\}^B(\alpha)$ is the characteristic function of a set weakly metarecursive in B. Kripke [10] proves Theorems 1 and 2 in a more general setting.

In [9] four notions of reducibility for sets of recursive ordinals are introduced and discussed. Two of these notions are "weakly metarecursive in" and "metarecursive in"; these two have the advantage of leading readily to generalizations of familiar arguments of recursion theory. The remaining two have the advantage of being justifiable model-theoretically; they are "invariantly definable from a set" and "invariantly definable from a property". In [17] it will be shown that the first two notions and the last coincide on a wide class of metarecursively enumerable sets.

4. A maximal set. One of the deepest results in the theory of recursively enumerable sets is Friedberg's construction of a maximal set [3]. A recursively enumerable set M is said to be maximal if its complement is infinite, and if for each recursively enumerable set R, either $R-M$ or $cR-M$ is finite. The transfer principle of section 2 suggests nine possible definitions of maximal, metarecursively enumerable set of recursive ordinals, since we can replace "infinite" by "infinite" or "not metafinite" or "unbounded", and since we can replace "finite" by "bounded" or "metafinite" or "finite". Throwing all caution to the winds, we choose the strongest of the nine possibilities. A metarecursively enumerable set M of recursive ordinals is said to be *maximal* if its complement is unbounded, and if for each metarecursively enumerable set R, either $R-M$ or $cR-M$ is finite. Note that a maximal set must be unbounded.

THEOREM 3. *There exists a maximal, metarecursively enumerable set of recursive ordinals.*

PROOF. Let Q be a Π^1_1 set of unique notations for the recursive ordinals. If e is a notation for α, let $|e| = \alpha$; the partial function $|e|$ is partial metarecursive. We claim Q is not metarecursive. Suppose it is. Then the function $h(\alpha)$, defined by

$$h(\alpha) = \begin{cases} |\alpha| \text{ if } \alpha \in Q \\ 0 \text{ otherwise,} \end{cases}$$

is metarecursive. But the range of h restricted to the natural numbers is unbounded. This contradicts (B^*). Thus Q is not regular.[2] $(B^*))$ is stated at the beginning of section 6.)

Let $r(\sigma)$ be a metarecursive function which enumerates Q without repetitions; it follows from (B^*) that any non-metafinite, metarecursively enumerable set can be enumerated without repetitions by some metarecursive function. Our construction is inspired by that of Friedberg [3]. One difference between our construction and Friedberg's consists of our use of the function $r(\sigma)$ to insure that the complement of M is unbounded. Without $r(\sigma)$ we would have only that cM is not metafinite. We are also indebted to C. E. M. Yates [20] for a modification of Friedberg's construction he devised in order to construct a complete, maximal set. Without Yates' modification, we would have for each metarecursively enumerable set R, either $R-M$ or $cR-M$ is bounded rather than finite.

We define a partial recursive function, $g(\sigma, e)$; for every recursive ordinal σ and every natural number e:

$$J(\sigma, e) = \{|r(\tau)| \mid \tau \leqq \sigma \;\&\; r(\tau) \leqq e\};$$

$$g(\sigma, e) = \begin{cases} \text{greatest member of } J(\sigma, e) \text{ if } J(\sigma, e) \neq \varnothing, \\ 0 \text{ otherwise.} \end{cases}$$

[2] The notion of regularity is defined near the end of section 6.

For any function $f(\sigma)$, we say $lim_\sigma f(\sigma)$ exists if $(E\gamma)(\sigma)\ (\sigma > \gamma \rightarrow f(\sigma) = f(\gamma))$. It is clear that $lim_\sigma g(\sigma, e)$ exists for each e; in fact, $lim_\sigma g(\sigma, e)$ is either 0 or the greatest ordinal which has a notation in Q less than or equal to e. Let $g(e) = lim_\sigma g(\sigma, e)$. Then the range of $g(e)$ is unbounded. In addition, if $d \geqq e$, then $g(d) \geqq g(e)$.

Let $\varphi_1(e, x)$ be the partial metarecursive function of Theorem 1 which enumerates all partial, metarecursive functions of one variable. For each natural number e, let R_e be the range of $\lambda x \mid \varphi_1(e, x)$. Thus $R_0, R_1, R_2, \ldots$ is a simultaneous, metarecursive enumeration of all metarecursively enumerable sets. There exist metarecursive functions, $u(\sigma)$ and $v(\sigma)$, such that for each σ, $u(\sigma) \epsilon R_{v(\sigma)}$, and such that for each e, $R_e = \{u(\tau) \mid v(\tau) = e\}$. For each σ and e, let $R_e^\sigma = \{u(\tau) \mid \tau \leqq \sigma \;\&\; v(\tau) = e\}$. In short, R_e^σ consists of all ordinals enumerated in R_e at or prior to stage σ of our simultaneous enumeration.

We will define a metarecursive function $M(\sigma)$ and a partial, metarecursive function $v(\sigma, e)$ simultaneously by recursion on σ. For each σ, $M(\sigma)$ will be a metafinite set of ordinals less than σ; M will equal $\bigcup\{M_\sigma \mid \sigma < w_1\}$. We will say η is in the i-th *e-state* at stage σ if $\eta \notin \bigcup\{M(\delta) \mid \delta < \sigma\}$ and

$$i = \sum \{2^{e-j} \mid \eta \in R_j^\sigma \;\&\; j \leqq e\}.$$

The e-state of η at stage σ, if it is defined, is a natural number less than 2^{j+1}. If $\eta \in \bigcup\{M(\delta) \mid \delta < \sigma\}$, then we will say η has no e-state at stage σ. The notion of e-state is due to Friedberg [3]. Principle (P) is the major reason for the use of the notion.

(P) *If $\alpha \epsilon$ i-th e-state at stage σ and $\beta \epsilon$ j-th e-state at stage σ and $i < j$, then for each $c \geqq e$, the c-state of α at stage σ is less than the c-state of β at stage σ.*

(P) is an immediate consequence of the fact that $1+2+2^2+\ldots+2^b < 2^{b+1}$.

We proceed at stage σ as follows. It is safe to assume $W(\sigma) = \bigcup\{M(\tau) \mid \tau < \sigma\}$ is metafinite, since we are defining a metarecursive function $\lambda\tau \mid M(\tau)$ with the property that for each τ, $M(\tau)$ is metafinite. (This last remark is a typical example of the role of (A*) and (B*), two principles stated at the beginning of section 6, in the definition of functions by metarecursion.) We define $v(\sigma, e)$ for all finite e by induction on e. Since $W(\sigma)$ is bounded, there must be an e-state at stage σ which contains an η greater than every member of $\{v(\sigma, i) \mid i<e\} \cup \{g(\sigma, e)\}$; let $j(\sigma, e)$ be the highest such e-state. We define:

$$v(\sigma, e) = \mu\eta[\eta \epsilon j(\sigma, e)\text{-th } e\text{-state} \;\&\; \eta > g(\sigma, e) \;\&\; (i)_{i<e}(v(\sigma, i) < \eta)];$$
$$M(\sigma) = \{\alpha \mid \alpha < \sigma\} - \{v(\sigma, e) \mid e \geqq 0\}.$$

Let $M = \bigcup\{M(\sigma) \mid \sigma < \omega_1\}$.

LEMMA 1. *For each e, $lim_\sigma v(\sigma, e)$ exists.*

PROOF. By induction on e. Suppose $lim_\sigma v(\sigma, i)$ exists for all $i<e$. Let γ

be the largest member of

$$\{lim_\sigma v(\sigma, i) \mid i<e\} \cup \{g(e)\}.$$

Let τ be so large that

$$(\sigma)_{\sigma \geqq \tau}[g(\sigma, e) = g(e) \,\&\, (i)_{i<e}(v(\sigma, i) = v(\tau, i))].$$

Then for each $\sigma>\tau$, $j(\sigma, e)$ is the highest e-state at stage σ which contains an $\eta>\gamma$. Let j be the greatest member of $\{j(\sigma, e) \mid \sigma>\tau\}$. Let α be the least member of

$$\{\eta \mid \eta>\gamma \,\&\, (E\sigma)_{\sigma>\tau}(\eta \epsilon j\text{-th } e\text{-state at stage } \sigma)\}.$$

Let $\sigma_1 > \tau$ be such that α is in the j-th e-state at stage σ_1. Then $j(\sigma_1, e) = j$ and $v(\sigma_1, e) = \alpha$.

We show $j(\sigma, e) = j$ and $v(\sigma, e) = \alpha$ for all $\sigma \geqq \sigma_1$ by induction on σ. Fix $\sigma_2 > \sigma_1$, and suppose $j(\sigma, e) = j$ and $v(\sigma, e) = \alpha$ when $\sigma_2 > \sigma \geqq \sigma_1$. Then $\alpha \notin \bigcup\{M(\delta) \mid \delta < \sigma_2\}$, and consequently, α is in some e-state at stage σ_2. Since $\sigma_2 > \sigma_1$, the e-state of α at stage σ_2 cannot be less than the e-state of α at stage σ_1. Since $\sigma_2 > \tau$ and $\alpha > \gamma$, the e-state of α at stage σ_2 cannot be greater than j. It follows $j(\sigma_2, e) = j$ and $v(\sigma_2, e) = \alpha$.

For each e, let $v(e) = lim_\sigma v(\sigma, e)$. Recall that $g(\sigma, e) < v(\sigma, e) < v(\sigma, e+1)$ for all σ. It follows that $cM = \{v(e) \mid e \geqq 0\}$, and that $v(e) > g(e)$ for all e. But then the complement of M is unbounded, since the range of $g(e)$ is unbounded.

Following Friedberg [3], we say α is a resident of the i-th e-state if α is in the i-th e-state at stage σ for all sufficiently large σ. Then $\alpha \epsilon cM$ if and only if α is a resident of some e-state. Since cM is infinite, it must be that for each e, some e-state has infinitely many residents. We say an e-state is well-resided if it has infinitely many residents.

LEMMA 2. *For each e, there is only one well-resided e-state.*

PROOF. By reductio ad absurdum. Fix e, and suppose both the i-th and the j-th e-states are well-resided, where $i < j$. It follows that there exist d and c such that $d > c \geqq e$, $v(c)$ is a resident of the i-th e-state, and $v(d)$ is a resident of the j-th e-state. Let σ be so large that $v(c)$ is in the i-th e-state at stage σ, $v(d)$ is in the j-th e-state at stage σ, and $v(\sigma, i) = v(i)$ for all $i \leqq c$. Let i^* be the c-state of $v(c)$ at stage σ, and let j^* be the c-state of $v(d)$ at stage σ. Then $i^* < j^*$ by principle (P); and $i^* = j(\sigma, c)$, since $v(c) = v(\sigma, c)$. It follows from the definition of $j(\sigma, c)$ that $v(d) < v(c)$, since $j(\sigma, c) < j^*$ and $v(d)$ is in the j^*-th c-state at stage σ. But $v(d) > v(c)$, because $d > c$.

Let R be a metarecursively enumerable set; then $R = R_e$ for some e. If $R-M$ were infinite, then one of the odd-numbered e-states would be well-resided. If $cR-M$ were infinite, then one of the even-numbered e-states would be well-resided. It follows from Lemma 2 that M is maximal.

5. Applications to Π_1^1 sets of natural numbers. In section 3 it was shown that each non-empty, Π_1^1 set is the range of a metarecursive function. (One effect of rule R3 is to make possible the enumeration of Π_1^1 sets.) The simultaneous, metarecursive enumeration of the metarecursively enumerable sets described in section 4 is easily changed to a simultaneous, metarecursive enumeration of the Π_1^1 sets; each R_e is replaced by S_e, where S_e consists of the finite members of R_e. By identifying the Π_1^1 sets with the metarecursively enumerable sets of natural numbers, we strengthen the well-known analogy between the Π_1^1 sets and the recursively enumerable sets. In fact, we can now apply the priority method to the Π_1^1 sets and obtain new results.

We call a Π_1^1 set M *maximal* if cM is infinite, and if for each Π_1^1 set S, either $S–M$ or $cS–M$ is finite.

(C) *There exists a maximal,* Π_1^1 *set.*

The proof of (C) can be given either as an immediate generalization of Friedberg's construction of a maximal set [3] or as a specialization of the proof of Theorem 3; the latter approach requires only that $g(\sigma, e)$ be replaced by e and R_e by S_e. Friedberg [3] also shows that each non-recursive, recursively enumerable set is the union of two disjoint, non-recursive, recursively enumerable sets. His argument readily generalizes, with the help of metarecursion theory, to a proof of (D).

(D) *Each non-hyperarithmetic,* Π_1^1 *set is the union of two disjoint, non-hyperarithmetic,* Π_1^1 *sets.*

(C) and (D) support the conjecture (due to H. Rogers, Jr., cf. [9]) that the ordering of recursively enumerable sets by set inclusion is ismorphic to the ordering of Π_1^1 sets by set inclusion, or at worst, isomorphic modulo the finite sets.

6. Post's problem. We now present a solution to Post's problem for the recursive ordinals. Our solution is analogous to Friedberg's solution of Post's problem for the natural numbers. Friedberg [2] makes use of the following combinatorial principle: (A) *the union of finitely many finite sets is finite.* A collection $\{K\}$ of metafinite sets is said to be *metafinite* if $\{K\} = \{K_\alpha \mid \alpha \epsilon H\}$ for some metafinite H. The metarecursion-theoretic analogue of (A) is (A*): *the union of the members of a metafinite collection of metafinite sets is metafinite.* (A*) follows readily from the definitions in section 1. We also need (B*): *the range of a metarecursive function restricted to a metafinite set is metafinite.* (B*) is the analogue of (B): *the range of a function restricted to a finite set is finite.*

A *requirement* $R = \{(F_a, H_a) \mid a \epsilon A\}$ is a collection (possibly void) of ordered pairs of disjoint metafinite sets of recursive ordinals. A set T of recursive ordinals *meets* requirement R if for some $a \epsilon A$, $F_a \subseteq T$ and $H_a \cap T = \varnothing$.

Let $t = (t_0, t_1, g)$ be a triple of functions defined on the recursive ordinals with recursive ordinals as values. We say t *enumerates requirements* if for each $\sigma \geqq 0$, $K_{t_0(\sigma)}$ and $K_{t_1(\sigma)}$ are disjoint, metafinite sets, and if all values of g are finite. The requirements enumerated by t are $R_0, R_1, R_2, \ldots$; for each natural number k,

$$R_k = \{(F^\sigma, H^\sigma) \mid g(\sigma) = k\},$$

where $F^\sigma = K_{t_0(\sigma)}$ and $H^\sigma = K_{t_1(\sigma)}$.

With each metarecursive t which enumerates requirements, we associate a set T called the *priority set* of t. T is metarecursively enumerable, and meets every member of a certain subclass of the class of requirements enumerated by t. T is defined by stages.

Stage $\sigma = 0$. $T_\sigma = \varnothing$.

Stage $\sigma > 0$. $T_\sigma = \bigcup\{T_\delta \mid \delta < \sigma\}$ if (a) or (b) or (c) is true:

(a) there is a $\psi < \sigma$ such that $g(\psi) < g(\sigma)$, $\psi > 0$, $F^\psi \nsubseteq \bigcup\{T_\delta \mid \delta < \psi\}$, $F^\psi \subseteq T_\psi$, $H^\psi \cap \bigcup\{T_\delta \mid \delta < \sigma\} = \varnothing$, and $H^\psi \cap F^\sigma \neq \varnothing$;

(b) there is a $\psi < \sigma$ such that $g(\psi) = g(\sigma)$, $\psi > 0$, $F^\psi \nsubseteq \bigcup\{T_\delta \mid \delta < \psi\}$, $F^\psi \subseteq T_\psi$, and $H^\psi \cap \bigcup\{T_\delta \mid \delta < \sigma\} = \varnothing$;

(c) $H^\sigma \cap \bigcup\{T_\delta \mid \delta < \sigma\} \neq \varnothing$.

Otherwise, $T_\sigma = \cup\{T_\delta \mid \delta < \sigma\} \cup F^\sigma$.

Let $T = \bigcup\{T_\delta \mid \delta < \omega_1\}$.

For each k, we say R_k is *met* at stage σ if $\sigma > 0$, $k = g(\sigma)$, $F^\sigma \nsubseteq \bigcup\{T_\delta \mid \delta < \sigma\}$ and $F^\sigma \subseteq T_\sigma$. If R_k is met at stage σ, then $H^\sigma \cap T_\sigma = \varnothing$, since clause (c) must be false at stage σ, and since $H^\sigma \cap F^\sigma = \varnothing$ for all σ.

For each k, we say R_k is *injured* at stage σ if there is a $\psi < \sigma$ such that R_k was met at stage ψ, $H^\psi \cap \bigcup\{T_\delta \mid \delta < \sigma\} = \varnothing$ and $H^\psi \cap T_\sigma \neq \varnothing$. If R_k is met at stage σ and is not injured at any later stage, then T meets R_k.

For each k, we say R_k is *t-dense* if for each metafinite set L, there is a $\sigma > 0$ such that $g(\sigma) = k$, $F^\sigma \bigcup\nsubseteq \{T_\delta \mid \delta < \sigma\}$, $H^\sigma \cap \bigcup\{T_\delta \mid \delta < \sigma\} = \varnothing$, and $L \cap F^\sigma = \varnothing$.

LEMMA 3. *If $\psi < \sigma$ and R_k is met at stage ψ and at stage σ, then there is an η such that $\psi < \eta < \sigma$ and R_k is injured at stage η.*

PROOF. We have $g(\psi) = g(\sigma) = k$, $F^\psi \subseteq T_\psi$ and $F^\psi \nsubseteq \bigcup\{T_\delta \mid \delta < \psi\}$. It follows that $H^\psi \cap \bigcup\{T_\delta \mid \delta < \sigma\} \neq \varnothing$, since otherwise clause (b) would be true at stage σ and R_k would not be met at stage σ. We know $H^\psi \cap \bigcup\{T_\delta \mid \delta < \psi\} = \varnothing$, since otherwise clause (c) would be at stage ψ and R_k would not be met at stage ψ. Then $H^\psi \cap T_\psi = \varnothing$, since $H^\alpha \cap F^\alpha = \varnothing$ for all α. Thus there is a unique η such that $\psi < \eta < \sigma$, $H^\psi \cap \bigcup\{T_\delta \mid \delta < \eta\} = \varnothing$, and $H^\psi \cap T_\eta \neq \varnothing$. But then R_k is injured at stage η.

LEMMA 4. *For each k, the set $\{\sigma \mid R_k$ is injured at stage $\sigma\}$ is finite.*

PROOF. By induction on k. Let $k \geqq 0$, and suppose that for each $i < k$, the set $\{\sigma \mid R_i$ is injured at stage $\sigma\}$ is finite. Then it follows from Lemma 3 that the set

$$S = \{\sigma | (\text{E}i)(i<k \;\&\; R_i \text{ is met at stage } \sigma)\}$$

is finite. We now show that the set $\{\sigma|R_k$ is injured at stage $\sigma\}$ is a subset of S. Fix σ, and suppose R_k is injured at stage σ. Then there is a $\psi < \sigma$ such that R_k was met at stage ψ, $H^\psi \cap \bigcup\{T_\delta \mid \delta < \sigma\} = \varnothing$, and $H^\psi \cap T_\sigma \neq \varnothing$; consequently, $0 < \psi$, $g(\psi) = k$, $F^\psi \nsubseteq \bigcup\{T_\delta \mid \delta < \psi\}$, and $F^\psi \nsubseteq T_\psi$. If $g(\psi)$ were less than or equal to $g(\sigma)$, then (a) or (b) would be true at stage σ, and R_k would not be injured at stage σ. Hence $g(\sigma) < k = g(\psi)$, and $R_{g(\sigma)}$ is met at stage σ.

LEMMA 5. *If t is metarecursive and enumerates requirements, then T, the priority set of T, is metarecursively enumerable and meets every t-dense requirement.*

PROOF. It follows from (A*) and (B*) that there exists a metarecursive function f such that $T_\sigma = K_{f(\sigma)}$ for all σ. But then $T = \bigcup\{K_{f(\sigma)} \mid \sigma < \omega_1\}$ is metarecursively enumerable. Fix k and let R_k be t-dense. We need a ψ such that R_k is met at stage ψ and is not injured at any later stage. By Lemmas 3 and 4, the set

$$K = \{\delta | (\text{E}i)_{i \leqq k} \,(R_i \text{ is met or injured at stage } \delta)\}$$

is finite; let η be the largest member of K. Let $L = \bigcup\{F^\delta \cup H^\delta \mid \delta \leqq \eta\}$; L is metafinite by (B*) and (A*). Since R_k is t-dense, there is a $\sigma > 0$ such that $g(\sigma) = k$, $F^\sigma \nsubseteq \bigcup\{T_\delta \mid \delta < \sigma\}$, $H^\sigma \cap \bigcup\{T_\delta \mid \delta < \sigma\} = \varnothing$, and $L \cap F^\sigma = \varnothing$. Clearly, $\sigma > \eta$, since $F^\sigma \neq \varnothing$ and $L \cap F^\sigma = \varnothing$. It follows that R_k is not met at stage σ and that $F^\sigma \nsubseteq T_\sigma$. Thus either (a) or (b) or (c) is true at stage σ. Clause (c) is false at stage σ. If clause (a) were true at stage σ, then there would be a ψ such that $g(\psi) < k$, $R_{g(\psi)}$ was met at stage ψ, and $R_{g(\psi)}$ was injured at stage σ; but $\sigma > \eta$. It follows clause (b) is true at stage σ. Thus there is a $\psi < \sigma$ such that $g(\psi) = g(\sigma) = k$, R_k was met at stage ψ, and R_k was not injured at stage δ for any δ such that $\psi \leqq \delta < \sigma$. Finally, R_k was not injured at stage $\delta \geqq \sigma$, since $\sigma > \eta$.

Let $p(\alpha, \beta)$ be a metarecursive function such that if $\alpha_1 \neq \alpha_2$ or $\beta_1 \neq \beta_2$, then $p(\alpha_1, \beta_1) \neq p(\alpha_2, \beta_2)$. Let $q(\eta)$ and $r(\eta)$ be metarecursive functions such that $q(\eta)$ is finite for all η and such that for each e and α, the set $\{\eta | q(\eta) = e \;\&\; r(\eta) = \alpha\}$ is unbounded.

THEOREM 4. *There exist two metarecursively enumerable sets such that neither is weakly metarecursive in the other.*

PROOF. By Lemma 5. We first define metarecursive functions T_σ, F^σ, H^σ and $g(\sigma)$ simultaneously by recursion on σ; for each σ, T_σ, F^σ and H^σ will be metafinite sets.

Stage $\sigma = 0$. $T_0 = F^0 = H^0 = \varnothing$, and $g(0) = 0$.

Stage $\sigma > 0$. Let $e = q(\sigma)$,

$$A_\sigma = \{\eta | 2\eta \in \bigcup\{T_\delta \mid \delta < \sigma\}\},$$

and

$$B_\sigma = \{\eta | 2\eta+1 \in \bigcup\{T_\delta \mid \delta < \sigma\}\}.$$

Case 1. σ is of the form $\tau+m$, where m is even and finite.

Case 1.1. There exist $\eta <\!\cdot\, \sigma$ and $\gamma < \sigma$ such that

(1) $T_1^1(A_\sigma, e, p(e, \eta), \gamma)$ & $U(e, \gamma) = 1$ & $p(e, \eta) \notin B_\sigma \cup K_{r(\sigma)}$;

let ψ be the least $\eta < \sigma$ such that (1) holds for some $\gamma < \sigma$. We set:

$$F^\sigma = \{2p(e, \psi)+1\};$$
$$H^\sigma = \{2\eta | \eta \notin A_\sigma\};$$
$$g(\sigma) = 2e+1.$$

Case 1.2. Otherwise. We set $F^\sigma = H^\sigma = \varnothing$ and $g(\sigma) = 0$.

Case 2. σ is of the form $\tau+m$, where m is odd and finite. Proceed as in case 1 with A and B interchanged, all members of F^σ even, all members of H^σ odd, and $g(\sigma)$ even.

Let T_σ be defined as in Lemma 5.

There exist metarecursive functions, t_0 and t_1, such that $F^\sigma = K_{t_0(\sigma)}$ and $H^\sigma = K_{t_1(\sigma)}$ for all σ. Let $t = (t_0, t_1, g)$ and $T = \bigcup\{T_\delta \mid \delta < \omega_1\}$ T is the priority set of t. Let $A = \{\eta | 2\eta \in T\}$ and $B = \{\eta | 2\eta + 1\} \in T$. By Lemma 5, T meets every t-dense requirement. We use this fact to show B is not weakly metarecursive in A. A similar argument shows A is not weakly metarecursive in B.

Fix e. Suppose $\{e\}^A(\alpha)$ is defined and less than 2 for all α. Let R_{2e+1}be t-dense. Thus T meets R_{2e+1}. It follows that there is a σ such that $g(\sigma)= 2e+1$, $F^\sigma \subseteq T$ and $H^\sigma \cap T = \varnothing$. Then case 1.1 must hold at stage σ, and there must be a ψ and a γ such that:

$$T_1^1(A_\sigma, e, p(e, \psi), \gamma) \text{ \& } U(e, \gamma) = 1;$$
$$F^\sigma = \{2p(e, \psi)+1\};$$
$$H^\sigma = \{2\eta | \eta \notin A_\sigma\}.$$

Since $F^\sigma \subseteq T$, we have $p(e, \psi) \in B$. Since $H^\sigma \cap T = \varnothing$, we have $\{e\}^A(p(e, \psi)) = 1$.

Now suppose R_{2e+1} is not t-dense. Note that $H^\sigma \cap \bigcup\{T_\delta \mid \delta < \sigma\} = \varnothing$ and $F^\sigma \cap \bigcup\{T_\delta \mid \delta < \sigma\} = \varnothing$ for all σ. It follows that there is a metafinite set L such that for all σ, either $g(\sigma) \neq 2e+1$ or $F^\sigma = \varnothing$ or $F^\sigma \cap L \neq \varnothing$. We use this fact to find a τ such that $\tau \notin B$ and $\{e\}^A(\tau) = 0$. Let η be such that $2p(e, \eta)+1 \notin L$; the existence of η follows from (B*). Let $\tau = p(e, \eta)$. If τ were a member of B, then there would be a σ such that $g(\sigma) = 2e+1$, $F^\sigma = \{2\tau+1\}$, and $F^\sigma \cap L = \varnothing$. Thus $\tau \notin B_\sigma$ for all σ. Suppose $\{e\}^A(\tau) = 1$. Then there is a $\sigma > \eta$ and a $\gamma < \sigma$ such that (1) holds and $K_{r(\sigma)} = \{\alpha | 2\alpha+1 \in L\}$. Then $g(\sigma) = 2e+1$, $F^\sigma \neq \varnothing$, and $F^\sigma \cap L = \varnothing$.

For each finite k, let $y(k)$ be the cardinality of $S_k = \{\sigma | R_k$ is met at stage σ of the construction of Theorem 4$\}$; by Lemma 4, $y(k)$ is finite. For each infinite α, let $y(\alpha) = 0$. We claim y is not a metarecursive function.

Suppose it is. For each finite k, let $w(k)$ be the greatest member of S_k; for each infinite α, let $w(\alpha) = 0$. Then w is metarecursive, since y is. It follows from (B*) that there is a γ such that $w(k) < \gamma$ for all finite k. Then $T = \bigcup\{F^\sigma \mid \sigma < \gamma\}$ is metafinite by (A*), and A and B are metarecursive.

Kripke [10] introduces the notion of ***admissible ordinal***, and then shows that for each admissible ordinal α, it is possible to develop a recursion theory for the ordinals less than α. He proves that Kleene's ω_1 is the first admissible ordinal after ω_0, and his notion of ω_1-recursive function is equivalent to our notion of metarecursive function. He obtains appropriate generalizations of Theorems 1 and 2 for all admissible α, and he proves Theorem 4 for a large class of admissible α which includes ω_1. In [17] it will be shown that the following solution of Post's problem generalizes to all admissible α: there exists recursively enumerable sets A and B such that A is not recursive in B, and B is not recursive.

The most striking phenomenon of metarecursion theory is the existence of bounded, metarecursively enumerable sets which are not metarecursive; we have an example of one at the beginning of the proof of Theorem 3. We call a set of recursive ordinals ***regular*** if its intersection with every metafinite set is metafinite. A metarecursively enumerable set is regular if every bounded, metarecursively enumerable subset of it is regular. More intuitively, a metarecursively enumerable A is regular if and only if some metarecursive function f enumerates it in such a way that for each α, f eventually stops enumerating new members of A less than α.

The automatic regularity of all sets in recursion theory is the main obstruction to lifting theorems of this theory up into metarecursion theory. We say that two sets have the same ***metadegree*** if each is metarecursive in the other. Dekker [1] shows: if f is a 1–1 recursive function, then the deficiency set of f has the same Turing-degree as f, and if the range of f is not recursive, then the deficiency set of f is simple (in fact, hypersimple). Dekker's result generalizes readily to metarecursion theory if it is assumed that the range of f is regular; without this assumption we are unable to show that the deficiency set of f has the same metadegree as the range of f. In [15] it was shown that each non-recursive, recursively enumerable A is the union of two disjoint, recursively enumerable sets whose Turing-degrees are incomparable. If we attempt to generalize this result to a metarecursively enumerable A, we appear to fail unless we assume A is regular.

The lack of regularity is not as alarming as it sounds, since it will be shown in [17] that every metadegree which contains a metarecursively enumerable set also contains a regular, metarecursively enumerable set. This last result makes it seem plausible that most, if not all, theorems about Turing-degrees of recursively enumerable sets can be lifted up into metarecursion theory. G. Driscoll plans to discuss this matter further.

It also will be shown in [17] that there exist infinitely many metadegrees of metarecursively enumerable sets which contain *only* regular sets.

One virtue of the lack of regularity is that it leads to an intricate theory of Π_1^1 sets of natural numbers. For example, it will be shown in [17] that there exist two Π_1^1 sets of natural numbers such that neither is weakly metarecursive in the other. This result can be taken to be the solution of Post's problem for Π_1^1 sets. Spector [18] showed that all Π_1^1 sets lie in two hyperdegrees, but in [17] it will be seen that they lie in infinitely many metadegrees.

REFERENCES

[1] J. C. E. DEKKER, *A theorem on hypersimple sets*, ***Proceedings of the American Mathematical Society***, 5 (1954), 791–796.

[2] R. M. FRIEDBERG, *Two recursively enumerable sets of incomparable degrees of unsolvability*, ***Proceedings of the National Academy of Science, U.S.A.***, 43 (1957), 236–238.

[3] R. M. FRIEDBERG, *Three theorems on recursive enumeration*, this JOURNAL, 23 (1958), 309–316.

[4] R. O. GANDY, *Proof of Mostowski's conjecture*, ***Bulletin de l'Académie Polonaise des Sciences***, 8 (1960, 571–574.

[5] K. GÖDEL, *Über formal unentscheidbare Sätze der Principia Mathematica und verwandter Systeme I*, ***Monatshefte für Mathematik und Physik***, 38 (1931), 173–178.

[6] S. C. KLEENE, ***Introduction to Metamathematics***, New York, Toronto, Amsterdam and Groningen, 1952.

[7] S. C. KLEENE, *On the forms of the predicates in the theory of constructive ordinals (second paper)*, ***American Journal of Mathematics***, 77 (1955), 405–428.

[8] G. KREISEL, *Set theoretic problems suggested by the notion of potential totality*, ***Infinitistic Methods***, Warsaw 1961, 103–140.

[9] G. KREISEL, *Model theoretic invariants: applications to recursive and hyperarithmetic operations*, ***Theory of Models Symposium***, Berkeley, forthcoming.

[10] S. KRIPKE, *Transfinite recursions on admissible ordinals*, in preparation.

[11] S. KRIPKE, *Admissible ordinals and the analytic hierarchy*, in preparation.

[12] A. MOSTOWSKI, ***Sentences Undecidable in Formalized Arithmetic***, Amsterdam, 1952.

[13] A. MOSTOWSKI, *Representability of sets in formal systems*, ***Recursive Function Theory, Proceedings of Symposia in Pure Mathematics***, 5 (1962), 29–48.

[14] E. L. POST, *Recursively enumerable sets of positive integers and their decision problems*, ***Bulletin of the American Mathematical Society***, 50 (1944), 284–316.

[15] G. E. SACKS, *On the degrees less than 0'*, ***Annals of Mathematics***, 77 (1963), 211–231.

[16] G. E. SACKS, *The recursively enumerable degrees are dense*, ***Annals of Mathematics***, 80 (1964), 300–312.

[17] G. E. SACKS, *Admissible ordinals, Post's problem and regularity*, in preparation.

[18] C. Spector, *Recursive well-orderings*, this Journal, 20 (1955), 151–163.
[19] R. M. Smullyan, ***Theory of Formal Systems***, Princeton, 1961.
[20] C. E. M. Yates, *Three theorems on the degrees of recursively enumerable sets*, ***Duke Mathematical Journal***, to appear.

UNIVERSITY OF PARIS
CORNELL UNIVERSITY

POST'S PROBLEM, ADMISSIBLE ORDINALS, AND REGULARITY

BY
GERALD E. SACKS(1)

1. **Introduction.** The basic notions of metarecursion theory were introduced in [7]. Metarecursion theory is an attempt to generalize the ideas and arguments of recursion theory from the natural numbers to the recursive ordinals. Ordinary recursion theory concerns itself with finite sets of natural numbers. Metarecursion theory deals in an analogous fashion with metafinite sets of recursive ordinals. In [7] it was seen that two of the deepest results of ordinary recursion theory, the solution of Post's problem [3] and the maximal set construction [4], generalize to the metarecursive case. Theorem 4 of [7] states that there exist two metarecursively enumerable sets of recursive ordinals such that neither is metarecursive in the other. Kreisel [6] casts doubt on the contention that Theorem 4 of [7] is the correct generalization of Post's problem. He wonders how to correctly formulate the notion of Turing reducibility for recursive ordinals and he discusses four possible choices. It is not our place to decide the correct notion of Turing reducibility for metarecursion theory; however, in §4 we show that there exist two metarecursively enumerable sets such that neither is reducible to the other by any of the four methods discussed by Kreisel.

In [7] two sets of recursive ordinals are said to have the same metadegree if each is metarecursive in the other. It is not known if the metadegrees of metarecursively enumerable sets are order-isomorphic to the Turing degrees of recursively enumerable sets, but all the existing evidence seems to favor an affirmative answer. Driscoll [2] has shown the metadegrees of metarecursively enumerable sets are dense; i.e., between any two comparable ones, there is a third. His proof is not a mere translation of the argument of [12]. There exists a formidable obstacle to the lifting up into metarecursion theory of results about recursively enumerable sets. Call a set of recursive ordinals regular if its intersection with every metafinite set is metafinite. In ordinary recursion theory, the intersection of any set of natural numbers with any finite set is finite; this platitude is essential in many arguments about recursively enumerable sets. Unfortunately, there exists a multitude of nonregular, metarecursively enumerable sets [7].

Received by the editors September 28, 1965 and, in revised form, February 8, 1966.

(1) The preparation of this paper was supported by U.S. Army Contract ARO-D-373. The author wishes to thank Professor G. Kreisel for invaluable help in the writing of this paper.

Nonetheless, Driscoll [2] was able to make a long detour around the obstacle of nonregularity with the aid of the principal result of §2: each metarecursively enumerable set has the same metadegree as some regular, metarecursively enumerable set.

Spector [14] showed that all nonhyperarithmetic, Π_1^1 sets have the same hyperdegree. It is possible to interpret his result as the answer to Post's problem for Π_1^1 sets. In this paper we offer what we hope is a happier alternative. Suppose we identify the natural numbers with the finite, recursive ordinals. Then, as was seen in [7], the Π_1^1 sets of natural numbers coincide with the metarecursively enumerable sets of finite, recursive ordinals. In §3 we show that there exist two Π_1^1 sets of natural numbers such that neither is metarecursive in the other.

Kripke [8], [9] has succeeded in generalizing a large portion of recursion theory by introducing the notion of admissible ordinal. An admissible ordinal has the property that the set of all lesser ordinals behaves very much like the natural numbers when one attempts to initiate the pre-Friedbergian arguments of recursion theory. Kripke's definition of admissible ordinal is expressed in terms of closure properties that refer to an equation calculus analogous to the one employed by Kleene [5] for ordinary recursion theory. Kripke's calculus is very convenient for the purposes of metarecursion theory and was utilized in [7]. (It turns out that ω_1, the least nonrecursive ordinal, is the first admissible ordinal occurring after ω.) In this paper, our first duty is metarecursion theory, but we cannot resist indicating which of our results generalize to every admissible ordinal. In particular, we give a solution to Post's problem which is uniform with respect to all admissible ordinals; i.e., we give a single argument in which the admissible ordinal α occurs as a parameter.

We conclude with a discussion of some open questions.

2. **Regularity and metarecursive enumerability.** In [7] a set of recursive ordinals was said to be regular if its intersection with every metafinite set was metafinite. Each metarecursive set is regular. Each nonhyperarithmetic, Π_1^1 set of natural numbers (regarded as a set of finite, recursive ordinals) is a nonregular, metarecursively enumerable set. In metarecursion theory we are forced to deal with bounded, metarecursively enumerable sets which are not metarecursive. This last remark is the reason that the phenomenon of nonregularity makes it difficult, if not impossible, to lift many results of ordinary recursion theory up into metarecursion theory. But the situation is not as gloomy as it appears, since we can show by an essentially elementary argument that each metadegree of a metarecursively enumerable set contains a regular, metarecursively enumerable set. In §4 we see that there exist metadegrees of metarecursively enumerable sets which contain *only* regular sets. The overall situation concerning regularity is not clear and is discussed in §7.

THEOREM 1. *Each metarecursively enumerable set has the same metadegree as some regular, metarecursively enumerable set.*

Proof. The main idea of what follows is inspired by Dekker's notion of deficiency set [1]. We use what might be called a double deficiency set. Let $\{K_\alpha\}$ be a standard indexing of the metafinite sets; i.e., there exist metarecursive functions j and k such that for each α, $K_\alpha = \{\beta \mid \beta < j(\alpha) \text{ and } k(\alpha,\beta) = 0\}$. Let Q be a Π_1^1 set of unique notations for the recursive ordinals; let η be a metarecursive function such that for each α, $\eta(\alpha)$ is the unique natural number in Q which is a notation for α. For convenience we assume that Q is a linearly ordered subset of Kleene's 0.

We write $A \leqq_M B$ to indicate A is metarecursive in B. Intuitively, $A \leqq_M B$ means there exists a method M for determining the truth or falsity of both of the following two assertions for an arbitrary metafinite K: $K \subseteq A$ and $K \subseteq cA$; the method M must be such that the status of K is decided by the fact that B satisfies some metafinite set of conditions such as those expressed by $K_1 \subseteq B$ and $K_2 \subseteq cB$. Thus M can be thought of as a metarecursively enumerable set of 5-tuples of the form $\langle 0,K,K_1,K_2,i\rangle$ and $\langle 1,K,K_1,K_2,i\rangle$, where $K \subseteq A$ if and only if for some $K_1 \subseteq B$ and $K_2 \subseteq cB$, we have $\langle 0, K,K_1,K_2,0\rangle \in M$, where $K \nsubseteq A$ if and only if for some $K_1 \subseteq B$ and $K_2 \subseteq cB$, we have $\langle 0,K,K_1,K_2,1\rangle \in M$, and where a similar relation holds between the 5-tuples of the form $\langle 1,K,K_1,K_2,i\rangle$, cA, and B.

We write $A \leqq_m B$ (read A is many-one reducible to B) to indicate that for some metarecursive function f, we have $\alpha \in A \leftrightarrow f(\alpha) \in B$. If $A \leqq_m B$, then $A \leqq_M B$. We write $A \leqq_w B$ to indicate A is weakly metarecursive in B. Let F_α be a standard enumeration of all the finite sets of recursive ordinals. We write $A \leqq_f B$ (read A is finitely reducible to B) to indicate $A \leqq_w B$ (there is an e such that $A = \{e\}^B$) and there is a metarecursive g such that for each α, the computation of the truth value of $\alpha \in A$ draws only on the information expressed by the membership of $F_{g(\alpha)} \subseteq B$ and $F_{g(\alpha)} \subseteq cB$. If $A \leqq_f B$ and $B \leqq_w C$, then $A \leqq_w C$. We need the notion of finite reducibility because Driscoll [2] has shown that the relation $\leqq_w$ is not transitive.

Suppose A is metarecursively enumerable and we wish to prove $A \leqq_M B$. It is sufficient to find a method M for showing $K \subseteq cA$ when and only when $K \subseteq cA$ is true. We merely combine the method M with an enumeration of A. If $K \subseteq A$, then eventually we will enumerate all of K in A. If $K \nsubseteq A$, then M will show that for some $\alpha \in K$, we have $\{\alpha\} \subseteq cA$. If $K \subseteq cA$, then M will show $K \subseteq cA$. If $K \nsubseteq cA$, then for some $\alpha \in K$, we will enumerate $\alpha \in A$. We will refer to the above observation as *principle* P.

Let A be a metarecursively enumerable set. We aim to find a metarecursively enumerable, regular B in the same metadegree as A. Let $A^* = \{\alpha \mid A \cap K_\alpha \neq \varnothing\}$. We check that A and A^* have the same metadegree. Observe that

$$(1)\qquad K_\alpha \subseteq cA^* \leftrightarrow \bigcup \{K_\gamma \mid \gamma \in K_\alpha\} \subseteq cA \leftrightarrow K_{h(\alpha)} \subseteq cA \leftrightarrow h(\alpha) \in cA^*,$$

where h is metarecursive. (We are using very heavily the fact that a metafinite union of metafinite sets is metafinite.) It follows from principle P and (1) that A^* is metarecursive in A. The definition of A^* implies that A is metarecursive in A^*.

Let f be a one-one, metarecursive function whose range is A^*. We introduce the double deficiency set of f:

$$D_f^2 = \{\alpha \mid (E\beta)(f(\beta) < f(\alpha) \;\&\; \eta(f(\beta)) < \eta(f(\alpha)) \;\&\; \beta > \alpha)\}.$$

We show $D_f{}^2$ is regular by showing $D_f{}^2 \cap L_\gamma$ is metafinite for every γ ($L_\gamma = \{\alpha \mid \alpha < \gamma\}$). Fix γ; we need only find a τ such that

$$(2) \qquad D_f{}^2 \cap L_\gamma = \{\alpha \mid (E\beta)_{\beta<\tau}(f(\beta) < f(\alpha) \;\&\; \eta(f(\beta)) < \eta(f(\alpha)) \;\&\; \beta > \alpha)\}.$$

We define a partial metarecursive function d by induction:

$$d(0) \cong \mu\beta(\beta > \gamma \;\&\; (E\alpha)(\alpha < \gamma \;\&\; f(\beta) < f(\alpha) \;\&\; \eta(f(\beta)) < \eta(f(\alpha))));$$

$$d(n+1) \cong \mu\beta(\beta > d(n) \;\&\; (m)_{m \leqq n}(f(\beta) < f(d(m)) \vee \eta(f(\beta)) < \eta(f(d(m))))$$

$$\& \; (E\alpha)(\alpha < \gamma \;\&\; f(\beta) < f(\alpha) \;\&\; \eta(f(\beta) < \eta(f(\alpha)))).$$

Clearly, the domain of d is a finite, initial segment of the natural numbers. Let τ be the least ordinal greater than γ and greater than any element of the range of d. Then τ satisfies (2). Note that τ, regarded as a function of γ, is weakly metarecursive in A^*. Then it follows from (3) and principle P that $D_f{}^2$ is metarecursive in A^*.

$$(3) \qquad K_\alpha \subseteq cD_f{}^2 \leftrightarrow K_\alpha \cap D_f{}^2 \cap L_{j(\alpha)} = \varnothing.$$

(Recall that $K_\alpha \subseteq L_{j(\alpha)}$ and j is metarecursive.)

Now we make an assumption about $D_f{}^2$ and show A^* is metarecursive in $D_f{}^2$. Let ρ be defined by

$$\rho(\alpha) \cong \mu\beta(f(\beta) > \alpha \;\&\; \eta(f(\beta)) > \eta(\alpha) \;\&\; \beta \notin D_f{}^2).$$

We assume ρ is total. Then ρ is weakly metarecursive in $D_f{}^2$, and (1) tells us

$$K_\alpha \subseteq cA^* \leftrightarrow h(\alpha) \notin A^* \leftrightarrow h(\alpha) \notin \{f(\beta) \mid \beta < \rho(h(\alpha))\}.$$

Finally, we are faced with the problem that ρ may not be total. Let γ be the least ordinal such that

$$(4) \qquad (\beta)(f(\beta) \leqq \gamma \vee \eta(f(\beta)) \leqq \eta(\gamma) \vee \beta \in D_f{}^2).$$

Then there must be a σ such that $\eta(f(\beta)) > \eta(\gamma)$ for all $\beta > \sigma$, and

$$(5) \qquad (\beta)_{\beta>\sigma}(f(\beta) \leqq \gamma \vee \beta \in D_f{}^2).$$

Let $A_\gamma{}^* = A^* \cap L_\gamma$. We claim $A_\gamma{}^*$ and A^* have the same metadegree. Let $A' = A^* - \{\alpha \mid \alpha \leqq \gamma \vee (E\delta)_{\delta \leqq \sigma}(\alpha = f(\delta))\}$. It suffices to show A' is metarecursive in $A_\gamma{}^*$. We make repeated applications of (5):

$$\alpha \in A' \leftrightarrow (E\beta)_{\beta > \sigma}(f(\beta) = \alpha \,\&\, \beta \in D_f{}^2),$$

$$\beta \in D_f{}^2 \leftrightarrow (E\delta_1)(\delta_1 > \beta \,\&\, f(\delta_1) < \alpha \,\&\, \eta(f(\delta_1)) < \eta(\alpha)),$$

$$\delta_1 > \sigma \,\&\, f(\delta_1) > \gamma \rightarrow \delta_1 \in D_f{}^2,$$

$$\delta_1 \in D_f{}^2 \leftrightarrow (E\delta_2)(\delta_2 > \delta_1 \,\&\, f(\delta_2) < f(\delta_1) \,\&\, \eta(f(\delta_2)) < \eta(f(\delta_1))),$$

$$\delta_2 > \sigma \,\&\, f(\delta_2) > \gamma \rightarrow \delta_2 \in D_f{}^2,$$

$$\vdots$$

$$(E\delta_m)(f(\delta_m) \leqq \gamma \,\&\, \eta(f(\delta_m)) < \eta(\alpha)).$$

But then we have

$$\alpha \in A' \leftrightarrow (E\beta)(E\delta)(\sigma < \beta < \delta \,\&\, \alpha = f(\beta) \,\&\, f(\delta) \leqq \gamma < \alpha \,\&\, \eta(f(\delta)) < \eta(\alpha)). \tag{6}$$

Let $F_{t(\alpha)}$ be the finite set $\{\varepsilon \mid \varepsilon \leqq \gamma \,\&\, \eta(\varepsilon) < \eta(\alpha)\}$; we can assume t is metarecursive, since our unique notations form a linearly ordered subset of Kleene's 0. Then (6) becomes

$$\alpha \in A' \leftrightarrow (E\beta)(E\delta)(\sigma < \beta < \delta \,\&\, \gamma < \alpha = f(\beta) \,\&\, f(\delta) \in F_{t(\alpha)}). \tag{7}$$

With the help of (1) we obtain

$$K_\alpha \subseteq cA' \leftrightarrow (K_\alpha - (A^* - A')) \subseteq cA^* \leftrightarrow q(\alpha) \in cA', \tag{8}$$

where q is metarecursive (we assume some repetition in our indexing of the metafinite sets). Principle P, (7) and (8) imply A' is metarecursive in $A_\gamma{}^*$.

Let $B = \{\eta(\alpha) \mid \alpha \in A_\gamma{}^*\}$. Since $A_\gamma{}^*$ is bounded, it must be that B and $A_\gamma{}^*$ have the same metadegree; in fact, $B \leqq_m A_\gamma{}^*$ and $A_\gamma{}^* \leqq_m B$. B is useful because it contains only finite ordinals. Let g be a one-one, metarecursive function whose range is B. Let

$$D_g = \{\alpha \mid (E\beta)(g(\beta) < g(\alpha) \,\&\, \beta > \alpha)\}.$$

D_g is seen to be regular by the same argument we used to see $D_f{}^2$ is regular.

We claim D_g and $A_\gamma{}^*$ have the same metadegree. D_g is metarecursive in B, and hence in $A_\gamma{}^*$, by the same argument used to show $D_f{}^2$ is metatecursive in A^*. It remains only to show $A_\gamma{}^*$ is metarecursuve in D_g. We note

$$K_\alpha \subseteq cA_\gamma^* \leftrightarrow (\tau)(\tau \in K_\alpha \,\&\, \tau < \gamma \rightarrow K_\tau \cap A = \varnothing)$$

(9)

$$\leftrightarrow K_{m(\alpha)} \subseteq cA \leftrightarrow h(m(\alpha)) \in cA^*,$$

where m is metarecursive. We can choose h and m so that

(10) $$(\alpha)(h(m(\alpha)) > \gamma \,\&\, \eta(h(m(\alpha)) > \eta(\gamma)) \,\&\, (\delta)_{\delta > \sigma}(h(m(\alpha)) \neq f(\delta))).$$

Then we have

(11) $$K_\alpha \subseteq cA_\gamma^* \leftrightarrow h(m(\alpha)) \in cA'.$$

Now $B \leqq_w D_g$ by an argument similar to the one used to show $A^* \leqq_M D_f{}^2$. (Actually, this is the argument used by Dekker [1] to prove the range of a one-one, recursive function is recursive in its deficiency set.) But then $A_\gamma^* \leqq_w D_g$, since $A_\gamma^* \leqq_m B$. It follows from (7) that $A' \leqq_f A_\gamma^*$; consequently, $A' \leqq_w D_g$. This last together with (11) and principle P give $A_\gamma^* \leqq_M D_g$.

The proof of Theorem 1 suffers from a certain lack of uniformity; this defect is discussed in §7.

Let α be an arbitrary admissible ordinal [8], [9]. Kripke calls a set of ordinals α-finite if it is γ-recursive and is bounded by some ordinal less than α. Call a set of ordinals (less than α) *regular* if its intersection with every α-finite set is α-finite. Define the notion of "α-recursive in" by strict analogy with the notion of "metarecursive in" as defined in [7]; simply replace "metarecursive" by "α-recursive" and "metafinite" by "α-finite". Let us say two sets have the same α-degree if each is α-recursive in the other.

THEOREM 1^α. *Each α-recursively enumerable set has the same α-degree as some regular, α-recursively enumerable set.*

Proof. We proceed as in Theorem 1, but some modifications are necessary. Let α^* be the projectum of α. (Kripke defines the *projectum* of α to be the least ordinal β such that there exists a one-one, α-recursive function whose range is bounded by β.) Note that an α-recursively enumerable set bounded by an ordinal less than α^* must be α-finite; this fact is analogous to the truism that every finite set of natural numbers is metafinite. Let t be a one-one, α-recursive function whose values are less than α^*. Then $t(\alpha)$ plays the same role that $\eta(\alpha)$ did in the proof of Theorem 1. The only difficulty lies in showing that A' is α-recursive in A_γ^* and that A_γ^* is α-recursive in D_g. Equation (6) becomes

(6^α) $$\eta \in A' \leftrightarrow (E\beta)(E\delta)(\sigma < \beta < \delta \,\&\, \eta = f(\beta) \,\&\, f(\delta) \leqq \gamma < \eta \,\&\, t(f(\delta)) < t(\eta)).$$

Let $K_{s(\eta)}$ be the α-finite set $\{\varepsilon \mid \varepsilon \leqq \gamma \,\&\, t(\varepsilon) < t(\eta)\}$, where s is α-recursive.

(7^α) $$\eta \in A' \leftrightarrow (E\beta)(E\delta)(\sigma < \beta < \delta \,\&\, \gamma < \eta = f(\beta) \,\&\, f(\delta) \in K_{s(\eta)}).$$

The set $\{t(f(\delta)) \mid f(\delta) \in K_{s(\eta)}\}$ is α-finite, since it is α-recursively enumerable and bounded by $t(\eta) < \alpha^*$. But then the set $\{f(\delta) \mid f(\delta) \in K_{s(\eta)}\}$ is α-finite. It follows from (7^α) and (8) that A' is α-recursive in A_γ^*. The computation of the truth-value of $\eta \in A'$ from A_γ^* uses only information about the membership of A_γ^* involving ordinals ε such that $t(\varepsilon) < t(\eta)$. But the computation of the truth-value of $\eta \in A'$ from B uses only information about the membership of B involving ordinals less than $t(\eta) < \alpha^*$. D_g is not α-finite, so there must be a τ such that $g(\tau) > t(\eta)$ and $\tau \notin D_g$. Any member of B less than $t(\eta)$ must be enumerated by g before $g(\tau)$. Thus the computation of the truth-value of $\eta \in A'$ from D_g is accomplished using only finitely much information about the membership of D_g. In short, A' is weakly α-recursive in D_g. This last together with (11) and principle P gives $A_\gamma{}^*$ is α-recursive in D_g.

We can strengthen Theorem 1^α as follows. Let α be an admissible ordinal. Let cB, the complement of B, denote $\{\beta \mid \beta < \alpha\} - B$. Call an α-recursively enumerable set unbounded if it is not bounded by any ordinal less than α. We say an α-recursively enumerable set B is *simple* if cB is unbounded, and if every α-recursively enumerable subset of cB is α-finite. Then we can prove: each non-α-recursive, α-recursively enumerable set has the same α-degree as some regular, simple, α-recursively enumerable set. This is a consequence of Theorem 1^α, since the ideas of Dekker [1] work quite nicely on regular sets, and since the deficiency set of a function which enumerates a regular set is regular.

3. **Metadegrees of Π_1^1 sets.** Before we prove Theorem 2, we recall some definitions of [7]. The T-predicate of [7] is analogous to that of Kleene [5]. Its definition is:

$$T_1^1\,(B, e, \alpha, \sigma) \leftrightarrow t(e,\sigma) \;=\; \langle e, M, N, \alpha, \gamma\rangle \,\&\, M \subseteq B \,\&\, N \subseteq cB;$$

$t(e,\sigma)$ is a special metarecursive function which indexes "deductions". Roughly speaking, $T_1^1\,(B,e,\alpha,\sigma)$ says: B is an arbitrary set of recursive ordinals; e is the Gödel number of a finite system of equations E; and $t(e,\sigma)$ is the Gödel number of a metafinite "deduction" from E and B whose last line gives the value γ of some function at argument α. $U(e,\sigma)$ is a metarecursive function such that if $t(e,\sigma) = \langle e, M, N, \alpha, \gamma\rangle$, then $U(e,\sigma) = \gamma$. Two useful partial functions are:

$$\{e\}_\sigma^B(\alpha) \;=\; \gamma \leftrightarrow T_1^1\,(B, e, \alpha, \sigma) \,\&\, U(e,\sigma) = \gamma;$$

$$\{e\}^B(\alpha) \;=\; \gamma \leftrightarrow (E\sigma)(\{e\}_\sigma^B(\alpha) = \gamma) \,\&\, (\sigma)(\{e\}_\sigma^B \text{ is defined } \rightarrow U(e,\sigma) = \gamma).$$

Theorem 2 of [7] states that if A is weakly metarecursive in B, then there exists a finite e such that $(\alpha)[a(\alpha) = \{e\}^B(\alpha)]$, where $a(\alpha)$ is the characteristic function of A. The converse is also true.

Let $M(e,\sigma)$ and $N(e,\sigma)$ be metarecursive functions such that if $t(e,\sigma) = \langle e, M, N, \alpha, \gamma\rangle$, then $M(e,\sigma) = M$ and $N(e,\sigma) = N$. (We are making use of a standard indexing of the metafinite sets which enables us to speak of partial metarecursive functions that take metafinite sets (M, N) as values.) We introduce two partial functions:

(12)
$$^{K}[e]^{B}_{\sigma}(\alpha) = \gamma \leftrightarrow \{e\}^{B}_{\sigma}(\alpha) = \gamma \,\&\, K - (N(e,\sigma) \cup B) \text{ is infinite;}$$
$$^{K}[e]^{B}(\alpha) = \gamma \leftrightarrow (E\sigma)(^{K}[e]^{B}_{\sigma}(\alpha) = \gamma) \,\&\, (\sigma)(^{K}[e]^{B}_{\sigma}(\gamma) \text{ is defined} \rightarrow U(e,\sigma) = \gamma).$$

The difference between $^{K}[e]^{B}$ and $\{e\}^{B}$ is the key that unlocks the proof of Theorem 2. The "deductions" associated with $^{K}[e]^{B}$ are restricted; each of them permits infinitely many members of K not to be in B or in $N(e,\sigma) \subseteq cB$. The B of Theorem 2 (as well as the A) have the property that $^{K}[e]^{B} \simeq \{e\}^{B}$ for all e and certain metafinite K's depending on e.

THEOREM 2. *There exist two Π^1_1 sets of natural numbers such that neither is weakly metarecursive in the other*(2).

Proof. We intend to metarecursively enumerate two sets, A and B, of natural numbers with an eye to meeting certain requirements. The first set of requirements will insure that A and B are "incomparable". The second set will insure that neither A nor B are metarecursive. The two sets of requirements are redundant, since if A is not weakly metarecursive in B, then A is not metarecursive. Sometimes, however, redundancy (overkill) is useful. We will simultaneously attempt to satisfy both sets of requirements, but we will be unable to prove directly that the first set is successfully met. Instead, a direct argument will be given that the second set is met, and then (and only then) it will be possible to show with the aid of the properties of $^{K}[e]^{B}$ that the first set is met.

We handle conflicting requirements by means of a system of priorities similar to that of Friedberg [3]. To be more precise, in the course of the construction, we make only finitely many attempts to satisfy each requirement. Driscoll [2] proves theorems about metarecursively enumerable sets by means of priority arguments in which infinitely many attempts are made to satisfy individual requirements.

We define eight partial metarecursive functions, $A(\sigma)$, $B(\sigma)$, $K(\sigma,e)$, $L(\sigma,e)$, $p(\sigma,e)$, $q(\sigma,e)$, $s(\sigma,e)$ and $v(\sigma,e)$ for all σ and all finite e by induction on σ; p,q,s and v will take finite ordinals as values, while the other four functions will take only metafinite sets as values.

Let r be a metarecursive function taking only finite values such that for each $c<\omega$ and $i<4$, the set $\{\sigma \mid r(\sigma) = c \,\&\, (E\tau)(\sigma = \tau + j$ and τ is a limit ordinal $\&\, j \equiv i\ (4))\}$ is unbounded.

Stage $\sigma = 0$. Let $A(0) = B(0) = \varnothing$, $K(0,e) = L(0,e) = \{m \mid m < \omega\}$, $p(0,e) = q(0,e) = 2^{2e+2}$, and $s(0,e) = v(0,e) = 2^{2e+1}$ for all e.

Stage $\sigma > 0$. If the partial function $\lambda\delta \mid K(\delta,e)$ changes value only finitely often as δ increases through the ordinals less than σ, let $K(\sigma^*,e)$ equal its "last" value; i.e.,

(2) It is not hard to show, using Spector [14], that if B is a nonhyperarithmetic, Π^1_1 set, then every Π^1_1 set is invariantly definable from B (regarded as a set). This last notion is defined in Kreisel [6].

$$(E\delta)_{\delta<\delta}(\rho)(\delta \leqq \rho < \sigma \rightarrow K(\rho, e) = K(\sigma^*, e)).$$

If the partial function changes value infinitely often, let $K(\sigma^*, e) = K(0, e)$. (We will see that the partial function $\lambda\sigma \mid K(\sigma, e)$ changes value only finitely often as σ increases through the recursive ordinals.) Define $L(\sigma^*, e)$, $p(\sigma^*, e)$, $q(\sigma^*, e)$, $s(\sigma^*, e)$ and $v(\sigma^*, e)$ similarly.

Let $c = r(\sigma)$ and $\sigma = \tau + j$, where τ is a limit ordinal and j is finite. Let $A(\sigma^*) = \bigcup \{A(\delta) \mid \delta < \sigma\}$ and $B(\sigma^*) = \bigcup \{B(\delta) \mid \delta < \sigma\}$.

Case 1. $j \equiv 0$ (4).

Case 1.1. $s(\sigma^*, c) \notin A(\sigma^*)$ and there is a $\gamma < \sigma$ such that $\{c\}_\gamma^\phi(s(\sigma^*, c)) = 1$. Define $A(\sigma) = A(\sigma^*) \cup \{s(\sigma^*, c)\}$. Set $B(\sigma) = B(\sigma^*)$, $K(\sigma, e) = K(\sigma^*, e)$, $L(\sigma, e) = L(\sigma^*, e)$, $q(\sigma, e) = q(\sigma^*, e)$ and $s(\sigma, e) = s(\sigma^*, e)$. If $L(\sigma, e) - B(\sigma)$ is finite for some e, then define $p(\sigma, e) = p(\sigma^*, e)$ and $v(\sigma, e) = v(\sigma^*, e)$ for all e. Suppose $L(\sigma, e) - B(\sigma)$ is infinite for all e; then set:

$$p(\sigma, e) = p(\sigma^*, e) \text{ if } e < c;$$

$$v(\sigma, e) = v(\sigma^*, e) \text{ if } e \leqq c;$$

$$p(\sigma, k) = \mu x[x \in L(\sigma, k) - B(\sigma) \,\&\, (Ey)(y \in L(\sigma, k) - B(\sigma) \,\&\, x > y > v(\sigma, k) + v(\sigma^*, k) + p(\sigma^*, k))] \text{ if } k \geqq c;$$

$$v(\sigma, k) = \mu x[x \in L(\sigma, k) - B(\sigma) \,\&\, x > p(\sigma, k-1)] \text{ if } k > c.$$

Case 1.2. $s(\sigma^*, c)$ does not satisfy the hypothesis of case 1.1. Change nothing and proceed to stage $\sigma + 1$.

Case 2. $j \equiv 1$ (4).

Case 2.1. $p(\sigma^*, c) \notin B(\sigma^*)$ and there is a $\gamma < \sigma$ such that ${}^{K(\sigma^*, c)}[c]_\gamma^{A(\sigma^*)}(p(\sigma^*, c)) = 1$; let δ be the least such γ. Define $B(\sigma) = B(\sigma^*) \cup \{p(\sigma^*, c)\}$. Set $A(\sigma) = A(\sigma^*)$. Let $L(\sigma, e) = L(\sigma^*, e)$, $p(\sigma, e) = p(\sigma^*, e)$ and $v(\sigma, e) = v(\sigma^*, e)$. Define $K(\sigma, e) = K(\sigma^*, e)$ for all $e < c$, $K(\sigma, c) = K(\sigma^*, c) - N(c, \delta)$, and $K(\sigma, e) = K(\sigma, c)$ for all $e > c$. If $K(\sigma, e) - A(\sigma)$ is finite for some e, then define $q(\sigma, e) = q(\sigma^*, e)$ and $s(\sigma, e) = s(\sigma^*, e)$ for all e. Suppose $K(\sigma, e) - A(\sigma)$ is infinite for all e; then set:

$$q(\sigma, e) = q(\sigma^*, e) \text{ if } e < c;$$

$$s(\sigma, e) = s(\sigma^*, e) \text{ if } e \leqq c;$$

$$q(\sigma, k) = \mu x[x \in K(\sigma, k) - A(\sigma) \,\&\, (Ey)(y \in K(\sigma, k) - A(\sigma) \,\&\, x > y > s(\sigma, k) + s(\sigma^*, k) + q(\sigma^*, k))] \text{ if } k \geqq c;$$

$$s(\sigma, k) = \mu x[x \in K(\sigma, k) - A(\sigma) \,\&\, x > q(\sigma, k-1)] \text{ if } k > c.$$

Case 2.2. $p(\sigma^*, c)$ does not satisfy the hypothesis of Case 2.1. Change nothing and proceed directly to stage $\sigma + 1$.

Case 3. $j \equiv 2$ (4). Similar to case 1 with A and B interchanged; we also interchange s and v, K and L, and p and q.

Case 4. $j \equiv 3$ (4). Similar to Case 2 with A and B interchanged. If case 4.1 holds and $L(\sigma, e) - B(\sigma)$ is infinite for all e, then set:

$$p(\sigma,e) = p(\sigma^*,e) \text{ if } e \leqq c;$$

$$v(\sigma,e) = v(\sigma^*,e) \text{ if } e \leqq c+1;$$

$$p(\sigma,k) = \mu x[x \in L(\sigma,k) - B(\sigma) \,\&\, (Ey)(y \in L(\sigma,k) - B(\sigma)$$

$$\& x > y > v(\sigma,k) + v(\sigma^*,k) + p(\sigma^*,k))] \text{ if } k > c;$$

$$v(\sigma,k) = \mu x[x \in L(\sigma,k) - B(\sigma) \,\&\, x > p(\sigma,k-1)] \text{ if } k > c+1.$$

That completes the construction. Note that we are using the fact that the predicate, α is the index of an *infinite*, metafinite set, is metarecursive.

The system of priorities we are following is buried in the above equations; we make it explicit. Our requirements are: $B \neq \{e\}^A$, $A \neq \{e\}^B$, $B \neq \{e\}^\Phi$, and $A \neq \{e\}^\Phi$, where $e \geqq 0$. $B \neq \{c\}^A$ has higher priority then $A \neq \{e\}^B$ for all $e \geqq c$. (Any conflict between two requirements is resolved in favor of the requirement of higher priority.) $A \neq \{c\}^B$ has higher priority than $A \neq \{e\}^\Phi$, $B \neq \{e\}^\Phi$, and $B \neq \{e\}^A$ for all $e > c$. $A \neq \{c\}^\Phi$ and $B \neq \{c\}^\Phi$ have higher priority than $B \neq \{e\}^A$ for all $e \geqq c$.

Let $f(\sigma)$ be a function from the recursive ordinals into the recursive ordinals. We say $f(\sigma)$ changes value only finitely often as σ increases through the recursive ordinals if there exists a finite sequence, $0 = \sigma_0 < \sigma_1 < \cdots < \sigma_m = \omega_1$, such that for each $i < m$, $f(\sigma)$ is constant when restricted to $\sigma_i \leqq \sigma < \sigma_{i+1}$.

LEMMA 1. *Let $f(\sigma,e)$ denote any one of the following six functions: $K(\sigma,e)$, $L(\sigma,e)$, $p(\sigma,e)$, $q(\sigma,e)$, $s(\sigma,e)$ or $v(\sigma,e)$. Then for each e, the function $\lambda\sigma | f(\sigma,e)$ changes value only finitely often as σ increases through the recursive ordinals.*

Proof. First we show $\lambda\sigma | p(\sigma,e)$ and $\lambda\sigma | v(\sigma,e)$ change value only finitely often for each e by an infinite descent on e. Fix e and suppose otherwise. Then Case 1.1 holds for infinitely many σ with $c \leqq e$ or Case 4.1 holds for infinitely σ with $c + 1 \leqq e$. Then $\lambda\sigma | s(\sigma,c')$ changes value infinitely often for some $c' \leqq e$ or $\lambda\sigma | q(\sigma,c')$ changes value infinitely often for some $c' < e$. Each of these last two alternatives implies that Case 2.1 or Case 3.1 holds for infinitely many σ and some $c < e$. If Case 2.1 holds for infinitely many σ and $c = c'' < e$, then $\lambda\sigma | p(\sigma,c'')$ changes value infinitely often. If Case 3.1 holds for infinitely many σ and $c = c'' < e$, then $\lambda\sigma | v(\sigma,c'')$ changes value infinitely often.

Infinitely many changes in $\lambda\sigma | s(\sigma,e)$ imply infinitely many changes in $\lambda\sigma | p(\sigma,c)$ for some $c < e$ or infinitely many changes in $\lambda\sigma | v(\sigma,c)$ for some $c < e$. A similar relationship holds between q and p and v. Infinitely many changes in $\lambda\sigma | K(\sigma,e)$

imply infinitely many changes in $\lambda\sigma \mid p(\sigma,c)$ for some $c < e$ by Case 2.1. A similar relationship holds between L and q.

LEMMA 2. *For each e and σ, $K(\sigma,e) - A(\sigma)$ and $L(\sigma,e) - B(\sigma)$ are infinite.*

Proof. Fix $e \geqq 0$ and $\sigma > 0$. First we show $L(\sigma^*,e) - B(\sigma^*)$ is infinite. By Lemma 1, there is a least $\gamma < \sigma$ such that $L(\sigma^*,e) = L(\sigma,e) = L(\gamma,e) = L(\tau,e)$ whenever $\sigma > \tau > \gamma$. Fix m; we locate a member of $L(\gamma,e) - B(\sigma^*)$ greater than m. Clearly, $f < v(\sigma^*,f) < p(\sigma^*,f) < v(\sigma^*,f+1)$ for all f. Let f be such that $e + m < f$. Then there is a y such that $m < v(\sigma^*,f) < y < p(\sigma^*,f)$ and $y \in L(\gamma,e) - B(\sigma^*)$. We prove this last assertion. Let τ be the least α such that $\alpha < \sigma$ and $p(\alpha,f) = p(\sigma^*,f) = p(\delta,f)$ when $\sigma > \delta > \alpha$; then $p(\tau,d) = p(\sigma^*,d) = p(\delta,d)$ when $\sigma > \delta > \tau$ and $d \leqq f$. Then $y \in L(\tau,f) - B(\tau)$, and so $y \in L(\tau,e) - B(\tau)$, since $e < f$. We claim $\gamma \leqq \tau$; if $\tau < \gamma$, then $L(\gamma,e) \neq L(\gamma^*,e)$ (by definition of γ) and $p(\gamma,f) \neq p(\gamma^*,f)$ (by Case 4.1); but $p(\gamma,f) = p(\gamma^*,f)$ if $\tau \leqq \gamma$. Since $\gamma \leqq \tau$, we have $y \in L(\gamma,f) - B(\tau)$. We still must see that $y \notin B(\sigma^*) - B(\tau)$. Suppose $y \in B(\delta) - B(\delta^*)$ where $\sigma > \delta > \tau$. Then $y = p(\delta^*,c)$ or $y = v(\delta^*,c)$. But $y \neq p(\delta^*,c)$, because $p(\delta^*,f-1) < v(\delta^*,f) < y < p(\delta^*,f)$; similarly, $y \neq v(\delta^*,c)$. Finally, $y \in L(\gamma,e) - B(\delta^*)$, since $e < f$.

Now we show $L(\sigma,e) - B(\sigma^*)$ is infinite. If $L(\sigma,e) \neq L(\sigma^*,e)$, Case 4.1 holds. It follows that ${}^{L(\sigma^*,c)}[c]_{\delta}^{B(\sigma^*)}(q(\sigma^*,c)) = 1$, and that $L(\sigma,c) = L(\sigma^*,c) - N(c,\delta)$. From (12) we obtain $L(\sigma^*,c) - (N(c,\delta) \cup B(\sigma^*))$ is infinite. Since $e \geqq c$, we have $L(\sigma,e) = L(\sigma,c)$ and $L(\sigma,e) - B(\sigma)$ is infinite. A similar argument shows $K(\sigma,e) - A(\sigma)$ is infinite.

LEMMA 3. *For each c, $A \neq \{c\}^{\Phi}$ and $B \neq \{c\}^{\Phi}$.*

Proof. Fix c and let $s(c) = \lim_{\sigma} s(\sigma,c)$ (the final value of $s(\sigma,c)$). Suppose $\{c\}^{\Phi}(s(c))$ is defined. Let γ be the least δ such that $s(c) = s(\delta,c) = s(\sigma,c)$ for all $\sigma > \delta$. Then $s(c) \notin A(\gamma)$ by Lemma 2. If there is a σ such that $\{c\}_{\sigma}^{\Phi}(s(c)) = 1$, then $s(c) \in A$ by Case 1. If $\{c\}^{\Phi}(s(c)) = 0$, then $s(c) \notin A$, since s and q have disjoint ranges. A similar argument shows $B \neq \{c\}^{\Phi}$.

LEMMA 4. *For each c, $B \neq \{c\}^{A}$ and $A \neq \{c\}^{B}$.*

Proof. Fix c and let $p(c) = \lim_{\sigma} p(\sigma,c)$. Suppose $\{c\}^{A}(p(c))$ is defined. Let ρ be the least δ such that $p(e) = p(\delta,e) = p(\sigma,e)$ for all $\sigma > \delta$ and $e \leqq c$. Then $p(c) \notin L(\rho,c) - B(\rho)$ by Lemma 2. Suppose $p(c) \in B$. Then Case 2.1 must hold at stage $\sigma > \rho$ and ${}^{K(\sigma^*,c)}[c]_{\delta}^{A(\sigma^*)}(p(c)) = 1$. It is enough to show $N(c,\delta) \cap A = \varnothing$, since $\{c\}^{A}(p(c))$ is defined. Now $K(\sigma,e) \cap N(c,\delta) = \varnothing$ for all $e \geqq c$. But $K(\sigma,c) = K(\alpha,c)$ for all $\alpha > \sigma$, since $\sigma > \rho$; consequently, $K(\tau,e) \subseteq K(\sigma,c)$ for all $\tau \geqq \sigma$ and $e \geqq c$. Thus $N(c,\delta) \cap K(\tau,e) = \varnothing$ for all $\tau \geqq \sigma$ and $e \geqq c$. Then $q(\tau,e)$ and $s(\tau,e) \notin N(c,\delta)$ for all $\tau \geqq \sigma$ and $e \geqq c$. So if $N(c,\delta) \cap A \neq \varnothing$, there must be an $f < c$ and a $\tau > \sigma$ such that $q(\tau^*,f) \in N(c,\delta) \cap A(\tau)$ or

$s(\tau^*,f) \in N(c,\delta) \cap A(\tau)$. Both of these alternatives are ruled out by the fact that $\sigma > \rho$.

Now suppose $p(c) \notin B$. We claim $\{c\}^A(p(c)) \neq 1$; suppose otherwise. Then there is a γ and a β such that $\gamma < \beta$ and $\{c\}_\gamma^{A(\sigma^*)}(p(c)) = 1$ for all $\sigma \geqq \beta$; in addition, $M(c,\gamma) \subseteq A$ and $N(c,\gamma) \subseteq cA$. Assume $\beta \geqq \rho$. Then $K(\sigma,c) = K(\beta,c)$ for all $\sigma \geqq \beta$. Any element added to A after stage β must be a member of $K(\beta,c)$ with finitely many exceptions (at most c). Consider $K(\beta,c) - (N(c,\gamma) \cup A(\sigma))$ for any $\sigma \geqq \beta$; it must be infinite, since otherwise $(A - A(\beta)) - (N(c,\gamma) \cup A(\sigma))$ would be finite and A would be metafinite. But Lemma 4 says A is not metafinite. But then ${}^{K(\sigma^*,c)}[c]_\gamma^{A(\sigma^*)}(p(c)) = 1$ for all $\sigma \geqq \beta$, and some instance of case 2.1 would put $p(c)$ in B. A similar argument shows $A \neq \{c\}^B$.

Note that it was not sufficient for us to know that various limits such as $\lim_\sigma p(\sigma,e)$ and $\lim_\sigma L(\sigma,e)$ existed. Our key lemma, Lemma 2, used Lemma 1 very heavily; we had to know that the function $\lambda_\sigma | L(\sigma,e)$ changed value only finitely often as σ increased through the recursive ordinals. This mode of argument seems to generalize only to those admissible ordinals which project into cardinals. We make a conjecture on this matter in §7.

THEOREM 3. *Every countable partial ordering is embeddable in the partial ordering of metadegrees of* Π_1^1 *sets.*

Proof. Combine the argument of Theorem 2 of §5 of [11] with the proof of Theorem 2 above. The proof of Lemma 2 above becomes more complicated, but no conceptual change is required.

4. **Subgeneric sets.** The notion of subgeneric set is due to Kreisel. Before we give the definition, we recall some features of Kripke's equation calculus introduced in [8] and sketched in [7]. Kripke's calculus is modeled on Kleene's [5]. The terms include function letters and numerals denoting recursive ordinals. The deduction rules are similar to the substitution rules of Kleene with one exception, rule (R3). By means of (R3) it is possible to deduce $(x)_{x<\beta}(f(x) = \mathbf{0})$ from the set of premises $\{f(\alpha) = \mathbf{0} \mid \alpha < \beta\}$, where $\boldsymbol{\alpha}$ and $\boldsymbol{\beta}$ are numerals denoting the recursive ordinals α and β. (Note: in Kripke's formalism, the formula $(x)_{x<\beta}\,(j(x) = 0)$ is rendered as the equation $(Ex < \boldsymbol{\beta})(f(x) = 1.)$ Kripke's calculus can be thought of as the usual means of determining the truth-value of formulas concerning function-values in which all quantifiers are bounded. Let E be a finite set of equations. Let $S_0^E = E$, and for each ordinal α (not necessarily recursive), let S_α^E be the set of all equations immediately deducible from $\bigcup\{S_\gamma^E | \gamma < \alpha\}$. Each member of S_α^E is an immediate consequence of finitely many premises in $\bigcup\{S_\gamma^E | \gamma < \alpha\}$ save those which result from (R3). If (R3) is applied to the members of $\bigcup\{S_\gamma^E | \gamma < \alpha\}$, then we require that the β mentioned in each use of (R3) be less than α; thus every member of S_ω^E is an immediate consequence of finitely many members of $\bigcup\{S_m^E | m < \omega\}$. Kripke shows $S_{\omega_1}^E = \bigcup\{S_\gamma^E \mid \gamma < \omega_1\}$, where ω_1 is the least nonrecursive ordinal.

In [7] it was observed that a function Φ is metarecursive if and only if for some E with principal function letter f, $\Phi(\gamma) = \delta \leftrightarrow f(\gamma) = \delta \in S^E_{\omega_1}$. We now characterize the relation, "weakly metarecursive in", in a similar fashion. Let B be an arbitrary set of recursive ordinals. Let $S_0^{E,g} = E \cup \{g(\alpha) = \mathbf{1} \mid \alpha \notin \beta\} \vee \{g(\alpha) = \mathbf{0} \mid \alpha \in \beta\}$; for each $\alpha > 0$, let $S_\alpha^{E,g}$ equal the set of all immediate consequences of $\bigcup\{S_\gamma^{E,g} \mid \gamma < \alpha\}$. (Do not forget the condition made in the definition of S_γ^E concerning the use of (R3).) It is not true that $S_{\omega_1}^{E,g} = \bigcup\{S_\gamma^{E,g} \mid \alpha < \omega_1\}$ for all B and E. It is possible to make deductions whose heights are not recursive ordinals. It is even possible to make deductions of recursive ordinal height which are not metafinite; i.e., for some B and recursive α, $S_\gamma^{E,g}$ may contain an equation deducible from B whose deduction-tree is not a metafinite object. Let

$$S^{E,g} = \bigcup\{S_\gamma^{E,g} \mid \alpha < \text{aleph-one}\}.$$

We define $A \leqq_{mc} B$ (A is metafinitely computable from B) to mean that for some finite E containing the principal function letter f and the given function letter g,

$$\alpha \notin A \leftrightarrow f(\alpha) = \mathbf{0} \in S^{E,g,m}$$

$$\alpha \notin A \leftrightarrow f(\alpha) = \mathbf{1} \in S^{E,g,m},$$

where $S^{E,g,m}$ is the set of all members of $S^{E,g}$ whose deduction-trees are metafinite. Clearly, the height of each member of $S^{E,g,m}$ is a recursive ordinal. $A \leqq_{mc} B$ means the membership relation of A is computable from B using the equation calculus and metafinite deductions only.

THEOREM 4. $A \leqq_w B \leftrightarrow A \leqq_{mc} B$.

Proof. If $A \leqq_w B$, then the computations used to decide A from B are metafinite and are metarecursively enumerable; consequently, $A \leqq_{mc} B$. Suppose $A \leqq_{mc} B$ by E. Each deduction permitted is a metafinite object; the roots of each deduction of $\alpha \in A$ or $\alpha \notin A$ consist of a metafinite set of membership conditions satisfied by B. The set of *all* metafinite deductions is metarecursively enumerable. It is not difficult to define the partial metarecursive functions needed to describe how each decision about the membership of A is determined by a metafinite neighborhood condition satisfied by B. Then $A \leqq_w B$.

We say B is *subgeneric* if for every E, we have $S_{\omega_1}^{E,g} = \bigcup\{S_\gamma^{E,g} \mid \alpha < \omega_1\}$, where $S_0^{E,g}$ includes the diagram of B. Intuitively, B is subgeneric if every equation deducible from E and B is deducible from E and B at some recursive ordinal. It is easily checked that every metarecursive set is subgeneric, and that every bounded, nonmetarecursive, metarecursively enumerable set is not subgeneric (first observed by Kreisel). In a letter to the author, Kreisel asked: do there exist subgeneric, nonmetarecursive, metarecursively enumerable sets? The reason for his interest lies in the light shed by the notion of subgenericity on the various notions of reducibility discussed in [6]. It turns out that one can prove the existence of nonmetarecursive, subgeneric sets by a standard measure-theoretic argument or by a "forcing" argument. (In fact, with probability one, a randomly chosen set of recursive ordinals is subgeneric.) However, a much stronger result, expressed by Theorem 6, can be obtained by the priority method. First we introduce a generally stronger notion. We say B is *hyperregular* if the range of each function weakly metarecursively in B, when restricted to a metafinite set, is bounded by a recursive ordinal. We will see that a set is both regular and hyperregular if and only if all possible deductions from the set are metafinite.

THEOREM 5. *A regular set of recursive ordinals is hyperregular if and only if it is subgeneric.*

Proof. Suppose B is regular and subgeneric. First we show B is completely regular: every bounded set weakly metarecursive in B is metafinite. Let C be bounded and weakly metarecursive in B. Let γ be the least upper bound of C. There exist functions $m(\delta)$ and $n(\delta)$ weakly metarecursive in B such that the truth-value of $\delta \in C$ is deducible from B (and some E not depending on δ) using only membership information about B expressed by $K_{m(\delta)} \subseteq B$ and $K_{n(\delta)} \subseteq cB$. Suppose $\rho = \{m(\delta) + n(\delta) \mid \delta < \gamma\}$ is unbounded. Then Q, a Π_1^1 set of unique notations, is computable from $\{|\tau| \mid \tau \in p\}$ by means of the unrestricted equation calculus. But then Q is computable from B, and consequently, Q is subgeneric. No nonmetarecursive Π_1^1 set is subgeneric, so ρ must be bounded. It follows from the regularity of B that C is metafinite.

Let f be weakly metarecursive in B, and let K be a metafinite set. Then $\{\langle |\alpha|, |f(\alpha)|\rangle \mid \alpha \in K\}$ is bounded and weakly metarecursive in B; by the previous paragraph, it must be metafinite.

Now suppose B is regular and hyperregular. Let E be a finite set of equations, and let β be a recursive ordinal. Suppose each member of $\{f(\gamma) = \mathbf{0} \mid \gamma < \beta\}$ is deducible from E and B by means of a metafinite deduction. Then there exist functions $s(\gamma)$ and $t(\gamma)$ weakly metarecursive in B such that some metafinite deduction of $f(\gamma) = \mathbf{0}$ from E and B uses only the membership information about B expressed by $K_{s(\gamma)} \subseteq B$ and $K_{t(\gamma)} \subseteq cB$. But $\{s(\gamma) \mid \gamma < \beta\}$ and $\{t(\gamma) \mid \gamma < \beta\}$ are bounded. It follows from the regularity of B that *all* members of $\{f(\gamma) = \mathbf{0} \mid \gamma < \beta\}$ are deducible from E and a fixed metafinite set of membership facts about B.

COROLLARY 1. *A metarecursively enumerable set is subgeneric if and only if it is regular and hyperregular.*

Proof. Any subgeneric metarecursively enumerable set is regular.

THEOREM 6. *There exists a nonmetarcursive, metarecursively enumerable, regular, hyperregular set of recursive ordinals.*

Proof. Let $R_0, R_1, \cdots, R_e, \cdots (e < \omega)$ be the standard enumeration of the metarecursively enumerable sets; let R_e^σ be all the members of R_e enumerated prior to stage σ; $\lambda\sigma e \mid R_e^\sigma$ is a metarecursive function which takes metafinite sets as values. Let $r(\sigma)$ be a metarecursive function taking only finite values such that for each c, the set $\{\sigma \mid r(\sigma) = c\}$ is unbounded. At stage σ we consider putting an element into A in order to insure that $A \cap R_e \neq \varnothing$ if R_e is unbounded; in this fashion we make A nonrecursive. At the same time we try to preserve all values of $\{e\}^A(m)$ $(m < \omega)$ as they develop. We do not add an element of R_c to A if it would change the apparent value of $\{e\}^A(m)$ for any $m < \omega$ and $e < c$. This system of priorities guarantees that any set of natural numbers metarecursive in A will be metarecursive, since for each e, $\{e\}^A$ will be "injured" only finitely often. It will follow that A is regular and hyper-regular.

Stage $\sigma \geqq 0$. Let $A(\sigma^*) = \bigcup\{A(\delta) \mid \delta < \sigma\}$. For each $e < \omega$, if for all $m < \omega$ there is a $\delta < \omega$ such that $\{e\}_\delta^{A(\sigma^*)}(m)$ is defined, let $\gamma(e,\sigma)$ be the least γ such that for all $m < \omega$, there is a $\delta < \sigma$ such that $\{e\}_\delta^{A(\sigma^*)}(m)$ is defined and $\gamma >$ all members of $N(e,\delta)$; otherwise, let $\gamma(e,\sigma) = 0$. We define

$$d(e,\sigma) = \mu x(x \geqq \gamma(e,\sigma) \,\&\, (\rho)(\rho < \sigma \rightarrow x \geqq d(e,\rho)))$$

for each e. Let $r(\sigma) = c$. If $R_c^\sigma \cap A(\sigma^*) = \varnothing$, and if there is a ρ such that $\rho \in R_c^\sigma \,\&\, (e)_{e<c}(\rho > d(e,\sigma)) \,\&\, \rho$ greater than any member of

$$\bigcup \{R_{r(\delta)}^\delta \cap (A(\delta) - A(\delta^*)) \mid \delta < \sigma \,\&\, r(\delta) < c\}$$

by at least ω, add the least such ρ to $A(\sigma^*)$ to obtain $A(\sigma)$; otherwise, let $A(\sigma) = A(\sigma^*)$.

Let $A = \bigcup \{A(\sigma) \mid \sigma < \omega_1\}$.

LEMMA 8. *For each* e, $\lim_\sigma d(e,\sigma)$ *exists.*

Proof. Fix e. For each $c \leqq e$, there is at most one σ such that at stage σ: A is augmented for the sake of R_c. (This is the purpose of the clause $R_c^\sigma \cap A(\sigma^*) = \varnothing$.) Let σ be so large that A is not augmented at stage ρ for the sake of R_c for any $\rho \geqq \sigma$ and $c \leqq e$. The set $\{\gamma(e,\rho) \mid \rho \geqq \sigma\}$ must have a maximum value, since if $\gamma(e,\delta) > 0$ for some $\delta \geqq \sigma$, then $\gamma(e,\rho) \leqq \gamma(e,\delta)$ for all $\rho > \delta$; suppose $\tau \geqq \sigma$ is such that $\gamma(e,\tau)$ is the maximum value. Then $d(e,\rho) = d(e,\tau)$ for all $\rho \geqq \tau$. (Note that if α is added to A for the sake of R_c at stage $\rho \geqq \sigma$, then $c > e$ and $\alpha > d(e,\rho)$.)

LEMMA 9. *A is not metarecursive.*

Proof. First we show cA is unbounded. Fix α; we claim $\{\beta \mid \alpha < \beta < \alpha+\omega\} \cap A$ is finite. Let τ and σ be such that $\alpha < \tau < \alpha + \omega$, $\tau \in A(\sigma)$, and $A(\sigma^*) \cap \{\beta \mid \alpha < \beta < \alpha + \omega\} = \varnothing$. Let $r(\sigma) = c$. If ρ is added to A at stage $\delta > \sigma$

for the sake of R_d, where $d > c$, then $\rho \geqq \alpha + \omega$, since ρ must be greater than τ by at least ω. Thus any member of $\{\beta \mid \alpha < \alpha + \omega\}$ added to A after stage σ must be added for the sake of R_d, where $d < c$; consequently, $\{\beta \mid \alpha < \beta < \alpha + \omega\} \cap A$ has at most $c + 1$ members.

Fix c and suppose R_c is unbounded. Let ρ have the following properties: $\rho \in R_c$, $(e)_{e<c}(\rho > \lim_\sigma d(e,\sigma))$, and ρ is greater by at least ω than any member of $\bigcup \{R^\delta_{r(\delta)} \cap (A(\delta) - A(\delta^*)) \mid \delta < \omega_1 \,\&\, r(\delta) < c\}$; such a ρ exists by Lemma 8 and the fact that for each e, there is at most one δ such that $R^\delta_{\sigma(\delta)} \cap (A(\delta) - A(\delta^*)) \neq \emptyset \,\&\, r(\delta) = e$. Then $A \cap R_c$ contains a member less than or equal to ρ.

LEMMA 10. *For each e, if $\{e\}^A(m)$ is defined for all $m < \omega$, then $\{e\}^A$ agrees with some metarecursive function on the natural numbers.*

Proof. Fix e and suppose $\{e\}^A(m)$ is defined for all $m < \omega$. Fix m; we show how to compute $\{e\}^A(m)$ metaeffectively. Let σ be so large that A is not augmented at stage ρ for the sake of R_c for any $\rho \geqq \sigma$ and $c \leqq e$, and such that $d(e,\rho) = d(e,\sigma)$ for all $\rho \geqq \sigma$. Let $\tau(m) \geqq \sigma$ have the property that for some $\delta < \tau(m)$, $\{e\}^{A(\tau(m)^*)}_\sigma(m)$ is defined; $\tau(m)$, as a function of m, is weakly metarecursive. Then $\{e\}^{A(\tau(m)^*)}_\rho(m)$ is defined and $d(e,\tau(m)) >$ all members of $N(e,\rho)$. But then $d(e,\gamma) >$ all members of $N(e,\rho)$ for all $\gamma \geqq \tau(m)$, and so $N(e,\rho) \subseteq cA$; consequently, $\{e\}^{A(\tau(m)^*)}_\rho(m) = \{e\}^A(m)$.

It follows from Lemma 10 that A is both regular and hyperregular.

If A and B are sets of recursive ordinals, let $A \leqq_c B$ mean that A is computable from B by means of Kripke's equation calculus; of course, deductions whose heights are nonrecursive ordinals are permitted. If B is regular and hyperregular, and A is computable from B, then A is weakly metarecursive in B.

Kreisel considers several notions of invariant definability in [6] and in private communications. In a future publication he will present the most appealing of these notions; for now, let us denote it by $\leqq_I$, and let us be content to hear some of Kreisel's observations concerning it; $\leqq_I$ is transitive; if A and B are sets of recursive ordinals and $A \leqq_I B$, then $A \leqq_c B$; if A and B are sets of natural numbers, then $A \leqq_I B \leftrightarrow A \leqq_c B \leftrightarrow A$ is hyperarithmetic in B.

It follows from Theorem 7 that there exist two metarecursively enumerable sets such that neither is invariantly definable from the other.

THEOREM 7. *There exist two metarecursively enumerable sets such that neither is computable from the other by means of Kripke's equation calculus.*

Proof. We lift Theorem 1 of §5 of [11] up into metarecursion theory as follows: each regular, nonmetarecursively enumerable set B is the disjoint union of two metarecursively enumerable sets of incomparable metadegrees. The regularity of B is all-important; we need to know that for each α we eventually finish enumerating all members of B less than α. Except for this point, the lifting up is routine. Let B be regular and hyperregular. Then $B = B_0 \cap B_I$, $B_0 \cap B_1 = \Phi$, and the metadegrees of B_0 and B_1 are incomparable. It follows that B_0 and B_1 are regular and hyperregular; but then neither is computable from the other.

We conclude the present section by outlining some generalizations to arbitrary admissible ordinals. At the beginning of this section we sketched Kripke's equation calculus for ω_1. The only change needed for an arbitrary admissible α is the introduction of numerals forordinals less than α. We define S_γ^E and $S_\gamma^{E,g}$ as before. By the definition of admissibility, we have $S_\alpha^E = \bigcup \{S_0^E \mid \gamma < \alpha\}$. Let B be a subset of α. We say B is *subgeneric* if for every E, $S_\alpha^{E,g} = \bigcup \{S_\gamma^{E,g} \mid \gamma < \alpha\}$, where $S_0^{E\,g}$ includes the diagram of B. We say B is *hyperregular* if the range of each function weakly α-recursive in B, when restricted to a set bounded by an ordinal less than α, is bounded by an ordinal less than α.

THEOREM 8. *The following three properties of a regular set B are equivalent to hyperregularity*:

(a) *every deduction from B is α-finite*;

(b) *the range of each function weakly α-recursive in B, when restricted to an α-finite set, is α-finite*;

(c) *the restriction of each function weakly α-recursive in B to an α-finite set is an α-finite partial function.*

Proof. We indicate only that hyperregularity plus regularity implies property (a). The second half of the proof of Theorem 5 uses no special properties of recursive ordinals, generalizes to an arbitrary admissible α with no change, and gives the desired result.

COROLLARY 2. *Each regular, hyperregular set is subgeneric.*

THEOREM 9. *For each admissible α, there exists a non-α-recursive, a-recursively enumerable, regular, hyperregular set.*

Proof. First we mimic the proof of Theorem 6; the argument splits into two cases distinguished by the truth-value of the statement: α^*, the projectum of α, is less than α. If $\alpha^* < \alpha$, the argument of Theorem 6 is followed closely, and the ordinals less than α^* take over the part of the natural numbers. If $\alpha^* = \alpha$, then a slight modification is necessary; instead of preserving values of $\{\varepsilon\}^A(\beta)$ for all $\beta < \alpha^*$, we preserve $\{\varepsilon\}^A(\beta)$ for all $\beta < \varepsilon$, and make use of the repetition inherent in any standard enumeration of the functions weakly α-recursive in A. Thus we obtain a non-α-recursive, α-recursively enumerable set A with the property that any function weakly α-recursive in A, when restricted to an α-finite set, is an α-finite partial function.

5. **The nonexistence of maximal sets.** Let α be admissible. We say an α- recursively enumerable set M is *maximal* if cM is not bounded by any ordinal less than α

and if for each α-recursively enumerable set R, either $R - M$ or cR is bounded by some ordinal less than α. Clearly, a maximal set is unbounded. The maximal set construction of [7] shows that if $\alpha^* = \omega$, then a maximal, α-recursively enumerable set exists. It follows that there exist uncountably many, countable, admissible α such that a maximal, α-recursively enumerable set exists. (In fact, it was shown in [7] that there exists a metarecursively enumerable set M with unbounded complement such that for any metarecursively enumerable R, either $R - M$ or cR is *finite*.) The purpose of the present section is to give an example of an important theorem about recursively enumerable sets which does not generalize to arbitrary admissible ordinals.

THEOREM 10. *There exists a countable, admissible ordinal α such that no α-recursively enumerable set is maximal.*

Proof. First we show that no maximal, aleph-one-recursively enumerable set exists. Assume that $V = L$. Let f be a one-one, aleph-one-recursive function which assigns an aleph-one-finite set of natural numbers to each countable ordinal; the equations defining f merely specify some diagonal procedure. Let g be the usual one-one function which assigns a member of the closed unit interval to each set of natural numbers. Then $h = g \circ f$ can be viewed as an aleph-one-recursive function, which associates a real number with each countable ordinal. Suppose M is a maximal, aleph-one-recursively enumerable set. Let $I_1, I_2, I_3, \cdots$ be the usual enumeration of all open subintervals of the unit interval with rational endpoints. For each n, let $J_n = \{\gamma \mid h(\gamma) \in I_n\}$; then J_n is aleph-one-recursively enumerable if a natural h is chosen; in fact, J_n is aleph-one-recursive uniformly in n. For each n, $J_n - M$ or $cJ_n - M$ is countable; let K_n be $J_n - M$ if $J_n - M$ is countable, and let K_n be $cJ_n - M$ otherwise. Let $K = \bigcup\{K_n \mid n < \omega\}$, and $N = cM - K$. Then for each n, $N \cap J_n = \emptyset$ or $N \cap cJ_n = \emptyset$. Let $\alpha, \beta \in N$ such that $h(\alpha) < h(\beta)$, and r_1, r_2 be rational numbers such that $h(\alpha) < r_1$ & $h(\beta) < r_2$. Let $I_m = (r_1, r_2)$. Then $\beta \in J_m$, $\alpha \notin J_m$, $N \cap J_m \neq \emptyset$, and $N \cap cJ_m \neq \emptyset$. (Of course, the aleph-one-recursive enumerability of M is irrelevant.)

Let P be a countable initial segment of the constructible sets such that P is a model of set theory and $V = L$. The above argument holds in P. Let γ be the countable ordinal which plays the part of aleph-one in P. The notion of admissibility is absolute, so γ is admissible, not only in the sense of P, but in actuality. There is no maximal γ-recursively enumerable set in P, so there is none anywhere, since all γ-recursively enumerable sets are in P. The existence of P cannot be proved in set theory, but P can safely be replaced by Q, a model for finitely many axioms of set theory and $V = L$ strong enough to carry out the above argument; and the existence of Q can be proved in set theory.

We do not know if there exists an uncountable, admissible ordinal α such that a maximal, α-recursively enumerable set exists. We conjecture that it is possible to give an elementary description of an ordinal satisfying Theorem 10.

Note that the argument of Theorem 10 shows there does not exist any maximal CA (complementary analytic) set of real numbers. (Call a CA set M maximal if its complement is uncountable and if for each CA set $R \supseteq M$, either $R \dot{-} M$ is countable or cR is countable.) This fact combined with the existence of maximal Π_1^1 sets provides strong evidence for the following assertion: the analogy between recursively enumerable sets and Π_1^1 sets is much stronger than the analog between Π_1^1 sets and CA sets. We will give other evidence in [13]. Of course, one is free to say that the above notion of maximality has no place in the theory of CA sets. The argument of Theorem 10 still works if one adopts the following definition: call a CA set M maximal if its complement is not Borel and if for each CA set $R \supseteq M$, either $R - M$ is Borel or cR is Borel.

6. **Post's problem for admissible ordinals.** Kripke [8] has obtained a solution of Post's problem for a large class of admissible ordinals. We give a solution of Post's problem which holds for every admissible α, but we leave open the structure of the upper semilattice of degrees of α-recursively enumerable sets.

THEOREM 11. *For each admissible ordinal α, there exist α-recursively enumerable sets A and B such that A is not α-recursive in B and B is not α-recursive.*

Proof. The introductory remarks of §3 readily generalize to an arbitrary admissible ordinal α; the natural numbers are replaced by the ordinals less than α^*, the projectum of α. Thus we have a T-predicate whose definition is:

$$T_1^1(B,\varepsilon,\alpha,\sigma) \leftrightarrow t(\varepsilon,\sigma) = \langle \varepsilon, M, N, \alpha, \gamma\rangle \,\&\, M \subseteq B \,\&\, N \subseteq cB,$$

where $\varepsilon < \alpha^*$, B is a set of ordinals less than α, and M and N are α-finite sets; $t(\varepsilon,\sigma)$ is a special α-recursive function which indexes "deductions." $T_1{}^1(B,\varepsilon,\alpha,\sigma)$ says: ε is the Gödel number of a finite system of equations E and $t(\varepsilon,\sigma)$ is the Gödel number of an α-finite "deduction" from E and B, whose last line gives the value γ of some function for argument α. $U(\varepsilon,\sigma)$ is an α-recursive function such that if $t(\varepsilon,\sigma) = \langle \varepsilon, M, N, \alpha, \gamma\rangle$, then $U(\varepsilon,\sigma) = \gamma$. Two useful partial functions are:

$$\{\varepsilon\}_\sigma^B(\alpha) = \gamma \leftrightarrow T_1{}^1(B,\varepsilon,\alpha,\sigma) \,\&\, U(\varepsilon,\sigma) = \gamma;$$

$$\{\varepsilon\}^B(\alpha) = \gamma \leftrightarrow (E\sigma)(\{\varepsilon\}_\sigma^B(\alpha) = \gamma) \,\&\, (\sigma)(\{\varepsilon\}_\sigma^B(\alpha) \text{ is defined} \rightarrow U(\varepsilon,\sigma) = \gamma).$$

If A is weakly α-recursively in B, then for some $\varepsilon < \alpha^*$, we have $(\gamma)(a(\gamma) = \{\varepsilon\}^B(\gamma))$, where $a(\gamma)$ is the characteristic function of A.

Let $\{R_\gamma \mid \gamma < \alpha^*\}$ be a standard simultaneous enumeration of all the α-recursively enumerable sets. Let R_γ^σ be the set of all members of R_γ enumerated prior to stage σ of the simultaneous enumeration. Let $r(\sigma)$ be an α-recursive function whose range is bounded by α^* such that for each $\varepsilon < \alpha^*$, the set $\{\sigma \mid r(\sigma) = \varepsilon\}$ is not bounded by any ordinal less than α.

We simultaneously define four α-recursive functions, $A(\sigma)$, $B(\sigma)$, $d(\varepsilon,\sigma)$ and $K(\varepsilon,\sigma)$ by induction on σ.

Stage $\sigma = 0$. $A(o) = B(o) = \varnothing$, $d(\varepsilon, o) = \varepsilon$, and $K(\varepsilon, o) = 0$ for all $\varepsilon < \alpha^*$.

Stage $\sigma > 0$. Let $\delta = r(\delta)$, $A(\sigma^*) = \bigcup\{A(\rho) \mid \rho < \sigma\}$, and $B(\sigma^*) = \bigcup\{B(\rho) \mid < \sigma^*\}$. Let $K(\varepsilon,\sigma^*) =$ least upper bound of $\{K(\varepsilon,\rho) \mid \rho < \sigma\}$ and $d(\varepsilon,\sigma^*) =$ least upper bound of $\{d(\varepsilon,\rho) \mid \rho < \sigma\}$.

If $R_\delta^\sigma \cap B(\sigma^*) = \varnothing$, and if there is a ρ such that $\rho \in R_\delta^\sigma \,\&\, (\varepsilon)_{\varepsilon<\delta}(\rho > K(\varepsilon,\sigma^*)) \,\&\, \rho$ greater than double any member of $\bigcup\{R_{r(\tau)}^\tau \cap (B(\tau) - B(\tau^*)) \mid \tau < \sigma \,\&\, r(\tau) < \delta\}$, then add the least such ρ to $B(\sigma^*)$ to obtain $B(\sigma)$; otherwise, $B(\sigma) = B(\sigma^*)$. We define $d(\varepsilon,\sigma)$ for all $\varepsilon < \alpha^*$:

$$d(\varepsilon,\sigma) = d(\varepsilon,\sigma^*) \text{ if } \varepsilon < \delta \text{ or if } B(\sigma) = B(\sigma^*);$$

$$d(\varepsilon,\sigma) = \mu x(\tau)(\tau < \varepsilon \rightarrow x > d(\tau,\sigma))$$
$$\& x > d(\varepsilon,\sigma^*) + \text{least upper bound of } A(\sigma^*) + 1) \text{ if}$$
$$\varepsilon \geqq \delta \,\&\, B(\sigma) \neq B(\sigma^*).$$

Now we consider $\{\delta\}_\gamma^{B(\sigma^*)}(d(\delta,\sigma))$ for all $\gamma < \sigma$. Suppose $d(\delta,\sigma) \notin A(\sigma^*)$ and there is a $\gamma < \sigma$ such that $\{\delta\}_\gamma^{B(\ ^*)}(d(\delta,\sigma)) = 1$; let τ be the least such γ. Define $A(\sigma) = A(\sigma^*) \cup \{d(\delta,\sigma)\}$. Let $K(\delta,\sigma) = K(\delta,\sigma^*) +$ least upper bound of $N(\delta,\tau)$, and let $K(\varepsilon,\sigma) = K(\varepsilon,\sigma^*)$ for all $\varepsilon \neq \delta$ and $< \alpha^*$. If our supposition is false, let $A(\sigma) = A(\sigma^*)$ and $K(\varepsilon,\sigma) = K(\varepsilon,\sigma^*)$ for all $\varepsilon < \alpha^*$.

LEMMA 11. *For each $\delta < \alpha^*$, there exists a τ such that for all $\varepsilon \leqq \delta$, $d(\varepsilon,\tau) = d(\varepsilon,\sigma)$ for all $\sigma \geqq \tau$.*

Proof. Fix $\delta < \alpha^*$. At the beginning of stage $\sigma > 0$, we consider adding an element to B for the sake of R_γ; this operation occurs at most once before the requirement that $B(\sigma^*) \cap R_\gamma^\sigma = \varnothing$. Let S be the set of all $\gamma \leqq \varepsilon$ such that at some stage σ, a number of R_γ^σ, where $\gamma = r(\sigma)$, is added to B. S is α-recursively enumerable and is bounded by $\delta < \alpha^*$. It follows that S is α-finite. Let τ be so large that all members of S have been enumerated prior to stage τ.

LEMMA 12. *B is not α-recursive.*

Proof. First we show cB is unbounded. Fix $\gamma < \alpha$ and suppose $\{\beta \mid \gamma \leqq \beta < \alpha\} \subseteq B$. Let α_0 be an infinite ordinal $> \gamma$ such that $\alpha_0 \in B(\sigma) - B(\sigma^*)$ and $\alpha_0 >$ every member of $B(\sigma^*)$. For each $n \geqq 0$, let α_{n+1} be the first member of B to be enumerated after α_n with the property that $\alpha_0 < \alpha_{n+1} < 2\alpha_n$; our supposition concerning γ implies α_n is well-defined for all $n < \omega$. Let $\delta(n)$ be such that α_n is added to B at stage $\sigma(n)$ for the sake of $R_{\delta(n)}$; in short, $r(\sigma(n)) = \delta(n)$. Clearly, any element added to B for the sake of R_δ at stage σ, where $\delta > \delta(n)$ and $\sigma > \sigma(n)$, must be greater than $2\alpha_n$; consequently, $\delta(n+1) < \delta(n)$ for all n, a contradiction.

Now fix δ and suppose R_δ is unbounded; we show $R_\delta \neq cB$. Let τ be as in the statement of Lemma 11. Let $\beta \geqq \tau$ be such that $\lim_\sigma d(\varepsilon,\sigma) \in A \leftrightarrow \lim_\sigma d(\varepsilon,\sigma) \in A(\beta^*)$ for all $\varepsilon \leqq \delta$. The existence of β follows from the properties of α^* and τ and the fact that $\delta < \alpha^*$. But then $K(\varepsilon,\beta) = K(\varepsilon,\sigma)$ for all $\varepsilon \leqq \delta$ and $\sigma \geqq \beta$. Let ρ be such that $\rho \in R_\delta$, $(\varepsilon)_{c<\delta}(\rho > \lim_\sigma K(\varepsilon,\sigma))$, and ρ is greater than double any member of $\bigcup \{R^\gamma_{r(\gamma)} \cap (B(\gamma) - B(\gamma^*)) \mid r(\gamma) < \delta\}$; the existence of ρ follows from the nature of α^*. Then B must contain some element of R .

LEMMA 13. *A is not weakly α-recursive in B.*

Proof. Fix δ. Suppose $\{\delta\}^B(\gamma)$ is defined for all $\gamma < \alpha$. Let $d = \lim_\sigma d(\delta,\sigma) \in A$. For some σ, we must have $d = d(\delta,\sigma) \notin A(\sigma^*)$ and $\{\delta\}^{B(\sigma^*)}_\tau(d) = 1$ for some $\tau < \sigma$. Also, $K(\delta,\sigma) \geqq$ every member of $N(\delta,\tau)$. We claim $\{\delta\}^B(d) = 1$; it is sufficient to show $N(\delta,\tau) \subseteq cB$. If some element of $N(\delta,\tau)$ is put in B at stage $\rho > \sigma$, then $r(\rho) = \beta \leqq \delta$ and $d(\delta,\rho) \neq d(\delta,\rho^*)$. But $\rho > \sigma$ and $\lambda\gamma \mid d(\delta,\gamma)$ is a nondecreasing function of γ, so $d(\delta,\rho) = d(\delta,\rho^*)$.

Now suppose $d \notin A$. Then $\{\varepsilon\}^B(d) \neq 1$, since otherwise the second half of stage σ would eventually compel us to put d in A.

The proof of Theorem 11 used only one combinatorial fact concerning all admissible ordinals; namely, an α-recursively enumerable set bounded by an ordinal less than α^*, the projectum of α, is α-finite. If $\alpha^* < \alpha$, then α^* behaves somewhat like ω does in metarecursion theory. If $\alpha^* = \alpha$, then all bounded, α-recursively enumerable sets are α-finite, and it is possible to initiate some familiar arguments of ordinary recursion theory. But we are concealing many combinatorial difficulties. The proof of Theorem 8 works because at most *one* attempt is made to satisfy each requirement of the form: if R_e is unbounded, then $R_e \cap B \neq \varnothing$. We have not found any combinatorial trick for attacking all admissible ordinals with priority arguments that permit more than one attempt to satisfy any requirement.

7. **Open questions and conjectures.** What follows is not a complete list of all interesting questions concerning the topics of this paper, but is, hopefully, a selection.

Q1. If α is admissible, is α^*, the projectum of α, admissible? This question was suggested by Kripke, who has shown that α^* must be the limit of admissible ordinals([3]).

Q2. Do there exist α-recursively enumerable subsets of α^* of incomparable α-degrees? We think yes. The answer to Q1 might be of some help.

Q3. Do there exist α-recursively enumerable sets such that neither is computable from the other by means of Kripke's equation calculus. We think yes.

Q4. For what α do maximal, α-recursively enumerable sets exist? In particular, is there a maximal, α-recursively enumerable set for some uncountable α?

If A is weakly metarecursive in B, then $A = \{e\}^B$ for some e as described at

(3) Kripke has recently shown the answer to Q1 is yes.

the beginning of §3. The reduction procedure used to compute A from B can be identified with the metarecursively enumerable set $\{t(e,\sigma) \mid \sigma < \omega_1\}$. A typical value of $t(e,\sigma)$ is $\langle e, M, N, \alpha, \gamma\rangle$; if $M \subseteq B$ and $N \subseteq cB$, then $\{e\}^B(\alpha) = \gamma$, where M and N are metafinite sets. We say e is intrinsically consistent, if for all α, σ_1 and σ_2, if $t(e,\sigma_1) = \langle e, M_1, N_1, \alpha, \gamma_1\rangle$, $t(e,\sigma_2) = \langle e, M_2, N_2, \alpha, \gamma_2\rangle$, and $(M_1 \cup M_2) \cap (N_1 \cup N_2) = \varnothing$, then $\gamma_1 = \gamma_2$.

Q5. For each e, does there exist an intrinsically consistent e^* such that $\{e\}^A \simeq \{e^*\}^A$ for all A? We think not. In ordinary recursion theory the answer is yes, since we can require that M and N define a finite, initial segment of a characteristic function rather than an arbitrary, basic neighborhood.

Q6. How does one define the jump operator for sets of recursive ordinals? A satisfactory definition should have the property that if A is metarecursive in B, then the jump of A is metarecursive in the jump of B. An answer to Q5 should help here.

Q7. Does there exist a metarecursive function f such that for all e, $R_{f(e)}$ is regular and has the same metadegree as R_e? The proof of Theorem 1 does not provide such an f, but we think it exists. An answer to Q8 might help.

Q8. If A is a nonregular, metarecursively enumerable set, does there exist a Π_1^1 set of the same metadegree as A? We conjecture yes(4).

In[13] we will give axioms for recursion theory in the sense that one gives axioms for group theory.Any object satisfying our axioms will be called a recursion system and will consist of an initial segment of the ordinaals and some partial functions on that segment and into it. It will be seen that most of the results of this paper hold for arbitrary recursion systems and that the notion of regular, hyperregular set (inspired by Kreisel's notion of subgeneric set) is the key to the classification of recursion systems.

References

1. J. C. E. Dekker, *A theorem on hypersimple sets*, Proc. Amer. Math. Soc. **5** (1954), 791–796.

2. G. Driscoll, *Contributions to metarecursion theory*, Ph. D. Thesis, Cornell University, Ithaca, N. Y., 1965.

3. R. M. Friedberg, *Two recursively enumerable sets of incomparable degrees of unsolvability*, Proc. Nat. Acad. Sci. U.S.A. **43** (1957), 236–238.

4. ———, *Three theorems on recursive enumeration*, J. Symbolic Logic **23** (1958), 309–316.

5. S. C. Kleene, *Introduction to metamathematics*, Van Nostrand, New York, 1952.

6. G. Kreisel, *Model-theoretic invariants*, Theory of Models Symposium, Berkeley, Calif., (to appear).

7. G. Kreisel and G. E. Sacks, *Metarecursive sets*, J. Symbolic Logic **31** (1966), 1–21.

8. S. Kripke, *Transfinite recursions on admissible ordinals*. I, II (abstracts), J. Symbolic Logic **29** (1964), 161–162.

9. ———, *Admissible ordinals and the analytic hierarchy* (abstract), J. Symbolic Logic **29** (1964), 162.

(4) *Note added in proof.* We can show the answer to Q8 is yes.

10. E. L. Post, *Recursively enumerable sets of positive integers and their decision procedures*, Bull. Amer. Math. Soc. **50** (1944), 284–316.

11. G. E. Sacks, *Degrees of unsolvability*, Annals of Mathematics Studies No. 55, Princeton Univ. Press, Princeton, N. J., 1963.

12. ———, *The recursively enumerable degrees are dense*, Ann. of Math. **80** (1964), 300–312.

13. ———, *Axioms for recursion theory*, (to appear).

14. C. Spector, *Recursive well-orderings*, J. Symbolic Logic **20** (1955), 151–163.

CORNELL UNIVERSITY,
ITHACA, NEW YORK

ON A THEOREM OF LACHLAN AND MARTIN

GERALD E. SACKS[1]

In [3] we raised the following question: does there exist a recursively enumerable degree $\boldsymbol{d}$ such that $\boldsymbol{0}^{(n)}<\boldsymbol{d}^{(n)}<\boldsymbol{0}^{(n+1)}$ for all $n\geqq 0$? This question was answered affirmatively by Lachlan [1] and by Martin [2]. Lachlan's proof combines a familiar priority argument with the fixed point theorem of Kleene. Martin's proof is a new form of priority argument based on Theorem 3 of section 6 of [3]. In this paper we give a vanishingly short proof of the Lachlan-Martin result without any use of priority. Our argument is an exercise in the fixed point theorem. We exploit, possibly for the first time, a uniformity concealed in most proofs of the fixed point theorem.

Let A be an arbitrary set of natural numbers, and let W_0^A, W_1^A, W_2^A, $\cdots$ be a standard simultaneous enumeration of all sets recursively enumerable in A. For each $e\geqq 0$, let

$$R_e^A = \{2n \mid n \in A\} \cup \{2n+1 \mid n \in W_e^A\}.$$

Then R_0^A, R_1^A, R_2^A, $\cdots$ is a standard simultaneous enumeration of sets recursively enumerable in A and of degree greater than or equal to that of A.

The proof of Theorem 3 of §6 of [3] makes clear that there exists a recursive function t such that

$$(A)(e)[\boldsymbol{A} < \boldsymbol{R}_{t(e)}^A < \boldsymbol{A}' \quad \text{and} \quad (\boldsymbol{R}_{t(e)}^A)' = \boldsymbol{R}_e^{A'}],$$

where $\boldsymbol{A}$ is the degree of A. By the fixed point theorem, there exists a c such that

$$(A)[R_{t(c)}^A = R_c^A].$$

The fact that c is independent of A is the key that unlocks the shiny little box containing our proof. It follows that

$$(A)[\boldsymbol{A} < \boldsymbol{R}_c^A < \boldsymbol{A}' \quad \text{and} \quad (\boldsymbol{R}_c^A)' = \boldsymbol{R}_c^{A'}].$$

But then

$$(A)[\boldsymbol{A} < \boldsymbol{R}_c^A < \boldsymbol{A}' < (\boldsymbol{R}_c^A)' < \boldsymbol{A}'']$$

and

Received by the editors December 27, 1965.

[1] The preparation of this paper was supported by U. S. Army Contract ARO-D-373.

$$(A)(n)[\boldsymbol{A}^{(n)} < (\boldsymbol{R}_c^A)^{(n)} < \boldsymbol{A}^{(n+1)}].$$

Let J and C be operators such that $J(A)=A'$ and $C(A)=R_c^A$. Then our argument (as well as that of Martin [2]) shows that $\boldsymbol{JC}=\boldsymbol{CJ}$. In other words we have found a solution of Post's problem which commutes with the jump operator up to degrees. A. Nerode pointed out that the version of the fixed point theorem used above is better formulated as a fixed point theorem for a class of effective transformations acting on a class of effective operators. From his point of view, it is not surprising that one should use a fixed point argument to study the equation $\boldsymbol{JX}=\boldsymbol{XJ}$.

We leave the reader with the following open question: does there exist a solution to Post's problem well defined on degrees? More precisely, does there exist a Gödel number c such that for all A and B, if $\boldsymbol{A}=\boldsymbol{B}$, then $\boldsymbol{R}_c^A=\boldsymbol{R}_c^B$ and $\boldsymbol{A}<\boldsymbol{R}_c^A<\boldsymbol{A}'$?

References

1. A. H. Lachlan, *On a problem of G. E. Sacks*, Proc. Amer. Math. Soc. **16** (1965), 972–979.

2. D. A. Martin, *On a question of G. E. Sacks*, J. Symbolic Logic 31 (1966), 66–69.

3. G. E. Sacks, *Degrees of unsolvability*, Annals of Mathematics Studies No. 55, Princeton Univ. Press, Princeton, N. J., 1963.

Cornell University

FUNDAMENTA
MATHEMATICAE
LXI (1967)

A minimal hyperdegree

by

R. O. Gandy and G. E. Sacks* (Cambridge Mass.)

Two sets of natural numbers have the same hyperdegree if each is hyperarithmetic in the other. A non-hyperarithmetic set is said to have *minimal hyperdegree* if all sets of lower hyperdegree are hyperarithmetic. In this paper we construct a set that has minimal hyperdegree, and we study a certain class of models of the hyperarithmetic comprehension axiom. We draw upon ideas occurring in Spector's construction of a minimal degree of unsolvability [9] and in Feferman's application of forcing to analysis [2]. Our argument mixes Cohen's forcing method [1] with classical truth considerations; however, the use of forcing is not essential to the construction of a set of minimal hyperdegree. Instead of forcing with finite conditions in the style of Feferman [2], we force with infinite, hyperarithmetic conditions. As one might expect, forcing with infinite conditions is much closer to truth than forcing with finite conditions. A set generic with respect to our notion of forcing must necessarily have minimal hyperdegree.

All of our forcing is with respect to a second order language L(S), which is virtually isomorphic to Feferman's language L*(S) ([2], p. 335). L(S) is the language of first order number theory augmented by the constant symbol **S**, some second order variables, and the membership symbol ϵ. Let O_1 be a π_1^1 subset of O [4], the set of all notations for recursive ordinals, such that each recursive ordinal has precisely one notation in O_1 [3]; if b is the unique notation in O_1 for the recursive ordinal β, we write $|b| = \beta$. In addition, the relation $|b| < |c|$ is the restriction of some recursively enumerable relation to O_1. For each $b \in O_1$, L(S) has set variables $X^b, Y^b, Z^b, \ldots$; L(S) also has set variables $X, Y, Z, \ldots$, number variables $x, y, z, \ldots$, a numeral $\bar{n}$ for each natural number n, and symbols for equality ($=$), successor ($'$), addition ($+$) and multiplication ($\cdot$).

For each $b \in O_1$, the variable X^b is said to be *ranked*; the variable X is said to be *unranked*. A formula $\mathfrak{F}$ of L(S) is called *ranked* if every set

* The second-named author was partially supported by the Guggenheim Foundation and by U. S. A. Contract ARO-D-373.

variable occurring in $\mathfrak{F}$ is ranked. A formula of $\mathbf{L}(\mathbf{S})$ is called *existential* if it is ranked or if it is of the form $(\mathrm{E}X)\mathfrak{F}$ with X the only unranked variable occurring in $\mathfrak{F}$. The *ordinal rank* of a ranked formula $\mathfrak{F}$ is the least ordinal a such that $a \geqslant |b|$ for every free variable X^b of $\mathfrak{F}$ and such that $a > |b|$ for every bound variable X^b of $\mathfrak{F}$. A formula $\mathfrak{F}$ is called *arithmetical* if no bound set variables occur in $\mathfrak{F}$.

Let X be an arbitrary set of natural numbers. Following Feferman [2], for each $b \in O_1$, we inductively define a structure $\mathcal{M}_b(X) = \mathcal{M}_b$ and truth in $\bigcup \{\mathcal{M}_a |\ |a| < |b|\}$:

(i) A sentence $\mathfrak{F}$ of ordinal rank $\leqslant |b|$ is true in $\bigcup \{\mathcal{M}_a |\ |a| < |b|\}$ if it is true when $\mathbf{S}$ is interpreted as X, the number variables of $\mathfrak{F}$ are restricted to ω, and each second order variable X^a of $\mathfrak{F}$ is restricted to $\mathcal{M}_a$.

(ii) For each formula $\mathfrak{G}(x)$ (with only x free) of ordinal rank $\leqslant |b|$, let $\hat{x}\mathfrak{G}(x) = \{n |\ \mathfrak{G}(\overline{n})$ is true in $\bigcup \{\mathcal{M}_a |\ |a| < |b|\}\}$; then $\mathcal{M}_b$ consists of all such sets $\hat{x}\mathfrak{G}(x)$.

We define $\mathcal{M}(X) = \bigcup \{\mathcal{M}_b(X) |\ b \in O_1\}$. A sentence $\mathfrak{F}$ of $\mathbf{L}(\mathbf{S})$ is true in $\mathcal{M}(X)$ ($\models_{\mathcal{M}(X)} \mathfrak{F}$) if it is true when each unranked variable of $\mathfrak{F}$ is restricted to $\mathcal{M}(X)$ and the remaining symbols of $\mathfrak{F}$ are interpreted according to (i) above.

Let U be a ranked or unranked set variable of $\mathbf{L}(\mathbf{S})$, and let t denote a number-theoretic term. We write $\mathfrak{F}(\hat{x}\mathfrak{G}(x))$ for the result of replacing each occurrence of $t \in U$ in $\mathfrak{F}(U)$ by $\mathfrak{G}(t)$.

If $a_0, a_1, \ldots, a_{m-1}$ is a finite sequence of natural numbers, then it is effectively represented by the sequence number $\pi\{p_i^{1+a_i} |\ i < m\}$, where p_i is the ith largest prime. Sequence numbers provide a means of referring to finite, initial segments of characteristic functions of sets of natural numbers as if they were natural numbers. We use $p, q, r, \ldots$ to denote finite, initial segments of characteristic functions. We write $p > r$ if $p \neq r$ and p is extended by r. Let P be a non-empty set of finite, initial segments; we say that P *defines* (or *is*) a *perfect, closed set* if

$$(p)_{p \in P}(\mathrm{E}q)_{q \in P}(\mathrm{E}r)_{r \in P}[q \neq r \;\&\; q \not> r \;\&\; r \not> q \;\&\; p > q \;\&\; p > r].$$

We write $X \in P$ if infinitely many initial segments of the characteristic function of X belong to P. (If P defines a perfect, closed set, then $\{X |\ X \in P\}$ is a perfect, closed set in the standard sense.) Let $P, Q, R, \ldots$ denote hyperarithmetic, perfect, closed (h.p.c.) sets; i.e., the set of sequence numbers of the initial segments in P is hyperarithmetic. We write $P \geqslant Q$ if $(X)(X \in Q \rightarrow X \in P)$. It is routine to assign indices to h.p.c. sets so that the following relation in x and Y is π_1^1: x is the index of a h.p.c. set P and $Y \in P$. For this reason it makes sense to say that the relation "P is a h.p.c. set and $Y \in P$" is π_1^1. Similarly, it makes sense to say that the set of all formulas of $\mathbf{L}(\mathbf{S})$ is π_1^1.

The forcing relation $P \Vdash_h \mathfrak{F}$, where P is a h.p.c. set and $\mathfrak{F}$ is a sentence of L(S), is defined by means of five closure conditions:

(i) $P \Vdash_h \mathfrak{F}$ if $\mathfrak{F}$ is ranked and $(X)(X \in P \to \models_{\mathcal{M}(X)} \mathfrak{F})$.

(ii) $P \Vdash_h (\mathrm{E}x)\mathfrak{F}(x)$ if $\mathfrak{F}(x)$ is unranked and $P \Vdash_h \mathfrak{F}(\bar{n})$ for some n.

(iii) $P \Vdash_h (\mathrm{E}X^b)\mathfrak{F}(X^b)$ if $\mathfrak{F}(X^b)$ is unranked and $P \Vdash_h \mathfrak{F}(\hat{x}\mathfrak{G}(x))$ for some $\mathfrak{G}(x)$ of ordinal rank $\leqslant |b|$.

(iv) $P \Vdash_h (\mathrm{E}X)\mathfrak{F}(X)$ if $(\mathrm{E}b)_{b \in O_1}[P \Vdash (\mathrm{E}X^b)\mathfrak{F}(X^b)]$.

(v) $P \Vdash_h \mathfrak{F}_1 \,\&\, \mathfrak{F}_2$ if $\mathfrak{F}_1 \,\&\, \mathfrak{F}_2$ is unranked, $P \Vdash_h \mathfrak{F}_1$, and $P \Vdash_h \mathfrak{F}_2$.

(vi) $P \Vdash_h \sim \mathfrak{F}$ if $\mathfrak{F}$ is unranked and $(Q)_{P \geqslant Q}[\sim Q \Vdash_h \mathfrak{F}]$.

We defined $\Vdash_h$ in terms of closure properties rather than transfinite induction in order to simplify the statements of our proofs (cf. Feferman [2], p. 336). A set S is said to be *generic* if for each sentence $\mathfrak{F}$ there is a P such that $S \in P$ and $[P \Vdash_h \mathfrak{F}$ or $P \Vdash_h \sim \mathfrak{F}]$. It is not immediate that generic sets exist, because the definition of $\Vdash_h$ contains some peculiarities, particularly the contrast between conditions (i) and (vi). It is precisely these peculiarities which make possible a quick proof of the all-important Lemma 1.

LEMMA 1. *The relation $P \Vdash_h \mathfrak{F}$, restricted to ranked $\mathfrak{F}$, is π_1^1.*

Proof. It is readily verified that the relation $\mathfrak{F}$ is true in $\mathcal{M}(X)$, restricted to ranked $\mathfrak{F}$, is π_1^1 in $\mathfrak{F}$ and X; one need only note that this relation can be defined by Σ_1^1 closure conditions (cf. Feferman [2], p. 337), The relation $X \in P$ is arithmetical *on* the π_1^1 set of all hyperarithmetic. perfect, closed P's. (This last point is elaborated in the comments immediately preceding Lemma 6.)

Kreisel's Lemma ([7], p. 307) states: if $P(x, y)$ is π_1^1 and $(x)(\mathrm{E}y)P(x, y)$, then there exists a hyperarithmetic function f such that $(x)P(x, f(x))$. Kreisel's argument also establishes a slightly stronger fact needed below: if $P(x, y)$ is π_1^1, then there exists a partial π_1^1 function f such that $(x)[(\mathrm{E}y)P(x, y) \to f(x)$ is defined $\& P(x, f(x))]$. (A partial function is π_1^1 if its graph is π_1^1).

The purpose of Lemma 2 (the "sequential" lemma) is to standardize a construction which we use repeatedly, and which is characteristic of forcing with closed, perfect sets.

LEMMA 2. *Let $\{\mathfrak{F}_m\}$ be a hyperarithmetic sequence of existential sentences If $(m)(Q)_{P \geqslant Q}(\mathrm{E}R)_{Q \geqslant R}[R \Vdash_h \mathfrak{F}_m]$, then $(\mathrm{E}Q)_{P \geqslant Q}(m)[Q \Vdash_h \mathfrak{F}_m]$.*

Proof, Suppose $(m)(Q)_{P \geqslant Q}(\mathrm{E}R)_{Q \geqslant R}[R \Vdash_h \mathfrak{F}_m]$. By Lemma 1 and condition (iv) of the definition of $\Vdash_h$, the following relation is π_1^1 in m, Q, and R: $P \geqslant Q \geqslant R \,\&\, R \Vdash_h \mathfrak{F}_m$. The argument of Kreisel's Lemma ([7], p. 307) shows that R can be regarded as a partial π_1^1 function f of m and Q: $R = f(m, Q) \to P \geqslant Q \geqslant R \,\&\, R \Vdash_h \mathfrak{F}_m$. Let us say that Q_1 and Q_2 are basic, disjoint h.p.c. subsets of Q if there exist $q_1, q_2 \in Q$ such that $q_1 \neq q_2$, $q_1 \not\geqslant q_2$, $q_2 \not\geqslant q_1$, $Q_1 = \{p|\ p \in Q \,\&\, q_1 \geqslant p\}$, and $Q_2 = \{p|\ p \in Q \,\&\, q_2 \geqslant p\}$. By iterating the partial π_1^1 function f, we can define a hyperarithmetic, partial function Q_i^m with the following properties: $Q_0^0 = P$; for each $m \geqslant 0$ and $i < 2^m$, Q_i^{m+1} and $Q_{i+2^m}^{m+1}$ are basic, disjoint, h.p.c. subsets of Q_i^m, $Q_i^{m+1} \Vdash_h \mathfrak{F}_m$, and $Q_{i+2^m}^{m+1} \Vdash_h \mathfrak{F}_m$. We define Q by

$$p \in Q \leftrightarrow (m)(\mathrm{E}_1 i)_{i<2^m}(p \in Q_i^m)\,.$$

Then Q is a h.p.c. set, $P \geqslant Q$, and

$$X \in Q \leftrightarrow (m)(\mathrm{E}_1 i)_{i<2^m}(X \in Q_i^m)\,.$$

Fix m and $X \in Q$. There exists an i such that $X \in Q_i^{m+1}$ and $Q_i^{m+1} \Vdash_h \mathfrak{F}_m$. Since $\mathfrak{F}_m$ is existential, $\models_{\mathcal{M}(X)} \mathfrak{F}_m$. But then $Q \Vdash_h \mathfrak{F}_m$.

LEMMA 3. $(\mathfrak{F})(P)(\mathrm{E}Q)_{P\geqslant Q}[Q \Vdash_h \mathfrak{F} \vee Q \Vdash_h \sim\mathfrak{F}]$.

Proof. Condition (vi) of the definition of $\Vdash_h$ makes it safe to assume that $\mathfrak{F}$ is ranked. Conditions (i) and (ii) of the definition of $\Vdash_h$ make it safe to assume that $\mathfrak{F}$ is in prenex normal form. By the full ordinal rank of $\mathfrak{F}$, we mean a function f defined on O_1 such that for each $b \in O_1$, $f(b)$ equals the number of occurrences of b in $\mathfrak{F}$. We say

$$f < g \quad \text{if} \quad (\mathrm{E}b)_{b\in O_1}\big[f(b) < g(b) \,\&\, (c)_{|c|>|b|}\big(f(c) = g(c)\big)\big]\,.$$

By the *arithmetical rank* of $\mathfrak{F}$, we mean the number m of occurrences of number-theoretic quantifiers in $\mathfrak{F}$. The rank of $\mathfrak{F}$ is (f, m). We say $(f, m) < (g, n)$ if $f < g$ or if $f = g$ and $m < n$. We prove the lemma for ranked $\mathfrak{F}$ by induction on the rank of $\mathfrak{F}$. Suppose that $\mathfrak{F}$ is $(\mathrm{E}X^b)\mathfrak{F}(X^b)$. Let $\{\mathfrak{G}_i(x)\}$ be a hyperarithmetic enumeration of all formulas (with just x free) of ordinal rank $\leqslant |b|$. For each i, $\mathfrak{F}\big(\hat{x}\mathfrak{G}_i(x)\big)$ has lower full ordinal rank than $(\mathrm{E}X^b)\mathfrak{F}(X^b)$. Our inductive hypothesis is: $(i)(P)(\mathrm{E}Q)_{P\geqslant Q}\big[Q \Vdash_h \mathfrak{F}\big(\hat{x}\mathfrak{G}_i(x)\big)$ or $Q \Vdash_h \sim\mathfrak{F}\big(\hat{x}\mathfrak{G}_i(x)\big)\big]$. Fix P. If $(\mathrm{E}i)(\mathrm{E}Q)_{P\geqslant Q}\big[Q \Vdash_h \mathfrak{F}\big(\hat{x}\mathfrak{G}_i(x)\big)\big]$, then all is well; suppose not. Then it follows from the inductive hypothesis that $(i)(Q)_{P\geqslant Q}(\mathrm{E}R)_{Q\geqslant R}\big[R \Vdash_h \sim\mathfrak{F}\big(\hat{x}\mathfrak{G}_i(x)\big)\big]$. By Lemma 2, there is a Q such that $P \geqslant Q$ and $(i)\big[Q \Vdash_h \sim\mathfrak{F}\big(\hat{x}\mathfrak{G}_i(x)\big)\big]$; but then $Q \Vdash_h \sim(\mathrm{E}X^b)\mathfrak{F}(X^b)$.

If $\mathfrak{F}$ is of the form $(\mathrm{E}x)\mathfrak{F}(x)$, we note that $\mathfrak{F}(\bar{n})$ has lower arithmetical rank than $(\mathrm{E}x)\mathfrak{F}(x)$ and then proceed as above. If $\mathfrak{F}$ has no quantifiers, then there is a $q \in P$ such that the desired $Q = \{p|\ p \in P \,\&\, q \geqslant p\}$.

The existence of generic sets follows from Lemma 3. A standard transfinite induction shows: if S is generic, then $(\mathfrak{F})[\models_{\mathcal{M}(X)}\mathfrak{F} \leftrightarrow$ $\leftrightarrow (\mathrm{E}P)(S \in P \,\&\, P \Vdash_h \mathfrak{F})]$.

LEMMA 4. *If S is genetric, $\mathfrak{F}(X)$ is arithmetical, and $\models_{\mathcal{M}(S)}(\mathrm{E}X)\mathfrak{F}(X)$, then $(\mathrm{E}X)[\mathfrak{F}(X) \,\&\, X$ is hyperarithmetic $\&\, X \in \mathcal{M}(S)]$.*

Proof. Since S is generic, there must be a P such that $S \in P$ and $P \Vdash_h (\mathrm{E}X^b)\mathfrak{F}(X^b)$ for some $b \in O_1$. Since P is a hyperarithmetic, perfect, closed set, there must exist a hyperarithmetic $H \in P$. It follows from the definition of $\Vdash_h$ for ranked sentences that $(\mathrm{E}X)[X \in \mathcal{M}_b(H) \,\&\, \mathfrak{F}(X)]$. A transfinite induction on O_1 in the style of Kleene ([4], p. 35) shows that every member of $\mathcal{M}_b(X)$ is hyperarithmetic in X for every X; consequently, $\mathfrak{F}(K)$ holds for some hyperarithmetic K. Following Kleene ([4], p. 35), we can find a ranked formula $\mathfrak{G}(x)$ such that $K = \{n \mid \models_{\mathcal{M}(X)} \mathfrak{G}(\bar{n})\}$ holds for every X.

Lemma 5. *For each π_1^1 set A, there exists an existential formula $P(x)$ such that for all generic S, $(n)[n \in A \leftrightarrow \models_{\mathcal{M}(S)} P(\bar{n})]$.*

Proof. By Gandy [3] there exists an arithmetical predicate $B(x, Y)$ such that for all n,

$$n \in A \to (\mathrm{E}Y)[B(n, Y) \,\&\, Y \text{ is hyperarithmetical}]\,.$$

At the end of the proof of Lemma 4, it was noted that every hyperarithmetic set belongs to $\mathcal{M}(X)$ for every X. It follows from Lemma 4 that for all n,

$$n \in A \leftrightarrow \models_{\mathcal{M}(S)} (\mathrm{E}Y) B(\bar{n}, Y)\,.$$

For the sake of Lemma 6, it is necessary to sharpen our previous observations concerning the assignment of indices to hyperarithmetic, closed sets. Kreisel ([7], p. 307) mentions an arithmetical formula $A(x, Y)$ with the following property: if $x \in O$, then $(\mathrm{E}_1 Y)A(x, Y)$ and $(Y)[A(x, Y) \to \to Y$ is a hyperarithmetic set of the same Turing degree as $H_x]$. Since each hyperarithmetic set is recursive in H_x for some x, it is possible to assign indices to hyperarithmetic sets and to obtain an arithmetical formula $B(x, Y)$ such that the set of indices I is π_1^1 and such that: if $x \in I$, then $(\mathrm{E}_1 Y)B(x, Y)$ and $(Y)[B(x, Y) \to (Y)_0 =$ the hyperarithmetic set whose index is $x]$, where $(Y)_0 = \{m \mid 2^m \in Y\}$.

I and B can be modified to obtain a π_1^1 set I_1 of indices for hyperarithmetic, perfect, closed sets and an arithmetical formula $B_1(x, Y)$ with the following properties: if $x \in I_1$, then $(\mathrm{E}_1 Y)B_1(x, Y)$ and $(Y)[B_1(x, Y) \to (Y)_0 =$ the h.p.c. set whose index is $x]$. Let $D(Y, Z)$ be an arithmetical formula which says: $(Y)_0$ contains infinitely many sequence numbers which represent initial segments of the characteristic function of Z. Then $x \in I_1 \,\&\, (Y)[B_1(x, Y) \to D(Y, Z)]$ says: x is the index of a h.p.c. P and $Z \in P$.

Lemma 6. *Let A be a π_1^1 set (of indices) of hyperarithmetic, perfect, closed sets, and let S be generic. If $(P)(\mathrm{E}Q)_{P \geqslant Q}(Q \in A)$, then $(\mathrm{E}Q)(Q \in A \,\&\, \,\&\, S \in Q)$.*

Proof. By Lemma 5 there is a formula $P(x)$ such that for all generic S, $(n)[n \in A \leftrightarrow \models_{\mathcal{M}(S)} P(\bar{n})]$. Let $\mathfrak{F}$ denote the following formula of L(S):

$$(\mathrm{E}x)\big[P(x)\ \&\ (Y)\big(B_1(x, Y) \rightarrow D(Y, \mathbf{S})\big)\big]\,.$$

Then for all generic S, $(\mathrm{E}Q)[Q \in A\ \&\ S \in Q] \leftrightarrow \models_{\mathcal{M}(S)} \mathfrak{F}$. Now fix the generic set S. To prove the lemma, it is enough to find a P such that $P \Vdash_h \mathfrak{F}$ and $S \in P$. Suppose (for the sake of a contradiction) that no such P exists. Then there must be a P such that $P \Vdash_h \sim\mathfrak{F}$ and $S \in P$. There must also be a Q_1 such that $P \geqslant Q_1$ and $Q_1 \in A$. By Lemma 3, there exists a generic set $S' \in Q_1$. Since $P \Vdash_h \sim\mathfrak{F}$ and $P \geqslant Q_1$, it follows $Q_1 \Vdash_h \sim\mathfrak{F}$. Thus $\models_{\mathcal{M}(S')} \sim\mathfrak{F}$, and consequently, $\sim(\mathrm{E}Q)[Q \in A\ \&\ S' \in Q]$. But $Q_1 \in A$ and $S' \in Q_1$.

LEMMA 7. *Let $\mathfrak{F}(x, Y)$ be a formula of* L(S) *whose only free variables are x and Y and whose only unranked variable is Y. If $P \Vdash_h (x)(\mathrm{E}Y)\mathfrak{F}(x, Y)$, then* $(\mathrm{E}Q)_{P \geqslant Q}(\mathrm{E}b)_{b \in O_1}[Q \Vdash_h (x)(\mathrm{E}Y^b)\mathfrak{F}(x, Y^b)]$.

Proof. Let $P \Vdash_h (x)(\mathrm{E}Y)\mathfrak{F}(x, Y)$. Then Lemma 3 and the definition of $\Vdash_h$ imply

$$(n)(Q)_{P \geqslant Q}(\mathrm{E}R)_{Q \geqslant R}[R \Vdash_h (\mathrm{E}Y)\mathfrak{F}(\bar{n}, Y)]\,.$$

It follows from Lemma 2 that there is a Q such that $P \geqslant Q$ and $(n)[Q \Vdash_h (\mathrm{E}Y)\mathfrak{F}(\bar{n}, Y)]$. This means that $(n)(\mathrm{E}b)_{b \in O_1}[Q \Vdash_h (\mathrm{E}Y^b)\mathfrak{F}(\bar{n}, Y^b)]$. By Lemma 1 and Kreisel's Lemma ([7], p. 307), there is a hyperarithmetic function f such that $(n)[Q \Vdash_h (\mathrm{E}Y^{f(n)})\mathfrak{F}(\bar{n}, Y^{f(n)})]$. By Spector [8], there is a $b \in O_1$ such that $(n)(|f(n)| \leqslant |b|)$. But then $Q \Vdash_h (x)(\mathrm{E}Y^b)\mathfrak{F}(x, Y^b)$.

LEMMA 8. *Let $\mathfrak{G}(x)$ be a ranked formula with only x free. For each P, there exists a Q such that $P \geqslant Q$ and either* (i) *or* (ii) *holds:*

(i) $(X)\big(X \in Q \rightarrow X$ *is hyperarithmetic in* $\{n \mid \models_{\mathcal{M}(X)} \mathfrak{G}(\bar{n})\}\big)$;

(ii) $(X)\big(X \in Q \rightarrow \{n \mid \models_{\mathcal{M}(X)} \mathfrak{G}(\bar{n})\}$ *is hyperarithmetic*$\}\big)$.

Proof. We proceed in the spirit of the proof of Lemma 2.

Case 1. $(Q)_{P \geqslant Q}(\mathrm{E}Q_1)_{Q \geqslant Q_1}(\mathrm{E}Q_2)_{Q \geqslant Q_2}(\mathrm{E}n)[Q_1 \Vdash_h \mathfrak{G}(\bar{n})\ \&\ Q_2 \Vdash_h \sim\mathfrak{G}(n)]$.

It follows from Lemma 1 and the argument of Kreisel's Lemma ([7], p. 307) that we can regard Q_1, Q_2, and n as partial π_1^1 functions of Q. By iterating these partial π_1^1 functions, we can define a hyperarithmetic, partial function Q_i^m with the following properties: $Q_0^0 = P$; for each $m \geqslant 0$ and $i < 2^m$, Q_i^{m+1} and $Q_{i+2^m}^{m+1}$ are basic, disjoint h.p.c. subsets of Q_i^m, and $(\mathrm{E}n)[Q_i^{m+1} \Vdash_h \mathfrak{G}(\bar{n})\ \&\ Q_{i+2^m}^{m+1} \Vdash_h \sim \mathfrak{G}(\bar{n})]$. We define Q by:

$$p \in Q \leftrightarrow (m)(\mathrm{E}_1 i)_{i<2^m}(p \in Q_i^m)\,.$$

Then Q is a h.p.c. set, $P \geqslant Q$, and

$$X \in Q \leftrightarrow (m)(\mathrm{E}_1 i)_{i<2^m}(X \in Q_i^m)\,.$$

We say that t puts X in Q if $(m)(X \in Q_{t(m)}^m)$. Each $X \in Q$ is put in Q by a unique t, and is hyperarithmetic in that t. We claim that Q satisfies condition (i) of the lemma. Fix $X \in Q$. Let $\mathcal{G} = \{n \mid \models_{\mathcal{M}(X)} \mathcal{G}(\bar{n})\}$. To see that X is hyperarithmetic in $\mathcal{G}$, it is enough to see that the unique t which puts X in Q is hyperarithmetic in $\mathcal{G}$. Consider the following definition of t:

$$t(0) = 0 \;\&\; \big(t(m+1) = t(m) \vee t(m+1) = t(m) + 2^m\big)\,;$$

$$t(m+1) = t(m) \leftrightarrow (\mathrm{E}n)[Q^{m+1}_{t(m)} \Vdash_h \mathcal{G}(\bar{n}) \;\&\; Q^{m+1}_{t(m)+2^m} \Vdash_h \sim \mathcal{G}(n) \;\&\; n \in \mathcal{G}]\,.$$

The above equations define t, because $\mathcal{G}(\bar{n})$ is a ranked formula, and consequently, $[X \in Q^{m+1}_i \;\&\; Q^{m+1}_i \Vdash_h \mathcal{G}(\bar{n})] \rightarrow n \in \mathcal{G}$. But then t is hyperarithmetic in $\mathcal{G}$, since the relation $P \Vdash_h \mathfrak{F}$ is hyperarithemtic when P is restricted to be a member of $\{Q^m_i \mid i < 2^m \;\&\; m \geqslant 0\}$ and $\mathfrak{F}$ is restricted to be a member of $\{\mathcal{G}(\bar{n}), \sim\mathcal{G}(\bar{n}) \mid n \geqslant 0\}$.

Case 2. $(\mathrm{E}R)_{P \geqslant R}(Q_1)_{R \geqslant Q_1}(Q_2)_{R \geqslant Q_2}(n) \sim [Q_1 \Vdash_h \mathcal{G}(\bar{n}) \;\&\; Q_2 \Vdash_h \sim \mathcal{G}(\bar{n})]$.

By Lemma 3, $(n)(\mathrm{E}Q)_{R \geqslant Q}[Q \Vdash_h \mathcal{G}(\bar{n}) \vee Q \Vdash_h \sim \mathcal{G}(\bar{n})]$. For each n, let

$$F_n = \begin{cases} \mathcal{G}(\bar{n}) & \text{if} \quad (\mathrm{E}Q)_{R \geqslant Q}[Q \Vdash_h \mathcal{G}(\bar{n})]\,, \\ \sim\mathcal{G}(\bar{n}) & \text{i-} \quad (\mathrm{E}Q)_{R \geqslant Q}[Q \Vdash_h \sim \mathcal{G}(\bar{n})]\,. \end{cases}$$

The defining property of R guarantees that F_n is well-defined; in addition $\{F_n\}$ is a hyperarithmetic sequence of ranked formulas. The nature of R also guarantees that

$$(n)(Q_1)_{R \geqslant Q_1}(\mathrm{E}Q_2)_{Q_1 \geqslant Q_2}[Q_2 \Vdash_h F_n]\,.$$

It follows from Lemma 2 that $(\mathrm{E}Q)_{R \geqslant Q}(n)[Q \Vdash_h F_n]$. If $X \in Q$, then $\models_{\mathcal{M}(X)} \mathcal{G}(\bar{n}) \leftrightarrow Q \Vdash_h \mathcal{G}(\bar{n})$. Thus Q satisfies condition (ii) of the lemma, since the relation $Q \Vdash_h \mathcal{G}(\bar{n})$ is hyperarithmetic in n.

LEMMA 9. *If S is generic, then the hyperarithmetic comprehension axiom holds in $\mathcal{M}(S)$.*

Proof. The argument of Feferman ([2], p. 339) makes clear that it is sufficient to show the following: let S be generic, and let $\mathfrak{F}(x, Y)$ be a formula of L(S) whose only free variables are x and Y and whose only unranked variable is Y; if $\models_{\mathcal{M}(S)}(x)(\mathrm{E}Y)\mathfrak{F}(x, Y)$, then for some $b \in O_1$,

$$\models_{\mathcal{M}(S)}(x)(\mathrm{E}Y^b)\mathfrak{F}(x, Y^b)\,.$$

Let P be such that $S \in P$ and $P \Vdash_h (x)(\mathrm{E}Y)\mathfrak{F}(x, Y)$. Let

$$A_1 = \{Q \mid (X)(X \in Q \rightarrow X \notin P)\},$$

and let

$$A_2 = \{Q \mid P \geqslant Q \;\&\; (\mathrm{E}b)_{b \in O_1}[Q \Vdash_h (x)(\mathrm{E}Y^b)\mathfrak{F}(x, Y^b)]\}\,.$$

Let A be the set of indices of all members of $A_1 \cup A_2$. It follows from Lemma 1 that A is π^1_1. It is our intention to apply Lemma 6 to A. We must first show $(Q)(\mathrm{E}R)_{Q \geqslant R}(R \in A)$. Fix Q. The last sentence of the proof of Lemma 4 implies there exists a ranked sentence $\mathfrak{F}$ such that $(X)[X \in P \leftrightarrow$

$\leftrightarrow \models_{\mathcal{M}(X)} \mathfrak{F}]$. By Lemma 3, there exists a Q_1 such that $Q \geqslant Q_1$, and $Q_1 \Vdash_h \mathfrak{F}$ or $Q_1 \Vdash_h \sim\mathfrak{F}$. If $Q_1 \Vdash_h \mathfrak{F}$, then $P \geqslant Q_1$, and by Lemma 7, $(\mathrm{E}R)_{Q_1 \geqslant R}(R \in A_2)$. If $Q_1 \Vdash_h \sim\mathfrak{F}$, then $Q_1 \in A_1$.

By Lemma 6, there is a $Q_2 \in A$ such that $S \in Q_2$. Since $S \in P$, it must be that $Q_2 \in A_2$. But then $Q_2 \Vdash_h (x)(\mathrm{E}Y^b)\mathfrak{F}(x, Y^b)$ for some $b \in O_1$, and consequently, $\models_{\mathcal{M}(S)}(x)(\mathrm{E}Y^b)\mathfrak{F}(x, Y^b)$.

LEMMA 10. *If S is generic, then S has minimal hyperdegree.*

Proof. Let S be generic. First we show that S is not hyperarithmetic. Suppose that S equals the hyperarithmetic set H. Following Kleene ([4], p. 35), we can find a ranked formula $\hat{x}\mathfrak{G}(x)$ such that $H = \{n \mid \models_{\mathcal{M}(X)} \mathfrak{G}(\bar{n})\}$ for all X. Since S is generic, there exists a P with the property that $P \Vdash_h \mathrm{S} = \hat{x}\mathfrak{G}(x)$. But then $(X)(X \in P \rightarrow X = H)$. This last is absurd, since P is uncountable.

Now suppose that K is hyperarithmetic in S. It follows from Lemma 9 and the argument of Kreisel ([6], p. 114) that $K \in \mathcal{M}(S)$. Let $\mathfrak{G}(x)$ be a ranked formula such that $n \in K \leftrightarrow \models_{\mathcal{M}(S)} \mathfrak{G}(\bar{n})$. Let

$$A_1 = \{Q \mid (X)(X \in Q \rightarrow X \text{ is hyperarithmetic in } \{n \mid \models_{\mathcal{M}(X)} \mathfrak{G}(\bar{n})\})\},$$

and let

$$A_2 = \{Q \mid (X)(X \in Q \rightarrow \{n \mid \models_{\mathcal{M}(X)} \mathfrak{G}(\bar{n})\} \text{ is hyperarithmetic})\}.$$

Let $A = A_1 \cup A_2$. Then the following observations imply that A is π_1^1: the relation "X is hyperarithmetic in Y" is π_1^1; the relation "$\models_{\mathcal{M}(X)} \mathfrak{G}(\bar{n})$" is hyperarithmetic (in X and n). By Lemma 8, $(P)(\mathrm{E}Q)_{P \geqslant Q}(Q \in A)$. By Lemma 6, $(\mathrm{E}Q)(Q \in A \;\&\; S \in Q)$. If $Q \in A_1$, then S is hyperarithmetic in K. If $Q \in A_2$, then K is hyperarithmetic.

THEOREM. *There exists a set of minimal hyperdegree less than the hyperdegree of O.*

Proof. The relation $P \Vdash_h \mathfrak{F}$, where P is a h.p.c. set and $\mathfrak{F}$ is an arbitrary sentence of L(S), is hyperarthmetic in O (cf. [2], p. 337). Let $\{\mathfrak{F}_m \mid m \geqslant 0\}$ be an enumeration hyperarithmetic in O of all sentences of L(S). By Lemma 3, we have $(m)(P)(\mathrm{E}Q)_{P \geqslant Q}[P \Vdash_h \mathfrak{F}_m \vee P \Vdash_h \sim \mathfrak{F}_m]$. The argument of Kreisel's Lemma ([7], p. 307), relativized to O, makes it possible to regard Q as a partial function of m and P whose graph is π_1^1 in O: $(m)(P)[P \geqslant f(m, P) \;\&\; (f(m, P) \Vdash_h \mathfrak{F}_m \vee f(m, P) \Vdash_h \sim \mathfrak{F}_m)]$. By iterating f, we can define a function P_m hyperarithmetic in O with the following properties: $P_m \geqslant P_{m+1} \;\&\; [P_m \Vdash_h \mathfrak{F}_m \vee P_m \Vdash_h \sim \mathfrak{F}_m]$. We define S by the formula:

$$n \in S \leftrightarrow (\mathrm{E}m)[P_m \Vdash_h \bar{n} \in \mathrm{S}].$$

Then S is generic and hyperarithmetic in O. (S is, in fact, the unique X such that $(m)(X \in P_m)$.) By Lemma 10, S has minimal hyperdegree. In [2], p. 340, it was observed that O is not of minimal hyperdegree.

We conclude with a word on the elimination of forcing from the construction of a set of minimal hyperdegree. The key lemmas underlying the proof of the theorem are Lemmas 7 and 8. Lemma 7 provides the means of insuring that ω_1^S, the least ordinal not recursive in S, equals ω_1, the least non-recursive ordinal. Lemma 8 provides the means of insuring that S has minimal hyperdegree, given that $\omega_1^S = \omega_1$. Both of the key lemmas are consequences of the nature of the forcing relation restricted to ranked sentences, where it coincides with the classical truth relation. These considerations, if pressed hard, lead to a construction of a set of minimal hyperdegree recursive in O without any recourse to the idea of forcing.

References

[1] P. J. Cohen, *The independence of the continuum hypothesis*, Proc. Nat. Acad. of Sci. U. S. A. 50 (1963), pp 1143-1148 (Part I) and 51 (1964), pp. 105-110 (Part II).

[2] S. Feferman, *Some applications of the notion of forcing and generic sets*, Fund. Math. 56 (1965), pp. 325-345.

[3] R. Gandy, *Proof of Mostowski's conjecture*, Bull. Acad. Polon. Sci., Ser. Math. 8 (1960), pp. 571-575.

[4] S. C. Kleene, *Hierarchies of number-theoretic predicates*, Bull. Amer. Math. Soc., 61 (1955), pp. 193-213.

[5] — *Quantification of number-theoretic functions*, Compos. Math. 15 (1959), pp. 23-40.

[6] G. Kreisel, *Set theoretic problems suggested by the notion of potential totality*, Proc. Symp. Infinitistic Methods, Warsaw 1961, pp. 103-140.

[7] — *The axiom of choice and the class of hyperarithmetic functions*, Indag. Math. 24 (1962), pp. 307-319.

[8] C. Spector, *Recursive well-orderings*, J. Symb. Log., 20 (1955), pp. 151-163.

[9] — *On degrees of recursive unsolvability*, Ann. of Math., 64 (1956), pp. 581-592.

UNIVERSITY OF MANCHESTER
and
MASSACHUSETTS INSTITUTE OF TECHNOLOGY

Reçu par la Rédaction le 17. 2. 1967

MEASURE-THEORETIC UNIFORMITY IN RECURSION THEORY AND SET THEORY

BY

GERALD E. SACKS(1)

1. **Introduction.** The notion of measure-theoretic uniformity was presented in [23], [24]. In this paper the notion is first utilized in an arithmetic setting and then extended to hyperarithmetic theory and set theory. The recursion-theoretic concepts touched on are: the arithmetic, hyperarithmetic and analytic hierarchies of Kleene [12]; ω-models of the hyperarithmetic comprehension axiom of Kreisel [16]; recursive ordinals [11]; and hyperdegrees [13]. One of the recursion-theoretic results we prove is: if $P(X)$ is a Π^1_1 predicate with free set variable X and the set $\hat{X}P(X)$ has positive Lebesgue measure, then $P(A)$ holds for some hyperarithmetic set A. A corollary of this result is: the set of all X such that the ordinals recursive in X coincide with the recursive ordinals has Lebesgue measure 1 (see footnote (9), below). In the area of set theory we are largely concerned with showing how relative consistency results follow in a natural manner from the notion of measure-theoretic uniformity. We develop Cohen's independence results [2] as well as a result of Solovay [29], [30] on the extendability of Lebesgue measure to all sets of reals. Solovay makes use of Cohen's forcing method, but he ingeniously replaces Cohen's finite forcing conditions by closed sets of positive measure. He shows: if ZF (Zermelo-Fraenkel set theory) is consistent, then ZF+ "there exists a translation-invariant, countably additive extension of Lebesque measure defined on all sets of reals"+"the dependent axiom of choice" is consistent(2). We reprove Solovay's theorem with emphasis on the notion of measure-theoretic uniformity and with the help of some elementary properties of uniformly distributed, independent random variables; however, the fine details of our argument are not substantially different from those originated by Solovay. §4 of this paper, which deals solely with set theory, can be read independently of §§2 and 3, but it is intended to be read as a natural continuation of the earlier sections.

Received by the editors March 13, 1968 and, in revised form, August 1, 1968.

(1) The preparation of this paper was partially supported by U.S. Army Contract ARO-D-373 and by NSF grant GP-6982. The author wishes to thank Professor A. Nerode for many helpful conversations on uniformity and definability, and Professor D. Lacombe for enabling the author to present the paper in a series of lectures [24] at the University of Paris in November, 1966.

(2) R. B. Jensen has shown that the dependent axiom of choice is stronger than the countable axiom of choice. Solovay has also shown: if ZF+ "there exists an inaccessible cardinal" is consistent, then so is ZF+ "the dependent axiom of choice" + "all sets of reals are Lebesque measurable".

Our interest in applying measure-theoretic ideas to questions of mathematical logic began with the reading of Spector's measure-theoretic construction of two incomparable hyperdegrees [32]. Spector showed, using Fubini's Theorem and the measurability of analytic sets, that the set of all pairs (X, Y) such that X and Y are hyperarithmetically incomparable has measure 1. Shoenfield extended Spector's argument to show there exists an uncountable set of pairwise incomparable hyperdegrees. Feferman [3] obtained two incomparable hyperdegrees by means of a forcing argument, but Spector's proof, although less constructive than Feferman's, is much shorter. In [22] we showed: if X is not recursive, then the set of all Y such that the Turing-degree of Y is incomparable with that of X has measure 1. In §3 we show: if X is not hyperarithmetic, then the set of all Y such that the hyperdegree of Y is incomparable with that of X has measure 1. There is a significant conceptual difference between these last two results. The proof of the former of the two relies on the countable additivity of Lebesgue measure and on the following fact: if X is Turing-reducible to (i.e., recursive in) Y, then X is reducible to Y by means of one of countably many reduction procedures whose nature does not depend on Y (i.e., $X=\{e\}^Y$). In contrast, if X is hyperarithmetic in Y, then X is reducible to Y by means of a hyperarithmetic reduction procedure whose nature depends on the ordinals recursive in Y. Since the ordinals recursive in Y can be arbitrarily large countable ordinals, the proof of the latter of the above two results has to take into account uncountably many hyperarithmetic reduction procedures. This difficulty is overcome in §3 by showing that if one ignores a set of Y's of measure 0, then one can assume the ordinals recursive in Y are the recursive ordinals.

It seems impractical at this point to give a general and formal account of the notion of measure-theoretic uniformity. Instead, we give a variety of examples and applications in the belief that the notion is simple enough to abstract from the examples. Let N be the natural numbers. We put the usual probability measure on 2^N as follows. Let m be the unbiased measure for $2=\{0, 1\}$: $m(\{0, 1\})=1$, $m(\{0\}) =m(\{1\})=\frac{1}{2}$, and $m(\varnothing)=0$, where $\varnothing$ denotes the empty set. Let μ be the product measure induced on 2^N by N. Of course μ is merely the product-space version of Lebesgue measure on the unit interval [0, 1] of the reals [10, p. 157]. What we need of measure theory can be found in Halmos [10].

Let $R(T, x, y)$ be a recursive predicate [12] of the set variable T and the natural number variables x and y. Thus the truth-value of $R(T, x, y)$ for any actual choice of $T\in 2^N$ and $x, y\in N$ is determined by finitely much information about the membership of T and the values of x and y. A familiar uniformity can be put as follows:

$$(T)[(x)(Ey)R(T, x, y)\rightarrow (Ef)(f \text{ recursive in } T \ \&\ (x)R(T, x, f(x)))].$$

(In fact, $f(x)\simeq \mu y R(T, x, y)$.) Another way of expressing this uniformity is: for each T, if the predicate $(x)(Ey)R(T, x, y)$ has a Skolem function, then it has a Skolem function recursive in T. Thus in our study of the predicate $(x)(Ey)R(T, x, y)$, we can reduce our attention from arbitrary Skolem functions to Skolem functions recursive in T. It is now quite natural to ask: can we reduce our attention still further to recursive Skolem functions? The answer is yes, if we ignore a set of T's of measure 0, and if we replace Skolem functions by *bounding functions*: f is a bounding function for $(x)(Ey)R(T, x, y)$ if $(x)(Ey)_{y \leqq f(x)}R(T, x, y)$. The existence of a bounding function is equivalent to the existence of a Skolem function.

PROPOSITION 1.1. *Let $R(T, x, y)$ be recursive. If the set of all T such that $(x)(Ey)R(T, x, y)$ has measure 1, then the set of all T such that $(Ef)[f$ recursive & $(x)(Ey)_{y \leqq f(x)}R(T, x, y)]$ also has measure 1.*

Proof. It is enough to show: for each rational $\delta > 0$, there exists a recursive f such that $\{T \mid (x)(Ey)_{y \leqq f(x)}R(T, x, y)\}$ has measure at least $1-\delta$. A nonempty, basic open subset of 2^N consists of all T satisfying a consistent conjunction of finitely many atomic conditions of the form $m \in T$ or their negations. The measure of a nonempty, basic open set is 2^{-n}, where n is the number of distinct atomic conditions that define the basic open set. Thus $\mu(\{T \mid m \in T \;\&\; m+1 \notin T\}) = 1/4$. Since R is recursive, there exists a recursive function $b_i^{m,n}$ such that for each m and n, $b_i^{m,n}$ is a basic open subset of 2^N, and such that for all T,

$$(m)(n)[R(T, m, n) \leftrightarrow (Ei)(T \in b_i^{m,n})].$$

For each m, let

$$f(m) = \mu t[\mu(\bigcup \{b_i^{m,n} \mid i \leqq t \;\&\; n \leqq t\}) \geqq 1 - \delta/2^{m+1}].$$

For each m, $f(m)$ is defined because $\mu(\bigcup \{b_i^{m,n} \mid i \geqq 0 \;\&\; n \geqq 0\}) = 1$. (Read "$\mu t$" as "the least t such that".)

A stronger and more attractive proposition than 1.1, and one closer to the idea of measure-theoretic uniformity would be: the set of all T such that

$$(x)(Ey)R(T, x, y) \rightarrow (Ef)[f \text{ recursive } \& \; (x)(Ey)_{y \leqq f(x)}R(T, x, y)]$$

has measure 1. Unfortunately, D. A. Martin has shown that for some recursive R this stronger proposition is false. We have to go to the language of arithmetic to realize the full flavor of measure-theoretic uniformity. Theorem 2.2 of §2 states: let $B(T, x, y)$ be arithmetic [12]; then the set of all T such that

$$(x)(Ey)B(T, x, y) \rightarrow (Ef)[f \text{ is arithmetic } \& \; (x)(Ey)_{y \leqq f(x)}B(T, x, y)]$$

has measure 1. In short, if we restrict T to a set of measure 1, then we can restrict ourselves, in our study of arithmetic predicates of T, to arithmetic bounding functions. The reduction *from* bounding functions arithmetic in T *to* arithmetic bounding functions is the essence of measure-theoretic uniformity in the arithmetic case.

2. **The arithmetic hierarchy.** Let $B(T)$ be an arithmetic predicate of T. The set $\hat{T}B(T)$ is a Borel subset of 2^N of finite rank and consequently is measurable. Let r

be a variable ranging over the rational numbers in the unit interval [0, 1]. By making use of some standard Gödel numbering, it is possible to regard the predicate $\mu(\hat{T}B(T)) \leqq r$ as a number-theoretic predicate whose free variables are B (ranging over the arithmetic predicates) and r (ranging over the rationals in [0, 1]). For each $n > 0$, an arithmetic predicate is said to belong to Σ_n^0 (Π_n^0) if it is expressible by means of a recursive matrix and a quantifier prefix beginning with an existential (universal) quantifier and passing through $n-1$ alternations of quantifiers. Recursive predicates are said to belong to both Σ_0^0 and Π_0^0.

LEMMA 2.1. (i) *For each* $n \geqq 0$: *the predicate* $\mu(\hat{T}B(T)) \leqq r$, *restricted to* $B \in \Sigma_n^0$, *is* Π_n^0.

(ii) *For each* $n \geqq 0$: *the predicate* $\mu(\hat{T}B(T)) \leqq r$, *restricted to* $B \in \Pi_n^0$, *is* Π_{n+1}^0.

(iii) *The predicate* $\mu(\hat{T}B(T)) \leqq r$, *where B is arithmetic and otherwise unrestricted, is hyperarithmetic.*

Proof. (i) and (ii) are proved simultaneously by induction on n. First let $n=0$. Then $B(T)$ is a recursive predicate, and there exists a finite union $\bigcup \{b_i \mid i<k\}$ of basic open subsets of 2^N such that

$$B(T) \leftrightarrow (Ei)_{i<k}(T \in b_i)$$

for all T; furthermore, the finite union $\bigcup \{b_i \mid i<k\}$ can be found effectively from a Gödel number of $B(T)$. Clearly, $\mu(\hat{T}B(T)) = \mu(\bigcup \{b_i \mid i<k\})$. Now suppose $n>0$ and $B(T) \in \Sigma_n^0$. Then $B(T)$ is of the form $(Ex)B(T, x)$, where $B(T, x) \in \Pi_{n-1}^0$, and

$$\mu(\hat{T}B(T)) \leqq r \leftrightarrow (m)[\mu(\hat{T}((Ex)_{x \leqq m}B(T, x))) \leqq r].$$

The inductive hypothesis for (ii) implies that the right side of the above formula is Π_n^0. Suppose instead that $B(T) \in \Pi_n^0$. Then $B(T)$ is of the form $(x)B(T, x)$, where $B(T, x) \in \Sigma_{n-1}^0$, and

$$\mu(\hat{T}B(T)) \leqq r \leftrightarrow (\delta)(Em)[\mu(\hat{T}((x)_{x \leqq m}B(T, x))) \leqq r+\delta],$$

where δ is a variable ranging over the positive rationals. The inductive hypothesis for (i) implies that the right side of this last formula is Π_{n+1}^0.

To see (iii) observe that the "effective" nature of the inductive step of the proof of (i) and (ii) implies the existence of a recursive function f with the following property: for each $n \geqq 0$, $f(n)$ is the Gödel number of a Π_n^0 predicate $P_{f(n)}(b, r)$ such that

$$P_{f(n)}(b, r) \leftrightarrow b \text{ is the Gödel number of } B(T) \;\&\; B(T) \in \Sigma_n^0 \;\&\; \mu(\hat{T}B(T)) \leqq r.$$

The predicate $P_{f(n)}(b, r)$, with n, b and r as free variables, is hyperarithmetic, since the truth set for arithmetic is hyperarithmetic.

In Theorem 2.2 we attempt to capture what we mean by measure-theoretic uniformity in the arithmetic case. A standard uniformity for arithmetic predicates of the form $B(T, x, y)$ can be put as follows: if $(x)(Ey)B(T, x, y)$, then $(Ef)[f$ is arithmetic in T & $(x)(Ey)_{y \leqq f(x)}B(T, x, y)$]. The corresponding measure-theoretic uniformity is obtained by demanding that the bounding function f be arithmetic rather than arithmetic in T. The cost of such a demand is a set of T's of measure 0.

THEOREM 2.2. *Let $B(T, x, y)$ be arithmetic. Then the set of all T's satisfying the following condition has measure* 1: *if $(x)(Ey)B(T, x, y)$, then $(Ef)[f$ is arithmetic & $(x)(Ey)_{y \leqq f(x)}B(T, x, y)]$.*

Proof. Let δ be a positive rational. We define an arithmetic f such that

$$\mu(\hat{T}(x)(Ey)_{y \leqq f(x)}B(T, x, y)) + \delta \geqq \mu(\hat{T}(x)(Ey)B(T, x, y)).$$

Let $b(x, r)$ be the predicate: $\mu(\hat{T}(Ey)B(T, x, y)) > r$. By Lemma 2.1, $b(x, r)$ is arithmetic. Then $f(x)$ is the least j such that for all r,

$$b(x, r) \to \mu(\hat{T}(Ey)_{y \leqq j}B(T, x, y)) > r - \delta/2^{x+1}.$$

COROLLARY 2.3. *If $B(T)$ is arithmetic and the set $\hat{T}B(T)$ has positive measure, then $B(A)$ holds for some arithmetic A*(3).

Proof. First we consider a special case, where $B(T)$ is Π_2^0, and then we show how to reduce the general case to the special case. Let $B(T)$ be $(x)(Ey)R(T, x, y)$, where $R(T, x, y)$ is recursive. By Theorem 2.2, there is an arithmetic function f such that

$$\mu(\hat{T}(x)(Ey)_{y \leqq f(x)}R(T, x, y)) > 0.$$

There exists an arithmetic function c_x such that for each x, c_x is a finite union of basic open subsets of 2^N and

$$(Ey)_{y \leqq f(x)}R(T, x, y) \leftrightarrow T \in c_x.$$

For each x, let $d_x = \bigcap \{c_i \mid i \leqq x\}$. Note that for each x, the closed set d_x is nonempty, because $\mu(\hat{T}B(T)) > 0$. If A is an arithmetic set such that $(x)(A \in d_x)$, then $B(A)$ holds. We define such an A by an induction on n:

$$n \in A \leftrightarrow \mu(\hat{T}((x)(T \in d_x) \ \& \ (i)_{i<n}(i \in T \leftrightarrow i \in A) \ \& \ n \in T)) > 0.$$

A is arithmetical, since it is arithmetic in the function d_x.

Now we show how to reduce the general case to the special case by considering an illustrative example. Let $B(T)$ be $(x)(Ey)(u)(Ev)R(T, x, y, u, v)$, where $R(T, x, y, u, v)$ is recursive. By Theorem 2.2, there is an arithmetic f such that

$$\hat{T}(x)(Ey)_{y \leqq f(x)}(u)(Ev)R(T, x, y, u, v)$$

has positive measure. By using a trick of Kleene [12], we can pass a bounded existential quantifier through a universal quantifier:

$$(Ey)_{y \leqq f(x)}(u)(Ev)R(T, x, y, u, v) \leftrightarrow (u)(Ey)_{y \leqq f(x)}(Ev)R(T, x, y, (u)_y, v).$$

(3) Corollary 2.3 was independently obtained by H. Tanaka [34].

By Theorem 2.2, there is an arithmetic g such that the set

$$\hat{T}(x)(u)(Ev)_{v \leqq g(x,u)}(Ey)_{y \leqq f(x)}R(T, x, y, (u)_y, v)$$

has positive measure. But now the argument of the special case can be applied.

Let us say that $B(T)$ holds for almost all T if $\mu(\hat{T}B(T))=1$. Then Corollary 2.3 is equivalent to the statement: if $B(T)$ is arithmetic and holds for all arithmetic T, then it holds for almost all T. It follows immediately that the predicate "T is arithmetic" is not arithmetic. This last result was first proved by Addison [1] by means of a forcing argument.

Corollary 2.3 is an example of what might be called a measure-theoretic basis result. Gandy's basis theorem [7] for arithmetic predicates states: if $B(T)$ is arithmetic and $(ET)B(T)$, then $B(A)$ holds for some A recursive in Kleene's O, the set of all notations for recursive ordinals. (Kleene [13] gives an example of an arithmetic predicate $B(T)$ such that $(ET)B(T)$ but $B(T)$ holds for no hyperarithmetic T.)

Call a set T *fundamental* (with respect to arithmetic) if $B(T)$ holds for every arithmetic B such that $\mu(\hat{T}B(T))=1$. It is possible to construct hyperarithmetic, fundamental (with respect to arithmetic) sets with the help of Lemma 2.1, but we prefer to deduce the existence of such sets as a corollary to Theorem 3.9. They behave very much like the generic (with respect to arithmetic) sets of Feferman [3]. In particular, they are not implicitly arithmetically definable; that is, they are not unique solutions of arithmetic predicates.

3. **The hyperarithmetic hierarchy.** An analytic predicate $P(T, x)$ (Kleene [12]) is said to be Π_1^1 if it is expressible in the form $(Y)A(Y, T, x)$, where $A(Y, T, x)$ is arithmetic. Let $P(T, x, y)$ be Π_1^1. A familiar uniformity due to Kreisel [16] can be put as follows: if $(x)(Ey)P(T, x, y)$, then

$$(Ef)[f \text{ is hyperarithmetic in } T \;\&\; (x)(Ey)_{y \leqq f(x)}P(T, x, y)].$$

The corresponding measure-theoretic uniformity is obtained by first discarding a set of T's of measure 0 and then requiring that the bounding function f be hyperarithmetic rather than hyperarithmetic in T. We develop this result (Corollary 3.12) and several related ones by studying a relativization of Kleene's ramified analytic hierarchy [14]. Very briefly, the ramified analytic hierarchy is obtained by starting with the empty set and iterating the process of analytic definability through the recursive ordinals(4). Kleene [14] showed that the ramified analytic hierachy is identical with the set of hyperarithmetic sets. Following Feferman [3], we relativize by starting with an arbitrary set T rather than the empty set(5).

(4) Many persons now use the term "ramified analytic hierarchy" to denote the result of iterating the process of analytic definability past the recursive ordinals to its endpoint β_0.

(5) It might seem more natural to relativize by replacing not only the empty set by T but also the recursive ordinals by the ordinals recursive in T; however, it follows from Lemma 3.5 that this relativization is identical with the one we adopted for almost all T.

The relativization is described by a ramified, second-order language $\mathscr{L}(\mathscr{T})$, which differs little from Feferman's language $\mathscr{L}^*(\mathscr{S})$ ([3, p. 335]). $\mathscr{L}(\mathscr{T})$ is the language of first-order number theory augmented by the constant symbol $\mathscr{T}$ denoting an arbitrary set of natural numbers, some second-order set variables, and the membership symbol ($\in$). Let O_1 be a Π_1^1 subset of O such that O_1 is linearly ordered by $<_O$ and has order-type ω_1 (the least nonrecursive ordinal). (The existence of O_1 is proved in [4] and in [6].) Thus each recursive ordinal has just one notation in O_1; if b is the unique notation in O_1 for the recursive ordinal β, we write $|b|=\beta$. For each $b \in O_1$, $\mathscr{L}(\mathscr{T})$ has ranked set variables X^b, $Y^b, Z^b, \ldots$; $\mathscr{L}(\mathscr{T})$ also has unranked set variables X, $Y, Z, \ldots$, a numeral $\bar{n}$ for each natural number n, and symbols for equality ($=$), successor ($'$), addition ($+$), and multiplication ($\cdot$).

A formula $\mathscr{F}$ of $\mathscr{L}(\mathscr{T})$ is said to be ranked if every set variable occurring in $\mathscr{F}$ is ranked. The *ordinal rank* of a ranked sentence $\mathscr{F}$ is the least α such that $\alpha > |b|$ for every variable X^b occurring in $\mathscr{F}$. Let T be an arbitrary set of natural numbers. Following Feferman [3], for each $b \in O_1$, we inductively define a structure $\mathscr{M}_b(T)$ and truth in the structure $\bigcup \{\mathscr{M}_a(T) \mid |a| < |b|\}$:

(i) A sentence $\mathscr{F}$ of ordinal rank $\leqq |b|$ is true in $\bigcup \{\mathscr{M}_a(T) \mid |a| < |b|\}$ if it is true when $\mathscr{T}$ is interpreted as T, the number variables of $\mathscr{F}$ are restricted to ω, and each set variable X^a of $\mathscr{F}$ is restricted to $\mathscr{M}_a(T)$.

(ii) For each formula $\mathscr{G}(x)$ (with only x free) of ordinal rank $\leqq |b|$, let $\hat{x}\mathscr{G}(x)$ denote the set $\{n \mid \mathscr{G}(\bar{n})$ is true in $\bigcup \{\mathscr{M}_a(T) \mid |a| < |b|\}\}$ $\mathscr{M}_b(T)$ consists of all sets of natural numbers defined in this manner.

We define $\mathscr{M}(T)=\bigcup \{\mathscr{M}_b(T) \mid b \in O_1\}$. Let $\mathscr{M}=\mathscr{M}(\phi)=$the set of all hyperarithmetic sets. A sentence $\mathscr{F}$ of $\mathscr{L}(\mathscr{T})$ is true (in symbols, $\mathscr{M}(T) \models \mathscr{F}$) if it is true when each unranked variable is restricted to $\mathscr{M}(T)$ and the remaining symbols of $\mathscr{F}$ are interpreted according to (i) and (ii). It is a routine matter to choose a Gödel numbering for the formulas of $\mathscr{L}(\mathscr{T})$ with the following properties: the predicate "e is the Gödel number of a ranked formula of $\mathscr{L}(\mathscr{T})$" is Π_1^1; for each $b \in O_1$, the predicate "e is the Gödel number of a ranked formula of $\mathscr{L}(\mathscr{T})$ of ordinal rank $< |b|$" is recursively enumerable (uniformly in b). The assignment of Gödel numbers makes it possible to classify various relations involving the formulas of $\mathscr{L}(\mathscr{T})$. We will occasionally blur the distinction between formulas and Gödel numbers of formulas.

The *full ordinal rank* of a ranked formula $\mathscr{F}$ is a function f: $O_1 \to \omega$ such that for each $b \in O_1, f(b)$ is the number of occurrences of (X^b), $(EY^b), \ldots$ in $\mathscr{F}$. If f and g are full ordinal ranks, then $f <_r g$ if

$$(Ec)[c \in O_1 \ \& f(c) < g(c) \ \& \ (d)(c <_{O_1} d \to f(d) = g(d))].$$

The relation $<_r$ is a well-ordering of the set of full ordinal ranks.

If $\mathscr{G}(x)$ is a ranked formula (with only x free), then $\mathscr{F}(\hat{x}\mathscr{G}(x))$ denotes the result of replacing each occurrence of $t \in Y$ in $\mathscr{F}(Y)$ by $\mathscr{G}(t)$. (t is a number-theoretic term.)

PROPOSITION 3.1. *Let $\mathscr{F}(Y^b)$ be a ranked formula in prenex normal form whose only free variable is Y^b. Let $\{\mathscr{G}_i(x) \mid i \leqq n\}$ be a finite sequence of formulas with only x free of ordinal rank $\leqq |b|$. Then the disjunction $\bigvee_{i \leqq n} \mathscr{F}(\hat{x}\mathscr{G}_i(x))$ is equivalent to a prenex normal sentence of full ordinal rank less than that of $(EY^b)\mathscr{F}(Y^b)$.*

Proof. By illustrative example. Let $\mathscr{F}(Y^b)$ be

$$(EX^c)(Y^d)(EZ^b)R(X^c, Y^d, Z^b, Y^b),$$

where R contains no quantifiers, and let $n=2$. The following formulas are equivalent:

$$\bigvee_{i \leqq 1} (EX^c)(Y^d)(EZ^b)R(X^c, Y^d, Z^b, \hat{x}\mathscr{G}_i(x));$$

$$(EX^c)\Big[\bigvee_{i \leqq 1} (Y^d)(EZ^b)R(X^c, Y^d, Z^b, \hat{x}\mathscr{G}_i(x))\Big];$$

$$(EX^c)(Y_0^d)(Y_1^d)\Big[\bigvee_{i \leqq 1} (EZ^b)R(X^c, Y_i^d, Z^b, \hat{x}\mathscr{G}_i(x))\Big];$$

$$(EX^c)(Y_0^d)(Y_1^d)(EZ^b)\Big[\bigvee_{i \leqq 1} R(X^c, Y_i^d, Z^b, \hat{x}\mathscr{G}_i(x))\Big].$$

The block of quantifiers $(Y_0^d)(Y_1^d)$ can be replaced by a single quantifier (Y^d) by means of a standard trick: each occurrence of $t \in Y_0^d (t \in Y_1^d)$ is replaced by $2t \in Y^d (2t+1 \in Y^d)$. The formula

$$\bigvee_{i \leqq 1} R(X^c, Y_i^d, Z^b, \hat{x}\mathscr{G}_i(x))$$

is equivalent to a prenex normal formula $Q(X^c, Y^d, Z^b)$ such that all of its quantifiers have superscripts which are notations for ordinals $<|b|$. Clearly,

$$(EY^b)(EX^c)(Y^d)(EZ^b)R(X^c, Y^d, Z^b, Y^b)$$

has greater full ordinal rank than

$$(EX^c)(Y^d)(EZ^b)Q(X^c, Y^d, Z^b).$$

For each sentence $\mathscr{F}$ of $\mathscr{L}(\mathscr{T})$, we define $p(\mathscr{F})$, the probability that $\mathscr{F}$ is true in $\mathscr{M}(T)$, to be the Lebesgue measure of $\{T \mid \mathscr{M}(T) \models \mathscr{F}\}$. (It is easily seen by transfinite induction that $\{T \mid \mathscr{M}(T) \models \mathscr{F}\}$ is a Borel subset of 2^N.)

LEMMA 3.2. *The predicate $p(\mathscr{F}) \geqq r$, restricted to ranked $\mathscr{F}$ and rational r, is Π_1^1.*

Proof. Spector observed that the intersection of all sets satisfying a Σ_1^1 condition is Π_1^1; i.e., if $A(X)$ is Σ_1^1, then $\bigcap \{X \mid A(X)\}$ is Π_1^1. We give an inductive definition of $p(\mathscr{F}) \geqq r$, restricted to ranked, *prenex normal* $\mathscr{F}$ and rational r, and then we note that the cases of our inductive definition can be expressed by means of Σ_1^1 closure conditions. The number-quantifier rank of $\mathscr{F}$ is the number of occurrences of $(x), (Ey), \ldots$ in $\mathscr{F}$. The rank of $\mathscr{F}$ is the ordered pair (f, m), where f is the full ordinal rank of $\mathscr{F}$ and m is the number-quantifier rank of $\mathscr{F}$. We say $(f, m) < (g, n)$

if $f <_r g$ or if $f = g$ and $m < n$. The inductive definition of $p(\mathscr{F}) \geqq r$ proceeds according to the rank of $\mathscr{F}$.

Case 0. $\mathscr{F}$ has no quantifiers. The predicate $p(\mathscr{F}) \geqq r$, restricted to $\mathscr{F}$ without quantifiers and rational r, is recursive; i.e., $p(\bar{m} \in \mathscr{T}) = \frac{1}{2}$ etc.

Case 1. $\mathscr{F}$ is of the form $(EX^b)\mathscr{F}_1(X^b)$. Let $\{\mathscr{G}_i(x) \mid i \geqq 0\}$ be the uniform (in b) recursive enumeration of formulas with only x free of ordinal rank $\leqq |b|$. Then $p(\mathscr{F}) \geqq r$ is defined to be

$$(\delta)(Em)\Big[p\Big(\bigvee_{i \leqq m} \mathscr{F}_1(\hat{x}\mathscr{G}_i(x))\Big) \geqq r - \delta\Big],$$

where δ is restricted to the positive rationals. By 3.1, $\bigvee_{i \leqq m} \mathscr{F}_1(\hat{x}\mathscr{G}_i(x))$ is equivalent to a prenex normal sentence of full ordinal rank less than that of $(EX^b)\mathscr{F}_1(X^b)$.

Case 2. $\mathscr{F}$ is of the form $(X^b)\mathscr{F}_1(X^b)$. Then $p(\mathscr{F}) \geqq r$ is defined to be

$$(m)\Big[p\Big(\bigwedge_{i \leqq m} \mathscr{F}_1(\hat{x}\mathscr{G}_i(x))\Big) \geqq r\Big].$$

By the dual of 3.1, $\bigwedge_{i \leqq m} \mathscr{F}_1(\hat{x}\mathscr{G}_i(x))$ is equivalent to a prenex normal sentence of full ordinal rank less than that of $(X^b)\mathscr{F}_1(X^b)$.

Case 3. $\mathscr{F}$ is of the form $(Ex)\mathscr{F}_1(x)$. Then $p(\mathscr{F}) \geqq r$ is defined to be

$$(\delta)(Em)\Big[p\Big(\bigvee_{i \leqq m} \mathscr{F}_1(\bar{i})\Big) \geqq r - \delta\Big],$$

where δ is restricted to the positive rationals. By the same argument used to establish 3.1, $\bigvee_{i \leqq m} \mathscr{F}_1(\bar{i})$ is equivalent to a prenex normal sentence of lower rank than that of $(Ex)\mathscr{F}_1(x)$; in fact, the full ordinal rank will be unchanged but the number-quantifier rank will be less.

Case 4. $\mathscr{F}$ is of the form $(x)\mathscr{F}_1(x)$. Dual to Case 3.

Case 2 can be rephrased as a Σ_1^1 closure condition as follows: If $\mathscr{F}$ is a ranked, prenex normal sentence of the form $(X^b)\mathscr{F}_1(X^b)$ and $(m)[p(\bigvee_{i \leqq m} \mathscr{F}_1(\hat{x}\mathscr{G}_i(x))) \geqq r]$, where $\{\mathscr{G}_i(x) \mid i \geqq 0\}$ is the uniform (in b) recursive enumeration of formulas with only x free of rank $\leqq |b|$, then $p(\mathscr{F}) \geqq r$. The hypothesis of Case 2 is Π_1^1 in variables $\mathscr{F}$ and r.

LEMMA 3.3. *Let $\mathscr{F}(x, Y)$ be a formula of $\mathscr{L}(\mathscr{T})$ whose only free variables are x and Y, and whose only unranked variable is Y. Then for each rational r, there is a $c \in O_1$ such that*

$$p((x)(EY)\mathscr{F}(x, Y)) \geqq r \rightarrow p((x)(EY^c)\mathscr{F}(x, Y^c)) \geqq r.\text{[6]}$$

Proof. Fix r and suppose $p((x)(EY)\mathscr{F}(x, Y)) \geqq r$. Define $(Y)_n$ to be $\{i \mid p_n^{1+i} \in Y\}$, where p_n is the nth smallest prime ($p_0 = 2, p_1 = 3, \ldots$). Then

$$(m)\Big[p\Big((EY)\Big(\bigwedge_{n \leqq m} \mathscr{F}(\bar{n}, (Y)_n)\Big)\Big) \geqq r\Big].$$

(6) It is not possible in general to improve 3.3 by finding a c that is independent of r.

It follows that

$$(m)(\delta)(Ec)\Big[c \in O_1 \;\&\; p\Big((EY^c)\Big(\bigwedge_{n \leqq m} \mathscr{F}(\bar{n}, (Y^c)_n)\Big)\Big) \geqq r-\delta\Big],$$

where δ is restricted to the positive rationals. It follows from 3.2 that the predicate

$$\Big[c \in O_1 \;\&\; p\Big((EY^c)\Big(\bigwedge_{n \leqq m} \mathscr{F}(\bar{n}, (Y^c)_n)\Big)\Big) \geqq r-\delta\Big]$$

is Π_1^1 in c, m and δ. By Kreisel's Lemma [16, p. 307], there exists a hyperarithmetic function $c(m, \delta)$ such that

$$(m)(\delta)\Big[p\Big((EY^{c(m,\delta)})\Big(\bigwedge_{n \leqq m} \mathscr{F}(\bar{n}, (Y^{c(m,\delta)})_n)\Big)\Big) \geqq r-\delta\Big].$$

By Spector [31] there is a $c \in O_1$ such that $(m)(\delta)(|c(m, \delta)| \leqq |c|)$. But then

$$(m)(\delta)\Big[p\Big((EY^c)\Big(\bigwedge_{n \leqq m} \mathscr{F}(\bar{n}, (Y^c)_n)\Big)\Big) \geqq r-\delta\Big],$$

and consequently, $p((x)(EY^c)\mathscr{F}(x, Y^c)) \geqq r$.

By an instance of the hyperarithmetic comprehension axiom, Kreisel [16] means the universal closure of any formula of the form

$$(x)[(EY)A(x, Y) \leftrightarrow (Z)B(x, Z)] \rightarrow (EX)(x)[x \in X \leftrightarrow (EY)A(x, Y)],$$

where $A(x, Y)$ and $B(x, Z)$ are arithmetic predicates that may contain free set variables other than Y and Z.

Lemma 3.4. *If $\mathscr{F}$ is an instance of the hyperarithmetic comprehension axiom, then* $p(\mathscr{F})=1$.

Proof. Let $A(x, Y)$ and $B(x, Y)$ be formulas of $\mathscr{L}(\mathscr{T})$ whose only free variables are x and Y and whose only unranked variable is Y. Let K be the set of all T such that $(x)[(EY)A(x, Y) \leftrightarrow (Z)B(x, Z)]$ is true in $\mathscr{M}(T)$. Then $(x)(EY)[A(x, Y) \vee \sim B(x, Y)]$ is true in $\mathscr{M}(T)$ for all $T \in K$. Fix $\delta > 0$. By 3.3 there is a $c \in O_1$ such that $(x)(EY^c)[A(x, Y^c) \vee \sim B(x, Y^c)]$ is true in $\mathscr{M}(T)$ for all $T \in J \subseteq K$, where $\mu(K-J) < \delta$. But then $(x)[(EY^c)A(x, Y^c) \leftrightarrow (EY)A(x, Y)]$ holds in $\mathscr{M}(T)$ for all $T \in J$. Since $(EY^c)A(x, Y^c)$ is a ranked formula with only x free, it follows that $(EX)(x)[x \in X \leftrightarrow (EY)A(x, Y)]$ holds in $\mathscr{M}(T)$ for all $T \in J$.

Lemma 3.4 suggests: if a set of T's of measure 0 is avoided, then $\mathscr{M}(T)$ is very similar to $\mathscr{M}$. The similarity is pursued further in Lemma 3.5, and is exploited to prove 3.7. The underlying cause of the similarity is expressed by Lemma 3.2, which says that the probability that a ranked sentence $\mathscr{F}$ is true in $\mathscr{M}(T)$ can be "computed" in $\mathscr{M}$. The use of the term "computed" is appropriate for two closely related reasons: the predicate $p(\mathscr{F}) \geqq r$, restricted to ranked $\mathscr{F}$ and rational r, is Σ_1^1 in the sense of $\mathscr{M}$ (i.e., the existential set quantifier is restricted to the sets of $\mathscr{M}$); and the predicate is metarecursive [17] as well.

LEMMA 3.5. *For almost all T, $\mathcal{M}(T)$ is the set of all sets hyperarithmetic in T*(7).

Proof. The argument of Kleene [14, p. 35], relativized to T, shows that the sets of $\mathcal{M}(T)$ are just those sets each of which is recursive in H_b^T for some $b \in O_1 \subseteq O^T$(8). According to Kreisel [15], if T belongs to an ω-model of the hyperarithmetic comprehension axiom, then every set hyperarithmetic in T also belongs to that ω-model. Now apply 3.4.

PROPOSITION 3.6. *Let $\{\mathscr{F}_i\}$ be a hyperarithmetic sequence of ranked sentences of $\mathscr{L}(\mathscr{T})$. Then there is a ranked sentence $\mathscr{F}$ such that for all T, $\mathcal{M}(T) \models \mathscr{F}$ if and only if $(i)[\mathcal{M}(T) \models \mathscr{F}_i]$.*

Proof. Let H be a hyperarithmetic set such that $n \in H \leftrightarrow n$ is the Gödel number of a member of $\{\mathscr{F}_i\}$. By Spector [31] there is a $c \in O_1$ such that the ordinal rank of $\mathscr{F}_i$ is $\leqq |c|$ for all i. Let V_c^T be the set of Gödel numbers of all formulas of ordinal rank $\leqq |c|$ which are true in $\mathcal{M}(T)$. By adapting the argument of Kleene [14, p. 36], one can show that V_c^T is recursive in $H_{c'}^T$ for some c' in $O_1 \subseteq O^T$ (see footnote (8)). Clearly, $(i)(i \in H \rightarrow i \in V_c^T)$ is equivalent to $(i)[\mathcal{M}(T) \models \mathscr{F}_i]$. Again by adapting the arguments of Kleene [14, p. 35], one can show the existence of ranked formulas $\mathscr{G}_1(x)$, $\mathscr{G}_2(x)$, and $\mathscr{G}_2(x)$ of $\mathscr{L}(\mathscr{T})$ such that for *all* T:

$$H = \{n \mid \mathcal{M}(T) \models \mathscr{G}_1(\bar{n})\};$$
$$H_{c'}^T = \{n \mid \mathcal{M}(T) \models \mathscr{G}_2(\bar{n})\};$$
$$V_c^T = \{n \mid \mathcal{M}(T) \models \mathscr{G}_3(\bar{n})\}.$$

The desired $\mathscr{F}$ is $(x)(\mathscr{G}_1(x) \rightarrow \mathscr{G}_3(x))$.

The Gödel number of $\mathscr{F}$ can be found effectively from the Gödel number of $\{\mathscr{F}_i\}$, since one can pass effectively from the Gödel number of $\{\mathscr{F}_i\}$ to c, c', and the Gödel numbers of $\mathscr{G}_1$ and $\mathscr{G}_3$.

LEMMA 3.7. *For each ranked sentence $\mathscr{F}$ and rational $\delta > 0$, there exists a ranked sentence $\mathscr{G}$ such that $p(\mathscr{F} \mathbin{\&} \sim\mathscr{G}) < \delta$ and such that $\{T \mid \mathcal{M}(T) \models \mathscr{G}\}$ is a closed subset of $\{T \mid \mathcal{M}(T) \models \mathscr{F}\}$.*

Proof. Let $P(\mathscr{F}, \delta, \mathscr{G})$ denote the following number-theoretic predicate: $\mathscr{F}$ is a ranked sentence, δ is a positive rational, $\mathscr{G}$ is a ranked sentence, $p(\mathscr{F} \mathbin{\&} \sim\mathscr{G}) < \delta$, and $\{T \mid \mathcal{M}(T) \models \mathscr{G}\}$ is a closed subset of $\{T \mid \mathcal{M}(T) \models \mathscr{F}\}$. With the help of 3.2 and the fact that $\mathcal{M}(T) \models \mathscr{F}$, restricted to ranked $\mathscr{F}$, is Π_1^1, it is not hard to check that $P(\mathscr{F}, \delta, \mathscr{G})$ is Π_1^1.

We prove by induction on the rank of $\mathscr{F}$ that $(\delta)(E\mathscr{G})P(\mathscr{F}, \delta, \mathscr{G})$. The cases are the same as those of 3.2. (We again assume that $\mathscr{F}$ is in prenex normal form.) The most

(7) Feferman [3] showed by means of a forcing argument with finite conditions that for all T outside a first category subset of 2^N, $\mathcal{M}(T)$ is the set of all sets hyperarithmetic in T.

(8) There is no harm in thinking of O_1 as a subset of O^T, since there exists a recursive function f with the following properties: for all T, $a <_{O_1} b \leftrightarrow f(a) <_{O^T} f(b)$, and $a \in O_1 \rightarrow |a| = |f(a)|_{O^T}$.

interesting case is when $\mathscr{F}$ is of the form $(X^b)\mathscr{F}_1(X^b)$. Let $\{\mathscr{G}_i(x) \mid i \geqq 0\}$ be the uniform (in b) recursive enumeration of formulas (with only x free) of ordinal rank $\leqq |b|$. For each i, $\mathscr{H}_1(\hat{x}\mathscr{G}_i(x))$ has lower full ordinal rank than $(X^b)\mathscr{F}_1(X^b)$. So by the induction hypothesis,

$$(i)(\delta)(E\mathscr{H})P(\mathscr{F}_1(\hat{x}\mathscr{G}_i(x)), \delta/2^{i+1}, \mathscr{H}).$$

Since P is Π_1^1, it follows from Kreisel's Lemma [16, p. 307] that there exists a hyperarithmetic sequence $\{\mathscr{H}_i \mid i \geqq 0\}$ of ranked formulas such that

$$(i)P(\mathscr{F}_1(\hat{x}\mathscr{G}_i(x)), \delta/2^{i+1}, \mathscr{H}_i).$$

By 3.6, there is a ranked formula $\mathscr{H}$ equivalent to the "infinite conjunction" of the $\mathscr{H}_i$'s. Then $P(\mathscr{F}, \delta, \mathscr{H})$.

LEMMA 3.8. *If $\mathscr{F}$ is a ranked sentence of $\mathscr{L}(\mathscr{T})$ and $p(\mathscr{F})>0$, then $\mathscr{M}(T) \models \mathscr{F}$ for some hyperarithmetic T.*

Proof. By 3.7 it is safe to assume that $\{T \mid \mathscr{M}(T) \models \mathscr{F}\}$ is a closed subset of 2^N. We inductively define a sequence $\{\mathscr{G}_n \mid n \geqq 0\}$ of atomic sentences:

$$\begin{aligned} \mathscr{G}_n &= \bar{n} \in \mathscr{T} \quad \text{if } p(\mathscr{F} \;\&\; (i)_{i<n}\mathscr{G}_i \;\&\; \bar{n} \in \mathscr{T}) > 0 \\ &= \bar{n} \notin \mathscr{T} \quad \text{otherwise.} \end{aligned}$$

By 3.2 the set $H=\{n \mid \mathscr{G}_n \models \bar{n} \in \mathscr{T}\}$ is hyperarithmetic. Clearly, $\mathscr{M}(H) \models \mathscr{F}$.

THEOREM 3.9. *If $P(X)$ is Π_1^1 and the Lebesgue measure of $\hat{X}P(X)$ is positive, then $P(H)$ holds for some hyperarithmetic H*(9).

Proof. By Gandy [6] or Spector [33], there is an arithmetic predicate $\mathscr{A}(X, Y)$ such that

$$(X)[P(X) \leftrightarrow (EY)(Y \text{ is hyperarithmetic in } X \;\&\; \mathscr{A}(X, Y)].$$

It follows from 3.5 that for almost all T,

$$[\mathscr{M}(T) \models (EY)\mathscr{A}(\mathscr{T}, Y)] \leftrightarrow P(T);$$

consequently,

$$p((EY)\mathscr{A}(\mathscr{T}, Y)) = \mu(\hat{X}P(X)) > 0.$$

By 3.8, $\mathscr{M}(H_1) \models (EY)\mathscr{A}(\mathscr{T}, Y)$ for some hyperarithmetic H_1. Since every member of $\mathscr{M}(H_1)$ is hyperarithmetic in H_1, there must be a hyperarithmetic $H \in \mathscr{M}(H_1)$ such that $\mathscr{A}(H_1, H)$, and hence $P(H)$, hold.

Theorem 3.9 is another example of what might be called a measure-theoretic basis theorem. Shoenfield's basis result [27] for Π_1^1 predicates states: if $P(X)$ is Π_1^1 and $(EX)P(X)$, then $P(C)$ holds for some C constructible in the sense of Gödel [9].

(9) Theorem 3.9 first appeared in Sacks [24]. Subsequently, H. Tanaka [34] independently developed another proof of 3.9 in which Corollary 5 of Sacks [23] (Corollary 3.10 of the present paper) is quoted but which avoids the formalism of the ramified analytic hierarchy.

Theorem 3.9 can be relativized in a reasonable manner. Let $P(A, X)$ be Π_1^1 with A and X as free set variables. For each A, if the Lebesgue measure of $\hat{X}P(A, X)$ is positive, then $P(A, H)$ holds for some H hyperarithmetic in A. It follows that if a complementary analytic set of reals [28] has positive Lebesgue measure, then it contains a real number hyperarithmetic in the determining system [28, p. 34] of its complement.

Another way of expressing Theorem 3.9 is: if $P(X)$ is Π_1^1 and holds for every hyperarithmetic X, then $P(X)$ holds for almost every X. We have suppressed a considerable amount of uniformity inherent in 3.9. It is not difficult to check: there is a partial recursive function f such that if e is the Gödel number of a Π_1^1 predicate $P(X)$ and the Lebesgue measure of $\hat{X}P(X)$ is positive, then $f(e)$ is defined and is the Gödel number of a hyperarithmetic set H such that $P(H)$ holds.

The least ordinal not recursive in X is denoted by ω_1^X, the least nonrecursive ordinal by ω_1.

Corollary 3.10. *The set* $\{X \mid \omega_1^X = \omega_1\}$ *has Lebesgue measure* 1.

Proof. First observe that $\omega_1^X > \omega_1$ is Π_1^1. By Gandy [6] or Feferman-Spector [4] there is a recursive linear ordering $<_R$ of the natural numbers whose maximal well-ordered initial segment has order-type ω_1. Then $\omega_1^X > \omega_1$ is equivalent to the following Π_1^1 predicate: $(Ee)[e$ is the Gödel number of a recursive-in-X well-ordering of the natural numbers $(<_e)]$ & $(f)[f$ is not a 1-1 order-preserving map of $<_e$ onto an initial segment of $<_R]$. Now by Spector [31], $\omega_1^H = \omega_1$ for all hyperarithmetic H. Then by 3.9, $\omega_1^X = \omega_1$ for almost all X.

Corollary 3.11. *If A is not hyperarithmetic, then the set* $\{X \mid A$ *is hyperarithmetic in* $X\}$ *has Lebesgue measure* 0([10]).

Proof. Suppose $\mu(\{X \mid A \leqq_h X\}) > 0$. By 3.5, $\mu(\{T \mid A \in \mathcal{M}(T)\}) > 0$. There must then exist a ranked formula $\mathcal{G}(x)$ such that

$$\mu(\{T \mid A = \{n \mid \mathcal{M}(T) \models \mathcal{G}(\bar{n})\}\}) > 0.$$

Let $K \subseteq \{T \mid A = \{n \mid \mathcal{M}(T) \models \mathcal{G}(\bar{n})\}\}$ be such that $\mu(K) = r > 0$, where r is rational. Let J be a finite union of basic open subsets of 2^N such that $\mu(K \Delta J) < r/3$. $(K \Delta J = (K-J) \cup (J-K).)$ Let $\mathcal{J}$ be a ranked sentence of $\mathcal{L}(\mathcal{T})$ with the property that

$$(T)[T \in J \leftrightarrow \mathcal{M}(T) \models \mathcal{J}].$$

Then

$$n \in A \leftrightarrow p(\mathcal{G}(\bar{n}) \;\&\; \mathcal{J}) \geqq 2r/3,$$

and consequently, A is hyperarithmetic by 3.2.

([10]) S. Kripke has greatly strengthened Corollary 3.11 in the following direction: let $K \subseteq 2^N$ be a set of measure 0 such that $\phi \notin K$ and such that any set of the same hyperdegree as some member of K is also a member of K; then the set of all X such that some $Y \in K$ is hyperarithmetic in X has measure 0.

COROLLARY 3.11′ (CARL JOCKUSCH). *Let* $P(X)$ *be* Π_1^1. *If* $(Y)(Y \in HYP \rightarrow \sim P(Y))$ *then the Lebesgue measure of* $\{X \mid (EY)(P(Y) \;\&\; Y \leq_h X)\}$ *is* 0.

Proof. By 3.9 it is enough to show $(EY)[P(Y) \;\&\; Y \leq_h X]$ is Π_1^1. This last predicate is equivalent to $(Eb)[b \in O^X \;\&\; Y$ is recursive in $H_b^X \;\&\; P(Y)]$.

A set A is said to be implicitly arithmetically definable if it is the unique solution of some arithmetic predicate; i.e., there is an arithmetic $B(X)$ such that $(E_1X)B(X)$ and $B(A)$. It is well known that each hyperarithmetic set is recursive in some implicitly arithmetically definable, hyperarithmetic set. Feferman [3] showed by means of a forcing argument with finite conditions that there exists a hyperarithmetic set which is not implicitly arithmetically definable. The notion of fundamental (with respect to arithmetic) was defined at the end of §2.

COROLLARY 3.12. *There exists a hyperarithmetic set which is fundamental (with respect to arithmetic). Each such set is not implicitly arithmetically definable.*

Proof. A set T is fundamental (with respect to arithmetic) if for every arithmetic predicate $B(X)$,

$$\mu(\hat{X}B(X)) = 1 \rightarrow B(T).$$

By 2.2 (iii) and 3.9, there exists a hyperarithmetic T which is fundamental with respect to arithmetic. If $B(X)$ is arithmetic and $B(T)$ holds, then $\hat{X}B(X)$ must have positive Lebesgue measure.

THEOREM 3.13. *Let* $P(T, x, y)$ *be* Π_1^1. *Then the set of all* T *satisfying the condition below has Lebesgue measure* 1:

$$(x)(Ey)P(T, x, y) \rightarrow (Ef)_{HYP}(x)(Ey)_{y \leq f(x)}P(T, x, y).$$

Proof. By Gandy [6] or Spector [33], there is an arithmetic formula $\mathscr{A}(T, x, y, Y)$ such that for all T, m, and n:

$$P(T, m, n) \leftrightarrow (EY)[Y \leq_h T \;\&\; \mathscr{A}(T, m, n, Y)].$$

By 3.5, we have for almost all T and all m, n:

$$P(T, m, n) \leftrightarrow \mathscr{M}(T) \models (EY)\mathscr{A}(\mathscr{T}, \bar{m}, \bar{n}, Y).$$

Then for almost all T:

$$(x)(Ey)P(T, x, y) \leftrightarrow \mathscr{M}(T) \models (x)(Ey)(EY)\mathscr{A}(\mathscr{T}, x, y, Y).$$

Let δ be a positive rational. We define a hyperarithmetic function f such that the set of all T satisfying the following condition has measure at least $1-\delta$:

$$(x)(Ey)P(T, x, y) \rightarrow (x)(Ey)_{y \leq f(x)}P(T, x, y).$$

Let $b(n, r)$ be the predicate

$$p((Ey)(EY)\mathscr{A}(\mathscr{T}, \bar{n}, y, Y)) > r,$$

where r is rational. By 3.2, $b(n, r)$ is hyperarithmetic. Then $f(n)$ is the least j such that for all r,

$$b(n, r) \rightarrow p((Ey)_{y \leqq j}(EY)\mathscr{A}(\mathscr{T}, n, y, Y)) > r - \delta/2^{n+1}.$$

Theorem 3.13 is the instance of measure-theoretic uniformity we associate with hyperarithmetic theory; by suppressing a set of T's of measure 0, we are able to restrict our attention from bounding functions hyperarithmetic in T (provided by Kreisel's Lemma [16]) to bounding functions that are actually hyperarithmetic. A sharpening of Theorem 3.13 will be useful.

COROLLARY 3.14. *Let $P(T, x, y)$ and $Q(x, y)$ be Π_1^1 formulas such that* $(x)(z)\cdot(Ey)_{z \leqq y}Q(x, y)$. *Then the set of all T satisfying the following condition has Lebesgue measure* 1:

$$(x)(Ey)[P(T, x, y) \ \& \ Q(x, y)] \rightarrow (Ef)_{HYP}(x)(Ey)_{y \leqq f(x)}[P(T, x, y) \ \& \ Q(x, f(x))].$$

Proof. By 3.13, for almost all T there exists a hyperarithmetic g such that

$$(x)(Ey)[P(T, x, y) \ \& \ Q(x, y)] \rightarrow (x)(Ey)_{y \leqq g(x)}[P(T, x, y) \ \& \ Q(x, y)].$$

Since $(x)(z)(Ey)_{z \leqq y}Q(x, y)$, it must be that

$$(x)(Ey)[g(x) \leqq y \ \& \ Q(x, y)].$$

By Kreisel's Lemma [16], there exists a hyperarithmetic f such that

$$(x)[g(x) \leqq f(x) \ \& \ Q(x, f(x))].$$

But then for almost all T there exists a hyperarithmetic f such that

$$(x)(Ey)[P(T, x, y) \ \& \ Q(x, y)] \rightarrow (x)(Ey)_{y \leqq f(x)}[P(T, x, y) \ \& \ Q(x, f(x))].$$

Let $T_0, T_1, T_2, \ldots$ be an arbitrary sequence of sets of natural numbers. We turn our attention to the structure $\mathscr{M}(T_0, T_1, T_2, \ldots)$ introduced by Feferman [3]. The language $\mathscr{L}(\mathscr{T}_0, \mathscr{T}_1, \mathscr{T}_2, \ldots)$ differs from $\mathscr{L}(\mathscr{T})$ (described at the beginning of §3) only in that the set constant $\mathscr{T}$ has been replaced by the set constants $\mathscr{T}_0, \mathscr{T}_1, \mathscr{T}_2, \ldots$; the structure $\mathscr{M}(T_0, T_1, T_2, \ldots)$ is defined in the same manner as $\mathscr{M}(T)$. Each set constant $\mathscr{T}_i$ is of course interpreted as T_i. Each member of $\mathscr{M}_b(T_0, T_1, T_2, \ldots)$ is defined by a formula $\mathscr{G}(x)$ of ordinal rank $\leqq |b|$, where $b \in O_1$. Note that each such formula $\mathscr{G}(x)$ can explicitly mention only finitely many $\mathscr{T}_i$'s; thus it is possible that $\mathscr{M}(T_0, T_1, T_2, \ldots)$ will not, in general, have a member in which all the T_i's are recursive.

We think of $(T_0, T_1, T_2, \ldots)$ as being a member of $(2^N)^N$. Lebesgue measure μ for 2^N was defined in §1 as the product measure induced by the unbiased measure m on each factor $2 = \{0, 1\}$. Lebesgue measure μ for $(2^N)^N$ is defined similarly. If $\mathscr{F}$ is a sentence of $\mathscr{L}(\mathscr{T}_0, \mathscr{T}_1, \mathscr{T}_2, \ldots)$, then the set

$$\{(T_0, T_1, \ldots) \mid \mathscr{M}(T_0, T_1, \ldots) \models \mathscr{F}\}$$

is a Borel subset of $(2^N)^N$ and so is Lebesgue-measurable; let $p(\mathscr{F})$, the probability that $\mathscr{F}$ is true in $\mathscr{M}(T_0, T_1, \ldots)$, be the measure of this set.

Let v and w be recursive functions such that the map $h: N \to N \times N$, defined by $h(n) = (v(n), w(n))$, is one-to-one and onto. Let $x \in (T)_y$ be defined by

$$(En)[n \in T \,\&\, v(n) = x \,\&\, w(n) = y].$$

Then the map $h: 2^N \to (2^N)^N$ defined by $h(T) = ((T)_0, (T)_1, (T)_2, \ldots)$ is one-to-one, onto, and measure-preserving; i.e., if $A \subseteq 2^N$, then A is a Borel subset of 2^N of measure m if and only if $h[A]$ is a Borel subset of $(2^N)^N$ of measure m (see footnote ([12])).

LEMMA 3.15. *The predicate $p(\mathscr{F}) \geq r$, where $\mathscr{F}$ is restricted to the ranked sentences of $\mathscr{L}(\mathscr{T}_0, \mathscr{T}_1, \mathscr{T}_2, \ldots)$ and r is rational, is Π_1^1.*

Proof. Same as 3.2.

By an instance of the Σ_1^1 axiom of choice, Kreisel [16] means the universal closure of any formula of the form:

$$(x)(EY)A(x, Y) \to (EY)(x)A(x, (Y)_x),$$

where $A(x, Y)$ is arithmetic and may contain free variables other than x and Y. Feferman [3] showed that the Σ_1^1 axiom of choice holds in $\mathscr{M}(T_0, T_1, \ldots)$ for all $(T_0, T_1, \ldots)$ outside a first category subset of $(2^N)^N$ by means of a Cohen-type forcing argument.

THEOREM 3.16. *If $\mathscr{F}$ is an instance of the Σ_1^1 axiom of choice, then $\mathscr{F}$ is true in $\mathscr{M}(T_0, T_1, T_2, \ldots)$ with probability* 1.

Proof. Let $\mathscr{A}(x, Y)$ be a formula of $\mathscr{L}(\mathscr{T}_0, \mathscr{T}_1, \ldots)$ whose only free variables are x and Y, and whose only set variable is Y. Let T be such that

$$\mathscr{M}((T)_0, (T)_1, \ldots) \vDash (x)(EY)\mathscr{A}(x, Y).$$

We intend to show that almost every such T has the property that

$$\mathscr{M}((T)_0, (T)_1, \ldots) \vDash (EY)(x)\mathscr{A}(x, (Y)_x).$$

The lemma then follows from the measure-preserving property of the map $h(T) = ((T)_0, (T)_1, \ldots)$. Let $\{\mathscr{G}_b(x) \mid b \in O_1\}$ be a Π_1^1 enumeration of all ranked formulas with only x free of $\mathscr{L}(\mathscr{T}_0, \mathscr{T}_1, \mathscr{T}_2, \ldots)$. Then

$$(n)(Eb)[b \in O_1 \,\&\, \mathscr{M}((T)_0, (T)_1, \ldots) \vDash \mathscr{A}(\bar{n}, \hat{x}\mathscr{G}_b(x))].$$

If T lies outside a certain set of measure 0, then it follows from 3.14 that there exists a hyperarithmetic function f such that

$$(n)(Eb)_{b \leq f(n)}[f(n) \in O_1 \,\&\, \mathscr{M}((T)_0, (T)_1, \ldots) \vDash \mathscr{A}(\bar{n}, \hat{x}\mathscr{G}_b(x))].$$

By Spector [31], there is a $c \in O_1$ such that $f(n) <_{O_1} c$ for all n. Thus

$$(n)(Eb)[b \leq_{O_1} c \,\&\, \mathscr{M}((T)_0, (T)_1, \ldots) \vDash \mathscr{A}(\bar{n}, \hat{x}\mathscr{G}_b(x))].$$

There exists a ranked formula $\mathscr{H}(x)$ of $\mathscr{L}(\mathscr{T}_0, \mathscr{T}_1, \mathscr{T}_2, \ldots)$ such that for all T, n, and $b \leqq_{O_1} c$:

$$\mathscr{M}((T)_0, (T)_1, \ldots) \models \mathscr{G}_b(\bar{n}) \leftrightarrow \mathscr{M}((T)_0, (T)_1, \ldots) \models \bar{n} \in (\hat{x}\mathscr{H}(x))_b.$$

The existence of $\mathscr{H}$ follows by the same argument used in 3.6. But then

$$\mathscr{M}((T)_0, (T)_1, \ldots) \models (x)(Ey)\mathscr{A}(x, (\hat{x}\mathscr{H}(x))_y).$$

It easily follows that

$$\mathscr{M}((T)_0, (T)_1, \ldots) \models (EY)(x)\mathscr{A}(x, (Y)_x).$$

The proof of Theorem 3.16 was little more than an application of Corollary 3.14, which was only a refinement of Theorem 3.13, the statement of measure-theoretic uniformity for the hyperarithmetic case. In the same manner, Lemma 3.4 and Theorem 3.9 can be viewed as consequences of Theorem 3.13. In short, if one keeps in mind the hyperarithmetic version of measure-theoretic uniformity, as expressed by Theorem 3.13, one can readily recover all the results of §3.

4. **The constructible hierarchy.** Let $\mathscr{M}$ be a countable initial segment of L, the class of constructible sets, such that $\mathscr{M}$ is a model of ZF, the axioms of Zermelo-Fraenkel set theory, and hence of $V=L$, the axiom of constructibility. ($\mathscr{M}=F''\alpha$ for some countable ordinal α, where F is defined in Gödel [9].) We utilize measure-theoretic ideas in the style of §3 to study the effect of adding an arbitrary set T of natural numbers to $\mathscr{M}$. The resulting structure is denoted by $\mathscr{M}(T)$. Lemma 4.2 is the instance of measure-theoretic uniformity we associate with the constructible hierachy of $\mathscr{M}$, and it readily implies that the replacement axiom holds in $\mathscr{M}(T)$ with probability 1. In order to give a complete proof of Lemma 4.1, it is necessary to give some details concerning the structure of $\mathscr{M}(T)$. For this purpose the symbolism of Tharp [36] is convenient. The symbols of the language $\mathscr{L}^0$ are: $\in$ (membership); unranked set variables $x, y, z, \ldots$; ranked set variables $x^\alpha, y^\alpha, z^\alpha, \ldots$ for each ordinal α of $\mathscr{M}$; propositional connectives, and existential and universal quantifiers for both ranked and unranked set variables. The atoms of $\mathscr{L}^0$ are of the form $t_1 \in t_2$, where t_1 and t_2 are variables. The formulas of $\mathscr{L}^0$ are constructed from the atoms and the logical symbols in the standard manner. We define a class $\mathscr{C}$ of constants (intended to name the members of $\mathscr{M}(T)$) by induction on the ordinals of $\mathscr{M}$.

$\mathscr{C}(0)=\{\bar{n} \mid n<\omega\}$.

$\mathscr{C}(\alpha+1)$: $\mathscr{T}$ is a member of $\mathscr{C}(\alpha+1)$. Let $\phi(x^\alpha, y_1, \ldots, y_n)$ be a formula of $\mathscr{L}^0$ whose only free variables are $x^\alpha, y_1, \ldots, y_n$ $(n \geqq 0)$, and whose only quantified variables are of the form x^β for β less than or equal to α. Then $\hat{x}^\alpha\phi(x^\alpha, c_1, \ldots, c_n)$, where each c_i $(1 \leqq i \leqq n)$ is either $\mathscr{T}$ or a member of $\bigcup \{\mathscr{C}(\beta) \mid \beta \leqq \alpha\}$, is a typical member of $\mathscr{C}(\alpha+1)$.

$\mathscr{C}(\lambda)=\bigcup \{\mathscr{C}(\beta) \mid \beta<\lambda\}$ for each limit ordinal λ.

Let $\mathscr{C}=\bigcup\{\mathscr{C}(\alpha)\mid\alpha\in\mathscr{M}\}$. The symbols of $\mathscr{L}(\mathscr{T})$ are those of $\mathscr{L}^0$ together with the members of $\mathscr{C}$. The atomic formulas of $\mathscr{L}(\mathscr{T})$ are of the form $t_1\in t_2$, where t_1 and t_2 are variables or members of $\mathscr{C}$.

A sentence $\mathscr{F}$ of $\mathscr{L}(\mathscr{T})$ is said to be ranked if all its variables are ranked. The *ordinal rank* of a ranked sentence $\mathscr{F}$, denoted by $o(\mathscr{F})$, is the least upper bound of $\{\beta\mid(x^\beta)$ or (Ey^β) occurs in $\mathscr{F}\}\cup\{\beta+1\mid\hat{x}^\beta$ occurs in $\mathscr{F}\}$. If $t\in\mathscr{C}$, then the ordinal rank of t, denoted by $o(t)$, is the ordinal rank of the sentence $\bar{0}\in t$. For each set T of natural numbers, we simultaneously define several concepts by induction on the ordinals of $\mathscr{M}$: $c\in\mathscr{C}(\alpha)$ and $b\in\mathscr{M}_\alpha(T)$ and c denotes b; $o(\mathscr{F})\leqq\alpha$ and $\mathscr{F}$ is true in $\mathscr{M}_\alpha(T)$.

1. $\alpha=0$. $\mathscr{M}_0(T)=\omega$ and $\bar{n}$ denotes n. $\mathscr{F}$ is true in $\mathscr{M}_0(T)$ if it is true when $\mathscr{T}$ is interpreted as T, $\bar{n}$ is interpreted as n, and the variables $x^0, y^0, z^0,\dots$ are restricted to ω.

2. $\alpha=\gamma+1$. Let $c\in\mathscr{C}(\alpha)$ be of the form $\hat{x}^\gamma\phi(x^\gamma)$. Then c denotes b, where

$$b=\{a\mid a\in\mathscr{M}_\gamma(T)\ \&\ (Ed)[d\in\mathscr{C}(\gamma)\ \&\ d\text{ denotes }a\ \&\ \mathscr{M}_\gamma(T)\models\phi(d)]\}.$$

$\mathscr{M}_\alpha(T)$ is the set of all such b's. Let $o(\mathscr{F})\leqq\alpha$. Then $\mathscr{F}$ is true in $\mathscr{M}_\alpha(T)$ if it is true when each quantified variable x^β of $\mathscr{F}$ is restricted to $\mathscr{M}_\beta(T)$, and each constant c of $\mathscr{C}(\beta)$ occurring in $\mathscr{F}$ is interpreted as the $b\in\mathscr{M}_\beta(T)$ it denotes.

3. $\alpha=\lambda$, *where* λ *is a limit ordinal.* $\mathscr{M}_\lambda(T)=\bigcup\{\mathscr{M}_\beta(T)\mid\beta<\lambda\}$.

Let $\mathscr{M}(T)=\bigcup\{\mathscr{M}_\alpha(T)\mid\alpha\in\mathscr{M}\}$. Let $\mathscr{F}$ be an arbitrary sentence of $\mathscr{L}(\mathscr{T})$. $\mathscr{F}$ is true in $\mathscr{M}(T)$ ($\mathscr{M}(T)\models\mathscr{F}$) if it is true when each unranked, quantified variable is restricted to $\mathscr{M}(T)$ and the remaining symbols are interpreted as above. If $\mathscr{F}$ is ranked, then $\mathscr{F}$ can be put in prenex normal form without any increase in ordinal rank. Assume $\mathscr{F}$ is in prenex normal form. The *quantifier alternation rank of* $\mathscr{F}$, denoted by $a(\mathscr{F})$, is the number of alternations of quantifiers in the prefix of $\mathscr{F}$. The *leading quantifier rank* of $\mathscr{F}$, denoted by $l(\mathscr{F})$, is the number of quantified variables occurring in the prefix of $\mathscr{F}$ on the left of the left-most alternation of quantifiers of $\mathscr{F}$ if $a(\mathscr{F})>0$. If $a(\mathscr{F})=0$, then $l(\mathscr{F})$ is the number of quantified variables occurring in the prefix of $\mathscr{F}$. Thus if $\mathscr{F}$ is $(x^\alpha)(x^\beta)(y^\alpha)(Ey^\gamma)\mathscr{G}$, where $\mathscr{G}$ has no quantifiers, then $l(\mathscr{F})=3$.

For each sentence $\mathscr{F}$, the countability of $\mathscr{M}$ implies that $\{T\mid\mathscr{M}(T)\models\mathscr{F}\}$ is a Borel subset of 2^N. For each $\mathscr{F}$, the probability that $\mathscr{F}$ is true in $\mathscr{M}(T)$, denoted by $p(\mathscr{F})$, is defined to be the Lebesgue measure of the Borel set corresponding to $\mathscr{F}$. We have just given what might be called the external definition of the function $p(\mathscr{F})$. In Lemma 4.1 we give the internal definition, i.e., a definition inside $\mathscr{M}$. If $\mathscr{G}$ is an instance of the replacement axiom, then the fact that the replacement axiom is true in $\mathscr{M}$ together with the fact the probability function $p(\mathscr{F})$ is $\mathscr{M}$-definable will suffice to show $p(\mathscr{G})=1$. Hopefully, we will see this matter as a typical instance of the phenomenon of measure-theoretic uniformity. Of course the $\mathscr{M}$-definability of $p(\mathscr{F})$ plays the same role here as the $\mathscr{M}$-definability of the forcing relation in Cohen [2]. The difference, if there is one, is that $p(\mathscr{F})$ has a trivial external definition.

Let A be an arbitrary set. We say A is $\mathcal{M}$-definable if there exists a formula $\mathcal{F}(x)$ of ZF with constants denoting elements of $\mathcal{M}$ such that A equals

$$\{a \mid a \in \mathcal{M} \;\&\; a \text{ satisfies } \mathcal{F}(x) \text{ in } \mathcal{M}\}.$$

LEMMA 4.1. *For each $n \geqq 0$: The function $p(\mathcal{F})$, restricted to sentences $\mathcal{F}$ of $\mathcal{L}(\mathcal{T})$ having at most n unranked quantifiers, is $\mathcal{M}$-definable.*

Proof. We consider only $n=0$, since Cases 1 and 2 below make clear how to handle unranked quantifiers. We assume $\mathcal{F}$ is in prenex normal form and define $p(\mathcal{F})$ by induction on the rank of $\mathcal{F}=r(\mathcal{F})$, where $r(\mathcal{F})$ is the triple $(o(\mathcal{F}), a(\mathcal{F}), l(\mathcal{F}))$. (The triples are ordered lexicographically.)

Case 1. $\mathcal{F}$ is $(x^\alpha)\mathcal{F}_1(x^\alpha)$. Then

$$p(\mathcal{F}) = \text{glb}\left\{p\Big(\bigwedge_{i \leqq n} \mathcal{F}_1(c_i)\Big) \;\Big|\; n < \omega,\ c_i \in \mathcal{C}(\alpha)\right\},$$

where glb denotes "greatest lower bound". If $c_i (i \leqq n) \in \mathcal{C}(\alpha)$, then $\bigwedge_{i \leqq n} \mathcal{F}_1(c_i)$ can be put in prenex normal form $\mathcal{G}$ such that $o(\mathcal{G}) \leqq o(\mathcal{F})$. We claim $r(\mathcal{G}) < r(\mathcal{F})$. If $l(\mathcal{F}) > 1$, then $a(\mathcal{G}) = a(\mathcal{F})$ and $l(\mathcal{G}) < l(\mathcal{F})$. If $l(\mathcal{F}) = 1$, then $a(\mathcal{G}) < a(\mathcal{F})$. The last two assertions follow from the kind of elementary quantifier manipulation occurring in 3.1.

Case 2. $\mathcal{F}$ is $(Ex^\alpha)\mathcal{F}_1(x^\alpha)$. Dual to Case 1.

Case 3. $\mathcal{F}$ has no quantifier prefix. Thus $\mathcal{F}$ is a propositional combination of atoms of the form $t_1 \in t_2$, where $t_1, t_2 \in \mathcal{C}$. If every atom of $\mathcal{F}$ has ordinal rank 0, then $p(\mathcal{F})$ is easily computed from the information given in Case 3a. If some atom of $\mathcal{F}$ has ordinal rank >0, then $\mathcal{F}$ is equivalent to a sentence of lower ordinal rank, as is indicated in Cases 3b, 3c and 3d.

Case 3a. $p(\bar{m} \in \bar{n}) = 1$ if $m < n$; $p(\bar{m} \in \mathcal{T}) = \frac{1}{2}$; $p(\mathcal{T} \in \bar{m}) = p(\mathcal{T} \in \mathcal{T}) = p(\bar{m} \in \bar{n}) = 0$ if $m \geqq n$.

Case 3b. $o(t_1) < o(t_2)$. t_2 must be of the form $\hat{x}^\beta \phi(x^\beta)$. $t_1 \in t_2$ is equivalent to $\phi(t_1)$. $o(t_1 \in t_2) = \beta + 1$ and $\beta \geqq o(\phi(t_1))$.

Case 3c. $o(t_1) > o(t_2)$. Then $t_1 \in t_2$ is equivalent to

$$(Ey^\alpha)[(x^\alpha)(x^\alpha \in y^\alpha \leftrightarrow \phi(x^\alpha)) \;\&\; y^\alpha \in t_2],$$

where t_1 is of the form $\hat{x}^\alpha \phi(x^\alpha)$. $o(t_1 \in t_2) = \alpha + 1$, but the ordinal rank of the above equivalent formula is $\leqq \alpha$.

Case 3d. $0 < o(t_1) = o(t_2)$. Similar to Case 3c.

Cases 1–3 constitute the definition of $p(\mathcal{F})$ "inside" $\mathcal{M}$. A routine induction shows the internal definition of the function $p(\mathcal{F})$ agrees with the external definition.

The proof of Lemma 4.1 relied heavily on the notion of prenex normal form. We comment further on this matter at the end of the paper. Lemma 4.2 is the instance of measure-theoretic uniformity we associate with the constructible hierarchy of $\mathcal{M}$. It is analogous to Theorems 2.2 and 3.13, but there is a difference worth noting.

Both 2.2 and 3.13 have the form: for almost every T there exists an f etc., but 4.1 has the form: there exists an f such that for almost all T etc. The difference is a consequence of the strong "closure" properties of the constructible hierarchy.

LEMMA 4.2. *Let $P(x, y)$ be a formula of $\mathcal{L}(\mathcal{T})$ whose only free variables are x and y. Then there exists an $\mathcal{M}$-definable function f such that for all $\alpha \in \mathcal{M}$ with probability 1:*

$$(x^\alpha)(Ey)P(x^\alpha, y) \leftrightarrow (x^\alpha)(Ey^{f(\alpha)})P(x^\alpha, y^{f(\alpha)}).$$

Proof. Let $g(\alpha, \beta) = p((x^\alpha)(Ey^\beta)P(x^\alpha, y^\beta))$; g is $\mathcal{M}$-definable by 3.1. For each α, $\lambda\beta \mid g(\alpha, \beta)$ is a nondecreasing $\mathcal{M}$-definable function from the ordinals into the unit interval. Since the replacement axiom holds in $\mathcal{M}$, there must exist $\mathcal{M}$-definable functions $h(\alpha)$ and $f(\alpha)$ such that $h(\alpha)$ is the least upper bound of $\{g(\alpha, \beta) \mid \beta \in \mathcal{M}\}$ and $g(\alpha, f(\alpha)) = h(\alpha)$.

LEMMA 4.3. *Let $\mathcal{F}$ be an instance of the replacement axiom. Then $p(\mathcal{F}) = 1$.*

Proof. The replacement axiom for $\mathcal{M}(T)$ can be split as follows: (i) the range of a function restricted to a set is contained in a set; (ii) let $P(x)$ be a formula of $\mathcal{L}(\mathcal{T})$ whose only free variable is x; then

$$(y)(Ez)(x)[x \in y \;\&\; P(x) \leftrightarrow x \in z].$$

By 4.2, (i) holds in $\mathcal{M}(T)$ for almost all T. But (ii) follows from (i) in $\mathcal{M}(T)$ by standard arguments implicit in Gödel [8, 9]. Let $y \in \mathcal{M}(T)$. Then it follows from (i) by means of a Skolem-Löwenheim argument that there exists a $v \in \mathcal{M}(T)$ such that $y \in v$, $y \subseteq v$, and in $\mathcal{M}(T)$,

$$(x)(x \in y \;\&\; P(x) \leftrightarrow x \in y \;\&\; P^v(x)),$$

where $P^v(x)$ is the result of restricting the quantifiers of P to v. Then $z \rightleftharpoons \hat{x}(x \in y \;\&\; P^v(x))$ is a first-order definable subset of v; and $z \in \mathcal{M}(T)$, since every first-order definable subset of a member of $\mathcal{M}(T)$ is a member of $\mathcal{M}(T)$.

PROPOSITION 4.4. *Let $\{\mathcal{F}_n \mid n \in \omega\}$ be a sequence of $\mathcal{M}$ of sentences of $\mathcal{L}(\mathcal{T})$ of countable (in the sense of $\mathcal{M}$) ordinal rank. Then there exists a sentence $\mathcal{F}$ of countable ordinal rank such that for all T,*

$$\mathcal{M}(T) \models \mathcal{F} \leftrightarrow (En)(\mathcal{M}(T) \models \mathcal{F}_n).$$

In addition, $\mathcal{F}$ regarded as a function of $\{\mathcal{F}_n \mid n \in \omega\}$, is $\mathcal{M}$-definable.

Proof. Let α be a countable (in the sense of $\mathcal{M}$) ordinal such that $o(\mathcal{F}_n) < \alpha$ for all n. Then

$$\mathcal{M}(T) \models \mathcal{F}_n \leftrightarrow \mathcal{M}_\alpha(T) \models \mathcal{F}_n$$

for all n and all T. By standard arguments there exist a countable ordinal $\beta > \alpha$ and a formula $P(x^0)$ of $\mathcal{L}(\mathcal{T})$ of ordinal rank β such that for all n and all T,

$$\mathcal{M}_\alpha(T) \models \mathcal{F}_n \leftrightarrow \mathcal{M}_\beta(T) \models P(\bar{n}).$$

The existence of $P(x^0)$ follows from two facts: (a) truth in $\mathscr{M}_\alpha(T)$ has a first-order definition (uniform in T) with quantifiers restricted to $\mathscr{M}_{\alpha+\omega}(T)$; (b) every first-order definable subset of a member of $\mathscr{M}(T)$ is a member of $\mathscr{M}(T)$. The desired $\mathscr{F}$ is $(Ex^0)P(x^0)$.

LEMMA 4.5. *For each $n \geqq 0$, there exists an $\mathscr{M}$-definable function $\lambda\mathscr{F} \mid \mathscr{F}^*$, defined for all sentences $\mathscr{F}$ of $\mathscr{L}(\mathscr{T})$ having at most n unranked quantifiers, such that $p(\mathscr{F} \leftrightarrow \mathscr{F}^*)=1$ and $\mathscr{F}^*$ has countable (in the sense of $\mathscr{M}$) ordinal rank.*

Proof. Similar to 4.1. We consider $n=0$ only, assume $\mathscr{F}$ is in prenex normal form, and define $\mathscr{F}^*$ by induction on the rank of $\mathscr{F}$. If $\mathscr{F}$ has countable ordinal rank, then $\mathscr{F}^*$ is $\mathscr{F}$. Let $\mathscr{F}$ have uncountable ordinal rank. Suppose $\mathscr{F}$ is $(Ex^\alpha)\mathscr{F}_1(x^\alpha)$. Then

$$p(\mathscr{F}) = \text{lub}\left\{p\Big(\bigvee_{i\leqq n} \mathscr{F}_1(c_i)\Big) \mid c_i(i \leqq n) \in C(\alpha)\right\}$$

where lub denotes "least upper bound". As in 4.1, $\bigvee_{i\leqq n} \mathscr{F}_1(c_i)$ is equivalent to a prenex normal sentence of lower rank than $(Ex^\alpha)\mathscr{F}_1(x^\alpha)$. Then

$$p\Big(\bigvee_{i\leqq n} \mathscr{F}_1(c_i) \leftrightarrow \Big(\bigvee_{i\leqq n} \mathscr{F}_1(c_i)\Big)^*\Big) = 1,$$

where $(\bigvee_{i\leqq n} \mathscr{F}_1(c_i))^*$ has countable ordinal rank. It follows from the $\mathscr{M}$-definability of $\lambda\mathscr{F} \mid p(\mathscr{F})$ (4.1) that there is an $\mathscr{M}$-definable sequence $\{c_i \mid i<\omega\} \subseteq C(\alpha)$ such that

$$p(\mathscr{F}) = \text{lub}\left\{p\Big(\bigvee_{i\leqq n} \mathscr{F}_1(c_i)\Big) \mid i < \omega\right\}.$$

Let $\mathscr{F}^*$ be equivalent to $\bigvee_{n<\omega} (\bigvee_{i\leqq n} \mathscr{F}(c_i))^*$. $\mathscr{F}^*$ exists by 4.4.

If $\mathscr{F}$ is $(x^\alpha)\mathscr{F}_1(x^\alpha)$, we proceed in a manner dual to the above.

If $\mathscr{F}$ has no quantifier prefix, then $\mathscr{F}$ is equivalent, as we saw in 4.1, to a sentence of lower ordinal rank.

LEMMA 4.6. *Let α be a cardinal of $\mathscr{M}$. Then $p(\alpha$ is a cardinal of $\mathscr{M}(T))=1$.*

Proof. We sharpen the argument of 4.2. Let α be an uncountable cardinal of $\mathscr{M}$, and let $\beta \in \alpha$. We show $p(\beta$ can be mapped 1-1 onto $\alpha)=0$. Let f be a constant $\mathscr{C}$ which denotes a function from β into α. For each $\gamma \in \beta$, let

$$k(\gamma) = \{\tau \mid \tau \in \alpha \;\&\; p(f(\gamma) = \tau) > 0\}.$$

By 4.1, $\lambda\gamma \mid k(\gamma)$ is $\mathscr{M}$-definable. Since f denotes a function and p is countably additive in the sense of $\mathscr{M}$, it follows that $k(\gamma)$ is a countable set of $\mathscr{M}$ for all $\gamma \in \beta$. But then $\bigcup \{k(\gamma) \mid \gamma \in \beta\}$ is a subset of α of cardinality (in the sense of $\mathscr{M}$) less than that of α, and consequently,

$$p(\{f(\gamma) \mid \gamma \in \beta\} \subseteq \bigcup \{k(\gamma) \mid \gamma \in \beta\} \neq \alpha) = 1.$$

LEMMA 4.7. *With probability 1: the power set axiom holds in $\mathscr{M}(T)$.*

Proof. Our argument is related to the one given by Gödel [9, p. 46] to show that the power set axiom holds in L. Let d be a constant of $\mathscr{C}$. It will suffice to find an ordinal γ of $\mathscr{M}$ such that for all $c \in \mathscr{C}$,

$$p(c \subseteq d \to c \in \mathscr{M}_\gamma(T)) = 1.$$

Let $d \in \mathscr{C}(\alpha+1) - \mathscr{C}(\alpha)$. Then the members of the set denoted by d can be denoted by constants of $\mathscr{C}(\alpha)$. By 4.5 there is an $\mathscr{M}$-definable function $\lambda bc \mid (b \in c)^*$ such that

$$p(b \in c \leftrightarrow (b \in c)^*) = 1$$

for all b, $c \in \mathscr{C}$, and such that $(b \in c)^*$ has countable ordinal rank. For each $c \in \mathscr{C}$, let $k(c) = \lambda b \mid (b \in c)^*$, where b is restricted to $\mathscr{C}(\alpha)$. If $k(c_1) = k(c_2)$, then

$$p(c_1 \subseteq d \;\&\; c_2 \subseteq d \to c_1 = c_2) = 1.$$

Since $\mathscr{C}(\alpha)$ is a set of $\mathscr{M}$ and the power set axiom holds in $\mathscr{M}$, it follows $\{k(c) \mid c \in \mathscr{C}\}$ is a set of $\mathscr{M}$. Since the replacement axiom holds in $\mathscr{M}$, there must exist a γ such that for all $c \in \mathscr{C}$,

$$(Ec')[c' \in C(\gamma) \;\&\; k(c') = k(c)].$$

LEMMA 4.8. *With probability* 1: *the generalized continuum hypothesis holds in* $\mathscr{M}(T)$.

Proof. We first sharpen the argument of 4.7, and then exploit the fact that the generalized continuum hypothesis holds in $\mathscr{M}$. Let α be an infinite ordinal of $\mathscr{M}$, and α^+ be the least ordinal greater than α of higher cardinality in the sense of $\mathscr{M}$. For each $c \in \mathscr{C}$, let $k(c) = \lambda b \mid (b \in c)^*$, where $b \in \alpha$, $p(b \in c \leftrightarrow (b \in c)^*) = 1$, and $(b \in c)^*$ has countable rank. As in 4.7, if $k(c_1) = k(c_2)$, then

$$p(c_1 \subseteq \alpha \;\&\; c_2 \subseteq \alpha \to c_1 = c_2) = 1.$$

As in 4.7, $\{k(c) \mid c \in \mathscr{C}\}$ is a set of $\mathscr{M}$. Let $d \subseteq \mathscr{C}$ be a set of $\mathscr{M}$ such that for all $c \in \mathscr{C}$,

$$(Ec')[c' \in d \;\&\; k(c') = k(c)],$$

and such that

$$(c_1)(c_2)[c_1 \in d \;\&\; c_2 \in d \to k(c_1) \neq k(c_2)].$$

Let d^* denote the set of sets of $\mathscr{M}(T)$ denoted by the constants of d. Then for all $c \in \mathscr{C}$,

$$p(c \subseteq \alpha \to c \in d^*) = 1.$$

Since $k(c)$ can be regarded as a function of $\mathscr{M}$ from α into ω_1, it follows that the cardinality of d (in $\mathscr{M}$) is at most $\omega_1^\alpha = 2^\alpha = \alpha^+$. But the cardinality of d^* (in $\mathscr{M}(T)$) is at most that of d with probability 1.

THEOREM 4.9 (R. SOLOVAY). *With probability* 1: $\mathscr{M}(T)$ *is a model of* $ZF + GCH + AC +$ "*cardinals are absolute*" $+ V \neq L$.

Proof. Apply 4.3, 4.6, 4.7, and 4.8. For all T, $\mathscr{M}(T) \models \mathscr{T} \in L$ if and only if $T \in \mathscr{M}$. Clearly $p(\mathscr{T} \in L) = 0$, since $\mathscr{M}$ is countable.

We now develop Cohen's results concerning the independence of the continuum hypothesis by making purely notational changes in the proof of Theorem 4.9. Let γ be an uncountable cardinal of $\mathscr{M}$. We study the structure $\mathscr{M}(T^\gamma)$, where T^γ is an *arbitrary* subset of γ. The language $\mathscr{L}(\mathscr{T}^\gamma)$ is similar to the language $\mathscr{L}(\mathscr{T})$ $(=\mathscr{L}(\mathscr{T}^\omega))$ except that $\mathscr{T}$ is replaced by $\mathscr{T}^\gamma$ and $\mathscr{C}(0)=\{\bar{n} \mid n \in \omega\}$ is replaced by $\{\bar{\beta} \mid \beta<\gamma\}$. $\mathscr{M}_\alpha(T^\gamma)$ is defined in the same manner as $\mathscr{M}_\alpha(T)$ save that $\mathscr{M}_0(T^\gamma)=\gamma$ rather than ω. Since γ is countable, the product measure on 2^γ closely resembles the product measure on 2^N. If $\mathscr{F}$ is a sentence of $\mathscr{L}(\mathscr{T}^\gamma)$, then the set

$$\{T^\gamma \mid T^\gamma \in 2^\gamma \ \&\ \mathscr{M}(T^\gamma) \models \mathscr{F}\}$$

is a Borel subset of 2^γ; its Lebesgue measure, denoted by $p(\mathscr{F})$, is the probability that $\mathscr{F}$ is true in $\mathscr{M}(T^\gamma)$.

THEOREM 4.10 (R. SOLOVAY). *Let γ be an uncountable cardinal of $\mathscr{M}$. With probability* 1: *$\mathscr{M}(T^\gamma)$ is a model of $ZF+AC+$"cardinals are absolute"$+V\neq L$. If γ is not cofinal with ω (in the sense of $\mathscr{M}$), then $2^\omega=\gamma$ holds with probability* 1 *in $\mathscr{M}(T^\gamma)$.*

Proof. Ranks are assigned to the sentences of $\mathscr{L}(\mathscr{T}^\gamma)$ in precisely the same way they were assigned to sentences of $\mathscr{L}(\mathscr{T})$. There is no difficulty encountered when $\mathscr{T}$ is replaced by $\mathscr{T}^\gamma$ in 4.1, 4.2, 4.3 and 4.4; 4.5, 4.6, and 4.7 are unaltered by the substitution of $\mathscr{T}^\gamma$ for $\mathscr{T}$, since they depend only on the countable additivity of the measure. Let $g:\gamma\times\omega\to\gamma$ be the canonical 1-1 correspondence between $\gamma\times\omega$ and γ. For each $\delta>\gamma$, let a_δ be a constant of $\mathscr{C}$ which denotes $\{n \mid g(\delta, n)\in T^\gamma\}$. If $\delta_1\neq\delta_2$, then $p(\bar{n}\in a_{\delta_1}\leftrightarrow \bar{n}\in a_{\delta_2})=\frac{1}{2}$ for each n, and consequently $p(a_{\delta_1}=a_{\delta_2})=0$. It follows that $p(2^\omega\geqq\gamma)=1$.

Now assume γ is not cofinal with ω. The argument of 4.8 serves to show $p(2^\omega=\gamma)$ $=1$. For each $c\in\mathscr{C}$, let $k(c)=\lambda n \mid (\bar{n}\in c)^*$, where $n\in\omega$ and $(\bar{n}\in c)^*$ has countable ordinal rank. If $k(c_1)=k(c_2)$, then

$$p(c_1 \subseteq \omega \ \&\ c_2 \subseteq \omega \to c_1 = c_2) = 1.$$

The cardinality of the set of sentences of $\mathscr{L}(\mathscr{T}^\gamma)$ of countable ordinal rank is γ; so $k(c)$ can be regarded as function from ω to γ. It follows $\{k(c) \mid c\in\mathscr{C}\}$ is a set (in $\mathscr{M}$) of cardinality at most $\gamma^\omega=\gamma$, since (in $\mathscr{M}$) γ is not cofinal with ω and the generalized continuum hypothesis holds. But the cardinality of 2^ω (in $\mathscr{M}(T^\gamma)$) is at most that of $\{k(c) \mid c\in\mathscr{C}\}$ with probability 1.

Let $(T_0, T_1, T_2, \ldots)$ be a sequence of arbitrary subsets of 2^N. We now study the effect of adjoining $(T_0, T_1, T_2, \ldots)$ to $\mathscr{M}$ in the same measure-theoretic fashion we studied the effect of adding an arbitrary T to $\mathscr{M}$. The language $\mathscr{L}(\mathscr{T}_0, \mathscr{T}_1, \mathscr{T}_2, \ldots)$ is obtained by altering $\mathscr{L}(\mathscr{T})$ as follows. Replace $\mathscr{T}$ by $\mathscr{T}_0, \mathscr{T}_1, \mathscr{T}_2, \ldots$; add primitive symbols $+$ (addition) and $\cdot$ (multiplication) with the proviso that they are to operate only on number-theoretic terms. $\mathscr{C}(0)$ is again the set of finite numerals. $\mathscr{C}(\alpha+1)$ is defined as at the beginning of §4 save that $\mathscr{T}$ is replaced by $\mathscr{T}_0, \mathscr{T}_1, \mathscr{T}_2, \ldots, \mathscr{T}_i, \ldots$ $(i<\omega)$ and the formula ϕ can have occurrences of $+$ and $\cdot$. Thus the constants of $\mathscr{C}(1)$ now denote all the subsets of ω which are arithmetical in some finite subsequence of $T_0, T_1, T_2, \ldots$; truth in the structure $\mathscr{M}(T_0, T_1, T_2, \ldots)$ and the relation "$c \in \mathscr{C}$ & $b \in \mathscr{M}(T_0, T_1, T_2, \ldots)$ & c denotes b" are defined as before.

Regard $(T_0, T_1, T_2, \ldots)$ as a point in the space $(2^N)^N$, and give $(2^N)^N$ the product measure induced by the measure m on each factor 2. ($m(\{0\})=m(\{1\})=\frac{1}{2}$.) Then for each sentence $\mathscr{F}$ of $\mathscr{L}(\mathscr{T}_0, \mathscr{T}_1, \mathscr{T}_2, \ldots)$, the set

$$\{(T_0, T_1, T_2, \ldots) \mid \mathscr{M}(T_0, T_1, T_2, \ldots) \models \mathscr{F}\}$$

is a Borel subset of $(2^N)^N$; the measure of this set, denoted by $p(\mathscr{F})$, is the probability that $\mathscr{F}$ is true in $\mathscr{M}(T_0, T_1, T_2, \ldots)$. The argument of 4.1 now shows: for each $n \geqq 0$, the function $p(\mathscr{F})$ restricted to sentences $\mathscr{F}$ of $\mathscr{L}(\mathscr{T}_0, \mathscr{T}_1, \mathscr{T}_2, \ldots)$, is $\mathscr{M}$-definable. The arguments of 4.2, 4.3, 4.4, 4.5, 4.6, and 4.7 still work, and together they show: with probability 1, $\mathscr{M}(T_0, T_1, T_2, \ldots)$ is a model of ZF+ "cardinals are preserved". Actually, it is necessary to restate and reprove 4.4, 4.5, and 4.7 as 4.17.1, 4.17.2, and 4.17.3. That task is postponed until 4.17 is established. It is instructive to consider why 4.8 now fails. The last line of the argument of 4.8 assumes the existence of an $\mathscr{M}(T)$-definable function which assigns to each set of $\mathscr{M}(T)$ a constant of $\mathscr{C}$ that denotes it; the function is obtained by assigning $\mathscr{T}$ to T and then exploiting an $\mathscr{M}$-definable well-ordering of the formulas and constants of $\mathscr{L}(\mathscr{T})$. In the present case, the language $\mathscr{L}(\mathscr{T}_0, \mathscr{T}_1, \mathscr{T}_2, \ldots)$ has an $\mathscr{M}$-definable well-ordering, but with probability 1 (as we shall see below) there is no $\mathscr{M}(T_0, T_1, T_2, \ldots)$-definable function which assigns $\mathscr{T}_i$ to T_i for all $i<\omega$.

$\mathscr{M}(T_0, T_1, \ldots)$ has certain symmetries studied by Cohen [1], Feferman [2], and Solovay [29], [30] which we shall study in terms of measure-preserving transformations of $(2^N)^N$. Let r be a function whose domain is $\{\mathscr{T}_i \mid i \in \omega\}$ and whose range is a subset of $\mathscr{C}(1)$; r can be extended to a function from $\mathscr{C}$ into $\mathscr{C}$ as follows:

$$r(c(\mathscr{T}_0, \ldots, \mathscr{T}_n)) = c(r(\mathscr{T}_0), \ldots, r(\mathscr{T}_n));$$

in the same manner, r can be extended to a function from the sentences of $\mathscr{L}(\mathscr{T}_0, \mathscr{T}_1, \ldots)$ into itself. If r is a map from $\mathscr{C}(1)$ into $\mathscr{C}(1)$, then r induces a map r^* from $(2^N)^N$ into $(2^N)^N$ as follows:

$$n \in r^*(T_i) \leftrightarrow \mathscr{M}(T_0, T_1, \ldots) \models \bar{n} \in r(\mathscr{T}_i);$$
$$r^*(T_0, T_1, \ldots) = (r^*(T_0), r^*(T_1), \ldots).$$

We say r is a *map from* $\mathscr{C}(1)$ *onto* $\mathscr{C}(1)$ if r is a map from $\mathscr{C}(1)$ into $\mathscr{C}(1)$ and if for each $i \in \omega$, there is a $u_i \in \mathscr{C}(1)$ such that $r(u_i)=\mathscr{T}_i$(11). If r is a map from $\mathscr{C}(1)$ onto $\mathscr{C}(1)$, then r^* is a 1-1 map from $(2^N)^N$ into $(2^N)^N$.

LEMMA 4.11. *Let r be a map of $\mathscr{C}(1)$ onto $\mathscr{C}(1)$. Then $\mathscr{M}(T_0, T_1, \ldots) \models r(\mathscr{F}) \leftrightarrow \mathscr{M}(r^*(T_0), r^*(T_1), \ldots) \models \mathscr{F}$, for all $(T_0, T_1, \ldots)$.*

---

(11) Let $c, d \in \mathscr{C}$. We say $c=d$ if c and d denote the same member of $\mathscr{M}(T_0, T_1, T_2, \ldots)$ for all $(T_0, T_1, T_2, \ldots)$. J. Rosenthal [21] has shown that this notion of equality is $\mathscr{M}$-definable. His result is not needed in the present paper.

Proof. By induction on the rank of $\mathcal{F}$. (We use the same notion of rank employed in 4.1.) It is sufficient to consider two cases.

Case 1. $\mathcal{F}$ is $\bar{n} \in \mathcal{T}_i$. Then $\mathcal{M}(T_0, T_1, \ldots) \models r(\bar{n} \in \mathcal{T}_i) \leftrightarrow \mathcal{M}(T_0, T_1, \ldots) \models \bar{n} \in r(\mathcal{T}_i) \leftrightarrow n \in r^*(T_i) \leftrightarrow \mathcal{M}(r(T_0), r(T_1), \ldots) \models \bar{n} \in \mathcal{T}_i$.

Case 2. $\mathcal{F}$ is $(Ex^\alpha)\mathcal{G}(x^\alpha, b)$, where b is the only member of $\mathcal{C}$ occurring in $\mathcal{F}$. It is easily checked that r maps $\mathcal{C}(\alpha)$ onto $\mathcal{C}(\alpha)$; i.e., for each $d \in \mathcal{C}(\alpha)$, there is a $c \in \mathcal{C}(\alpha)$ such that $r(c) = d$. The following formulas are equivalent:

$$\mathcal{M}(r^*(T_0), r^*(T_1), \ldots) \models (Ex^\alpha)\mathcal{G}(x^\alpha, b);$$
$$(Ec)[c \in \mathcal{C}(\alpha) \ \& \ \mathcal{M}(r^*(T_0), r^*(T_1), \ldots) \models \mathcal{G}(c, b)];$$
$$(Ec)[c \in \mathcal{C}(\alpha) \ \& \ \mathcal{M}(T_0, T_1, \ldots) \models \mathcal{G}(r(c), r(b))];$$
$$(Ed)[d \in \mathcal{C}(\alpha) \ \& \ \mathcal{M}(T_0, T_1, \ldots) \models \mathcal{G}(d, r(b))];$$
$$\mathcal{M}(T_0, T_1, \ldots) \models r((Ex^\alpha)\mathcal{G}(x^\alpha, b)).$$

The second formula is equivalent to the third by the induction hypothesis.

LEMMA 4.12. *Let r be a map of $\mathcal{C}(1)$ onto $\mathcal{C}(1)$ such that the induced map $r^*: (2^N)^N \to (2^N)^N$ is measure-preserving. Then $p(\mathcal{F}) = p(r(\mathcal{F}))$.*

Proof. r^* is 1-1 and maps $(2^N)^N$ onto a set whose complement has measure 0. Now apply 4.11.

Feferman's transformation lemma [3, 4.8] is a forcing-theoretic version of our Lemma 4.12.

Let $\mathcal{C}(\alpha; \mathcal{T}_0, \mathcal{T}_1, \ldots, \mathcal{T}_{k-1})$ $(k \geqq 0)$ be the set of all constants of $\mathcal{C}(\alpha)$ whose $\mathcal{T}_i$-symbols are included in the sequence $\mathcal{T}_0, \ldots, \mathcal{T}_{k-1}$. Let $\mathcal{M}(\alpha; T_0, \ldots, T_{k-1})$ be the set of all sets of $\mathcal{M}_\alpha(T_0, T_1, T_2, \ldots)$ denoted by members of

$$\mathcal{C}(\alpha; \mathcal{T}_0, \ldots, \mathcal{T}_{k-1}).$$

It is clear that $\mathcal{C}$ has a constant which denotes $\mathcal{M}(\alpha; T_0, \ldots, T_{k-1})$; let us adopt the convention of denoting $\mathcal{M}(\alpha; T_0, \ldots, T_{k-1})$ by $\mathcal{M}(\alpha; \mathcal{T}_0, \ldots, \mathcal{T}_{k-1})$. Similarly, let

$$\mathcal{C}(\mathcal{T}_0, \ldots, \mathcal{T}_{k-1}) = \bigcup \{\mathcal{C}(\alpha; \mathcal{T}_0, \ldots, \mathcal{T}_{k-1}) \mid \alpha \in \mathcal{M}\},$$

and let $\mathcal{M}(\mathcal{T}_0, \ldots, \mathcal{T}_{k-1})$ denote $\bigcup \{\mathcal{M}(\alpha; T_0, \ldots, T_{k-1}) \mid \alpha \in \mathcal{M}\}$.

PROPOSITION 4.13 (R. SOLOVAY). *For each $k \geqq 0$: $\mathcal{M}(\mathcal{T}_0, \ldots, \mathcal{T}_{k-1})$ is well-ordered with probability* 1.

Proof. Let $\mathcal{N} = \mathcal{M}(T_0, T_1, T_2, \ldots)$ be a model of ZF with the allowable exception of the power set axiom. Fix $k \geqq 0$. In $\mathcal{N}$ we can define truth in $\mathcal{N}$ for all ranked sentences whose $\mathcal{T}_i$-symbols are included in the list $\mathcal{T}_0, \ldots, \mathcal{T}_{k-1}$. It follows by standard arguments that there exists a 1-1 $\mathcal{N}$-definfible map h_k of $\mathcal{M}(T_0, \ldots, T_{k-1})$ into $\mathcal{C}(\mathcal{T}_0, \ldots, \mathcal{T}_{k-1})$, i.e., each set of $\mathcal{M}(T_0, \ldots, T_{k-1})$ is mapped to a constant of $\mathcal{C}(\mathcal{T}_0, \ldots, \mathcal{T}_{k-1})$ which denotes that set. But there exists an $\mathcal{M}$-definable (hence $\mathcal{N}$-definable) well-ordering of $\mathcal{C}$. The definition of h_k has $T_0, \ldots, T_{k-1}$ as parameters.

LEMMA 4.14 (R. SOLOVAY). *Let $\mathscr{F}(x^\alpha)$ be a formula of $\mathscr{L}(\mathscr{T}_0, \mathscr{T}_1, \ldots)$ whose only free variable is x^α, and whose $\mathscr{T}_i$-symbols are included among $\mathscr{T}_0, \ldots, \mathscr{T}_{k-1}$. Then with probability* 1,

$$(Ex^\alpha)\mathscr{F}(x^\alpha) \leftrightarrow (Ex^\alpha)[\mathscr{F}(x^\alpha) \ \&\ x^\alpha \in \mathscr{M}(\alpha; \mathscr{T}_0, \ldots, \mathscr{T}_k)].$$

Similarly, with probability 1,

$$(Ex)\mathscr{F}(x) \leftrightarrow (Ex)[\mathscr{F}(x) \ \&\ x \in \mathscr{M}(\mathscr{T}_0, \ldots, \mathscr{T}_k)].$$

Proof. Let $k=1$. To prove the first half of the lemma, it is enough to prove: for each sequence $c_0, c_1, \ldots, c_n \in \mathscr{C}(\alpha)$, there exists a sequence $r(c_0), r(c_1), \ldots, r(c_n) \in \mathscr{C}(\alpha; \mathscr{T}_0, \mathscr{T}_1)$ such that

$$p\Big(\bigvee_{i \leq n} \mathscr{F}(c_i)\Big) = p\Big(\bigvee_{i \leq n} \mathscr{F}(r(c_i))\Big).$$

Let m be so large that $c_0, c_1, \ldots, c_n \in \mathscr{C}(\alpha; \mathscr{T}_0, \ldots, \mathscr{T}_m)$. Consider the following map r of $\mathscr{C}(1)$ into $\mathscr{C}(1)$:

$$r(\mathscr{T}_0) = \mathscr{T}_0; r(\mathscr{T}_{i+m+1}) = \mathscr{T}_{i+2} \qquad (i \geq 0);$$
$$r(\mathscr{T}_{j+1}) = \hat{x}_0(\overline{m}x_0 + j \in \mathscr{T}_1) \qquad (j < m).$$

It is not difficult to check that r maps $\mathscr{C}(1)$ onto $\mathscr{C}(1)$ and that r^* is measure-preserving([12]). Clearly, $r(c_i) \in \mathscr{C}(\alpha; \mathscr{T}_0, \mathscr{T}_1)$ and $r(\mathscr{F}(c_i))$ is $\mathscr{F}(r(c_i))$ for each $i \leq n$. Now apply 4.12.

We specify some maps of $\mathscr{C}(1)$ into $\mathscr{C}(1)$ needed for the proof of Lemma 4.15. For each $n \geq 0$ and $i \geq 0$, let

$$(\mathscr{T}_m)_i = \hat{x}_0[(2^{i+1}x^0 + 2^i - 1) \in \mathscr{T}_m];$$
$$(\mathscr{T}_m)^n = \hat{x}_0[(2^n x^0 + 2^n - 1) \in \mathscr{T}_m].$$

For each $m, n \geq 0$, we define a map r_n^m of $\mathscr{C}(1)$ into $\mathscr{C}(1)$:

$$r_n^m(\mathscr{T}_j) = \mathscr{T}_j \qquad (j < m);$$
$$r_n^m(\mathscr{T}_{m+i}) = (\mathscr{T}_m)_i \qquad (i < n);$$
$$r_n^m(\mathscr{T}_{m+n}) = (\mathscr{T}_m)^n;$$
$$r_n^m(\mathscr{T}_{m+n+k}) = \mathscr{T}_{m+k} \qquad (k \geq 1).$$

It is not hard to see that $(r_n^m)^*: (2^N)^N \to (2^N)^N$ is measure-preserving (see footnote ([12])). We claim that r_n^m maps $\mathscr{C}(1)$ onto $\mathscr{C}(1)$. Consider the constant $d_n^m \in \mathscr{C}(1)$ defined by

$$x^0 \in d_n^m \leftrightarrow (Ey^0)(y^0 \in \mathscr{T}_{m+n} \ \&\ x^0 = 2^n y^0 + 2^n - 1)$$
$$\vee\ (Ei)_{0 \leq i < n}(Ey^0)(y^0 \in \mathscr{T}_{m+i} \ \&\ x^0 = 2^{i+1}y^0 + 2^i - 1).$$

([12]) Cf. Halmos [10, p. 159 (7)].

Clearly, $r_n^m(d_n^m)=\mathscr{T}_m$. Thus r_n^m is a map of $\mathscr{C}(1)$ into $\mathscr{C}(1)$ which meets the hypotheses of Lemma 4.12.

LEMMA 4.15 (R. SOLOVAY). *With probability* 1, *the dependent axiom of choice holds in* $\mathscr{M}(T_0, T_1, T_2, \ldots)$.

Proof. Let $\mathscr{F}(x, y)$ be a formula of $\mathscr{L}(\mathscr{T}_0, \mathscr{T}_1, \ldots)$ whose only free variables are x and y and all of whose $\mathscr{T}_i$-symbols are included among $\mathscr{T}_0, \mathscr{T}_1, \ldots, \mathscr{T}_{m-1}$ $(m \geqq 0)$. We intend to find an $f \in \mathscr{C}(\mathscr{T}_0, \ldots, \mathscr{T}_m)$ such that f denotes a function on ω and such that with probability 1,

$$(x)(Ey)\mathscr{F}(x, y) \to (x^0)\mathscr{F}(f(x^0), f(x^0+\bar{1})).$$

For the sake of notational simplicity, let $m=1$. By 4.14, for each $n \geqq 0$, the following two sentences are equivalent with probability 1:

$$(x)(Ey)[x \in \mathscr{M}(\mathscr{T}_0, \ldots, \mathscr{T}_n) \to \mathscr{F}(x, y)];$$
$$(x)(Ey)[x \in \mathscr{M}(\mathscr{T}_0, \ldots, \mathscr{T}_n) \to \mathscr{F}(x, y) \ \&\ y \in \mathscr{M}(\mathscr{T}_0, \ldots, \mathscr{T}_{n+1})].$$

With the aid of 4.12, we apply r_{n+1}^1 to the above pair of sentences to obtain another pair equivalent with probability 1:

$$(x)(Ey)[x \in \mathscr{M}(\mathscr{T}_0, (\mathscr{T}_1)_0, \ldots, (\mathscr{T}_1)_{n-1}) \to \mathscr{F}(x, y)];$$
$$(x)(Ey)[x \in \mathscr{M}(\mathscr{T}_0, (\mathscr{T}_1)_0, \ldots, (\mathscr{T}_1)_{n-1}) \to \mathscr{F}(x, y) \ \&\ y \in \mathscr{M}(\mathscr{T}_0, (\mathscr{T}_1)_0, \ldots, (\mathscr{T}_1)_n)].$$

(*Note*: $\mathscr{M}(\mathscr{T}_0, (\mathscr{T}_1)_0, \ldots, (\mathscr{T}_1)_n)$ denotes the collection of all sets of $\mathscr{M}(T_0, T_1, T_2, \ldots)$ denoted by constants of the form $r_{n+1}^1(c)$ for some $c \in \mathscr{C}(\mathscr{T}_0, \mathscr{T}_1, \ldots, \mathscr{T}_{n+1})$.) It follows from 4.13 that the sequence of sets $\mathscr{M}(\mathscr{T}_0, (\mathscr{T}_1)_0, \ldots, (\mathscr{T}_1)_n)$ $(n \geqq 0)$ can be well-ordered uniformly in n with probability 1. But then for some $f \in \mathscr{C}$, we have with probability 1: if $(x)(Ey)\mathscr{F}(x, y)$, then

$$f(\bar{0}) \in \mathscr{M}(\mathscr{T}_0),$$
$$f(\bar{n}+\bar{1}) \in \mathscr{M}(\mathscr{T}_0, (\mathscr{T}_1)_0, \ldots, (\mathscr{T}_1)_n) \qquad (n \geqq 0),$$
$$(x^0)\mathscr{F}(f(x^0), f(x^0+\bar{1})).$$

We can insist that $f \in \mathscr{C}(\mathscr{T}_0, \mathscr{T}_1)$, since the constants that denote the well-orderings needed in the definition of f are members of $\mathscr{C}(\mathscr{T}_0, \mathscr{T}_1)$.

The argument of Lemma 4.15 can be greatly simplified, if one merely wishes to establish the countable axiom of choice (see footnote (2)). In fact, Proposition 4.13 and Lemma 4.14 readily imply: with probability 1, the product of a family of nonempty sets, indexed by a well-ordered set, is nonempty.

A very simple notion of conditional probability of truth in $\mathscr{M}(T_0, T_1, T_2, \ldots)$ will make it possible to strengthen Lemma 4.12. Let $\{\mathscr{T}_i \mid i \in K\}$ be a finite set of $\mathscr{T}_i$-symbols; let $A_i \subseteq \omega$ for each $i \in K$. The *conditional probability that* $\mathscr{F}$ *is true in* $\mathscr{M}(T_0, T_1, T_2, \ldots)$, *given that* $T_i = A_i$ *for all* $i \in K$, is denoted by

$$p(\mathscr{F} \mid \mathscr{T}_i = A_i, i \in K),$$

and is defined as follows. Let $(2^N)^N_K$ be the set of all sequences $(T_0, T_1, T_2, \ldots)$ such that $T_i = A_i$ for all $i \in K$. Think of $(2^N)^N_K$ as the product of countably many copies of 2^N and finitely many one-point spaces, one one-point space $\{A_i\}$ for each $i \in K$. Give each copy of 2^N the usual Lebesgue measure, and give each one-point space measure 1. Let $(2^N)^N_K$ have the product measure induced by the measures assigned to its factors. Then $p(\mathscr{F} \mid \mathscr{T}_i = A_i, i \in K)$ is the measure of the set of all sequences $(T_0, T_1, T_2, \ldots)$ of $(2^N)^N_K$ such that $\mathscr{F}$ is true in $\mathscr{M}(T_0, T_1, T_2, \ldots)$.

Let r be a map of $\mathscr{C}(1)$ into $\mathscr{C}(1)$. Then for each i, $r(\mathscr{T}_i)$ is a member of $\mathscr{C}(1)$ in which only finitely many $\mathscr{T}_j$-symbols occur. Suppose for each i, either all or none of the $\mathscr{T}_j$-symbols occurring in $r(\mathscr{T}_i)$ also occur in the set $\{\mathscr{T}_i \mid i \in K\}$. We describe this state of affairs by saying that *fixing* $\{\mathscr{T}_i \mid i \in K\}$ *has the effect of fixing* $\{r(\mathscr{T}_i) \mid i \in K_r\}$, where K_r is the set of all i such that every $\mathscr{T}_j$-symbol occurring in $r(\mathscr{T}_i)$ also occurs in $\{\mathscr{T}_i \mid i \in K\}$. We will always assume, for convenience only, that K and K_r are finite. Observe that fixing $T_i = A_i$ for all $i \in K$ has the effect of fixing $r^*(T_i) = B_i$ for all $i \in K_r$, where B_i is the set in $\mathscr{M}(T_0, T_1, T_2, \ldots)$ denoted by $r(\mathscr{T}_i)$ when $T_i = A_i$ for all $i \in K$.

LEMMA 4.16. *Let r be a map of $\mathscr{C}(1)$ onto $\mathscr{C}(1)$ such that $r^*: (2^N)^N \to (2^N)^N$ is measure-preserving. Suppose that fixing $\{\mathscr{T}_i \mid i \in K\}$ has the effect of fixing*

$$\{r(\mathscr{T}_i) \mid i \in K_r\}.$$

Then for each sentence $\mathscr{F}$ of $\mathscr{L}(\mathscr{T}_0, \mathscr{T}_1, \mathscr{T}_2, \ldots)$,

$$p(r(\mathscr{F}) \mid \mathscr{T}_i = A_i, i \in K) = p(\mathscr{F} \mid \mathscr{T}_i = B_i, i \in K_r),$$

where for each $i \in K_r$, B_i is the set denoted by $r(\mathscr{T}_i)$ when $T_i = A_i$ for all $i \in K$.

Proof. We extend the argument of 4.12. The map $r^*: (2^N)^N \to (2^N)^N$ is 1-1. Let r^*_K be the restriction of r^* to $(2^N)^N_K$. Then r^*_K is a 1-1 map of $(2^N)^N_K$ into $(2^N)^N_{K_r}$. The lemma will now follow from 4.11, after we show r^*_K is measure-preserving. Let $\{c_i \mid i \in \omega\}$ be a sequence of constants of $\mathscr{C}$ which denote subsets of ω. We say $\{c_i \mid i \leqq n\}$ is a sequence of *independent, uniformly distributed ω-terms* if

$$p(\overline{m}_1 \in c_{i1} \;\&\; \overline{m}_2 \in c_{i2} \;\&\cdots\&\; \overline{m}_k \in c_{ik}) = 1/2^k$$

whenever $0 \leqq i1 \leqq i2 \leqq \cdots \leqq ik \leqq n$ and $m_j = m_p \to ij \neq ip$ for $1 \leqq j < p \leqq k$. We say $\{c_i \mid i \in \omega\}$ is a sequence of independent, uniformly distributed ω-terms if $\{c_i \mid i \leqq n\}$ is such a sequence for all $n \geqq 0$. It follows from 4.12 that $\{r(\mathscr{T}_i) \mid i \in \omega\}$ is a sequence of independent, uniformly distributed ω-terms. This last fact implies r^*_K is measure-preserving (cf. Halmos [10, p. 191–192]).

COROLLARY 4.17. *Let $\{\mathscr{T}_i \mid i \in K\}$ include all the $\mathscr{T}_i$-symbols occurring in the sentence $\mathscr{F}$ of $\mathscr{L}(\mathscr{T}_0, \mathscr{T}_1, \mathscr{T}_2, \ldots)$. Then for every sequence $\{A_i \mid i \in K\}$ of subsets of ω, $p(\mathscr{F} \mid \mathscr{T}_i = A_i, i \in K) = 0$ or 1*(13).

(13) S. Kochen pointed out that Corollary 4.17 is an instance of the 0–1 law [10, p. 201].

Proof. Let b denote an arbitrary propositional combination of sentences of the form $\bar{m} \in \mathcal{T}_i$, $\bar{n} \notin \mathcal{T}_j$, where $i, j \notin K$. Then $p(b)$ can be thought of as the measure of an arbitrary finite union of basic open subsets of $(2^N)^N_K$. ($(2^N)^N_K$ is defined just before 4.16.) It follows from 4.16 that

$$p(\mathcal{F} \;\&\; b \mid \mathcal{T}_i = A_i, i \in K) = p(b)p(\mathcal{F} \mid \mathcal{T}_i = A_i, i \in K).$$

For example, if $K = \{0\}$, b is $\bar{m} \in \mathcal{T}_1$, and r is the map defined by $r(\mathcal{T}_1) = \omega - \mathcal{T}_1$ and $r(\mathcal{T}_i) = \mathcal{T}_i$ $(i \neq 1)$, then

$$p(\mathcal{F} \;\&\; \bar{m} \in \mathcal{T}_1 \mid \mathcal{T}_0 = A_0) = p(\mathcal{F} \;\&\; \bar{m} \notin \mathcal{T}_1 \mid \mathcal{T}_0 = A_0)$$

by 4.16, and consequently,

$$p(\mathcal{F} \;\&\; m \in \mathcal{T}_1 \mid \mathcal{T}_0 = A_0) = \tfrac{1}{2}p(\mathcal{F} \mid \mathcal{T}_0 = A_0).$$

For each $\delta > 0$, there exists a b such that

$$p(\mathcal{F} \leftrightarrow b \mid \mathcal{T}_i = A_i, i \in K) \geqq 1 - \delta.$$

But then $p(\mathcal{F} \mid \mathcal{T}_i = A_i, i \in K)$ must be a solution of $x = x^2$.

Suppose $\{\mathcal{T}_i \mid i \in K\}$ includes all the $\mathcal{T}_i$-symbols occurring in $\mathcal{F}$, where $\mathcal{F}$ is a sentence of $\mathcal{L}(\mathcal{T}_0, \mathcal{T}_1, \mathcal{T}_2, \ldots)$. Then Corollary 4.17 tells us that the truth-value of $\mathcal{F}$ in the structure $\mathcal{M}(T_0, T_1, T_2, \ldots)$ is determined by the choice of $\{T_i \mid i \in K\}$ for almost all sequences $(T_0, T_1, T_2, \ldots)$. Let $(2^N)^K$ be the product of K-many copies of 2^N, one copy of 2^N for each $i \in K$. Give $(2^N)^K$ the usual product measure. Identify the sequences $\{A_i \mid i \in K\}$ with the points of $(2^N)^K$. Then by Fubini's theorem $p(\mathcal{F})$ is the measure of the set of all sequences $\{A_i \mid i \in K\}$ such that $p(\mathcal{F} \mid \mathcal{T}_i = A_i, i \in K) = 1$.

PROPOSITION 4.17.1 (CF. 4.4). *Let $\{\mathcal{F}_n \mid n \in \omega\}$ be a sequence of $\mathcal{M}$ of sentences of $\mathcal{L}(\mathcal{T}_0, \mathcal{T}_1, \ldots)$ of countable (in the sense of $\mathcal{M}$) ordinal rank such that for some k, no $\mathcal{T}_i$-symbol with $i \geqq k$ occurs in any $\mathcal{F}_n$. Then there exists a sentence $\mathcal{F}$ of countable ordinal rank containing no $\mathcal{T}_i$-symbols not occurring in at least one $\mathcal{F}_n$ and such that for almost all $(T_0, T_1, \ldots)$,*

$$\mathcal{M}(T_0, T_1, \ldots) \models \mathcal{F} \leftrightarrow (En)(\mathcal{M}(T_0, T_1, \ldots) \models \mathcal{F}_n).$$

In addition, $\mathcal{F}$ regarded as a function of $\{\mathcal{F}_n \mid n \in \omega\}$ is $\mathcal{M}$-definable.

Proof. The argument of 4.4 establishes the first part of the proposition, namely, that the desired $\mathcal{F}$ exists. The $\mathcal{M}$-definability of $\mathcal{F}$ as a function of $\{\mathcal{F}_n \mid n \in \omega\}$ follows from the $\mathcal{M}$-definability of the probability function p restricted to sentences of $\mathcal{L}(\mathcal{T}_0, \mathcal{T}_1, \ldots)$ of countable ordinal rank and from the following equations which characterize $\mathcal{F}$ up to a set of measure 0:

$$(i)\Big[p\Big(\bigvee_{n \leqq i} \mathcal{F}_n \rightarrow \mathcal{F}\Big) = 1\Big];$$

$$(\delta)(Ei)\Big[p\Big(\mathcal{F} \rightarrow \bigvee_{n \leqq i} \mathcal{F}_n\Big) \geqq 1 - \delta\Big];$$

δ is a variable ranging over the positive rationals.

LEMMA 4.17.2 (CF. 4.5). *For each $n \geq 0$: there exists an $\mathcal{M}$-definable function $\lambda\mathcal{F} \mid \mathcal{F}^*$, defined for all sentences $\mathcal{F}$ of $\mathcal{L}(\mathcal{T}_0, \mathcal{T}_1, \ldots)$ having at most n unranked quantifiers, such that $p(\mathcal{F} \leftrightarrow \mathcal{F}^*) = 1$ and such that $\mathcal{F}^*$ has countable (in the sense of $\mathcal{M}$) ordinal rank and contains no $\mathcal{T}_i$-symbols not occurring in $\mathcal{F}$.*

Proof. We proceed by an induction on rank after the fashion of 4.5. Let $\mathcal{F}$ be a prenex normal sentence of uncountable ordinal rank of the form $(Ex^\alpha)\mathcal{F}_1(x^\alpha)$ such that $\mathcal{T}_0$ is the only $\mathcal{T}_i$-symbol occurring in $\mathcal{F}$. By 4.14,

$$p(\mathcal{F}) = \text{lub}\left\{p\Big(\bigvee_{i\leq n} \mathcal{F}_1(c_i)\Big) \mid c_i \in \mathcal{C}(\alpha; \mathcal{T}_0, \mathcal{T}_1)\right\},$$

where lub denotes "least upper bound". As in 4.1, $\bigvee_{i\leq n} \mathcal{F}_1(c_i)$ is equivalent to a prenex normal sentence of lower rank than $(Ex^\alpha)\mathcal{F}_1(x^\alpha)$. Then

$$p\Big(\bigvee_{i\leq n} \mathcal{F}_1(c_i) \leftrightarrow \Big(\bigvee_{i\leq n} \mathcal{F}_1(c_i)\Big)^*\Big) = 1,$$

where $(\bigvee_{i\leq n} \mathcal{F}_1(c_i))^*$ has countable ordinal rank and contains no $\mathcal{T}_i$-symbols other than $\mathcal{T}_0$ or $\mathcal{T}_1$. It follows from the $\mathcal{M}$-definability of p that there exists an $\mathcal{M}$-definable sequence $\{c_i \mid i<\omega\} \subseteq \mathcal{C}(\alpha; \mathcal{T}_0, \mathcal{T}_1)$ such that

$$p(\mathcal{F}) = \text{lub}\left\{p\Big(\bigvee_{i\leq n} \mathcal{F}_1(c_i)\Big) \mid n < \omega\right\}.$$

Let $\mathcal{G}$ be a sentence of countable ordinal rank equivalent with probability 1 to $\bigvee_{n<\omega} (\bigvee_{i\leq n} \mathcal{F}_1(c_i))^*$. The existence of $\mathcal{G}$, as well as the fact that $\mathcal{G}$ contains no $\mathcal{T}_i$-symbols other than $\mathcal{T}_0$ or $\mathcal{T}_1$, is a consequence of 4.17.1.

If $\mathcal{G}$ contains only $\mathcal{T}_0$, then the desired $\mathcal{F}^*$ is $\mathcal{G}$. Suppose $\mathcal{G}$ contains both $\mathcal{T}_0$ and $\mathcal{T}_1$. We know that $p(\mathcal{F} \leftrightarrow \mathcal{G}) = 1$ and that the only $\mathcal{T}_i$-symbol occurring in $\mathcal{F}$ is $\mathcal{T}_0$. It follows from 4.17 that $p(\mathcal{G} \mid \mathcal{T}_0 = T_0)$ equals 0 or 1 for almost all T_0. The function $\lambda T_0 \mid p(\mathcal{G} \mid \mathcal{T}_0 = T_0)$ is $\mathcal{M}(T_0, T_1, \ldots)$-definable for almost all $(T_0, T_1, \ldots)$; its definition contains no $\mathcal{T}_i$-symbol other than $\mathcal{T}_0$, and has countable ordinal rank because $\mathcal{G}$ does. Thus there is a formula $\mathcal{H}(\mathcal{T}_0, x^0)$ of countable ordinal rank such that for almost all $(T_0, T_1, \ldots)$:

$$\mathcal{M}(T_0, T_1, \ldots) \models \mathcal{H}(\mathcal{T}_0, \bar{0}) \leftrightarrow p(\mathcal{G} \mid \mathcal{T}_0 = T_0) = 0;$$
$$\mathcal{M}(T_0, T_1, \ldots) \models \mathcal{H}(\mathcal{T}_0, \bar{1}) \leftrightarrow p(\mathcal{G} \mid \mathcal{T}_0 = T_0) = 1.$$

Then the desired $\mathcal{F}^*$ is $\mathcal{H}(\mathcal{T}_0, \bar{1})$.

Note that the above argument merely establishes the existence of $\mathcal{F}^*$ and not its $\mathcal{M}$-definability. The $\mathcal{M}$-definability of $\mathcal{F}^*$ is a consequence of the $\mathcal{M}$-definability of the probability function p (restricted to sentences of $\mathcal{L}(\mathcal{T}_0, \mathcal{T}_1, \ldots)$ having at most n unranked quantifiers) and the following characterization of $\mathcal{F}^*$ up to a set of measure 0: $p(\mathcal{F} \leftrightarrow \mathcal{F}^*) = 1$; $\mathcal{F}^*$ has countable ordinal rank and contains no $\mathcal{T}_i$-symbol not occurring in $\mathcal{F}$.

LEMMA 4.17.3 (CF. 4.7). *With probability 1: the power set axiom holds in $\mathcal{M}(T_0, T_1, \ldots)$.*

Proof. Same as 4.7 save that 4.5 is replaced by 4.17.2.

Feferman [3] showed that if $T_0, T_1, T_2, \ldots$ is a sequence generic in the sense of Cohen (i.e., generic in the sense of forcing with finite conditions), then in $\mathcal{M}(T_0, T_1, T_2, \ldots)$, the Boolean algebra of all subsets of ω has no nonprincipal, maximal ideal. The proof of 4.18 is a measure-theoretic analogue of his argument.

THEOREM 4.18. *With probability* 1: *The Boolean algebra of all subsets of* ω *has no nonprincipal, maximal ideal in the structure* $\mathcal{M}(T_0, T_1, T_2, \ldots)$.

Proof. Let $c \in \mathcal{C}$ be a constant which denotes a set of subsets of ω. Suppose $p(c$ is a nonprincipal, maximal ideal of $2^\omega) > 0$. Let $\mathcal{T}_0, \ldots, \mathcal{T}_{n-1}$ include all the $\mathcal{T}_i$-symbols occurring in c. By 4.17 and Fubini's theorem, there must exist sets $A_0, \ldots, A_{n-1}$ such that

$$p(c \text{ is a nonprincipal, maximal ideal of } 2^\omega \mid \mathcal{T}_i = A_i, i < n) = 1.$$

Let d' denote $\omega - d$. We intend to show

$$p(\mathcal{T}_n \in c \;\&\; \mathcal{T}_n' \in c \mid \mathcal{T}_i = A_i, i < n) = 1.$$

Either $p(\mathcal{T}_n \in c \mid \mathcal{T}_i = A_i, i<n)$ or $p(\mathcal{T}_n' \in c \mid \mathcal{T}_i = A_i, i<n)$ must be positive. Suppose the former is positive; we claim it must be equal 1. To see the claim, let B be the set of all A such that

$$p(\mathcal{T}_n \in c \mid \mathcal{T}_i = A_i, i < n, \mathcal{T}_n = A) = 1.$$

By 4.17 and Fubini's theorem, $p(\mathcal{T}_n \in c \mid \mathcal{T}_i = A_i, i<n)$ equals the measure of $B \subseteq 2^N$. Let fin (B) be the collection of all subsets of ω that differ finitely from some member of B. If fin $(B) = B$, then the measure of B must be 0 or 1 by a standard zero-one law argument [10]. Let $f(A)$ be a subset of ω differing finitely from A, and let $f(\mathcal{T}_n)$ be a term that differs from $\mathcal{T}_n$ in the same way that $f(A)$ differs from A. By 4.16,

$$\begin{aligned} p(f(\mathcal{T}_n) \in c \mid \mathcal{T}_i = A_i, i < n, \mathcal{T}_n = A) \\ = p(\mathcal{T}_n \in c \mid \mathcal{T}_i = A_i, i < n, \mathcal{T}_n = f(A)). \end{aligned}$$

But

$$p(\mathcal{T}_n \in c \leftrightarrow f(\mathcal{T}_n) \in c \mid \mathcal{T}_i = A_i, i < n) = 1,$$

since c denotes a *nonprincipal* maximal ideal with probability 1 when $\mathcal{T}_i$ is fixed at A_i for all $i<n$. That establishes the claim that $p(\mathcal{T}_n \in c \mid \mathcal{T}_i = A_i, i<n) = 1$. Another application of 4.16 gives

$$p(\mathcal{T}_n' \in c \mid \mathcal{T}_i = A_i, i < n) = 1.$$

In order to develop Solovay's relative consistency result on the extendability of Lebesgue measure, we shall replace $\mathcal{C}$ by $\mathcal{C}^*$, $\mathcal{L}(\mathcal{T}_0, \mathcal{T}_1, \mathcal{T}_2, \ldots)$ by $\mathcal{L}^*(\mathcal{T}_0, \mathcal{T}_1, \mathcal{T}_2, \ldots)$, and $\mathcal{M}(T_0, T_1, T_2, \ldots)$ by $\mathcal{M}^*(T_0, T_1, T_2, \ldots)$. These changes will be small but important; our sole purpose is to insure that $\mathcal{C}^*(1)$ is as symmetrical as possible from a measure-theoretic point of view. This device was originated by R. Solovay.

Let $c_0 = c_0(\mathcal{T}_0, \ldots, \mathcal{T}_n)$ be an ω-term of $\mathscr{C}$, that is to say, a constant of $\mathscr{C}$ which denotes a subset of ω. (We can assume that the $\mathcal{T}_i$-symbols occurring in c_0 are $\mathcal{T}_0, \mathcal{T}_1, \ldots, \mathcal{T}_n$ for all sufficiently large n, since we can always add $\mathcal{T}_i$-symbols to c_0 in a spurious fashion.) Recall the notion of "sequence of independent, uniformly distributed ω-terms" introduced at the end of the proof of 4.16. We say $c_0 = c_0(\mathcal{T}_0, \ldots, \mathcal{T}_n)$ is a *projective, uniformly distributed* ω-term if there exist ω-terms $c_i(\mathcal{T}_0, \ldots, \mathcal{T}_n)$ $(1 \leqq i \leqq n)$ and $k_i(\mathcal{T}_0, \ldots, \mathcal{T}_n)$ $(i \leqq n)$ of $\mathscr{C}$ such that $\{c_i \mid i \leqq n\}$ and $\{k_i \mid i \leqq n\}$ are sequences of independent, uniformly distributed ω-terms and such that

$$\mathcal{T}_i = k_i(c_0, \ldots, c_n) = c_i(k_0, \ldots, k_n)$$

for all $i \leqq n$. ($k_i(c_0, \ldots, c_n)$ is the result of substituting c_i for $\mathcal{T}_i$ in k_i.) Note that if c_0 is a projective, uniformly distributed ω-term by virtue of the existence of c_i $(1 \leqq i \leqq n)$ and k_i $(i \leqq n)$, then c_i $(1 \leqq i \leqq n)$ and k_i $(i \leqq n)$ also are projective, uniformly distributed ω-terms. It is tedious but routine to check that $(\mathcal{T}_m)_i$, $(\mathcal{T}_m)^n$ and d_n^m, which were defined immediately preceding the proof of Lemma 4.15, are projective, uniformly distributed ω-terms. The collection of all such terms is $\mathscr{M}$-definable, since $p(\mathscr{F})$, restricted to the ranked sentences of $\mathscr{L}(\mathcal{T}_0, \mathcal{T}_1, \mathcal{T}_2, \ldots)$, is $\mathscr{M}$-definable.

Let $\mathscr{C}^*(0) = \mathscr{C}(0) = \{\bar{n} \mid n \in \omega\}$. Let $\mathscr{C}^*(1)$ be the set of all projective, uniformly distributed ω-terms. In the previous paragraph it was remarked that $\mathscr{C}^*(1)$ is $\mathscr{M}$-definable. It is convenient to make $\mathscr{C}^*(1)$ a set of $\mathscr{M}$ by the following device. The proof for the power set axiom for $\mathscr{M}(T_0, T_1, T_2, \ldots)$ (essentially the argument of 4.7) shows that there exists a set $b \in \mathscr{M}$ of ω-terms such that with probability 1, each ω-term of $\mathscr{C}$ is equal to some ω-term of b. Thus there is no harm in restricting the concept of ω-term to b. $\mathscr{C}^*(\gamma+1)$ $(\gamma > 0)$ and $\mathscr{C}^*(\lambda)$ (λ a limit ordinal) are defined in the same manner as $\mathscr{C}(\gamma+1)$ and $\mathscr{C}(\lambda)$. The constants of $\mathscr{C}^*(\gamma+1)$ $(\gamma>0)$ denote the first-order definable subsets of $\mathscr{M}^*_{\gamma+1}(T_0, T_1, T_2, \ldots)$. The atomic formulas of $\mathscr{L}^*(\mathcal{T}_0, \mathcal{T}_1, \mathcal{T}_2, \ldots)$ are of the form $t_1 \in t_2$, where t_1 and t_2 are variables or constants of $\mathscr{C}^*$.

Let $\mathscr{M}^*_0(T_0, T_1, T_2, \ldots) = \omega$. Let $t \in \mathscr{C}^*(1)$; suppose $\{\mathcal{T}_i \mid i \in K\}$ is the set of all $\mathcal{T}_i$-symbols occurring in t. By Lemma 4.17,

$$p(\bar{n} \in t \mid \mathcal{T}_i = T_i, i \in K) = 0 \text{ or } 1.$$

We say that t denotes $\{n \mid p(\bar{n} \in t \mid \mathcal{T}_i = T_i, i \in K) = 1\}$. $\mathscr{M}^*_1(T_0, T_1, T_2, \ldots)$ is the set of all sets denoted by some $t \in \mathscr{C}^*(1)$. $\mathscr{M}^*_{\gamma+1}(T_0, T_1, T_2, \ldots)$ $(\gamma > 0)$ is defined in the same inductive fashion as $\mathscr{M}_{\gamma+1}(T_0, T_1, T_2, \ldots)$. Thus the only conceptual difference between $\mathscr{M}(T_0, T_1, T_2, \ldots)$ and $\mathscr{M}^*(T_0, T_1, T_2, \ldots)$, or between $\mathscr{L}(\mathcal{T}_0, \mathcal{T}_1, \mathcal{T}_2, \ldots)$ and $\mathscr{L}^*(\mathcal{T}_0, \mathcal{T}_1, \mathcal{T}_2, \ldots)$, is the difference between $\mathscr{C}(1)$ and $\mathscr{C}^*(1)$.

A routine transfinite induction establishes the existence of an $\mathscr{M}$-definable $f: \mathscr{C}^* \to \mathscr{C}$ with the following properties for each $c \in \mathscr{C}^*$: if c is an ω-term, then $f(c)$

is an ω-term; c and $f(c)$ have the same occurrences of $\mathcal{T}_j$-symbols; the set denoted by c in $\mathcal{M}^*(T_0, T_1, \ldots)$ equals the set denoted by $f(c)$ in $\mathcal{M}(T_0, T_1, \ldots)$ for all $(T_0, T_1, \ldots)$.

For each sentence $\mathcal{F}$ of $\mathcal{L}^*(\mathcal{T}_0, \mathcal{T}_1, \mathcal{T}_2, \ldots)$, the probability that $\mathcal{F}$ is true in $\mathcal{M}^*(T_0, T_1, T_2, \ldots)$, denoted by $p^*(\mathcal{F})$, is the measure of the Borel set

$$\{(T_0, T_1, T_2, \ldots) \mid \mathcal{M}^*(T_0, T_1, T_2, \ldots) \models \mathcal{F}\}.$$

The argument of 4.1 shows: for each $n \geqq 0$, the function $p^*(\mathcal{F})$, restricted to sentences of $\mathcal{L}^*(\mathcal{T}_0, \mathcal{T}_1, \mathcal{T}_2, \ldots)$ having at most n unranked quantifiers, is $\mathcal{M}$-definable. Note that the peculiar definition of $\mathcal{C}^*(1)$ requires the use of the $\mathcal{M}$-definability of p for ranked sentences of $\mathcal{L}(\mathcal{T}_0, \mathcal{T}_1, \mathcal{T}_2, \ldots)$ in the proof of the $\mathcal{M}$-definability of p^*. The arguments of 4.17.1–3 show that with probability 1, $\mathcal{M}^*(T_0, T_1, T_2, \ldots)$ is a model of ZF. Observe that if $t \in \mathcal{C}^*(1) \subseteq \mathcal{C}$, then $p^*(\bar{n} \in t) = p(\bar{n} \in t)$ for all $n \in \omega$.

LEMMA 4.19. *Let r be a map of $\mathcal{C}^*(1)$ onto $\mathcal{C}^*(1)$ such that $r^*: (2^N)^N \to (2^N)^N$ is measure-preserving. Suppose that fixing $\{\mathcal{T}_i \mid i \in K\}$ has the effect of fixing*

$$\{r(\mathcal{T}_i) \mid i \in K_r\}.$$

Then for each sentence $\mathcal{F}$ of $\mathcal{L}^*(\mathcal{T}_0, \mathcal{T}_1, \mathcal{T}_2, \ldots)$,

$$p(r(\mathcal{F}) \mid \mathcal{T}_i = A_i, i \in K) = p(\mathcal{F} \mid \mathcal{T}_i = B_i, i \in K_r),$$

where for each $i \in K_r$, B_i is the set denoted by $r(\mathcal{T}_i)$ in $\mathcal{M}^*(T_0, T_1, T_2, \ldots)$ when $T_i = A_i$ for all $i \in K$.

Proof. Same as 4.16. The arguments of 4.11 and 4.12 are correct for $\mathcal{M}^*(T_0, T_1, T_2, \ldots)$. The differences between $\mathcal{C}(1)$ and $\mathcal{C}^*(1)$ are irrelevant.

LEMMA 4.20. *Let $\{\mathcal{T}_i \mid i \in K\}$ include all the $\mathcal{T}_i$-symbols occurring in the sentence $\mathcal{F}$ of $\mathcal{L}^*(\mathcal{T}_0, \mathcal{T}_1, \mathcal{T}_2, \ldots)$. Then for every sequence $\{A_i \mid i \in K\}$ of subsets of ω, $p(\mathcal{F} \mid \mathcal{T}_i = A_i, i \in K) = 0$ or 1.*

Proof. Same as 4.17. But it is necessary to check that the maps of $\mathcal{C}(1)$ onto $\mathcal{C}(1)$ employed in the proof of 4.17 are also maps of $\mathcal{C}^*(1)$ onto $\mathcal{C}^*(1)$. This is a routine matter, since all the relevant terms of $\mathcal{C}$ ($\omega - \mathcal{T}_1$, etc.) are projective, uniformly distributed ω-terms.

LEMMA 4.21. *With probability* 1, *the dependent axiom of choice holds in $\mathcal{M}^*(T_0, T_1, T_2, \ldots)$.*

Proof. Same as 4.15. But it is necessary to check that the maps of $\mathcal{C}(1)$ onto $\mathcal{C}(1)$ employed in the proofs of 4.14 and 4.15 are also maps of $\mathcal{C}^*(1)$ onto $\mathcal{C}^*(1)$.

From now on we will identify elements of 2^N with real numbers in the closed unit interval [0, 1] by means of the usual dyadic expansions. Thus if $A, B \in 2^N$, then $A + B$ (mod 1) will correspond to the arithmetic sum (mod 1) of the real numbers corresponding to A and B. A similar remark applies to ω-terms of $\mathcal{C}^*$.

LEMMA 4.22. (a) *Let* $c(\mathscr{T}_0, \ldots, \mathscr{T}_n) \in \mathscr{C}^*(1)$; *then there exists a map* r *of* $\mathscr{C}^*(1)$ *onto* $\mathscr{C}^*(1)$ *such that* $r(\mathscr{T}_0)=c(\mathscr{T}_0, \ldots, \mathscr{T}_n)$, $r(\mathscr{T}_1)=\mathscr{T}_{n+1}$, *and the induced map* $r^*: (2^N)^N \to (2^N)^N$ *is measure-preserving.*

(b) *If* $t(\mathscr{T}_0, \ldots, \mathscr{T}_{n-1})$ *is an* ω*-term of* $\mathscr{C}^*$, *then there exists a* $c_n \in \mathscr{C}^*(1)$ *such that* $c_n = \mathscr{T}_n + t(\mathscr{T}_0, \ldots, \mathscr{T}_{n-1})$ (mod 1) *with probability* 1. *In addition, there exists a map* r *of* $\mathscr{C}^*(1)$ *onto* $\mathscr{C}^*(1)$ *such that* $r(\mathscr{T}_i)=\mathscr{T}_i$ $(i<n)$, $r(\mathscr{T}_n)=c_n$, *and the induced map* $r^*: (2^N)^N \to (2^N)^N$ *is measure-preserving.*

Proof. Since $c_0=c(\mathscr{T}_0, \ldots, \mathscr{T}_n) \in \mathscr{C}^*(1)$, there must exist $c_i(\mathscr{T}_0, \ldots, \mathscr{T}_n)$ and $k_i(\mathscr{T}_0, \ldots, \mathscr{T}_n)$ $(i \leq n)$ such that $\{c_i \mid i \leq n\}$ and $\{k_i \mid i \leq n\}$ are sequences of independent, uniformly distributed ω-terms of $\mathscr{C}^*(1)$ and such that $\mathscr{T}_i=k_i(c_0, \ldots, c_n)$ $(i \leq n)$. Then the map r, defined by

$$r(\mathscr{T}_0) = c(\mathscr{T}_0, \ldots, \mathscr{T}_n) = c_0,$$
$$r(\mathscr{T}_1) = \mathscr{T}_{n+1}, r(\mathscr{T}_{n+k}) = \mathscr{T}_{n+k} \qquad (k > 1),$$
$$r(\mathscr{T}_{i+1}) = c_i \qquad (1 \leq i \leq n),$$

maps $\mathscr{C}^*(1)$ onto $\mathscr{C}^*(1)$. The induced map $r^*: (2^N)^N \to (2^N)^N$ is measure-preserving, since $\{r(\mathscr{T}_i) \mid i \in \omega\}$ is a sequence of independent, uniformly distributed ω-terms. (Cf. end of proof of 4.16.)

By the remarks immediately preceding 4.19, there exists an ω-term $t_1(\mathscr{T}_0, \ldots, \mathscr{T}_{n-1}) \in \mathscr{C}$ such that $\mathscr{T}_n+t(\mathscr{T}_0, \ldots, \mathscr{T}_{n-1})$ denotes the same set in $\mathscr{M}^*(T_0, T_1, \ldots)$ that $\mathscr{T}_n+t_1(\mathscr{T}_0, \ldots, \mathscr{T}_{n-1})$ denotes in $\mathscr{M}(T_0, T_1, \ldots)$. We must show $c_n=\mathscr{T}_n+t_1$ is a projective, uniformly distributed ω-term. For this purpose, it is convenient to think of the ω-terms of $\mathscr{C}$ as random variables [5] defined on the sample space $(2^N)^N$. An ω-term is uniformly distributed in our sense if and only if it is uniformly distributed in the standard sense when it is interpreted as a random variable. Thus if $c_0, \ldots, c_n$ is a sequence of uniformly distributed ω-terms such that c_i and c_j $(0 \leq i<j \leq n)$ have no $\mathscr{T}_k$-symbols in common, then $c_0, \ldots, c_n$ is a sequence of *independent*, uniformly distributed ω-terms (cf. Halmos [10, p. 194]). We claim

$$\mathscr{T}_0, \ldots, \mathscr{T}_{n-1}, \mathscr{T}_n+t_1(\mathscr{T}_0, \ldots, \mathscr{T}_{n-1}), \mathscr{T}_{n+1}, \ldots$$

is a sequence of independent, uniformly distributed ω-terms. The claim is a consequence of the following type of standard argument. Let X_0, X_1, Y be random variables on $(2^N)^N$ such that X_0, X_1 are independent and uniformly distributed, and such that X_0, Y are independent. Then X_0+Y (mod 1), X_1 are independent and uniformly distributed, since

$$\begin{aligned} p(X_0+Y=a \;\&\; X_1=c) &= \sum_b p(Y=a-b \;\&\; X_0=b \;\&\; X_1=c) \\ &= p(X_0=a)p(X_1=c)\sum_b p(Y=a-b \mid X_0=b \;\&\; X_1=c) \\ &= p(X_0=a)p(X_1=c). \end{aligned}$$

To check that $c_n=\mathscr{T}_n+t_1(\mathscr{T}_0, \ldots, \mathscr{T}_{n-1})$ is projective and uniformly distributed, let $c_i = k_i = \mathcal{T}_i$ $(i<n)$ and let $k_n = \mathcal{T}_n - t_1(\mathcal{T}_0, \ldots, \mathcal{T}_{n-1})$ (mod 1). The desired map r is defined by $r(\mathcal{T}_i) = \mathcal{T}_i$ $(i \neq n)$ and $r(\mathcal{T}_n) = c_n$.

Let x^Ω be a restricted variable of $\mathcal{L}^*(\mathcal{T}_0, \mathcal{T}_1, \ldots)$, restricted to the subsets of ω in $\mathcal{M}^*(T_0, T_1, \ldots)$. Thus we can denote each subset of 2^N in $\mathcal{M}^*(T_0, T_1, \ldots)$ by a constant of the form $\hat{x}^\Omega \mathcal{F}(x^\Omega)$, where $\mathcal{F}(x^\Omega)$ is a formula of $\mathcal{L}^*(\mathcal{T}_0, \mathcal{T}_1, \ldots)$ whose only free variable is x^Ω. For each $(T) = (T_0, T_1, T_2, \ldots)$, we define the *absolute* (in T) *measure* of $\hat{x}^\Omega \mathcal{F}(x^\Omega)$, denoted by $\mu_a^T(\hat{x}^\Omega \mathcal{F}(x^\Omega))$, as follows:

$$\mu_a^T(\hat{x}^\Omega \mathcal{F}(x^\Omega)) = p^*(\mathcal{F}(\mathcal{T}_j) \mid \mathcal{T}_i = T_i,\ i \in K),$$

where $j \notin K$ and $\{\mathcal{T}_i \mid i \in K\}$ is the set of all $\mathcal{T}_i$-symbols occurring in $\mathcal{F}$. (By 4.19, the value of j is irrelevant so long as $j \notin K$.) For each $(T) = (T_0, T_1, T_2, \ldots)$, we say $\mathcal{M}^*(T_0, T_1, T_2, \ldots)$ is *measure-complete* if for every $\mathcal{F}(x^\Omega) \in \mathcal{L}^*(\mathcal{T}_0, \mathcal{T}_1, \ldots)$, we have

$$\mu_a^T(\hat{x}^\Omega \mathcal{F}(x^\Omega)) > 0 \to \mathcal{M}^*(T_0, T_1, \ldots) \models (Ex^\Omega)\mathcal{F}(x^\Omega).$$

The concept of absolute (in T) measure has been defined for definitions of subsets of 2^N in $\mathcal{M}^*(T_0, T_1, \ldots)$ rather than subsets of 2^N; for each $(T) = (T_0, T_1, \ldots)$, let us say μ_a^T *is well-defined on all subsets of* 2^N if all definitions of any given subset of 2^N in $\mathcal{M}^*(T_0, T_1, \ldots)$ have the same absolute (in T) measure.

LEMMA 4.23. *With probability* 1: (a) $\mathcal{M}^*(T_0, T_1, T_2, \ldots)$ *is measure-complete*; (b) *if* $\mathcal{M}^*(T_0, T_1, T_2, \ldots)$ *is measure-complete, then* μ_a^T *is well-defined on all subsets of* 2^N *in* $\mathcal{M}^*(T_0, T_1, T_2, \ldots)$.

Proof. First we do (a). Let $\mathcal{T}_0$ be the only $\mathcal{T}_i$-symbol occurring in $\mathcal{F}(x^\Omega, \mathcal{T}_0)$. We must show: for almost all $(T) = (T_0, T_1, T_2, \ldots)$, if $\mu_a^T(\hat{x}^\Omega \mathcal{F}(x^\Omega, \mathcal{T}_0)) > 0$, then $\mathcal{M}^*(T_0, T_1, T_2, \ldots) \models (Ex^\Omega)\mathcal{F}(x^\Omega, \mathcal{T}_0)$. Suppose not. Then by 4.20 and Fubini's theorem, there exists a set $K \subseteq 2^N$ of positive measure such that for all $T_0 \in K$,

$$p^*(\sim(Ex^\Omega)\mathcal{F}(x^\Omega, \mathcal{T}_0) \mid \mathcal{T}_0 = T_0) = 1,$$
$$p^*(\mathcal{F}(\mathcal{T}_1, \mathcal{T}_0) \mid \mathcal{T}_0 = T_0) > 0.$$

Now we do (b). Let $\mathcal{F}_0(x^\Omega, \mathcal{T}_0)$ and $\mathcal{F}_1(x^\Omega, \mathcal{T}_1)$ be formulas such that $\mathcal{T}_i$ is the only $\mathcal{T}_i$-symbol occurring in $\mathcal{F}_i(x^\Omega, \mathcal{T}_i)$ $(i<2)$. By 4.20 and Fubini's theorem,

$$m_0 = \mu_a^T(\hat{x}^\Omega \mathcal{F}_0(x^\Omega, \mathcal{T}_0)) = p^*(\mathcal{F}_0(\mathcal{T}_2, \mathcal{T}_0) \mid \mathcal{T}_0 = T_0, \mathcal{T}_1 = T_1)$$

for almost all $(T) = (T_0, T_1, T_2, \ldots)$. Similarly,

$$m_1 = \mu_a^T(\hat{x}^\Omega \mathcal{F}_1(x^\Omega, \mathcal{T}_1)) = p^*(\mathcal{F}_1(\mathcal{T}_2, \mathcal{T}_1) \mid \mathcal{T}_0 = T_0, \mathcal{T}_1 = T_1)$$

for almost all (T). Let $\mathcal{F}_0 \bigtriangleup \mathcal{F}_1$ denote $(\mathcal{F}_0 \mathbin{\&} \sim\mathcal{F}_1) \vee (\mathcal{F}_1 \mathbin{\&} \sim\mathcal{F}_0)$. Then for almost all (T), if

$$m_\bigtriangleup = \mu_a^T(\hat{x}^\Omega(\mathcal{F}_0(x^\Omega, \mathcal{T}_0) \bigtriangleup \mathcal{F}_1(x^\Omega, \mathcal{T}_1))) = 0,$$

then $m_0=m_1$. Finally, for all $(T)=(T_0, T_1, T_2, \ldots)$, if $\mathcal{M}^*(T_0, T_1, T_2, \ldots)$ is measure-complete and

$$\mathcal{M}^*(T_0, T_1, T_2, \ldots) \models (x^\Omega)(\mathcal{F}_0(x^\Omega, \mathcal{T}_0) = \mathcal{F}_1(x^\Omega, \mathcal{T}_1)),$$

then $m_\Delta=0$.

Lemma 4.23 strongly suggests that μ_a^T is a good candidate for a countably additive, translation-invariant extension of Lebesgue measure to all subsets of 2^N in $\mathcal{M}^*(T_0, T_1, T_2, \ldots)$ for almost all $(T)=(T_0, T_1, T_2, \ldots)$. We shall of course carry out this suggestion, but there is a delicate problem, easily overlooked, that we must face immediately. The definition of μ_a^T was given by means of terms denoting subsets of 2^N in $\mathcal{M}^*(T_0, T_1, T_2, \ldots)$. With probability 1, there is no map in $\mathcal{M}^*(T_0, T_1, T_2, \ldots)$ which associates with each subset of 2^N a term denoting that subset; this follows from the argument of Theorem 4.18. Nonetheless, we are able to define μ_a^T in $\mathcal{M}(T_0, T_1, \ldots)$ for almost all $(T)=(T_0, T_1, \ldots)$ by exploiting the measure-theoretic symmetries of $\mathscr{C}^*(1)$. In fact, Lemma 4.24 is the sole reason we insisted that every ω-term of $\mathscr{C}^*(1)$ be projective and uniformly distributed.

LEMMA 4.24. *With probability* 1: μ_a^T *is* $\mathcal{M}(T_0, T_1, \ldots)$*-definable.*

Proof. Let s be an arbitrary subset of 2^N in $\mathcal{M}^*(T_0, T_1, T_2, \ldots)$. For almost all $(T)=(T_0, T_1, T_2, \ldots)$, there exist $\mathcal{M}^*(T_0, T_1, T_2, \ldots)$-definable maps f and b such that $f(s)=R(x^\Omega, y)$ is a ranked formula of $\mathcal{L}^*(\mathcal{T}_0, \mathcal{T}_1, \ldots)$ in which no $\mathcal{T}_i$-symbols occur, $b(s)=b$ is a subset of ω in $\mathcal{M}^*(T_0, T_1, T_2, \ldots)$, and $\hat{x}^\Omega R(x^\Omega, y)$ denotes s in $\mathcal{M}^*(T_0, T_1, T_2, \ldots)$ when y is interpreted as b. The set b is merely an encoding of the finitely many T_i's that are denoted by $\mathcal{T}_i$-symbols in some term denoting s; it is easily checked that all such encodings can be accomplished by projective, uniformly distributed ω-terms (cf. d_n^m defined immediately preceding 4.15). Thus b is denoted by some term $c(\mathcal{T}_0, \mathcal{T}_1, \mathcal{T}_2) \in \mathscr{C}^*(1)$, and

$$\mu_a^T(s) = p^*(R(\mathcal{T}_3, c(\mathcal{T}_0, \mathcal{T}_1, \mathcal{T}_2)) \mid \mathcal{T}_i = T_i, i \leqq 2).$$

Unfortunately there is (with probability 1) no $\mathcal{M}^*(T_0, T_1, \ldots)$-definable map h such that $h(r)=c(\mathcal{T}_0, \mathcal{T}_1, \mathcal{T}_2)$. But, thanks to the symmetry of $\mathscr{C}^*(1)$, we don't need c. We claim:

$$\mu_a^T(s) = p^*(R(\mathcal{T}_1, \mathcal{T}_0) \mid \mathcal{T}_0 = b).$$

The claim follows from 4.22 (a), 4.19, and Fubini's theorem. Finally, note that the conditional probability function $p^*(\mathcal{F} \mid \mathcal{T}_0=b)$, restricted to ranked sentences $\mathcal{F} \in \mathcal{L}^*(\mathcal{T}_0, \mathcal{T}_1, \ldots)$ and sets $b \in \mathcal{M}^*(T_0, T_1, \ldots)$, is $\mathcal{M}^*(T_0, T_1, \ldots)$-definable for almost all $(T_0, T_1, \ldots)$ by a routine modification of the argument of 4.1.

LEMMA 4.25. *With probability* 1: μ_a^T *is countably additive, translation-invariant, and agrees with Lebesgue measure on all Lebesgue-measurable sets.*

Proof. Let $\mathcal{F}(x^\Omega, \mathcal{T}_0, x^0)$ be a formula of $\mathcal{L}^*(\mathcal{T}_0, \mathcal{T}_1, \ldots)$ whose only $\mathcal{T}_i$-symbol is $\mathcal{T}_0$. Let $A \subseteq \omega$ be such that for all $m \neq n$,

$$p^*((x^\Omega) \sim [\mathcal{F}(x^\Omega, \mathcal{T}_0, \overline{m}) \,\&\, \mathcal{F}(x^\Omega, \mathcal{T}_0, \overline{n})] \mid \mathcal{T}_0 = A) = 1.$$

Let $(T)=(A, T_1, T_2, \ldots)$. Then

$$\sum_n \mu_a^T(\hat{x}^\Omega \mathscr{F}(x^\Omega, \mathscr{T}_0, \bar{n})) = \sum_n p^*(\mathscr{F}(\mathscr{T}_1, \mathscr{T}_0, \bar{n}) \mid \mathscr{T}_0 = A)$$
$$= p^*((Ex^0)\mathscr{F}(\mathscr{T}_1, \mathscr{T}_0, x^0) \mid \mathscr{T}_0 = A)$$
$$= \mu_a^T(\hat{x}^\Omega (Ex^0)\mathscr{F}(x^\Omega, \mathscr{T}_0, x^0)).$$

The countable additivity of μ_a^T for almost all $(T)=(T_0, T_1, T_2, \ldots)$ now follows from 4.20 and Fubini's theorem.

We establish the translation-invariance of μ_a^T for almost all (T) by a variation of the argument of 4.24. Let $\mathscr{F}(x^\Omega, \mathscr{T}_0)$ be a formula of $\mathscr{L}^*(\mathscr{T}_0, \mathscr{T}_1, \ldots)$ whose only $\mathscr{T}_i$-symbol is $\mathscr{T}_0$, and let $c(\mathscr{T}_0, \mathscr{T}_1)$ be an ω-constant whose only $\mathscr{T}_i$-symbols are $\mathscr{T}_0$ and $\mathscr{T}_1$. Then

$$\mu_c = \mu_a^T(\hat{x}^\Omega \mathscr{F}(x^\Omega + c(\mathscr{T}_0, \mathscr{T}_1), \mathscr{T}_0))$$
$$= p^*(\mathscr{F}(\mathscr{T}_2 + c(\mathscr{T}_0, \mathscr{T}_1), \mathscr{T}_0) \mid \mathscr{T}_i = T_i, i \leqq 1)$$

for all $(T)=(T_0, T_1, \ldots)$; and

$$\mu = \mu_a^T(\hat{x}^\Omega \mathscr{F}(x^\Omega, \mathscr{T}_0)) = p^*(\mathscr{F}(\mathscr{T}_2, \mathscr{T}_0) \mid \mathscr{T}_i = T_i, i \leqq 1)$$

for almost all (T) by 4.20 and Fubini's theorem. By 4.22 (b) and 4.19,

$$p^*(\mathscr{F}(\mathscr{T}_2 + c(\mathscr{T}_0, \mathscr{T}_1), \mathscr{T}_0) \mid \mathscr{T}_i = T_i, i \leqq 2)$$
$$= p^*(\mathscr{F}(\mathscr{T}_2, \mathscr{T}_0) \mid \mathscr{T}_i = T_i, i \leqq 1, \mathscr{T}_2 = T_2 + c(T_0, T_1))$$

for almost all (T). But then by Fubini's theorem $\mu_c = \mu$ for almost all T.

For almost all (T), μ_a^T agrees with Lebesgue measure on all Lebesgue-measurable sets because μ_a^T is countably additive and agrees with Lebesgue measure on all basic open subsets of 2^N. (Of course we are using the fact that every countably additive measure on 2^N is regular (Halmos [10]); this fact requires the countable axiom of choice in its proof; fortunately, 4.21 provides us with the countable axiom of choice for almost all (T).)

THEOREM 4.26 (SOLOVAY [29], [30]). *If ZF is consistent, then ZF+"there exists a countably additive, translation-invariant extension of Lebesgue measure to all sets of reals"+"dependent axiom of choice" is consistent.*

Proof. By 4.21, 4.24 and 4.25, $\mathscr{M}^*(T_0, T_1, T_2, \ldots)$ is the desired model with probability one.

Scott and Solovay [25], [26] have devised a very valuable way of viewing forcing arguments in terms of homomorphisms of Boolean algebras. Consider what happens from their point of view when a Cohen-generic $T \subseteq N$ is added to $\mathscr{M}$. Cohen [1] defined a forcing relation, $q \Vdash \mathscr{F}$, where q is a finite set of conditions on T and $\mathscr{F}$ is a sentence of $\mathscr{L}(\mathscr{T})$. Suppose q is interpreted as a basic open subset of 2^N. Then

$$\phi(\mathscr{F}) = \bigcup \{q \mid q \Vdash \sim\sim \mathscr{F}\}$$

defines a map from the sentences of $\mathscr{L}(\mathscr{T})$ into RO, the Boolean algebra of regular open subsets of 2^N. RO is a complete, Boolean algebra, and ϕ is a homomorphism in the sense of complete Boolean algebras. By that we mean, $\phi(\mathscr{F}_1 \vee \mathscr{F}_2)=\phi(\mathscr{F}_1) \vee \phi(\mathscr{F}_2)$. $\phi(\sim\mathscr{F})=$ interior of $2^N-\phi(\mathscr{F})$, and $\phi((Ex^\alpha)\mathscr{F}(x^\alpha))=\bigcup\{\phi(c) \mid c \in \mathscr{C}(\alpha)\}$. Cohen's key lemma is: for each $n \geqq 0$, the map ϕ, restricted to sentences having at most n unranked quantifiers, is $\mathscr{M}$-definable. The $\mathscr{M}$-definability of ϕ, together with the fact that ϕ is a homomorphism in the sense of complete Boolean algebras, is all that is used to show $\phi(\mathscr{F})=1$ for each instance $\mathscr{F}$ of the replacement axiom.

Our probability function $p(\mathscr{F})$ is a map from the sentences of $\mathscr{L}(\mathscr{T})$ into the usual linear ordering of the reals in [0, 1]. Now p is not a homomorphism, since $p(\mathscr{F}_1 \vee \mathscr{F}_2)$ need not equal

$$\max(p(\mathscr{F}_1), p(\mathscr{F}_2)).$$

But p is a "near-homomorphism"; it preserves negations and monotonic unions: $p(\sim\mathscr{F})=1-p(\mathscr{F})$, and $p((Ex^\alpha)\mathscr{F}(x^\alpha))=$ least upper bound of $\{p(\mathscr{F}(c_0) \vee \cdots \vee \mathscr{F}(c_n)) \mid c_i \in \mathscr{C}(\alpha)\}$. The fact that p is not a homomorphism is exactly the reason that prenex normal form was exploited so heavily in the proof (Lemma 4.1) of the $\mathscr{M}$-definability of p. The $\mathscr{M}$-definability of ϕ, together with the fact that p is a "near-homomorphism", was enough to show (Lemmas 4.2 and 4.3) that $p(\mathscr{F})=1$ for each instance $\mathscr{F}$ of the replacement axiom. Thus the measure-theoretic uniformity approach seems to suggest that the Scott-Solovay Boolean-valued model point of view can be generalized still further, possibly by dropping homomorphisms in favor of "near-homomorphisms".

Solovay [29] significantly extended Cohen's forcing method by using closed subsets of 2^N of positive measure as forcing conditions. His forcing relation, $q \Vdash \mathscr{F}$, is identical with Cohen's save that q is a set of natural numbers of $\mathscr{M}$ which encodes a closed subset of 2^N of positive measure. The proof of Solovay's Theorem (4.26) is based on this notion of forcing. Call $T \subseteq N$ Solovay-generic if it is generic in his sense of forcing. Call T *fundamental* if for all sentences $\mathscr{F}$ of ZF, $p(\mathscr{F})=1 \leftrightarrow \mathscr{M}(T) \models \mathscr{F}$. (By Corollary 4.17, $p(\mathscr{F})=1$ or 0 if $\mathscr{F}$ has no $\mathscr{T}_i$-symbols.) Solovay has observed that T is fundamental if and only if T has the same degree of nonconstructibility as some Solovay-generic T'. (In this case, T and T' have the same degree if each is constructible from the other via the ordinals of $\mathscr{M}$.) Thus the transition described in the previous paragraph from homomorphisms to "near-homomorphisms" might correspond to a transition from sets to degrees of sets.

References

1. J. W. Addison, *The undefinability of the definable*, Abstract 622–71, Notices Amer. Math. Soc. **12** (1965), 347–348.

2. P. J. Cohen, *The independence of the continuum hypothesis*. I, II, Proc. Nat. Acad. Sci. U.S.A. **50** (1963), 1143–1148; **51** (1964), 105–110.

3. S. Feferman, *Some applications of the notions of forcing and generic sets*, Fund. Math. **56** (1965), 325–345.

4. S. Feferman and C. Spector, *Incompleteness along paths in progressions of theories*, J. Symbolic Logic **27** (1962), 383–390.

5. W. Feller, *An introduction to probability theory and its applications*, Wiley, New York, 1957.

6. R. O. Gandy, *Proof of Mostowski's conjecture*, Bull. Acad. Polon. Sci. Ser. Math. **8** (1960), 571–575.

7. ———, *On a problem of Kleene*, Bull. Amer. Math. Soc. **66** (1960), 501–502.

8. K. Gödel, *Consistency proof of the generalized continuum hypothesis*, Proc. Nat. Acad. Sci. U.S.A. **25** (1939), 220–224.

9. ———, *The consistency of the axiom of choice and of the generalized continuum hypothesis with the axioms of set theory*, 4th printing, Princeton Univ. Press, Princeton, N. J., 1958.

10. P. R. Halmos, *Measure theory*, Van Nostrand, Princeton, N. J., 1950.

11. S. C. Kleene, *On the forms of the predicates in the theory of constructive ordinals* (*second paper*), Amer. J. Math. **77** (1955), 405–428.

12. ———, *Arithmetical predicates and function quantifiers*, Trans. Amer. Math. Soc. **79** (1955), 312–340.

13. ———, *Hierarchies of number-theoretic predicates*, Bull. Amer. Math. Soc. **61** (1955), 193–213.

14. ———, *Quantification of number-theoretic functions*, Compositio Math. **14** (1959), 23–40.

15. G. Kreisel, *Set theoretic problems suggested by the notion of potential totality*, Proc. Sympos. Infinitistic Methods, Warsaw, 1961, pp. 103–140.

16. ———, *The axiom of choice and the class of hyperarithmetic functions*, Indag. Math. **24** (1962), 307–319.

17. G. Kreisel and G. E. Sacks, *Metarecursive sets*, J. Symbolic Logic **30** (1965), 318–338.

18. S. Kripke, *Transfinite recursions on admissible ordinals.* I and II (abstracts), J. Symbolic Logic **29** (1964), 161–162.

19. R. A. Platek, *Foundations of recursion theory*, Ph.D. Thesis, Stanford Univ., Stanford, Calif., 1965.

20. H. Rogers, *Theory of recursive functions and effective computability*, McGraw-Hill, New York, 1967.

21. J. Rosenthal, *Addition of relations in model theory*; *Truth in all of certain well-founded countable models arising in set theory*, Ph.D. Thesis, Massachusetts Institute of Technology, Cambridge, Mass., 1968.

22. G. E. Sacks, *Degrees of unsolvability*, Princeton Univ. Press, Princeton, N. J., 1963.

23. ———, *Measure-theoretic uniformity*, Bull. Amer. Math. Soc. **73** (1967), 169–174.

24. ———, *Higher recursion theory*, Lecture Notes, Massachusetts Institute of Technology, Cambridge, Mass., 1966.

25. D. Scott, *Lectures on Boolean-valued models for set theory*, Amer. Math. Soc. Summer Institute on Axiomatic Set Theory, UCLA, 1967. Lecture Notes.(14)

26. D. Scott and R. Solovay, "Boolean-valued models for set theory" in *Axiomatic set theory*, Proc. Sympos. Pure Math., vol. 13, Amer. Math. Soc., Providence, R. I. (to appear).

27. J. R. Shoenfield, *The problem of predicativity*, Essays on the Foundations of Mathematics, Magnes Press, Hebrew Univ., Jerusalem, 1961, pp. 132–139.

28. W. Sierpiński, *Les ensembles projectifs et analytiques*, Mémor. Sci. Math., no. 112, Gauthier-Villars, Paris, 1950.

(14) This reference is a preliminary version of [26].

29. R. Solovay, *The measure problem*, Abstract 65T-62, Notices Amer. Math. Soc. **12** (1965), 217.

30. ———, *The measure problem*, (to appear).

31. C. Spector, *Recursive well-orderings*, J. Symbolic Logic **20** (1955), 151–163.

32. ———, *Measure-theoretic construction of incomparable hyperdegrees*, J. Symbolic Logic **23** (1958), 280–288.

33. ———, *Hyperarithmetical quantifiers*, Fund. Math. **48** (1959), 312–320.

34. H. Tanaka, *Some results in the effective descriptive set theory*, Publ. RIMS, Kyoto Univ. Ser A **3** (1967), 11–52.

35. ———, *A basis result for Π_1^1-sets of positive measure*, Hosei University, 1967.

36. L. Tharp, *Set theory*, Lecture Notes, Massachusetts Institute of Technology, Cambridge, Mass., 1965.

MASSACHUSETTS INSTITUTE OF TECHNOLOGY,
CAMBRIDGE, MASSACHUSETTS

Actes, Congrès intern. math., 1970. Tome 1, p. 251 à 254.

RECURSION IN OBJECTS OF FINITE TYPE

by GERALD E. SACKS

My hope here in Nice is to draw attention to the work of S. C. Kleene [7] on recursion in objects of finite type. In pursuit of that hope I will touch lightly on some related developments in generalized recursion theory. My Nicene creed is: Kleene's notion of recursive object of finite type and Gödel's notion of constructible set are of similar, but not of the same, substance. An Athanasian might see them as the same after reading Shoenfield [20] on hierarchies, but the Arian view is more balanced in the light of Moschovakis [11, 12] on hyperprojective sets.

I owe much to R. Gandy, T. Grilliot, and P. Hinman, who patiently explained to me the concept of recursion in objects of finite type, and to G. Kreisel [8], who taught me that such things as " concepts " exist in the context of recursion theory.

An object of type 0 is a natural number. An *object of type* $n > 0$ is a total function whose arguments and values are of type $< n$. $U, V, \ldots$ denote objects of finite type. Kleene [7] introduced a transitive relation $U \leq V$ (to be read U is *recursive in* V). If U and V are objects of type 1, then $\leq$ coincides with Turing reducibility. For each finite type, 0 ambiguously denotes the function of that type which is everywhere equal to 0. If $U \leq 0$, then U is said to be recursive. If $U \leq V$ and $V \leq U$, then $U \equiv V$ (to be read U and V have the same *degree*). $X, Y, \ldots$ denote members of 2^ω called reals, and $F, G, H, \ldots$ denote total functions from the reals into the reals.

For each $n > 0$, $\underline{^nE}$ is the characteristic function of equality for objects of type $< n$. Thus $^2E(X, Y) = 0$ if $X = Y$, and $= 1$ otherwise. ^2E has the same degree as the Turing jump operator. A result of great internal beauty obtained by Kleene [7] is: the objects of type ≤ 2 recursive in ^2E are just the hyperarithmetic ones. $\underline{S_kU}$, the *k-section* of U, is the set of all objects of type k recursive in U. Kleene [7] asked: do there exist F's such that (1) S_1F consists of the arithmetic reals?, (2) S_1F consists of the Δ_2^1 reals? Recently Grilliot [5] answered (1) negatively by showing: if S_1F is closed under the Turing jump, then $^2E \leq F$. (2) is answered affirmatively below.

Platek [13] calls a transitive set A *admissible* if A is closed under finitary set operations and all instances of the Σ_1 reflection and Δ_0 comprehension axiom schemas are true in A. A function f from A into A is called A-recursive if the graph of f is a Σ_1 subset of A. For every F it is possible to construe S_1F as a countable transitive set $\underline{AS_1F}$ by exploiting the standard encoding of hereditarily countable sets by reals. An immediate consequence of Shoenfield [20], Hinman [6], and Grilliot [5] is: AS_1F is admissible if and only if $^2E \leq F$. It follows from Gandy's work [3] on selection operators that if $^2E \leq F$, then AS_1F satisfies the Σ_1 dependent choice axiom schema.

THEOREM 1 [17]. — (*i*) and (*ii*) are equivalent.

(*i*) A is a countable admissible set that satisfies the Σ_1 dependent choice axiom schema, and every member of A is countable in A.

(*ii*) There exists an F of type 2 such that ${}^2E \leq F$ and $A = AS_1F$.

For each ordinal α let L_α be the set of all constructible sets of constructible order $<\alpha$. Kripke [10] and Platek [13] call α an *admissible ordinal* if L_α satisfies the Σ_1 replacement axiom schema. A function is α-recursive if its graph is a Σ_1 subset of L_α. A set is α-recursively enumerable if it is the range of an α-recursive function. If α is admissible, then by Gödel there is an α-recursive well-ordering of all the computations needed to compute all the α-recursive functions. It follows that a great deal of classical recursion theory can be generalized from ω to α. For example, it can be shown that the Friedberg-Muchnik solution of Post's problem holds for every admissible α. The recursive functions of ordinals were first defined by Takeuti [22]. One of his early results restated in current terms says that every cardinal is an admissible ordinal. His proof is an application of *Gödel's Skolem-Lowenheim principle for L*: Σ_1 subsystems of L are isomorphic to initial segments of L. Gödel's principle (with L replaced by L_α) is central to current work on admissible ordinals; it plays an unexpected part in the solution of Post's problem [19]. I say " unexpected " because the use of model-theoretic ideas in recursion theory was at one time a surprise to me. On the other hand Kreisel's approach [8, 9] to generalized recursion theory was based from the beginning on the model-theoretic notion of *implicit invariant definability*. Later Barwise showed by means of a compactness argument: if A is a countable admissible set, then the implicitly invariantly definable functions from A into A are equivalent to the A-recursive functions (The equivalence fails for most uncountable A's). I would like to recommend the joint paper [1] of Barwise, Gandy, and Moschovakis as a starting point for any one curious about the great variety of ideas now current in generalized recursion theory.

COROLLARY 2. — If α is a countable admissible ordinal, then there exists an F of type 2 such that $L_\alpha \cap 2^\omega = S_1F$, and such that for every G of type 2 and of lower degree than F, $L_\alpha \cap 2^\omega \neq S_1G$.

COROLLARY 3. — If $n > 0$, then there exists an F of type 2 such that the reals recursive in F are just the Δ^1_n reals.

THEOREM 4. — If U is of type n and ${}^nE \leq U$, then $S_1U = S_1F$ for some F of type 2.

The above four results are proved with the aid of Gödel's Skolem-Lowenheim principle for L, Cohen's forcing method, and Grilliot's hierarchies based on objects of finite type [4]. The next theorem combines forcing with the Friedberg-Muchnik priority method. Platek [13] calls X F-recursive in Y if $X \# F, Y$. Two reals have the same *F-degree* if each is F-recursive in the other. Hinman calls a real *F-recursively enumerable* if it is the range of a partial function of type 1 recursive in F. A well-known result of Spector [21] can be extended to show: if ${}^2E \leq F$, then all non-F-recursive, F-recursively enumerable reals have the same F-degree. I say *X is Σ_1 in Y over $\underline{AS_1F}$* if X is a Σ_1 subset of $AS_1F(Y)$, where $AS_1F(Y)$ is the result of adjoining Y to $\underline{AS_1F}$ and closing under Δ_0 comprehension.

THEOREM 5. — If ${}^2E \leq F$, then there exist two F-recursively enumerable reals such that neither is Δ_1 in the other over AS_1F.

Kleene [7] showed that the ${}^2E$-recursively enumerable reals were just the Π^1_1 reals. Theorem 5 for $F = {}^2E$ was proved in [15].

The *superjump* is a fundamental object of type 3 introduced by Gandy [3]; it lifts F to F^1. Let $\{e\}^F(X)$ denote the value (possibly undefined) of the e-th partial function of type 2 recursive in F for real argument X. The value of $F^1(e, X)$ is 0 if $\{e\}^F(X)$ is defined and 1 otherwise. ${}^2E^1$ is the hyperjump and has the same degree as E_1, an object of type 2 associated with the Souslin operation and introduced by Tugué [23]. Gandy [3] showed: if $F \leq G$, then $F^1 \leq G^1$. Hinman has asked: is there a condition on G that implies the existence of an F such that $F^1 \equiv G$? Hinman's question was inspired by Friedberg's classic result [2]: if $JO \leq X$, then there exists a Y such that $JY \equiv X$, where 0 is the empty set and J is the Turing jump.

THEOREM 6 [18]. — Assume the continuum hypothesis. Then there exists an H such that $(G)(EF)[H \leq G \rightarrow F^1 \equiv G]$.

The F's of Theorems 1 through 5 are constructed in countably many steps, but the F of Theorem 6 is constructed in uncountably many steps. If the continuum hypothesis is dropped, then Theorem 6 can be approximated in the sense of Theorem 7. The continuum hypothesis is needed to make the approximations cohere with one another.

THEOREM 7. — If S_1G is closed under hyperjump, then there exists an F such that $S_1F^1 = S_1G$.

The next theorem is intended to suggest that the Tugué hierarchy for S_1E_1 is similar to the Shoenfield hierarchy for S_1F^1 whenever ${}^2E \leq F$; it was proved in [16] for the case of $F = {}^2E$.

THEOREM 8. — If ${}^2E \leq F$, then the F-degrees of S_1F^1 have a minimal, but no least, upper bound.

Most of the results of this paper have the following form: a structure B associated with some generalization of recursion theory is given; then an object U of type n is constructed such that the members of B coincide with the objects of type $< n$ that are recursive in U. Since Kleene's definition of relative recursiveness is inductive, it follows that B can be defined by an induction based on U. If enough results of the above form can be found, it may be possible (as Kreisel has suggested) to prove theorems about structures occurring in generalizations of recursion theory by thinking of them as having been built up by inductive definitions based on objects of finite (or higher) type. Among the means to that end would be various sharpenings of Theorem 4. The superjump ${}^3S$ is an object of type 3 of lower degree than ${}^3E$, but an application of Corollary 2 above to Platek [14] provides an F of type 2 such that $S_1{}^3S = S_1F$. So it seems likely that the hypothesis " ${}^nE \leq U$ " of Theorem 4 can be replaced by something of wider scope. Theorem 4 can be extended from 1-sections to k-sections as follows. For each n there is a V of type n such that for all U of type n: if $V \leq U$ and $k < n$, then $S_kU = S_kW$ for some W of type $k + 1$. It is possible that ${}^nE$ may suffice for V, but at the moment I need a V whose degree appears to be higher than ${}^nE$ save when $k = 1$ (Added in proof: if Gödel's axiom of constructibility holds, then $V = {}^nE$.)

BIBLIOGRAPHY

[1] J. Barwise, R. Gandy and Y. Moschovakis. — The next admissible set, *Jour. Symb. Log.*, to appear.

[2] R. Friedberg. — A cirterion for completeness of degrees of unsolvability, *Jour. Symb. Log.*, 22 (1957), pp. 159-160.

[3] R. Gandy. — *General recursive functionals of finite type and hierarchies of functions*, University of Clermont-Ferrand (1962).

[4] T. Grilliot. — Hierarchies based on objects of finite type, *Jour. Symb. Log.*, 34 (1969), pp. 177-182.

[5] —. — On effectively discontinuous type 2 objects, *Jour. Symb. Log.*, to appear.

[6] P. Hinman. — Ad astra per aspera: hierarchy schemata in recursive function theory, Ph. D. Thesis, Berkeley (1966).

[7] S. C. Kleene. — Recursive functionals and quantifiers of finite type, *Trans. Amer. Math. Soc.*, 91 (1959), pp. 1-52; 108 (1963), pp. 106-142.

[8] G. Kreisel. — Model theoretic invariants: applications to recursive and hyperarithmetic operations, *The Theory of Models*, North Holland (1965), pp. 190-205.

[9] — and G. E. Sacks. — Metarecursive sets, *Jour. Symb. Log.*, 30 (1965), pp. 318-338.

[10] S. Kriple. — Transfinite recursions on admissible ordinals, *Jour. Symb. Log. (Abstracts)*, 29 (1964), pp. 161-162.

[11] Y. Moschovakis. — Hyperanalytic predicates, *Trans. Amer. Math. Soc.*, 129 (1967), pp. 249-282.

[12] —. — Abstract first order computability, *Trans. Amer. Math. Soc.*, 138 (1969), pp. 427-504.

[13] R. Platek. — Foundations of recursion theory, Ph. D. Thesis, Stanford (1966).

[14] —. — A countable hierarchy for the superjump, *Proceedings of the* 1969 *Manchester conference*, to appear.

[15] G. E. Sacks. — On the reducibility of Π_1^1 sets, *Advances in Math.*, to appear.

[16] —. — Countable admissible ordinals and hyperdegrees, to appear.

[17] —. — The 1-section of a type *n* object, to appear.

[18] —. — Inverting the superjump, to appear.

[19] —. — and S. Simpson. — The α-finite injury method, to appear.

[20] J. Shoenfield. — A hierarchy based on a type 2 object, *Trans. Amer. Math. Soc.*, 134 (1968), pp. 103-108.

[21] C. Spector. — Recursive well-orderings, *Jour. Symb. Log.*, 20 (1955), pp. 151-163.

[22] G. Takeuti. — On the recursive functions of ordinals, *Jour. Math. Soc. Japan*, 12 (1960), pp. 119-128.

[23] T. Tugué. — Predicates recursive in a type 2 object and Kleene hierarchies, *Comment. Math. Univ. St. Paul* (Tokyo), 8 (1960), pp. 97-117.

P-P69
Department of Mathematics,
MIT
Cambridge, Mass 02139
(U. S. A.)

FORCING WITH PERFECT CLOSED SETS

GERALD E. SACKS[1]

0. **Introduction.** Forcing with perfect closed sets has its origin in Cohen's invention of forcing with finite conditions [**1**] and in Spector's construction of a minimal Turing degree [**28**]. Cohen taught that truth can be approximated more easily and more completely in the intermediate stages of a construction than any recursion theorist would have believed possible, and Spector revealed that infinite approximation by means of perfect closed conditions is radically different from finite approximation. The difference, it seems to us, is manifested in the sequential lemma, Lemma 1.4 of § 1 (see Footnote 2, p. 335).

In § 1 we study perfect closed forcing in the context of set theory. We show: if ZF (Zermelo-Fraenkel axioms for set theory) is consistent, then so is ZF plus there exist precisely two degrees of nonconstructibility. Two sets of natural numbers have the same degree of nonconstructibility if each is constructible from the other. All the constructible subsets of ω constitute one degree of nonconstructibility denoted by 0. A degree c is said to be less than or equal to a degree d if some set of degree c is constructible from some set of degree d. A nonzero degree is called minimal if the only degree less than it is 0. We lift Spector's construction of a minimal Turing degree from recursion theory to set theory in order to force all the nonconstructible subsets of ω to occupy the same degree. We also show by means of a transformation lemma that sets which are generic with respect to perfect closed forcing have the following curious property: if T is generic, then every set which has the same degree of nonconstructibility as T is generic.

In § 2 we examine perfect closed forcing in the hyperarithmetic case. The

[1] The preparation of this paper was supported in part by U.S. Army Contract DAHCO-4-67-C-0052. The author wishes to thank Professor J. Barkley Rosser for introducing him to Spector's minimal degree argument.

principal result of [5] is reviewed in order to bring out certain technical difficulties which crop up in the hyperarithmetic case and which have no counterpart in the set-theoretic case. It has to be shown in the hyperarithmetic case that the forcing relation, restricted to sentences of recursive ordinal rank, is Π_1^1, and this is seen to be the case only after the forcing relation is defined with certain precautions; if it is defined by routine analogy with the set-theoretic case, then its classification becomes uncertain. The corresponding problem of definability of the forcing relation in the set-theoretic case presents no problem at all.

In § 3 we touch on the arithmetic case and introduce the notion of pointed, perfect closed set. Using that notion, we obtain an order-theoretic characterization of the Turing degree of the truth-set for arithmetic in terms of the degrees of the arithmetic sets and the jump operator. We also briefly describe Spector's minimal Turing degree construction in terms of perfect closed forcing as well as several results based on forcing with pointed perfect closed sets whose detailed proofs will appear elsewhere [23].

Let $2 = \{0, 1\}$ have the discrete topology and let 2^ω have the product topology [8]. Let $P, Q, R, \ldots$ be perfect closed subsets of 2^ω. We conceive of a typical P as a tree with binary branching; each path of the tree is an infinite sequence of 0's and 1's and is viewed as the characteristic function of a subset of ω. Every path of P branches infinitely often; this is equivalent to saying that P is perfect. Suppose we adopt some standard recursive one-to-one correspondence between all finite sequences of 0's and 1's and ω. Then P can be encoded as a set of sequence numbers corresponding to finite, initial segments of paths of P. We will use $P, Q, R, \ldots$ ambiguously to denote perfect closed subsets of 2^ω and their encodings as subsets of ω. Whenever we classify P, we refer to the encoding of P. Thus if we say P is arithmetic, we mean that the encoding of P as a subset of ω is arithmetic. Let T be an arbitrary subset of ω. If we say $T \in P$, then clearly we are thinking of P as a subset of 2^ω. If we say $P \subseteq Q$, then we mean $(T)(T \in P \rightarrow T \in Q)$.

Let $p, q, r, \ldots$ ambiguously denote sequence numbers and the finite, initial segments of characteristic functions they encode. We say $p \geq r$ if $p = r$ or if p is extended by r. Let P be a nonempty set of sequence numbers. Then P is an encoding of a perfect closed subset of 2^ω if and only if

$$(p)_{p \in P}(\exists q)_{q \in P}(\exists r)_{r \in P}[p \geq q \;\&\; p \geq r \;\&\; q \ngeq r \;\&\; r \ngeq q].$$

We say $T \in p$ (or T satisfies p) if p is an initial segment of the characteristic function of T.

1. **The set-theoretic case.** Let $\mathscr{M}$ be a countable, transitive set which is a model of ZF plus V = L (Gödel's axiom of constructibility [6]). Let $T \subseteq \omega$. $\mathscr{M}(T)$ is the set of all sets constructible from T via the ordinals of $\mathscr{M}$. In order to make the definition of $\mathscr{M}(T)$ more precise, we adopt the symbolism of Tharp [30]. The symbols of $\mathscr{L}^0$ are: $\in$ (membership); unranked set variables $x, y, z, \ldots$; ranked set variables $x^\alpha, y^\alpha, z^\alpha, \ldots$ for each ordinal $\alpha \in \mathscr{M}$; propositional connectives; and quantifiers for both ranked and unranked variables. The atoms of $\mathscr{L}^0$ are of the form $t_1 \in t_2$, where t_1 and t_2 are variables. The formulas of $\mathscr{L}^0$ are

constructed from the atoms and the logical symbols in the usual fashion. A class $\mathscr{C}$ of constants intended to name the members of $\mathscr{M}(T)$ is defined by induction on the ordinals of $\mathscr{M}$.

$\mathscr{C}(0) = \{\bar{n} \mid n < \omega\}$.

$\mathscr{C}(\alpha + 1)$: $\mathscr{T} \in \mathscr{C}(\alpha + 1)$; let $\phi(x^\alpha, y_1, \ldots, y_n)$ be a formula of $\mathscr{L}^0$ whose only free variables are $x^\alpha, y_1, \ldots, y_n$ $(n \geq 0)$ and whose quantified variables are all of the form x^β for β's $\leq \alpha$; then $\hat{x}^\alpha\phi(x^\alpha, c_1, \ldots, c_n)$, where each c_i $(1 \leq i \leq n)$ is either $\mathscr{T}$ or a member of $\bigcup\{\mathscr{C}(\beta) \mid \beta \leq \alpha\}$, is a typical member of $\mathscr{C}(\alpha + 1)$.

$\mathscr{C}(\lambda) = \bigcup\{\mathscr{C}(\beta) \mid \beta < \lambda\}$ for each limit ordinal λ.

Let $\mathscr{C} = \bigcup\{\mathscr{C}(\alpha) \mid \alpha \in \mathscr{M}\}$. The symbols of $\mathscr{L}(\mathscr{T})$ are those of $\mathscr{L}^0$ together with the members of $\mathscr{C}$. The atomic formulas of $\mathscr{L}(\mathscr{T})$ are of the form $t_1 \in t_2$, where t_1 and t_2 are variables or members of $\mathscr{C}$.

A sentence $\mathscr{F}$ of $\mathscr{L}(\mathscr{T})$ is said to be *ranked* if all its variables are ranked. The ordinal rank of a ranked sentence $\mathscr{F}$, denoted by $o(\mathscr{F})$, is the least upper bound of $\{\beta \mid (x^\beta) \text{ or } (\exists x^\beta) \text{ occurs in } \mathscr{F}\} \cup \{\beta + 1 \mid \hat{x}^\beta \text{ occurs in } \mathscr{F}\}$. If $t \in \mathscr{C}$, then $o(t) = o(\bar{0} \in t)$. For each $T \subseteq \omega$ it is a routine matter to simultaneously inductively define the following concepts: $c \in \mathscr{C}(\alpha)$ and $b \in \mathscr{M}_\alpha(T)$ and c denotes b; $o(\mathscr{F}) \leq \alpha$ and $\mathscr{F}$ is true in $\mathscr{M}_\alpha(T)$. Of course $\mathscr{T}$ is interpreted at T, x^α ranges over $\mathscr{M}_\alpha(T)$, and x ranges over $\mathscr{M}(T) = \bigcup\{\mathscr{M}_\alpha(T) \mid \alpha \in \mathscr{M}\}$. We write $\mathscr{M}(T) \models \mathscr{F}$ to mean $\mathscr{F}$ is true in $\mathscr{M}(T)$. Let 0 denote the empty set, and let $\mathscr{M}_\alpha$ be $\mathscr{M}_\alpha(0)$. Then $\mathscr{M} = \bigcup\{\mathscr{M}_\alpha \mid \alpha \in \mathscr{M}\}$.

Now we define the forcing relation $P \Vdash \mathscr{F}$, where $P \in \mathscr{M}$ and $\mathscr{F}$ is a sentence of $\mathscr{L}(\mathscr{T})$. Of course $P \in \mathscr{M}$ means P is a perfect closed subset of 2^ω whose standard encoding as a subset of ω is a member of $\mathscr{M}$. The definition is orthodox save for the interpretation of P.

(1) $P \Vdash (\exists x)\mathscr{F}(x)$ if $P \Vdash \mathscr{F}(c)$ for some $c \in \mathscr{C}$.

(2) $P \Vdash (\exists x^\alpha)\mathscr{F}(x^\alpha)$ if $P \Vdash \mathscr{F}(c)$ for some $c \in \mathscr{C}(\alpha)$.

(3) $P \Vdash \mathscr{F}\ \&\ \mathscr{G}$ if $P \Vdash \mathscr{F}$ and $P \Vdash \mathscr{G}$.

(4) $P \Vdash \sim\mathscr{F}$ if $(Q)[P \supseteq Q \rightarrow \sim(Q \Vdash \mathscr{F})]$.

(5) $P \Vdash \bar{n} \in \mathscr{T}$ if $(T)[T \in P \rightarrow \mathscr{M}(T) \models \bar{n} \in \mathscr{T}]$.

(6) $P \Vdash \bar{m} \in \bar{n}$ if $m < n$.

(7) $P \Vdash t_1 \in t_2$ if $o(t_1) < o(t_2)$, t_2 is $\hat{x}^\beta\phi(x^\beta)$ and $P \Vdash \phi(t_1)$.

(8) $P \Vdash t_1 \in t_2$ if $o(t_1) > o(t_2)$, t_1 is $\hat{x}^\alpha\phi(x^\alpha)$ and $P \Vdash (\exists y^\alpha)[(x^\alpha)(x^\alpha \in y^\alpha \leftrightarrow \phi(x^\alpha))\ \&\ y^\alpha \in t_2]$.

(9) $P \Vdash t_1 \in t_2$ if $o(t_1) = o(t_2) > 0$. Similar to (8).

We say T is *generic* if for each sentence $\mathscr{F}$ of $\mathscr{L}(T)$ there is a perfect closed $P \in \mathscr{M}$ such that $T \in P$ and either $P \Vdash \mathscr{F}$ or $P \Vdash \sim\mathscr{F}$. Standard arguments [7] suffice to show: (i) $(\mathscr{F})(P) \sim [P \Vdash \mathscr{F}$ and $P \Vdash \sim\mathscr{F}]$; (ii) $(P)(\mathscr{F})(\exists Q)[P \supseteq Q\ \&\ Q \Vdash \mathscr{F}$ or $Q \Vdash \sim\mathscr{F}]$; (iii) $(P)(\mathscr{F})(Q)[P \supseteq Q\ \&\ P \Vdash \mathscr{F} \rightarrow Q \Vdash \mathscr{F}]$; (iv) generic T's exist. A delicate but not troublesome point arises in the course of the proof of (v): if T is generic, then $\mathscr{M}(T) \models \mathscr{F}$ if and only if $(\exists P)[T \in P\ \&\ P \Vdash \mathscr{F}]$. The difficulty can be put as follows. Suppose T is generic, $T \in P \cap Q$, $P \Vdash \mathscr{F}$, and

$Q \Vdash \mathscr{S}$; then there should be an $R \subseteq P \cap Q$ such that $T \in R$ and $R \Vdash \mathscr{F} \,\&\, \mathscr{S}$. In the case of Cohen-forcing with finite conditions, it was safe to let R be $P \cap Q$, but now a different argument is required, because the intersection of two overlapping, perfect closed sets need not be perfect.

It is not difficult to show by a Cantor-Bendixson analysis that if $T \notin \mathscr{M}$ and $T \in P \cap Q$, then $T \in R \subseteq P \cap Q$, where R is the perfect kernel of $P \cap Q$. In short, if a perfect closed set is encoded by a subset of ω belonging to $\mathscr{M}$, then so is its perfect kernel. But Kreisel **[13]** has shown that this type of argument will not work in the hyperarithmetic case of § 2. His result is: there is a hyperarithmetically encodable, closed set whose perfect kernel is not hyperarithmetically encodable. The proof of proposition 1.1 (ii) does work in the hyperarithmetic case.

PROPOSITION 1.1. (i) *If T is generic, then $T \notin \mathscr{M}$.* (ii) *If T is generic and $T \in P \cap Q$, then $T \in R$ for some $R \subseteq P \cap Q$.*

PROOF. Let $c \in \mathscr{C}$ denote a subset of ω, and let $\bar{c}(x^0)$ denote the sequence number which represents the characteristic function of c restricted to arguments $< x^0$.

To prove (i) let T be generic and suppose $T = d \in \mathscr{M}$. Then there is a P such that $T \in P$ and (a) $P \Vdash (\exists x^0)(\bar{\mathscr{T}}(x^0) \neq \bar{d}(x^0))$ or (b) $P \Vdash \sim(\exists x^0)(\bar{\mathscr{T}}(x^0) \neq \bar{d}(x^0))$. If (a) holds, then for some n, $P \Vdash \bar{\mathscr{T}}(\bar{n}) \neq \bar{d}(\bar{n})$, and consequently $\bar{T}(n) \neq \bar{d}(n)$. If (b) holds, then $(T)(T \in P \rightarrow T = d)$.

To prove (ii) let $\mathscr{F}$ be $(\exists x^0)(\bar{\mathscr{T}}(x^0) \notin P \cap Q)$. Then there is an R such that $T \in R$ and $R \Vdash \mathscr{F}$ or $R \Vdash \sim\mathscr{F}$. If $R \Vdash \mathscr{F}$, then $T \notin P \cap Q$. So $R \Vdash \sim\mathscr{F}$. This means that

$$(R_1)_{R \supseteq R_1}(n) \sim [R_1 \Vdash \bar{\mathscr{T}}(n) \notin P \cap Q].$$

But then $R \subseteq P \cap Q$.

Let A be an arbitrary set. We say A is $\mathscr{M}$-definable if there exists a formula $\mathscr{F}(x)$ of ZF with constants denoting elements of $\mathscr{M}$ such that $A = \{b \mid b \in \mathscr{M} \,\&\, \mathscr{M} \models \mathscr{F}(b)\}$.

LEMMA 1.2. *For each $n \geq 0$: the relation $P \Vdash \mathscr{F}$, restricted to sentences $\mathscr{F}$ of $\mathscr{L}(\mathscr{T})$ having at most n unranked quantifiers, is $\mathscr{M}$-definable.*

The proof of 1.2 is standard [7]. It follows from 1.2 that if T is generic, then the replacement axiom holds in $\mathscr{M}$. It suffices to prove: let $\mathscr{F}(x, y)$ be a formula of $\mathscr{L}(\mathscr{T})$ whose only free variables are x and y: if $P \Vdash (x^\alpha)(\exists y)\mathscr{F}(x^\alpha, y)$, then $P \Vdash (x^\alpha)(\exists y^\beta)\mathscr{F}(x^\alpha, y^\beta)$ for some β. P has the property that

$$(c)_{c \in \mathscr{C}(\alpha)}(Q)_{P \supseteq Q}(\exists R)_{Q \supseteq R}(\exists d)_{d \in \mathscr{C}(\alpha)}[R \Vdash \mathscr{F}(c, d)].$$

By 1.2, d can be regarded as an $\mathscr{M}$-definable function of c and Q. Since $\mathscr{C}(\alpha)$ and $\{Q \mid P \supseteq Q\}$ are sets of $\mathscr{M}$, the range of d, restricted to $\mathscr{C}(\alpha) \times \{Q \mid P \supseteq Q\}$, is a set of $\mathscr{M}$ and thus a subset of $\mathscr{C}(\beta)$ for some sufficiently large β. But this means $P \Vdash (x^\alpha)(\exists y^\beta)\mathscr{F}(x^\alpha, y^\beta)$.

LEMMA 1.3. *For each $n \geq 0$: There exists an $\mathscr{M}$-definable function $\lambda\mathscr{F} \mid \mathscr{F}^*$. defined for all sentences $\mathscr{F}$ of $\mathscr{L}(\mathscr{T})$ having at most n unranked quantifiers, such that*

$$(P)[P \Vdash \mathscr{F} \leftrightarrow P \Vdash \mathscr{F}^*]$$

and the ordinal rank of $\mathscr{F}^*$ is less than ω_2 (in the sense of $\mathscr{M}$).

PROOF. Standard. Let $K = \{P \mid P \Vdash \mathscr{F}\}$. By 1.2, K is a subset of 2^ω in $\mathscr{M}$. By Gödel [6], $K \in \mathscr{M}_\alpha$ for some $\alpha < \omega_2$, where ω_2 is the ω_2 of $\mathscr{M}$. Let $\mathscr{F}^*$ be $(\exists P)(P \in K \ \& \ \mathscr{T} \in P)$. Clearly the ordinal rank of $\mathscr{F}^* < \omega_2$. Suppose $P \Vdash \mathscr{F}^*$. Then for some Q, $P \Vdash Q \in K \ \& \ \mathscr{T} \in Q$. It follows $Q \in K$, since Q and K are members of $\mathscr{M}$ and no falsehood about $\mathscr{M}$ can ever be forced. $P \subseteq Q$, because $P \Vdash \mathscr{T} \in Q$ implies $(p)(p \in P \rightarrow p \in Q)$. But then $P \Vdash \mathscr{F}$.

If one forces with finite conditions in the manner of Cohen, then the ω_2 in Lemma 1.3 is replaced by ω_1. On the other hand it can be shown that if one is forcing with perfect closed conditions, then the ω_2 of Lemma 1.3 cannot be reduced to ω_1.

We turn now to the problem of showing for generic T that the cardinals of $\mathscr{M}(T)$ are the same as those of $\mathscr{M}$. In the case of Cohen-forcing with finite conditions [1] or Solovay-forcing with closed sets of positive measure [25], [26], the preservation of cardinals in generic extensions is a consequence of the countable chain condition: if J is a set of pairwise incomparable conditions (i.e., P, $Q \in J$ $\& \ P \neq Q \rightarrow P \nsubseteq Q \ \& \ Q \nsubseteq P$), then J is countable. It is not difficult to see that the countable chain condition fails for perfect closed conditions. Nonetheless, cardinals are preserved, thanks to Lemma 1.4, the sequential lemma.[2]

LEMMA 1.4. *Let $\{\mathscr{F}_i \mid i \in \omega\}$ be a countable (in the sense of $\mathscr{M}$) set of sentences of $\mathscr{L}(\mathscr{T})$. Let P have the property that*

$$(i)(Q)_{P \supseteq Q}(\exists R)_{Q \supseteq R}[R \Vdash \mathscr{F}_i].$$

Then there exists an $R \subseteq P$ such that for each i,

$$R = \bigcup_{j < 2^i} R_j^i, \quad \text{and} \quad R_j^i \Vdash \mathscr{F}_i$$

for all $j < 2^i$. In addition, $R_j^i \cap R_k^i = 0$ if $j \neq k$.

PROOF. Suppose $(i)(Q)_{P \supseteq Q}(\exists R)_{Q \supseteq R}[R \Vdash \mathscr{F}_i]$. By 1.2, there is an $\mathscr{M}$-definable function $f(i, Q)$ such that if $P \supseteq Q$, then $Q \supseteq f(i, Q)$ and $f(i, Q) \Vdash \mathscr{F}_i$. Let us say Q_1 and Q_2 are basic, disjoint perfect closed subsets of Q if there exist sequence numbers q_1 and $q_2 \in Q$ such that $q_1 \ngeq q_2$, $q_2 \ngeq q_1$, $Q_1 = \{p \mid p \in Q \ \& \ q_1 \geq p\}$, and $Q_2 = \{p \mid p \in P \ \& \ q_2 \geq p\}$. By iterating $f(i, Q)$, it is not difficult to

[2] Lemma 1.4 is called the sequential lemma in order to be consistent with [5], where a hyperarithmetic version (Lemma 2.2 of the present paper) of 1.4 occurs. Mathias [15] formulates 1.4 more abstractly and calls the result the fusion lemma. Shoenfield [24] invokes the term "splitting" to describe the proof of 1.4.

develop an $\mathcal{M}$-definable function $\lambda ij \mid Q^i_j$ with the following properties: $Q^0_0 \subseteq P$ and $Q^0_0 \Vdash \mathcal{F}_0$; for each i and $j < 2^i$, Q^{i+1}_j and $Q^{i+1}_{j+2^i}$ are basic, disjoint perfect closed subsets of Q^i_j; $Q^{i+1}_j \Vdash \mathcal{F}_{i+1}$, and $Q^{i+1}_{j+2^i} \Vdash \mathcal{F}_{i+1}$. Let $R = \bigcap_i \bigcup_{j<2^i} Q^i_j$, and let $R^i_j = R \cap Q^i_j$.

We say P weakly forces $\mathcal{F}$ ($P \Vdash^* \mathcal{F}$) if $P \Vdash \sim\sim\mathcal{F}$. An immediate corollary of Lemma 1.4 is

$$(i)(Q)_{P\supseteq Q}(\exists R)_{Q\supseteq R}[R \Vdash \mathcal{F}_i] \rightarrow (\exists R)_{P\supseteq R}(i)[R \Vdash^* \mathcal{F}_i].$$

Another way of expressing the message of 1.4 is: for all P and all sequences $\{\mathcal{F}_i \mid i \in \omega\}$ (in the sense of $\mathcal{M}$) of sentences of $\mathcal{L}(\mathcal{T})$,

$$(\exists R)_{P\supseteq R}(i)[R \Vdash^* \mathcal{F}_i \vee R \Vdash^* \sim\mathcal{F}_i].$$

Thus perfect closed forcing gives us the power to "decide" every one of a countable set of sentences by means of a single forcing condition. If we concentrate on sentences of countable ordinal rank, then we can do even more; this additional power is important in the hyperarithmetic case (§ 2), where the sentences of recursive ordinal rank behave very much like the sentences of countable ordinal rank in the set-theoretic case.

LEMMA 1.5. *Let $\mathcal{F}$ be a sentence of $\mathcal{L}(\mathcal{T})$ of countable (in the sense of $\mathcal{M}$) ordinal rank. Then for each P there is a $Q \subseteq P$ such that either* (1) *or* (2) *holds:*

(1) $(T)(T \in Q \rightarrow \mathcal{M}(T) \models \mathcal{F})$;

(2) $(T)(T \in Q \rightarrow \mathcal{M}(T) \models \sim\mathcal{F})$.

PROOF. By transfinite induction on the rank of $\mathcal{F}$. We give only the principal inductive step. Let $\mathcal{F}$ be $(\exists x^\alpha)\mathcal{F}(x^\alpha)$, where α is countable in the sense of $\mathcal{M}$. Let $\{c_i \mid i \in \omega\}$ be an enumeration of $\mathcal{C}(\alpha)$; $\mathcal{C}(\alpha)$ is countable (in the sense of $\mathcal{M}$) because α is. For each i, $\mathcal{F}(c_i)$ has lower rank than $(\exists x^\alpha)\mathcal{F}(x^\alpha)$. It follows from the inductive hypothesis that for each i and each P, there is a $Q \subseteq P$ such that either (1) or (2) holds:

(1) $(T)(T \in Q \rightarrow \mathcal{M}(T) \models \mathcal{F}(c_i))$;

(2) $(T)(T \in Q \rightarrow \mathcal{M}(T) \models \sim\mathcal{F}(c_i))$.

Fix P. If there is a $Q \subseteq P$ such that (1) holds, then the argument is complete; suppose there is no such Q. Then it follows that

$$(i)(Q)_{P\supseteq Q}(\exists R)_{Q\supseteq R}(T)[T \in R \rightarrow \mathcal{M}(T) \models \sim\mathcal{F}(c_i)].$$

To obtain a $Q \subseteq P$ such that $(R)[T \in Q \rightarrow \mathcal{M}(T) \models (x^\alpha) \sim \mathcal{F}(x^\alpha)]$, one repeats the construction used to prove Lemma 1.4 with one small change. The $\mathcal{M}$-definable function $f(i, Q)$ now has the property that if $P \supseteq Q$, then $Q \supseteq f(i, Q)$ and

$$(T)(T \in f(i, Q) \rightarrow \mathcal{M}(T) \models \sim\mathcal{F}(c_i)).$$

LEMMA 1.6. *Let* $\{\mathscr{F}_i\}$ *be a countable (in the sense of* $\mathscr{M}$*) set of sentences of* $\mathscr{L}(\mathscr{T})$ *of countable ordinal rank. Let P have the property that*

$$(i)(Q)_{P \supseteq Q}(\exists R)_{Q \supseteq R}[R \Vdash \mathscr{F}_i].$$

Then there exists a $Q \subseteq P$ *such that*

$$(i)(T)[T \in Q \rightarrow \mathscr{M}(T) \models \mathscr{F}_i].$$

PROOF. We claim that P has the property that

$$(i)(Q)_{P \supseteq Q}(\exists R)_{Q \supseteq R}(T)[T \in R \rightarrow \mathscr{M}(T) \models \mathscr{F}_i].$$

Fix i and Q. Then there is an $R \subseteq Q$ such that $R \Vdash \mathscr{F}_i$. By Lemma 1.5, there is an $R_1 \subseteq R$ such that either

(1) $(T)(T \in R_1 \rightarrow \mathscr{M}(T) \models \mathscr{F}_i)$ or

(2) $(T)(T \in R_1 \rightarrow \mathscr{M}(T) \models \sim\mathscr{F}_i)$.

Since there is a generic $T \in R_1 \subseteq R$ and $R \Vdash \mathscr{F}_i$, it is clear that (1) holds. But the existence of the desired Q follows from our claim and a repetition of the argument occurring at the end of the proof of Lemma 1.5.

LEMMA 1.7. *If T is generic, then* $\mathscr{M}$ *and* $\mathscr{M}(T)$ *have the same cardinals.*

PROOF. Let α be an uncountable cardinal in the sense of $\mathscr{M}$, and let $\beta < \alpha$. Let $f \in \mathscr{C}$ denote a function which maps β into α. For each $\gamma < \beta$, let

$$k(\gamma) = \{\tau \mid \tau < \alpha \ \& \ (\exists P)(P \Vdash f(\gamma) = \tau)\}.$$

By Lemma 1.2, $k(\gamma) \in \mathscr{M}$; and the cardinality of $k(\gamma)$ is at most ω_1, since the set of all P's has cardinality equal to ω_1. If $\alpha > \omega_1$, then $K = \bigcup\{k(\gamma) \mid \gamma < \beta\}$ has cardinality less than α. If T is generic, then in $\mathscr{M}(T)$ the range of f is a subset of K.

Suppose $\alpha = \omega_1$ and $\beta = \omega_0$. Now we need the sequential lemma, 1.4. Suppose for the sake of a contradiction $P \Vdash f$ is a 1-1 map of ω_0 onto ω_1. It suffices to find an $R \subseteq P$ and a $\gamma < \omega_1$ such that $R \Vdash^*$ range of f is bounded above by γ. P has the property that

$$(i)(Q)_{P \supseteq Q}(\exists R)_{Q \supseteq R}[R \Vdash (\exists \tau)(f(i) = \tau < \omega_1)].$$

By Lemma 1.4 there is an $R \subseteq P$ such that for each i,

$$R = \bigcup_{j < 2^i} R^i_j, \quad \text{and} \quad R^i_j \Vdash f(i) = \tau^i_j < \omega_1$$

for all $j < 2^i$, in addition, $R^i_j \cap R^i_k = 0$ if $j \neq k$. Let $\gamma = \text{lub}\,\{\tau^i_j \mid i \in \omega \ \& \ j < 2^i\}$. Then $\gamma < \omega_1$ since $\lambda ij \mid \tau^i_j$ is $\mathscr{M}$-definable; and $R \Vdash^*$ range of f is bounded above by γ.

The next lemma lifts the heart of Spector's minimal degree argument **[28]** from recursion theory to set theory. Let $c \in \mathscr{C}$ denote a subset of ω. For each T, generic or not generic, let

$$c(T) = \{n \mid \mathscr{M}(T) \models \bar{n} \in c\}.$$

Lemma 1.8. *Let $c \in \mathscr{C}$ denote a subset of ω. For each P there exists a $Q \subseteq P$ such that either* (1) *or* (2) *holds:*

(1) $(T)[T \in Q \ \&\ T \text{ is generic} \to T \in \mathscr{M}(c(T))]$

(2) $(\exists b)(T)[T \in Q \ \&\ T \text{ is generic} \to c(T) = b \in \mathscr{M}]$.

If in addition c has countable (in the sense of $\mathscr{M}$) ordinal rank, then either (1′) *or* (2′) *holds:*

(1′) $(T)[T \in Q \to T \in \mathscr{M}_{\omega_1}(c(T))]$;

(2′) $(\exists b)(T)[T \in Q \to c(T) = b \in \mathscr{M}]$.

Proof. We continue to exploit the construction occurring in the proof of the sequential lemma, 1.4.

Case 1. $(Q)_{P \supseteq Q}(\exists Q_1)_{Q \supseteq Q_1}(\exists Q_2)_{Q \supseteq Q_2}(\exists n)[Q_1 \Vdash \bar{n} \in c \ \&\ Q_2 \Vdash \bar{n} \notin c]$. By Lemma 1.2, Q_1, Q_2 and n can be regarded as $\mathscr{M}$-definable functions of Q. Iteration of these functions in the style of the proof of Lemma 1.4 leads to an $\mathscr{M}$-definable function $\lambda im \mid Q_i^m$ with the following properties: $Q_0^0 \subseteq P$; for each $m \geq 0$ and $i < 2^m$, Q_i^{m+1} and $Q_{i+2^m}^{m+1}$ are basic, disjoint perfect closed subsets of Q_i^m, and

$$(\exists n)[Q_i^{m+1} \Vdash \bar{n} \in c \ \&\ Q_{i+2^m}^{m+1} \Vdash \bar{n} \notin c].$$

We define Q by

$$T \in Q \leftrightarrow (m)(\exists_1 i)_{i<2^m}(T \in Q_i^m).$$

We say that t puts T in Q if $(m)(T \in Q_{t(m)}^m)$. Each $T \in Q$ is put in Q by a unique t and is arithmetic in that t, $\lambda im \mid Q_i^m$. We claim Q satisfies condition (1) of the lemma. Fix a generic $T \in Q$. To see that $T \in \mathscr{M}(c(T))$, it is enough to see that the unique t which puts T in Q is a member of $\mathscr{M}(c(T))$. Consider the following definition of t:

$$t(0) = 0 \ \&\ [t(m+1) = t(m) \vee t(m+1) = t(m) + 2^m];$$

$$t(m+1) = t(m) \leftrightarrow (\exists n)[Q_{t(m)}^{m+1} \Vdash \bar{n} \in c \ \&\ Q_{t(m)+2^m}^{m+1} \Vdash \bar{n} \notin c \ \&\ n \in c(T)].$$

Clearly, t is arithmetic in $c(T)$, $\lambda im \mid Q_i^m$, and the forcing relation $R \Vdash \mathscr{F}$ restricted to R's of the form Q_i^m and $\mathscr{F}$'s of the form $\bar{n} \in c$ and $\bar{n} \notin c$. Since $\lambda im \mid Q_i^m$ and $R \Vdash \mathscr{F}$ restricted both belong to $\mathscr{M} \subseteq \mathscr{M}(c(T))$, it follows that $T \in \mathscr{M}(c(T))$. We had to assume that T was generic in order to be sure that

$$T \in Q_{t(m)}^{m+1} \ \&\ Q_{t(m)}^{m+1} \Vdash \bar{n} \in c \to n \in c(T).$$

The assumption of genericity can be dropped if c has countable ordinal rank. Because then Lemma 1.5 makes it possible to choose Q_i^{m+1} and $Q_{i+2^m}^{m+1}$ so that there exists an n such that for all T,

$$T \in Q_i^{m+1} \to \mathscr{M}(T) \models \bar{n} \in c,$$

$$T \in Q_{i+2^m}^{m+1} \to \mathscr{M}(T) \models \bar{n} \notin c.$$

Note that if X, $Y \in Q$ and $X \neq Y$, then $c(X) \neq c(Y)$.

Case 2. Case 1 fails; then there is an $R \subseteq P$ such that

$$(Q_1)_{R \supseteq Q_1}(Q_2)_{R \supseteq Q_2}(n) \sim [Q_1 \Vdash \bar{n} \notin c \ \& \ Q_2 \Vdash \bar{n} \in c].$$

For each $n \in \omega$, let

$$\mathscr{F}_n \text{ be } \begin{cases} \bar{n} \in c & \text{if } (\exists Q)_{R \supseteq Q}[Q \Vdash n \in c] \\ \bar{n} \notin c & \text{if } (\exists Q)_{R \supseteq Q}[Q \Vdash n \notin c]. \end{cases}$$

The defining property of R guarantees that $\mathscr{F}_n$ is well defined and that

$$(n)(Q_1)_{R \supseteq Q_1}(\exists Q_2)_{Q_1 \supseteq Q_2}[Q_2 \Vdash \mathscr{F}_n].$$

By Lemma 1.4 there is a $Q \subseteq R \subseteq P$ such that $(n)[Q \Vdash^* \mathscr{F}_n]$. If $T \in Q$ and T is generic, then

$$\mathscr{M}(T) \models \bar{n} \in c \leftrightarrow Q \Vdash^* \bar{n} \in c.$$

So Q satisfies condition (2) of the lemma, since $Q \Vdash^* \bar{n} \in c$ is $\mathscr{M}$-definable. Suppose c has countable ordinal rank. By Lemma 1.6 there is a $Q \subseteq R$ such that

$$(n)(T)[T \in Q \rightarrow \mathscr{M}(T) \models \mathscr{F}_n].$$

It follows that

$$(n)(T)[T \in Q \rightarrow (n \in c(T) \leftrightarrow \mathscr{F}_n \text{ is } \bar{n} \in c)].$$

So Q satisfies condition (2′) of the lemma, since $\lambda n \mid \mathscr{F}_n$ is $\mathscr{M}$-definable.

THEOREM 1.9. *If T is generic, then $\mathscr{M}(T)$ is a model of* ZF *which satisfies the following sentences: every constructible cardinal is a cardinal; there exist exactly two degrees of nonconstructibility.*

PROOF. We noted in the remarks subsequent to Lemma 1.2 that the replacement axiom holds in $\mathscr{M}(T)$. Clearly, the following also holds in $\mathscr{M}(T)$: there is a set $T \subseteq \omega$ such that every set is constructible from T. By Lemma 1.7 the cardinals of $\mathscr{M}(T)$ are the same as the cardinals of $\mathscr{M}$. But then the argument of the last chapter of Gödel [6] shows that the generalized continuum hypothesis holds in $\mathscr{M}(T)$. So the power set axiom must hold in $\mathscr{M}(T)$. (Of course one can show the power set axiom holds in $\mathscr{M}(T)$ without using 1.7 or the sequential lemma.)

By Proposition 1.1 T is not constructible in the sense of $\mathscr{M}(T)$.

Suppose $K \subseteq \omega$ and $K \in \mathscr{M}(T)$. Let $c \in \mathscr{C}$ define K: $\bar{n} \in K \leftrightarrow \mathscr{M}(T) \models \bar{n} \in c$. Consider the sentence $\mathscr{F}$:

$$\mathscr{T} \in \mathscr{M}(c(\mathscr{T})) \vee c(\mathscr{T}) \in \mathscr{M}.$$

By Lemma 1.8 there is no P that forces the negation of $\mathscr{F}$. Since T is generic, it follows that

$$T \in \mathscr{M}(c(T)) \quad \text{or} \quad c(T) \in \mathscr{M}.$$

In other words, $T \in \mathscr{M}(K)$ or $K \in \mathscr{M}$.

THEOREM 1.10. *If constructible ω_2 is countable, then there exists a minimal degree of nonconstructibility.*

PROOF. By Lemma 1.3 T is generic if and only if T is generic with respect to all sentences of ordinal rank less than the ω_2 of $\mathcal{M}$. So if constructible ω_2 is countable, then generic T's actually exist and are of minimal degree by Theorem 1.9.

It would be interesting to find some axiom of infinity which, when added to ZF, decides every elementary question about the partial ordering of degrees of nonconstructibility. R. Solovay suggested that a measurable cardinal might suffice for this purpose.

If T is Cohen-generic (i.e., generic in the sense of forcing with finite conditions), then it is easy to see that T does not have minimal degree in $\mathcal{M}(T)$ by looking at the even and odd parts of T. R. Solovay has shown: if T is Cohen-generic and $T' \in \mathcal{M}(T)$, then there exists a Cohen-generic T'' such that $\mathcal{M}(T') = \mathcal{M}(T'')$. It follows that minimal degrees of nonconstructibility do not occur in Cohen-generic extensions of $\mathcal{M}$.

Thus far we have been a little vague about how to encode a perfect closed $P \subseteq 2^\omega$ as a set of sequence numbers. Now we choose a particular method of encoding for the sake of the proof of Lemma 1.11, a transformation lemma associated with perfect closed forcing. As in § 0, $p, q, r, \ldots$ ambiguously denote sequence numbers and the finite, initial segments of characteristic functions represented by those sequence numbers. We encode P in the form of a function $\lambda ji \mid p_i^j$ with the following properties: p_0^0 is the null initial segment; for each $j \geq 0$ and $i < 2^j$, $p_i^j \geq p_i^{j+1}$, $p_i^j \geq p_{i+2^j}^{j+1}$, $p_i^{j+1} \not\geq p_{i+2^j}^{j+1}$ and $p_{i+2^j}^{j+1} \not\geq p_i^{j+1}$. Then

$$(T)[T \in P \leftrightarrow (j)(\exists_1 i)(i < 2^j \;\&\; T \in p_i^j)].$$

As in § 0, $T \in p$ means p is an initial segment of the characteristic function of T.

Suppose Q is similarly encoded by $\lambda ji \mid q_i^j$. Then we define:

$$h_{P,Q}(p_i^j) = q_i^j.$$

For each $T \in P$, let $h_{P,Q}(T)$ be the unique member of $\bigcap\{h_{P,Q}(p_i^j) \mid i < 2^j \;\&\; T \in p_i^j\}$. Then $h_{P,Q}$ is a 1-1 map of P onto Q. We will call $h_{P,Q}$ a *canonical homeomorphism of P onto Q*. If $R \subseteq P$, then $h_{P,Q}(R) = \{h_{P,Q}(T) \mid T \in R\}$ is a perfect closed subset of Q. The map $h_{P,Q}$ induces a map $h_{P,Q}: \mathcal{L}(\mathcal{T}) \to \mathcal{L}(\mathcal{T})$ as follows. Let $\mathcal{T} \in p$ denote a finite conjunction of sentences of the form $\bar{n} \in \mathcal{T}$ and $\bar{m} \notin \mathcal{T}$ with the defining property that

$$(T)(T \in p \leftrightarrow \mathcal{M}(T) \models \mathcal{T} \in p).$$

For each n, let $\{r_i^n\}$ be the finite set

$$\{p_i^{n+1} \mid i < 2^{n+1} \;\&\; (T)(T \in p_i^{n+1} \to n \in T)\},$$

and similarly, let $\{s_i^n\}$ be

$$\{p_i^{n+1} \mid i < 2^{n+1} \;\&\; (T)(T \in p_i^{n+1} \to n \notin T)\}.$$

Then for every $T \in P$:

$$n \in T \leftrightarrow (\exists i)(T \in r_i^n) \;\&\; (i)(T \notin s_i^n).$$

Let c be a number-theoretic term of $\mathscr{L}(\mathscr{T})$. Then $h_{P,Q}(c \in \mathscr{T})$ is defined to be

$$(\exists i)[\mathscr{T} \in h_{P,Q}(r_i^c)] \;\&\; (i)[\mathscr{T} \notin h_{P,Q}(s_i^c)];$$

if $\mathscr{F}$ is a formula of $\mathscr{L}(\mathscr{F})$, then $h_{P,Q}(\mathscr{F})$ is the result of replacing each occurrence of $c \in \mathscr{T}$ in $\mathscr{F}$ by $h_{P,Q}(c \in \mathscr{T})$; $h(d)$, where $d \in \mathscr{C}$, is defined similarly. If $h_{Q,P}(\mathscr{T}) \in \mathscr{C}(1)$, then the following transformation lemma (cf. Feferman **[2]**) can be established by induction on the rank of $\mathscr{F}$: for all $T \in P$, $\mathscr{M}(T) \models \mathscr{F}$ if and only if $\mathscr{M}(h_{P,Q}(T)) \models h_{P,Q}(\mathscr{F})$.

LEMMA 1.11. *Let h be a canonical homeomorphism of P_1 onto Q_1. Then for each $T \in P_1$, T is generic if and only if $h(T)$ is generic.*

PROOF. Let K be an $\mathscr{M}$-definable set of P's. We say K is *dense* if

$$(R)(\exists P)[R \supseteq P \;\&\; P \in K].$$

It is standard to observe that T is generic if and only if $T \in \bigcup\{P \mid P \in K\}$ for every $\mathscr{M}$-definable, dense K. Clearly, if K is $\mathscr{M}$-definable and dense, then there cannot be an R that forces

$$\mathscr{T} \notin \bigcup\{P \mid P \in K\};$$

and for each sentence $\mathscr{F}$ of $\mathscr{L}(\mathscr{T})$, the set

$$\{P \mid P \Vdash \mathscr{F} \quad \text{or} \quad P \Vdash \sim\mathscr{F}\}$$

is $\mathscr{M}$-definable and dense.

Fix P_1. We say K_1 is *dense in* P_1 if $(P)[P \in K_1 \rightarrow P \subseteq P_1]$ and

$$(R)(\exists P)[R \subseteq P_1 \rightarrow R \supseteq P \;\&\; P \in K_1].$$

The above standard observation can be relativized to P_1: $T \in P_1$ is generic if and only if $T \in \bigcup\{P \mid P \in K_1\}$ for every $\mathscr{M}$-definable K_1 dense in P_1. Suppose that $T \in P_1$ is generic, and that K_1 is dense in P_1. There cannot be an R such that $T \in R$ and

$$R \Vdash \mathscr{T} \in P_1 \quad \& \quad \mathscr{T} \notin \bigcup\{P \mid P \in K_1\}.$$

If there were such an R, then by Proposition 1.1 there would be an $R_1 \subseteq R \cap P_1$. But then the density of K_1 in P_1 would provide a $P \subseteq R$, such that

$$P \Vdash \mathscr{T} \in \bigcup\{P \mid P \in K_1\}.$$

Since h is a canonical homeomorphism of P_1 onto Q_1, $\bigcup\{P \mid P \in K_1\}$ is dense in P_1 if and only if $\bigcup\{h(P) \mid P \in K_1\}$ is dense in Q_1; furthermore, if K_2 is dense in Q_1, then $K_2 = \bigcup\{h(P) \mid P \in K_1\}$ for some K_1 dense in P_1.

THEOREM 1.12. *If T_1 is generic and $T_2 \in \mathscr{M}(T_1) - \mathscr{M}$, then T_2 is generic.*

PROOF. Since $\mathscr{M}$ and $\mathscr{M}(T_1)$ have the same cardinals, $T_2 = c(T_1) = \{n \mid \mathscr{M}(T_1) \models \bar{n} \in c\}$ for some $c \in \mathscr{C}(\omega_1)$. By Lemma 1.8 there is a Q such that $T_1 \in Q$ and

$$(T)(T \in Q \rightarrow T \in \mathscr{M}_{\omega_1}(c(T)).$$

Recall the end of Case 1 of the proof of Lemma 1.8. Q has the property that

$$(X)(Y)[X, Y \in Q \;\&\; X \neq Y \rightarrow c(X) \neq c(Y)].$$

Let $R = \{c(T) \mid T \in Q]$; then c is a 1-1 map of Q onto R. We claim a small addition to Case 1 of Lemma 1.8 guarantees that c can be construed as a canonical homeomorphism of Q onto R. The addition is: for all m and $i < 2^{m+1}$, require that (i) or (ii) hold:

(i) $(T)(T \in Q_i^{m+1} \rightarrow \mathscr{M}(T) \models \bar{m} \in c)$;
(ii) $(T)(T \in Q_i^{m+1} \rightarrow \mathscr{M}(T) \models \bar{m} \notin c)$.

Then Lemma 1.11 implies that $T_2 = c(T_1)$ is generic.

For more information concerning the properties of $\mathscr{M}(T)$ for generic T see Prikry **[17]** and Mathias **[15]**.

The minimality of the degree of a generic T was established in the proof of Theorem 1.9 by clauses (1) and (2) of Lemma 1.8. If one wishes to build a more complicated model of ZF in which the degrees of subsets of ω are controlled by perfect closed forcing, one cannot in general prove Lemma 1.8 (1) or Lemma 1.8 (2). In that emergency one falls back on clauses (1′) and (2′) of Lemma 1.8, which suffice if cardinals are preserved.

Since the relation A is constructible from B is analytical (in fact, Σ_2^1) in the sense of Kleene **[9]**, it can be used to study the independence of various questions concerning the analytical hierarchy. A. Lévy has shown by means of his cardinal-collapsing method that if ZF is consistent, then ZF plus the negation of the axiom of choice for some Π_2^1 predicate is consistent. The axiom of choice for a predicate $P(n, X)$ is:

$$(n)(\exists X)P(n, X) \rightarrow (\exists X)(n)P(n, (X)_n),$$

where $(X)_n = \{m \mid 2^m \cdot 3^n \in X\}$. In Lévy's model, constructible ω_n is countable for all finite n. His result limiting the provable versions of the axiom of choice can be obtained by extending the arguments of this section to build a model $\mathscr{N}$ in which the degrees of nonconstructibility are isomorphic to ω and in which every constructible cardinal is a cardinal. Thus it is true in $\mathscr{N}$ that: for each $n > 0$, there is an X such that $(X)_0, (X)_1, \ldots, (X)_n$ have distinct degrees, but there is no X such that for all $n > 0$, $(X)_0, (X)_1, \ldots, (X)_n$ have distinct degrees.

S. Kripke and D. A. Martin carried on in the above vein and built a model of ZF in which a Δ_4^1 subset of ω failed to be constructible.[3] Mathias **[15]** gives an intuitive account of how this is done. The general idea is as follows. Choose an appropriate sequence $\langle \mathscr{S}_n \rangle$ of logically independent sentences from the theory of partial order. Than a set $A \subseteq \omega$ can be encoded as an analytical set in a model $\mathscr{N}$

[3] The question: is every analytical subset of ω constructible? was first suggested to us by G. Kreisel.

by taking steps so that $n \in A$ if and only if $\mathscr{S}_n$ is true in the partial ordering of degrees of nonconstructibility of $\mathscr{N}$. Jensen succeeded in using these ideas to obtain the best possible result: he lowered Δ_4^1 to Δ_3^1 and arranged for the axiom of choice to be true in his model. Jensen's result was first obtained by Solovay, who managed it without any use of degrees.

Silver has found another way of building a model in which minimal degrees of nonconstructibility occur. He makes use of a special kind of perfect closed condition he calls coinfinite. A coinfinite condition $P = (A, B)$ is given by a pair $A, B \subseteq \omega$ such that $A \cap B = 0$ and $\omega - (A \cup B)$ is infinite; we say $T \in P$ if $A \subseteq T$ and $T \cap B = 0$. The sequential lemma (1.4), mildly altered, holds for coinfinite conditions. (For more information about coinfinite conditions, see Mathias [**15**].) Coinfinite conditions were discussed but not used in Spector's paper on minimal Turing degrees [**28**]. They were used to prove Theorem 2 of [**21**].

2. **The hyperarithmetic case.** In the present section our purpose is to expound forcing with perfect closed sets in a context where the definability of the forcing relation is a more delicate matter than it is in the set-theoretic context of § 1. With this end in mind we review the minimal hyperdegree construction of [**5**] and add some intuitive remarks that were unfortunately left out of [**5**].

The language $\mathscr{L}(\mathscr{T})$ is little different from Feferman's language $\mathscr{L}^*(\mathscr{S})$ [**2**, p. 335]. $\mathscr{L}(\mathscr{T})$ is first order number theory augmented by a constant set symbol $\mathscr{T}$, certain set variables, and the membership symbol $\in$. Let O_1 be a Π_1^1 set of unique notations [**3**] for recursive ordinals: if $b \in O_1$ is the unique notation for β, then we write $|b| = \beta$. In addition we suppose there exists a recursively enumerable relation $R(c, b)$ such that for all $b \in O_1$, $\{c \mid R(c, b)\} = \{c \mid |c| < |b|\}$. For each $b \in O_1$, $\mathscr{L}(\mathscr{T})$ has ranked set variables $X^b, X^b, Z^b, \ldots$; $\mathscr{L}(\mathscr{T})$ also has unranked set variables $X, Y, Z, \ldots$, number variables $x, y, z, \ldots$, a numeral $\bar{n}$ for each natural number n, and symbols for equality, successor, addition and multiplication.

A formula $\mathscr{F}$ of $\mathscr{L}(\mathscr{T})$ is said to be *ranked* if every set variable occurring in $\mathscr{F}$ is ranked. A formula of $\mathscr{L}(\mathscr{T})$ is said to be existential (or Σ_1^1) if it is ranked or it is of the form $(\exists X)\mathscr{F}$ with X the only unranked variable occurring in $\mathscr{F}$. The *ordinal rank* of a ranked formula $\mathscr{F}$ is least ordinal α such that $\alpha \geq |b|$ for every free variable X^b of $\mathscr{F}$ and such that $\alpha > |b|$ for every bound variable X^b of $\mathscr{F}$.

Let T be an arbitrary subset of ω. Following Feferman [**2**], for each $b \in O_1$ we inductively define a structure $\mathscr{M}_b(T)$ and truth in $\bigcup\{\mathscr{M}_a(T) \mid |a| < |b|\}$.

(i) A sentence $\mathscr{F}$ of ordinal rank $\leq |b|$ is true in $\bigcup\{\mathscr{M}_a(T) \mid |a| < |b|\}$ if it is true when $\mathscr{T}$ is interpreted as T, the number variables of $\mathscr{F}$ are restricted to ω, and each ranked variable X^a of $\mathscr{F}$ is restricted to $\mathscr{M}_a(T)$.

(ii) For each formula $\mathscr{G}(x)$ (with only x free) of ordinal rank $\leq |b|$, let $\hat{x}\mathscr{G}(x)$ be $\{n \mid \mathscr{G}(\bar{n})$ is true in $\bigcup\{\mathscr{M}_a(T) \mid |a| < |b|\}\}$; then $\mathscr{M}_b(T)$ consists of all such sets $\hat{x}\mathscr{G}(x)$.

Let $\mathscr{M}(T)$ be $\bigcup\{\mathscr{M}_b(T) \mid b \in O_1\}$. Then $\mathscr{F}$ is true in $\mathscr{M}(T)$ (written $\mathscr{M}(T) \models \mathscr{F}$) if it is true when each unranked variable of $\mathscr{F}$ is restricted to $\mathscr{M}(T)$ and the remaining symbols of $\mathscr{F}$ are interpreted as in (i) above. If T is hyperarithmetic, then $\mathscr{M}(T)$ is the set of all hyperarithmetic sets, which we denote by $\mathscr{M}$.

$\mathscr{F}(\hat{x}\mathscr{G}(x))$ denotes the result of replacing each occurrence of $t \in \mathscr{U}$ in $\mathscr{F}(\mathscr{U})$ by $\mathscr{G}(t)$, where t is a number-theoretic term and $\mathscr{U}$ is a set variable.

As in § 0 we use $P, Q, R, \ldots$ to ambiguously denote perfect closed subsets of 2^ω and their encodings as subsets of ω. But now we insist that $P, Q, R, \ldots$, be hyperarithmetic. Thus we can write, as we did in § 1, $P \in \mathscr{M}$. It is routine to assign indices to hyperarithmetic, perfect closed sets so that the following relation in x and Y is Π_1^1: x is the index of a hyperarithmetic, perfect closed set P and $Y \in P$. For this reason it makes sense to regard the set of all hyperarithmetic, perfect closed (h.p.c.) sets as a Π_1^1 subset of ω. Similarly, one regards the set of all formulas of $\mathscr{L}(\mathscr{T})$ as a Π_1^1 subset of ω.

It is instructive to see what happens when the analogy between the constructible hierarchy of § 1 and the hyperarithmetic hierarchy of the present section is pursued. In both cases a perfect closed set is acceptable as a forcing condition only if it can be encoded by a subset of ω belonging to $\mathscr{M}$. The analogy suggests that the forcing relation $P \Vdash \mathscr{F}$, where P is h.p.c. set and $\mathscr{F}$ is a sentence of $\mathscr{L}(\mathscr{T})$, should be defined by the following inductive clauses:

(1) $P \Vdash (\exists X)\mathscr{F}(X)$ if $(\exists b)_{b \in O_1}[P \Vdash (\exists X^b)\mathscr{F}(X^b)]$.

(2a) $P \Vdash (\exists X^b)\mathscr{F}(X^b)$ if $P \Vdash \mathscr{F}(\hat{x}\mathscr{G}(x))$ for some $\mathscr{G}(x)$ of ordinal rank $\leq |b|$.

(2b) $P \Vdash (\exists x)\mathscr{F}(x)$ if $P \Vdash \mathscr{F}(\bar{n})$ for some $n \in \omega$.

(3) $P \Vdash \mathscr{F} \vee \mathscr{G}$ if $P \Vdash \mathscr{F}$ or if $P \Vdash \mathscr{G}$.

(4) $P \Vdash \sim\mathscr{F}$ if $(Q)_{P \supseteq Q} \sim [Q \Vdash \mathscr{F}]$.

(5) $P \Vdash \bar{n} \in \mathscr{T}$ if $(T)[T \in P \rightarrow \mathscr{M}(T) \models \bar{n} \in \mathscr{T}]$.

(6) $P \Vdash \mathscr{F}$ if $\mathscr{F}$ is a true, quantifier-free sentence of arithmetic.

The key property of the forcing relation in § 1 is its $\mathscr{M}$-definability as expressed by Lemma 1.2. The analogous fact for the present section (Lemma 2.1) is: the forcing relation restricted to Σ_1^1 $\mathscr{F}$'s is Π_1^1. As we shall see, Lemma 2.1 suffices to establish the sequential lemma and the hyperarithmetic comprehension axiom in $\mathscr{M}(T)$ for all generic T. (To see the analogy between Lemmas 1.2 and 2.1, recall that a Π_1^1-relation becomes Σ_1^1 [**4**, 29] when the universe is restricted to $\mathscr{M}$ = the hyperarithmetic sets. Thus 2.1 says that the forcing relation restricted to Σ_1^1 $\mathscr{F}$'s is $\mathscr{M}$-definable by means of a Σ_1^1 formula.) But we are unable to show that the forcing relation $\Vdash$, defined by (1)—(6) above and restricted to Σ_1^1 sentences, is Π_1^1. The difficulty is in clause (4): even for a ranked $\mathscr{F}$, the definition of $P \Vdash \sim\mathscr{F}$ involves quantification over the set of all Q's $\subseteq P$, that is to say, quantification over a Π_1^1 set of indices. We would be very interested to know the precise classification of $\Vdash$ restricted to Σ_1^1 $\mathscr{F}$'s. A more intuitive way of stating the difficulty is: the collection of all P's is a class of $\mathscr{M}$ but not a set of $\mathscr{M}$. If we restricted P to be recursive in some fixed hyperarithmetic set H, then the difficulty would vanish, but a generic set would not in general be of minimal hyperdegree, and the sequential lemma would hold only for sentences below a certain ordinal rank determined by H.

So we drop the analogy and define $\Vdash_h$ in the manner of [5].

(i) $P \Vdash_h (\exists X)\mathscr{F}(X)$ if $(\exists b)_{b \in O_1}[P \Vdash_h (\exists X^b)\mathscr{F}(X^b)]$.

(iia) $P \Vdash_h (\exists X^b)\mathscr{F}(X^b)$ if $\mathscr{F}(X^b)$ is unranked and $P \Vdash_h \mathscr{F}(\hat{x}\mathscr{G}(x))$ for some $\mathscr{G}(x)$ of ordinal rank $\leq |b|$.

(iib) $P \Vdash_h (\exists x)\mathscr{F}(x)$ if $\mathscr{F}(x)$ is unranked and $P \Vdash_h \mathscr{F}(\bar{n})$ for some $n \in \omega$.

(iii) $P \Vdash_h \mathscr{F} \vee \mathscr{G}$ if $\mathscr{F} \vee \mathscr{G}$ is unranked and $P \Vdash_h \mathscr{F}$ or $P \Vdash_h \mathscr{G}$.

(iv) $P \Vdash_h \sim\mathscr{F}$ if $\mathscr{F}$ is unranked and $(Q)_{P \supseteq Q} \sim [Q \Vdash_h \mathscr{F}]$.

(v) $P \Vdash_h \mathscr{F}$ if $\mathscr{F}$ is ranked and $(T)[T \in P \rightarrow \mathscr{M}(T) \models \mathscr{F}]$.

LEMMA 2.1 [5]. *The relation $P \Vdash_h \mathscr{F}$, restricted to Σ_1^1 $\mathscr{F}$'s of $\mathscr{L}(\mathscr{T})$, is Π_1^1.*

The proof of Lemma 2.1 is nothing more than inspection of clauses (v) and (i) of the definition of $\Vdash_h$. But we must pay a price, fortunately within our means, for the use of clause (v). It is no longer obvious as it was in § 1 that

$$(\mathscr{F})(P)(\exists Q)_{P \supseteq Q}[Q \Vdash_h \mathscr{F} \text{ or } Q \Vdash_h \sim\mathscr{F}],$$

because clause (iv) does not apply to ranked sentences. But this difficulty can be overcome by a sequential lemma (2.2) argument.

LEMMA 2.2 [5]. *Let $\{\mathscr{F}_i\}$ be a hyperarithmetic sequence of Σ_1^1 sentences of $\mathscr{L}(\mathscr{T})$. Let P have the property that*

$$(i)(Q)_{P \supseteq Q}(\exists R)_{Q \supseteq R}[R \Vdash_h \mathscr{F}_i].$$

Then there exists a $Q \subseteq P$ such that

$$(i)[Q \Vdash_h \mathscr{F}_i].$$

Note that Lemma 2.2 is closer in flavor to Lemma 1.6 than to Lemma 1.4. But the proof of 2.2 is essentially the same as that of 1.4. The only real difference is that the $\mathscr{M}$-definable function $f(i, Q)$ of 1.4 becomes a partial Π_1^1 function defined for all i and all hyperarithmetic $Q \subseteq P$. (A function is called partial Π_1^1 if its graph is Π_1^1.) The existence of f follows from Lemma 2.1 and an important lemma of Kreisel [14] which says: if $P(x, y)$ is Π_1^1, then there exists a partial Π_1^1 function g such that

$$(x)[(\exists y)P(x, y) \rightarrow g(x) \text{ is defined } \& \ P(x, g(x))].^4$$

The desired $Q \subseteq P$ is built, as it was in Lemma 1.4, by iterating f. To see that Q is hyperarithmetic one has to use the fact that a partial Π_1^1 function, restricted to a hyperarithmetic subset of its domain, is hyperarithmetic.

LEMMA 2.3 [5]. $(\mathscr{F})(P)(\exists Q)_{P \supseteq Q}[Q \Vdash_h \mathscr{F} \vee Q \Vdash_h \sim\mathscr{F}]$.

The proof of Lemma 2.3 is very much like that of Lemma 1.5 and proceeds by transfinite induction on the rank, appropriately defined, of $\mathscr{F}$. The only interesting

[4] Kreisel's lemma generalizes a familiar lemma of recursion theory which says the same thing with Π_1^1 replaced by recursively enumerable.

inductive step occurs when $\mathscr{F}$ is a ranked sentence of the form $(\exists X^b)\mathscr{F}(X^b)$. If there is a $\mathscr{G}(x)$ of ordinal rank $\leq |b|$ and a $Q \subseteq P$ such that $Q \Vdash_h \mathscr{F}(\hat{x}\mathscr{G}(x))$, then $Q \Vdash_h (\exists X^b)\mathscr{F}(X^b)$. Suppose there is no such $\mathscr{G}(x)$ and Q. Let $\{\mathscr{G}_i(x)\}$ be a hyperarithmetic enumeration of all $\mathscr{G}(x)$'s of ordinal rank $\leq |b|$. Then the induction hypothesis implies

$$(i)(Q)_{P \supseteq Q}(\exists R)_{Q \supseteq R}[R \Vdash_h \sim\mathscr{F}(\hat{x}\mathscr{G}_i(x))].$$

Now the sequential lemma (2.2) can be applied, as it was in Lemma 1.5, to obtain a $Q \subseteq P$ such that $Q \Vdash (X^b) \sim \mathscr{F}(X^b)$.

T is said to be *generic* if for every sentence $\mathscr{F}$ of $\mathscr{L}(\mathscr{T})$ there is a P such that $T \in P$ and either $P \Vdash_h \mathscr{F}$ or $P \Vdash_h \sim\mathscr{F}$. The existence of generic T's follows from Lemma 2.3. The minimal degree argument of Lemma 1.8 is easily modified to prove the next lemma.

LEMMA 2.4 [5]. *Let $\mathscr{G}(x)$ be a ranked formula of $\mathscr{L}(\mathscr{T})$ with only x free. For each P there exists a $Q \subseteq P$ such that either* (1) *or* (2) *holds:*

(1) $(T)[T \in Q \rightarrow T$ *is hyperarithmetic in* $\{n \mid \mathscr{M}(T) \models \mathscr{G}(\bar{n})\}]$;

(2) $(\exists H)(T)[T \in Q \rightarrow \{n \mid \mathscr{M}(T) \models \mathscr{G}(\bar{n})\} = H$ *is hyperarithmetic*].

LEMMA 2.5 [5]. *If T is generic, then every set hyperarithmetic in T belongs to* $\mathscr{M}(T)$.

Before we discuss the proof of Lemma 2.5, we make some related observations without proof. For every T it is the case that every member of $\mathscr{M}(T)$ is hyperarithmetic in T. If T is generic, then $\omega_1^T = \omega_1$, where ω_1^T is the least ordinal not recursive in T.

According to Kreisel [13] the conclusion of Lemma 2.5 is equivalent to the statement that the hyperarithmetic comprehension axiom (h.c.a.) holds in $\mathscr{M}(T)$. A typical instance of the h.c.a. is the universal closure of a formula of the form

$$(x)[(\exists Y)\mathscr{A}(x, Y) \leftrightarrow (Z)\mathscr{B}(x, Z)] \rightarrow (\exists X)(x)[x \in X \leftrightarrow (\exists Y)\mathscr{A}(x, Y)],$$

where $\mathscr{A}(x, Y)$ and $\mathscr{B}(x, Z)$ are arithmetic formulas which may contain free set variables other than Y or Z. By an argument of Feferman [2, p. 339] which holds for *all* T, the hyperarithmetic comprehension axiom is true in $\mathscr{M}(T)$ if the following Σ_1^1-bounding principle is true in $\mathscr{M}(T)$: let $\mathscr{F}(x, Y)$ be a formula of $\mathscr{L}(\mathscr{T})$ whose only free variables are x and Y and whose only unranked variable is Y; if $(x)(\exists Y)\mathscr{F}(x, Y)$, then $(x)(\exists Y^b)\mathscr{F}(x, Y^b)$ for some $b \in O_1$. The principle holds in $\mathscr{M}(T)$ for all generic T by Lemmas 2.6 and 2.7.

LEMMA 2.6. *Let $\mathscr{F}(x, Y)$ be a formula of $\mathscr{L}(\mathscr{T})$ whose only free variables are x and Y and whose only unranked variable is Y. If $P \Vdash_h (x)(\exists Y)\mathscr{F}(x, Y)$, then $Q \Vdash_h (x)(\exists Y^b)\mathscr{F}(x, Y^b)$ for some $Q \subseteq P$ and $b \in O_1$.*

The proof of Lemma 2.6 is another application of the sequential lemma (2.2). Suppose $P \Vdash_h (x)(\exists Y)\mathscr{F}(x, Y)$, where $\mathscr{F}(x, Y)$ is as in Lemma 2.6. Then

$$(n)(Q)_{P \supseteq Q}(\exists R)_{Q \supseteq R}[R \Vdash_h (\exists Y)\mathscr{F}(\bar{n}, Y)].$$

By Lemma 2.2 there is a $Q \subseteq P$ such that

$$(n)(\exists b)[b \in O_1 \ \& \ Q \Vdash_h (\exists Y^b)\mathscr{F}(\bar{n}, Y)].$$

By Lemma 2.1 and Kreisel's lemma (as stated just before 2.2), there is a hyperarithmetic function f such that

$$(n)[f(n) \in O_1 \ \& \ Q \Vdash_h (\exists Y^{f(n)})\mathscr{F}(n, Y^{f(n)})].$$

By Spector [27] there is a $b \in O_1$ such that $(n)(|f(n)| \leq |b|)$. So

$$Q \Vdash_h (x)(\exists Y^b)\mathscr{F}(x, Y^b).$$

Lemma 2.6 almost establishes that the Σ_1^1-bounding principle holds in $\mathscr{M}(T)$ for all generic T. What is lacking are some technical details concerning definability that are immediate in the set-theoretic context of § 1 but that require proof in the hyperarithmetic case.

Let A be a set of indices of hyperarithmetic perfect closed sets. We write $P \in A$ to mean that the index of the perfect closed set P, encoded as a hyperarithmetic set of sequence numbers, belongs to A. A is said to be *dense* if $(P)(\exists Q)[Q \in P \ \& \ Q \subseteq P]$.

A set A is said to be $\mathscr{M}$-definable if there is a formula $\mathscr{F}(x)$ of $\mathscr{L}(\mathscr{T})$ in which $\mathscr{T}$ does not occur such that $A = \{n \mid \mathscr{M} \models \mathscr{F}(\bar{n})\}$. For each T: a set A is said to be $\mathscr{M}(T)$-definable if there is a formula $\mathscr{F}(x)$ of $\mathscr{L}(\mathscr{T})$ such that

$$A = \{n \mid \mathscr{M}(T) \models \mathscr{F}(\bar{n})\}.$$

A is called *generically persistent*[5] if there is a formula $\mathscr{F}(x)$ of $\mathscr{L}(\mathscr{T})$ such that for all generic T, $A = \{n \mid \mathscr{M}(T) \quad \mathscr{F}(\bar{n})\}$.

LEMMA 2.7. *T is generic if and only if $T \in \bigcup\{P \mid P \in A\}$ for every dense A arithmetic in Kleene's O.*

LEMMA 2.8. *A is arithmetic in Kleene's O if and only if A is $\mathscr{M}$-definable if and only if A is generically persistent.*

Let T be generic; we use Lemmas 2.6 and 2.7 to show the Σ_1^1-bounding principle holds in $\mathscr{M}(T)$. Let $\mathscr{F}(x, Y)$ be as in the hypothesis of 2.6. Let A be the set of all Q's such that for some $b \in O_1$,

$$Q \Vdash_h (x)(\exists Y)\mathscr{F}(x, Y) \rightarrow (x)(\exists Y^b)\mathscr{F}(x, Y^b).$$

A is arithmetic in Kleene's O, since both O_1 and the indexing of all P's is Π_1^1. A is dense by 2.6. Then 2.7 provides a $Q \in A$ such that $T \in Q$.

To complete the proof of Lemma 2.5 we need only prove Lemmas 2.7 and 2.8.

---

[5] Persistence is a term used by H. Putnam [18] to describe a well-known property of the hyperarithmetic sets. Kreisel [13] formulates the persistence property as follows: a set is hyperarithmetic if and only if it belongs to every ω-model of the hyperarithmetic comprehension axiom. Another way of putting it is: a set A is hyperarithmetic if and only if there is a ranked formula $\mathscr{G}(x)$ of $\mathscr{L}(\mathscr{T})$ such that for all T, $A = \{n \mid \mathscr{M}(T) \models \mathscr{F}(\bar{n})\}$. This formulation of persistence is a consequence of a slight modification of Kleene [10, p. 35].

Suppose A is dense and arithmetic in Kleene's O. By 2.8 A is generically persistent. Let $\mathscr{A}(x)$ be a formula such that for *all* generic T, $A = \{n \mid \mathscr{M}(T) \models \mathscr{A}(\bar{n})\}$. We claim there is an $\mathscr{F}$ such that for all generic T,

$$\mathscr{M}(T) \models \mathscr{F} \leftrightarrow T \in \bigcup\{Q \mid Q \in A\}.$$

The desired $\mathscr{F}$ is $(\exists x)[\mathscr{A}(x) \;\&\; (Y)(\mathscr{B}(x, Y) \rightarrow \mathscr{D}(Y, \mathscr{T}))]$, where $\mathscr{B}(x, Y)$ and $\mathscr{D}(Y, Z)$ are arithmetic formulas with the following properties. Let I be the Π_1^1 set of indices of hyperarithmetic perfect closed sets. If $n \in I$, let P_n be the hyperarithmetic perfect closed set whose index is n. For each $n \in I$, $(\exists_1 Y)\mathscr{B}(n, Y)$ and $(Y)[\mathscr{B}(n, Y) \rightarrow (Y)_0 = P_n]$.[6] $\mathscr{D}(P, T)$ says $T \in P$. To see that every generic $T \in \bigcup\{Q \mid Q \in A\}$, it suffices to see there is no P such that $P \Vdash_h \sim\mathscr{F}$. If there were such a P, then by the density of A there would be a $Q \subseteq P$ such that $Q \in A$ and $Q \Vdash_h \sim\sim\mathscr{F}$.

Suppose T is arbitrary but that $T \in \bigcup\{Q \mid Q \in A\}$ for every dense A arithmetic in Kleene's O. T is generic if $T \in \bigcup\{Q \mid Q \Vdash_h \mathscr{F}$ or $Q \Vdash_h \sim\mathscr{F}\} = K_{\mathscr{F}}$ for every sentence $\mathscr{F}$ of $\mathscr{L}(\mathscr{T})$. But it is easy to check that $K_{\mathscr{F}}$ is dense and arithmetic in Kleene's O.

We turn to Lemma 2.8. It is enough to show: (i) O is $\mathscr{M}$-definable; (ii) O is generically persistent; (iii) if A is generically persistent, then A is arithmetic in O. We dispose immediately of (iii). Suppose $A = \{n \mid \mathscr{M}(T) \models \mathscr{A}(\bar{n})\}$ for all generic T. Then $n \in A \leftrightarrow (\exists P)(P \Vdash_h \mathscr{A}(\bar{n}))$. A is arithmetic in O since $P \Vdash_h \mathscr{F}$, restricted to $\mathscr{F}$'s of the form $\mathscr{A}(\bar{n})$, is arithmetic in O. Mostowski's conjecture [4] provides an arithmetic formula $C(X, n)$ such that for all n, $n \in O \leftrightarrow (\exists X)[X \in \text{HYP} \;\&\; C(X, n)]$. Thus (i) O is $\mathscr{M}$-definable by a Σ_1^1 formula. O will then be generically persistent if every arithmetic formula $C(X)$ and every generic T are such that if $\mathscr{M}(T) \models (\exists X)C(X)$, then $(\exists X)[X \in \text{HYP} \;\&\; C(X)]$. Suppose $P \Vdash_h (\exists X)C(X)$. Then $P \Vdash_h C(\hat{x}\mathscr{G}(x))$ for some ranked $\mathscr{G}(x)$. Let $H \in P$ be hyperarithmetic. Then $J = \{n \mid \mathscr{M}(H) \models \mathscr{G}(\bar{n})\}$ is hyperarithmetic and $C(J)$.

THEOREM 2.9 [5]. *If T is generic, then T has minimal hyperdegree.*

If T is generic, then T is not hyperarithmetic by the same mode of argument employed in Proposition 1.1 to show that a set generic in the sense of § 1 is not a constructible set of $\mathscr{M}$. The fact that T has minimal hyperdegree follows from Lemmas 2.4, 2.5 and 2.7.

THEOREM 2.10. *If T is generic, then every set of the same hyperdegree as T is generic.*

Let S have the same hyperdegree as some generic T. Then

$$S = \{n \mid \mathscr{M}(T) \models \mathscr{G}(\bar{n})\},$$

where $\mathscr{G}(x)$ is a ranked formula. By Lemmas 2.4 and 2.7 there is a Q such that $T \in Q$ and Q satisfies clause (1) of 2.4. Then the argument of Theorem 1.12,

[6] The existence of $\mathscr{B}$ follows from the uniform implicit arithmetic definability of hyperarithmetic sets of arbitrarily high Turing degree, cf. Kreisel [7, p. 307].

routinely modified, establishes that S is generic. The modification includes Lemmas 2.7 and 2.8.

3. **The arithmetic case.** In accord with § 0, $P, Q, R, \ldots$ ambiguously denote arithmetic perfect closed subsets of 2^ω and their encodings as arithmetic subsets of ω. One of the principal objectives of the present section is to introduce pointed, perfect closed sets and to apply them to a problem concerning the Turing degrees of certain hyperarithmetic sets and in particular the truth set for arithmetic.

The arithmetic language $\mathscr{L}(\mathscr{T})$ is the language of first order number theory together with a constant set symbol $\mathscr{T}$ and a predicate letter $\mathscr{R}$ denoting a certain recursive predicate. $\mathscr{L}(\mathscr{T})$ has symbols for membership ($\in$), addition ($+$), multiplication ($\cdot$), and successor ($'$), and a numeral $\bar{n}$ for each natural number n. $\mathscr{L}(\mathscr{T})$ is not essentially different from Feferman's language $L^*(S)$ [**2**, p. 328]. The predicate letter $\mathscr{R}$ is associated with the jump operator [**19**] for Turing degrees. By the enumeration theorem there is a recursive predicate $\mathscr{R}_2(T, x_1, x_2, y)$ such that for each recursive predicate $\mathscr{P}(T, x_2, y)$ there is an e with the property that

$$(T)(x_2)[(\exists y)\mathscr{R}_2(T, e, x_2, y) \leftrightarrow (\exists y)\mathscr{P}(T, x_2, y)].$$

The *jump* of T, denoted by T', can be defined by

$$2^m \cdot 3^n \in T' \leftrightarrow (\exists y)\mathscr{R}_2(T, m, n, y)$$

with the understanding that every member of T' is of the form $2^m \cdot 3^n$. T' can be thought of as the disjoint recursive union of all sets recursively enumerable in T. $T^{(n)}$, the n-jump of T, is defined by: $T^{(0)} = T$; $T^{(n+1)} = (T^{(n)})'$. 0 is the empty set. A set is arithmetic if and only if it is recursive in $0^{(n)}$ for some n. $T^{(\omega)}$, the ω-jump of T, is the disjoint recursive union of all n-jumps of T:

$$2^m \cdot 3^n \in T^{(\omega)} \leftrightarrow m \in T^{(n)}.$$

We call $0^{(\omega)}$ the truth set for arithmetic.

It is essential to note the existence of a recursive predicate $\mathscr{R}(T, n, x, y)$ such that for all T and n:

$$n \in T^{(2)} \leftrightarrow (\exists x)(y)\mathscr{R}(T, n, x, y).$$

The language $\mathscr{L}(\mathscr{T})$ includes $\mathscr{R}$; thus the 2-jump of T is definable by a Σ_2 formula of $\mathscr{L}(\mathscr{T})$. It is convenient to assume that the truth or falsity of $\mathscr{R}(T, n, x, y)$ for each choice of T, n, x and y is determined by an initial segment of the characteristic function of T of length at most y.

Let $\mathscr{F}$ be a sentence of $\mathscr{L}(\mathscr{T})$, and let T be an arbitrary subset of ω. We say T makes $\mathscr{F}$ true (or more briefly, $\mathscr{F}$ is true) when $\mathscr{F}$ is true subject to the interpretation of $\mathscr{T}$ as T and to the standard interpretation of the arithmetic symbols and $\mathscr{R}$. The forcing relation, $P \Vdash \mathscr{F}$, where $\mathscr{F}$ is a sentence of $\mathscr{L}(\mathscr{T})$ and P is an arithmetic perfect closed set, is defined by: $P \Vdash \mathscr{F} \leftrightarrow (T)[T \in P \rightarrow \mathscr{F}]$. Since we require every $T \in P$ to make $\mathscr{F}$ true, it is perhaps not appropriate to say P forces $\mathscr{F}$. On the other hand, if forcing is thought of as an approximation of truth, it seems reasonable to expect perfect closed forcing to provide a perfect approximation of truth.

We say T is *generic* if for each $\mathscr{F}$ there is a P such that $T \in P$ and either $P \Vdash \mathscr{F}$ or $P \Vdash \sim\mathscr{F}$. The existence of generic T's is an immediate consequence of Lemma 3.1. We could prove 3.1 by the sequential lemma approach of § 1 and 2, but it is more instructive to invoke what might be termed the *local forcing* approach. Let $p, q, r, \ldots$ be finite, initial segments of characteristic functions of subsets of ω. ($p \geq q$ and $T \in p$ are defined in § 0.) The definition of $p \Vdash \mathscr{F}$, where $\mathscr{F}$ is a sentence of $\mathscr{L}(\mathscr{T})$, is given by four inductive clauses:

(i) $p \Vdash \mathscr{F}$ if $\mathscr{F}$ is a true equation of arithmetic, or if $\mathscr{F}$ is of the form $\bar{n} \in \mathscr{T}$ or $\mathscr{R}(\mathscr{T}, \bar{m}, \bar{n}, \bar{c})$ and every $T \in p$ makes $\mathscr{F}$ true.

(ii) $p \Vdash (\exists x)\mathscr{F}(x)$ if $(\exists n)[p \Vdash \mathscr{F}(\bar{n})]$.

(iii) $p \Vdash \sim\mathscr{G}$ if $(q)_{p \geq q} \sim [q \Vdash \mathscr{G}]$.

(iv) $p \Vdash \mathscr{F}_1 \,\&\, \mathscr{F}_2$ if $p \Vdash \mathscr{F}_1$ and $p \Vdash \mathscr{F}_2$.

In short, $p \Vdash \mathscr{F}$ follows the rules of Cohen's forcing with finite conditions as formulated by Feferman **[2]** in the arithmetic case. Let P be an arithmetic perfect closed set. The forcing relation $p \Vdash^P \mathscr{F}$, where $p \in P$ and $\mathscr{F}$ is a sentence of $\mathscr{L}(\mathscr{T})$, is defined exactly as $p \Vdash \mathscr{F}$ was defined above except that clause (iii) becomes:

$$(\text{iii})^P \quad p \Vdash^P \sim\mathscr{G} \quad \text{if} \quad (q)_{p \geq q}[q \in P \rightarrow \sim q \Vdash^P \mathscr{G}].$$

We call $p \Vdash^P \mathscr{F}$ the *localization* of $p \Vdash \mathscr{F}$ to P. The idea of local forcing was used by Spector **[28]** in his construction of a minimal Turing degree. Spector's fundamental lemma [**28**, p. 588] was in essence the following: let $\mathscr{F}$ be a Σ_2 sentence of $\mathscr{L}(\mathscr{T})$, and let P be a recursive perfect closed set; then there exists a recursive perfect closed $Q \subseteq P$ such that either $(T)(T \in Q \rightarrow \mathscr{F})$ or $(T)(T \in Q \rightarrow \sim\mathscr{F})$. The details of Spector's proof coincide with the details of the local forcing argument given in Lemma 3.1 for the case $\mathscr{F} \in \Sigma_2$. Spector uses his lemma to construct a sequence $P_0 \supseteq P_1 \supseteq P_2 \supseteq \cdots$ of recursive perfect closed sets with the following properties: for all $T \in P_e$, if $\{e\}^T$ is total, then $\{e\}^T$ is recursive or T is recursive in $\{e\}^T$ ($\{e\}^T$ is the eth function partial recursive in T according to the standard enumeration); for all $T \in P_e$, T is not the eth partial recursive function. Thus if $T \in \bigcap\{P_e \mid e \geq 0\}$, it is clear that T has minimal Turing degree.

The properties of $\Vdash^P$ are virtually the same as those of $\Vdash$; intuitively, this is the case because P and 2^ω are homeomorphic by means of an arithmetic map. T is *generic for P* if for each sentence $\mathscr{F}$ of $\mathscr{L}(\mathscr{T})$ there is a $p \in P$ such that $T \in p$ and either $p \Vdash^P \mathscr{F}$ or $p \Vdash^P \sim\mathscr{F}$. If T is generic for P, then $T \in P$, and T makes $\mathscr{F}$ true if and only if there is a $p \in P$ such that $T \in p$ and $p \Vdash^P \mathscr{F}$.

LEMMA 3.1. *Let P be an arithmetic perfect closed set, and let $\mathscr{F}$ be a sentence of the arithmetic language $\mathscr{L}(\mathscr{T})$. Then there exists an arithmetic perfect closed $Q \subseteq P$ such that either* (a) *or* (b) *holds:*

(a) $(T)(T \in Q \rightarrow \mathscr{F})$;

(b) $(T)(T \in Q \rightarrow \sim\mathscr{F})$.

Furthermore, if $\mathscr{F}$ is Σ_2, then Q is recursive in P.

PROOF. There must be a $p \in P$ such that either (a) $p \Vdash^P \mathscr{F}$ or (b) $p \Vdash^P \sim\mathscr{F}$. Suppose (a). We develop a $Q \subseteq P$ satisfying clause (a) of the lemma. Let $\{\mathscr{F}_m \mid m \geq 0\}$ be an enumeration of all sentences of $\mathscr{L}(\mathscr{T})$ having no more quantifiers than $\mathscr{F}$ does. The relation $q \Vdash^P \mathscr{F}_m$, viewed as a relation on q and m, is arithmetic, because a finite bound has been imposed on the number of quantifiers that can occur in $\mathscr{F}_m$. It follows there must exist a partial arithmetic function $f(q, m)$ such that if $q \in P$, then $f(q, m) \in P$, $q \geq f(q, m)$, and either $f(q, m) \Vdash^P \mathscr{F}_m$ or $f(q, m) \Vdash^P \sim\mathscr{F}_m$. By iteration of f, it is possible to define a partial arithmetic function $\lambda im \mid q_i^m$ with the following properties: $q_0^0 = p$; for each $m \geq 0$ and $i < 2^m$, $q_i^m \geq q_i^{m+1}$, $q_i^m \geq q_{i+2^m}^{m+1}$, $q_i^{m+1} \ngeq q_{i+2^m}^{m+1}$, $q_{i+2^m}^{m+1} \ngeq q_i^{m+1}$, $q_i^{m+1} \Vdash^P \mathscr{F}_m$ or $q_i^{m+1} \Vdash^P \sim\mathscr{F}_m$, and $q_{i+2^m}^{m+1} \Vdash^P \mathscr{F}_m$ or $q_{i+2^m}^{m+1} \Vdash^P \sim\mathscr{F}_m$. Then $Q = \{q_i^m \mid m \geq 0 \;\&\; i < 2^m\}$ is an arithmetic perfect closed set. If $T \in Q$, then T is generic for P with respect to all sentences in $\{\mathscr{F}_m\}$, and consequently, T makes $\mathscr{F}$ true since $T \in p = q_0^0$ and $p \Vdash^P \mathscr{F}$.

Now suppose $\mathscr{F}$ is Σ_2. Let $\mathscr{F}$ be $(\exists x)(y)\mathscr{S}(x, y)$, where $\mathscr{S}(x, y)$ is a quantifier-free formula of $\mathscr{L}(\mathscr{T})$. Suppose again that (a) $p \Vdash^P \mathscr{F}$. This means $p \Vdash^P (y)\mathscr{S}(\bar{n}, y)$ for some n, and so

$$(m)(q)_{q \in P}(\exists r)_{r \in P}[p \geq q \rightarrow q \geq r \;\&\; r \Vdash^P \mathscr{S}(\bar{n}, \bar{m})].$$

The relation $r \Vdash^P \mathscr{S}(\bar{n}, \bar{m})$, viewed as a relation on r, n and m, is recursive in P, because $\mathscr{S}$ has no quantifiers. It follows there exists a function $f(m, q)$ partial recursive in P such that if $q \in P$ and $p \geq q$, then $f(m, q) \in P$, $q \geq f(m, q)$, and $f(m, q) \Vdash^P \mathscr{S}(\bar{n}, \bar{m})$. Let $\mathscr{F}_m$ be $\mathscr{S}(\bar{n}, \bar{m})$, and define $\lambda im \mid q_i^m$ by iterating $f(m, q)$ as in the previous paragraph. Then $Q = \{q_i^m \mid m \geq 0 \;\&\; i < 2^m\}$ is recursive in P, and every $T \in Q$ makes $(\exists x)(y)\mathscr{S}(x, y)$ true.

Finally, suppose (b) $p \Vdash^P (x)(\exists y) \sim\mathscr{S}(x, y)$. Then

$$(n)(q)_{q \in P}(\exists m)(\exists r)_{r \in P}[p \geq q \rightarrow q \geq r \;\&\; r \Vdash^P \sim\mathscr{S}(\bar{n}, \bar{m})].$$

As above, we can make m and r into functions of n and q partial recursive in P, functions which are defined whenever $q \in P$ and $p \geq q$. Then as above r is iterated to obtain a Q recursive in P such that every $T \in Q$ makes $(x)(\exists y) \sim \mathscr{S}(x, y)$ true.

One of the advantages of using local forcing instead of an appropriate sequential lemma to prove Lemma 3.1 is that local forcing makes it easier to see Q is arithmetic.

We say P is a *pointed*, arithmetic perfect closed set if

$$(T)(T \in P \rightarrow P \text{ is recursive in } T).$$

Every perfect closed P has T's such that T is recursive in P; let T_P be the canonical one obtained by traveling up P viewed as a tree and turning left at every branch point. If P is pointed, then P and T_P have the same Turing degree, and that degree is less than or equal to the degree of every $T \in P$. Pointed, perfect closed sets are useful for forcing the degree of a generic T to be an upper bound for some given countable set of degrees. The notion of pointedness is used in **[23]** to show among other things: (1) every countable admissible **[12]**, **[16]** ordinal $> \omega$ is of the form ω_1^T for some $T \subseteq \omega$; (2) if α is a countable admissible ordinal $>\omega$, then $\omega_1^T = \alpha$ for some T with the minimal property that for every T' of lower hyperdegree than T, $\omega_1^{T'} < \alpha$. Recently Friedman and Jensen found a very interesting proof of (1) which replaces forcing by a compactness theorem of Barwise for infinitary languages. Their compactness approach does not seem to yield the minimal property of (2). A typical consequence of (2) is: the set of hyperdegrees of the Δ_2^1 sets has a minimal upper bound. The idea of pointedness goes back to the Kleene-Post proof [**11**] that there exists a set of lower degree than that of $0^{(\omega)}$ but of higher degree than that of $0^{(m)}$ for all m, i.e., the degree of the truth set for arithmetic is not a minimal upper bound for the degrees of the arithmetic sets. We say T is *pointedly generic* if for every sentence $\mathscr{F}$ of $\mathscr{L}(\mathscr{T})$ there is a pointed, arithmetic perfect closed P such that $T \in P$ and either $P \Vdash \mathscr{F}$ or $P \Vdash \sim\mathscr{F}$. The existence of a pointedly generic T follows from 3.1 and 3.2 and the mode of argument occurring in Lemma 3.3. We state without proof the fact that the Turing degree of any pointedly generic T is a minimal upper bound for the Turing degrees of the arithmetic sets, since it was shown in [**20**] that every countable set of Turing degrees has a minimal upper bound.

PROPOSITION 3.2. *If P is a pointed, arithmetic perfect closed set recursive in the arithmetic set A, then there exists a pointed, arithmetic perfect closed set $Q \subseteq P$ such that A and Q have the same Turing degree.*

PROOF. Although P may be encoded in a rather arbitrary way as an arithmetic set of sequence numbers, we can always re-encode P as $\{p_i^m \mid m \geq 0 \;\&\; i < 2^m\}$, where $\lambda im \mid p_i^m$ is recursive in the given encoding of P and is such that for all m and $i < 2^m$: $p_i^m \geq p_i^{m+1}$, $p_i^m \geq p_{i+2^m}^{m+1}$, $p_i^{m+1} \ngeq p_{i+2^m}^{m+1}$ and $p_{i+2^m}^{m+1} \ngeq p_i^{m+1}$. Since P is pointed, the re-encoding of P has the same Turing degree as the given encoding of P; i.e., the given encoding is recursive in the canonical path T_P, and T_P is recursive in $\lambda im \mid p_i^m$.

We define Q inductively: $p_0^0 \in Q$; if $m+1$ is odd and $p_i^m \in Q$, then p_i^{m+1}, $p_{i+2^m}^{m+1} \in Q$; if $m+1$ is even, $p_i^m \in Q$ and $(m+1)/2 \in A$, then $p_i^{m+1} \in Q$; if $m+1$ is even, $p_i^m \in Q$ and $(m+1)/2 \notin A$, then $p_{i+2^m}^{m+1} \in Q$. Clearly, $Q \subseteq P$, and Q is recursive in P, A. But P is recursive in A, so Q is recursive in A. Consider the canonical path $T_Q \in Q$. T_Q is recursive in Q, and $T_Q \in P$; so P by virtue of its pointedness is recursive in T_Q and hence in Q. But then A is recursive in Q, since A is recursive in P, Q as a consequence of

$$(m+1)/2 \in A \leftrightarrow (\exists i)(i < 2^m \;\&\; p_i^{m+1} \in Q).$$

To see that Q is pointed, fix $T \in Q$. A is recursive in T, P as a consequence of

$$(m+1)/2 \in A \leftrightarrow (\exists i)(i < 2^m \;\&\; T \in p_i^{m+1}).$$

Then Q is recursive in T, P, since Q is recursive in A. But P is recursive in T because $T \in P$; so Q is recursive in T.

THEOREM 3.3. *There exists a T such that $T^{(2)}$ has the same Turing degree as $0^{(\omega)}$, and such that $0^{(n)}$ is recursive in T for all n.*

PROOF. Let $\{\mathscr{F}_m\}$ be an enumeration of all Σ_2 sentences of $\mathscr{L}(\mathscr{T})$. We intend to define two functions, $\lambda m \mid P_m$ and $\lambda m \mid t(m)$, each recursive in $0^{(\omega)}$ and with the following properties: $P_0 = 2^\omega$; $P_m \supseteq P_{m+1}$; P_m has the same Turing degree as $0^{(m)}$; $P_{m+1} \Vdash \mathscr{F}_m$ or $P_{m+1} \Vdash \sim\mathscr{F}_m$; $t(m) = 0 \leftrightarrow P_{m+1} \Vdash \mathscr{F}_m$.

Assume that P_m has been defined and that P_m is a pointed, perfect closed set which has the same Turing degree as $0^{(m)}$. By Proposition 3.2, there is a pointed $R_m \subseteq P_m$ whose Turing degree is $0^{(m+1)}$. By Lemma 3.1, there is a $P_{m+1} \subseteq R_m$ such that P_{m+1} is recursive in R_m and either $P_{m+1} \Vdash \mathscr{F}_m$ or $P_{m+1} \Vdash \sim\mathscr{F}_m$. Since R_m is pointed, it follows that every $P \subseteq R_m$ and recursive in R_m is pointed and recursive in R_m. (This fact was demonstrated in 3.2.) So P_{m+1} is pointed and has the same Turing degree as $0^{(m+1)}$. If $P_{m+1} \Vdash \mathscr{F}_m$, put $t(m) = 0$; otherwise, put $t(m) = 1$.

We obtain $\lambda m \mid P_m$ and $\lambda m \mid t(m)$ recursive in $0^{(\omega)}$ by appealing to various uniformities. Examination of the proof of Proposition 3.2 reveals that R_m is recursive uniformly in P'_m; i.e., R_m is the result of applying a fixed recursive reduction procedure to P'_m, a procedure independent of the value of P_m. To develop P_{m+1} and $t(m)$ from Lemma 3.1, we must first decide if there is a $p \in R_m$ such that $p \Vdash^{R_m} \mathscr{F}_m$. Since $\mathscr{F}_m$ is Σ_2, the answer as a function of m is uniformly recursive in $R_m^{(2)}$. Thus P_{m+1} is uniformly recursive in $P_m^{(3)}$, and $\lambda m \mid t(m)$ is recursive in $\lambda m \mid R_m^{(2)}$. Since P_0 is recursive and $\lambda m \mid 0^{(m)}$ is recursive in $0^{(\omega)}$, it follows that $\lambda m \mid P_m$ and $\lambda m \mid t(m)$ are too.

Since $\{\mathscr{F}_m\}$ includes all sentences of the form $\bar{n} \in \mathscr{T}$, it must be that

$$\bigcap\{P_m \mid m \geq 0\}$$

has a unique member T recursive in $0^{(\omega)}$. And for each m, $0^{(m)}$ is recursive in T since $T \in P_m$ and P_m is pointed and of the same degree as $0^{(m)}$. Since the recursive predicate $\mathscr{R}$ is part of $\mathscr{L}(\mathscr{T})$, $\bar{m} \in \mathscr{T}^{(2)}$ is a Σ_2 formula of $\mathscr{L}(\mathscr{T})$. Assume $\mathscr{F}_{2m}$ is $\bar{m} \in \mathscr{T}^{(2)}$ for all m. Then $m \in T^{(2)} \leftrightarrow t(2m) = 0$, and consequently, $T^{(2)}$ is recursive in $0^{(\omega)}$. Putnam [18] has shown: for all T, if $(m)(0^{(m)}$ is recursive in $T)$, then $0^{(\omega)}$ is recursive in $T^{(2)}$.

For each set A, let $\mathbf{A}$ be the Turing degree of A. After Kleene and Post **[11]** showed that $\mathbf{0}^{(\omega)}$ is not a minimal upper bound for $\{\mathbf{0}^{(m)} \mid m \geq 0\}$, Spector **[28]**, proved that $\{\mathbf{0}^{(m)} \mid m \geq 0\}$ has no least upper bound. Putnam **[18]** calls $\mathbf{A}$ the *n-least upper bound* for $\{\mathbf{B}_i \mid i \geq 0\}$ if $\mathbf{A}$ is the least member of $\{\mathbf{C}^{(n)} \mid (i)[\mathbf{B}_i \leq \mathbf{C}]\}$. Thus Spector ruled out a 0-least upper bound for $\{\mathbf{0}^{(m)} \mid m \geq 0\}$. It is not difficult to show in the style of Theorem 3.3 that $\{\mathbf{0}^{(m)} \mid m \geq 0\}$ lacks a 1-least upper bound. And of course 3.3 says that $\mathbf{0}^{(\omega)}$ is the 2-least upper bound of $\{\mathbf{0}^{(m)} \mid m \geq 0\}$. By Spector's uniqueness theorem [27], it is acceptable to define $\mathbf{0}^{(\alpha)}$ for each recursive ordinal α to be the Turing degree of the hyperarithmetic set H_b, where b is any notation for α. Then the argument of Theorem 3.3 is readily extended to give the following result.

THEOREM 3.4. *If α is a recursive limit ordinal, then $\mathbf{0}^{(\alpha)}$ is the 2-least upper bound of $\{\mathbf{0}^{(\beta)} \mid \beta < \alpha\}$.*

It can be shown that the set of all Turing degrees of hyperarithmetic sets has no 2-least upper bound.[7] This fact together with Theorem 3.4 leads to a tidy characterization of the hyperarithmetic hierarchy in terms of the jump operator and the notion of 2-least upper bound. H. Putnam **[18]** has extended Theorem 3.4 in a nontrivial fashion to the entire constructible hierarchy.

A set T is said to be *implicitly arithmetically definable* if there exists an arithmetic predicate $\mathscr{A}(X)$ such that $\mathscr{A}(T)$ and $(\exists_1 X)\mathscr{A}(X)$. Every implicitly arithmetically definable (i.a.d.) set is hyperarithmetic, and every hyperarithmetic set is recursive in some i.a.d. set. Feferman **[2]** found a hyperarithmetic set (in fact recursive in $0^{(\omega)}$) that was not i.a.d. We say two sets have the same arithmetic degree if each is arithmetic in the other.

THEOREM 3.5. *There exists a nonarithmetic, hyperarithmetic set T such that every nonarithmetic set arithmetic in T fails to be implicitly arithmetically definable.*

PROOF. Let $\{\mathscr{F}_n\}$ be an enumeration of the sentences of the arithmetic language $\mathscr{L}(\mathscr{T})$. We claim there is a function $\lambda n \mid P_n$ recursive in $0^{(\omega)}$ such that for each n, $P_n \supseteq P_{n+1}$, P_n is an arithmetic perfect closed set, and either $P_n \Vdash \mathscr{F}_n$ or $P_n \Vdash \sim\mathscr{F}_n$. The claim follows from some uniformities implicit in the proof of Theorem 3.1. Assume $\mathscr{F}$ is in prenex normal form and let $r(\mathscr{F}) = 1$ plus the number of alternations of quantifiers in the prenex of $\mathscr{F}$. Then the set $\{p \mid p \Vdash^P \mathscr{F}\}$ is recursive (uniformly) in the set $P^{(r(\mathscr{F}))}$. It follows that the first conclusion of Lemma 3.1 can be made to read: there exists a $Q \subseteq P$ such that either $P \Vdash \mathscr{F}$ or $P \Vdash \sim\mathscr{F}$ and such that Q is recursive (uniformly) in $P^{(1+r(\mathscr{F}))}$. But then the sequence $P_0 \supseteq P_1 \supseteq P_2 \supseteq \cdots$ can be constructed recursively from $0^{(\omega)}$. Let T be the unique member of $\bigcap\{P_n \mid n \geq 0\}$. T is generic, and the argument of Theorem 1.12, routinely modified, establishes that every set of the same arithmetic degree is generic. (T is of course of minimal arithmetic degree.) So it remains only to see that T is not implicitly arithmetically definable. This last is clear since if $\mathscr{A}(T)$ holds, then there is an arithmetic, perfect closed P such that $T \in P$ and

$$(T_1)(T_1 \in P \rightarrow \mathscr{A}(T_1)).$$

REFERENCES

1. P. J. Cohen, *The independence of the continuum hypothesis*, Proc. Nat. Acad. Sci. U.S.A. **50** (1963), 1143–1148 and **51** (1964), 105–110.

2. S. Feferman, *Some applications of the notion of forcing and generic sets*, Fund. Math. **56** (1965), 325–345.

3. S. Feferman and C. Spector, *Incompleteness along paths in progressions of theories*, J. Symbolic Logic **27** (1962), 383–390.

4. R. Gandy, *Proof of Mostowski's conjecture*, Bull. Acad. Polon. Sci. Sér. Math. **8** (1960), 571–575.

5. R. Gandy and G. E. Sacks, *A minimal hyperdegree*, Fund. Math. **61** (1967), 215–223.

[7] In fact it has no β-least upper bound for any recursive β.

6. K. Gödel, *The consistency of the axiom of choice and of the generalized continuum hypothesis*, Princeton Univ. Press, Princeton, N.J., 1966.

7. R. B. Jensen, "Concrete models of set theory" in *Sets, models and recursion theory*, Amsterdam, 1967, 44–74.

8. J. L. Kelley, *General topology*, New York, 1955.

9. S. C. Kleene, *Hierarchies of number-theoretic predicates*, Bull. Amer. Math. Soc. **61** (1955), 193–213.

10. ——, *Quantification of number-theoretic predicates*, Compositio Math. **15** (1959), 23–40.

11. S. C. Kleene and E. L. Post, *The upper semi-lattice of degrees of recursive unsolvability*, Ann. of Math. **59** (1954), 379–407.

12. S. Kripke, *Transfinite recursions on admissible ordinals*, I and II (abstracts), J. Symbolic Logic **29** (1964), 161–162.

13. G. Kreisel, *Set theoretic problems suggested by the notion of potential totality*, Proc. Sympos. Infinitistic Methods, Warsaw, 1961, 103–140.

14. ——, *The axiom of choice and the class of hyperarithmetic functions*, Indag. Math. **24** (1962), 307–319.

15. A. R. D. Mathias, *A survey of recent results in set theory*, these Proceedings, part II.

16. R. A. Platek, Ph.D. Thesis, Stanford University, 1965.

17. K. Prikry, *Models constructed using perfect sets*, Lectures Notes Axiomatic Set Theory (University of California, Los Angeles, 1967) pp. IV-K-1—IV-K-3.

18. H. Putnam, *Collected papers on hierarchy theory*, to appear.

19. H. Rogers, Jr., *Theory of recursive functions and effective computability*, McGraw-Hill, New York, 1967.

20. G. E. Sacks, *Degrees of unsolvability*, 2nd ed., Princeton Univ. Press, Princeton, N.J., 1966.

21. ——, *Post's problem, admissible ordinals, and regularity*, Trans. Amer. Math. Soc. **124** (1966), 1–23.

22. ——, *Measure-theoretic uniformity in recursion theory and set theory*, Trans. Amer. Math. Soc., **142**(1969), 381–420.

23. ——, *Countable admissible ordinals and hyperdegrees*, to appear.

24. J. R. Shoenfield, 1967 UCLA Recursion Theory Seminar Notes, Appendix B.

25. R. Solovay, *The measure problem*, abstract 65T-62, Notices Amer. Math. Soc. **12** (1965), 217.

26. ——, *The measure problem*, to appear.

27. C. Spector, *Recursive well-orderings*, J. Symbolic Logic, **20** (1955), 151–163.

28. ——, *On degrees of recursive unsolvability*, Ann. of Math. **64** (1956), 581–592.

29. ——, *Hyperarithmetical quantifiers*, Fund. Math., **48** (1959), 313–320.

30. L. Tharp, Set theory lecture notes, Massachusetts Institute of Technology, 1965.

MASSACHUSETTS INSTITUTE OF TECHNOLOGY

Reprinted from Annals of Mathematical Logic 4 (1972) 343–367 with kind permission of Elsevier Science - NL, Sara Burgerhartstraat 25, 1055 KV Amsterdam, The Netherlands.

THE α-FINITE INJURY METHOD

G.E. SACKS and S.G. SIMPSON *

Massachusetts Institute of Technology, Cambridge, Mass. 02139, U.S.A.

Received 10 April 1972

§0. Introduction

Recursive functions of ordinals were invented by Takeuti [1] and subsequently specialized by Kripke [2] and Platek [3] to Σ_1 admissible initial segments of the ordinals, and by Kreisel and Sacks [4] to the recursive ordinals. Takeuti's principal result was: if f is recursive and β is an infinite ordinal, then

$$\text{cardinality } f\beta \leq \text{cardinality } \beta .$$

The above inequality follows from a downward Skolem–Löwenheim argument similar to the one originated by Gödel [5] in order to show the axiom of constructibility implies the generalized continuum hypothesis. Takeuti's argument succeeds because the graph of a recursive function is Σ_1; his inequality fails dramatically if f is replaced by the Σ_2 function that enumerates the constructible cardinals in increasing order. Thus right from the start, the generalization of recursion theory from the integers to the ordinals was concomitant with the model theoretic properties of the constructible sets. [1]

In this paper the Friedberg–Muchnik solution of Post's problem is

* The preparation of this paper was partially supported by NSF GP29079.

[1] Browse in (rather than read) the monumental survey of generalized recursion theory by Kreisel [8] in order to grasp the ontological questions ignored here.

lifted from ω to α, where α is any Σ_1 admissible ordinal, with the help of some downward Skolem–Löwenheim arguments.

Let L_α be the set of all sets constructible via ordinals less than α. α is Σ_1 admissible if L_α satisfies the Σ_1 replacement axiom schema of **ZF**. A partial function $f: \alpha \to \alpha$ is α-recursive if its graph has a Σ_1 definition over L_α (with parameters in L_α). A set A is α-recursively enumerable if it is the range of a partial α-recursive function. Theorem 4.1 furnishes two α-recursively enumerable sets such that neither is α-recursive in the other.

It is hoped that the proof of 4.1 reifies the α-finite injury method. (An abstract account of the method would be premature, hence abortive, at this time.) A set H is α-finite if $H \subset \alpha$ and $H \in L_\alpha$. α-finite sets interact with α-recursive functions, when α is Σ_1 admissible, in much the same way finite sets interact with recursive functions, but not for the same reasons. The α-finite injury method specializes, when $\alpha = \omega$, to the finite injury (or priority) method of Friedberg and Muchnik. They reasoned along the following lines.

A recursively enumerable set A is sought which will satisfy a recursively enumerable sequence $\{R_m \mid m < \omega\}$ of requirements. Each R_m is a recursively enumerable sequence of neighborhood conditions, i.e. pairs $\langle H, K\rangle$ of disjoint finite subsets of ω. A is said to satisfy R_m if there is an $\langle H, K\rangle \in R_m$ such that $H \subset A$ and $K \subset \omega - A$. All difficulties arise from the possibility of conflict between requirements. Suppose at stage s of the enumeration of A, it is the case that $\langle H, K\rangle \in R_m$ and no member of K is yet in A, and that consequently all of H is added to A with the intent of satisfying R_m. It can happen that at some later stage t, $\langle H^*, K^*\rangle \in R_n$ and no member of K^* is yet in A. In that event it is tempting to add all of H^* to A at stage t so as to satisfy R_n. But suppose $H^* \cap K$ is nonempty. Then the addition of H^* to A will injure what was done at stage s to satisfy R_m. The situation is summarized by saying R_m and R_n conflict at stage t.

Friedberg and Muchnik resolved the conflict in favor of R_m if $m < n$, and in favor of R_n otherwise. In other words they assigned higher priority to the requirement with lesser subscript (or index): if $m < n$, then H^* is not added to A at stage t; if $m > n$, then H^* is added and R_m is injured in order to satisfy R_n. The principal Friedberg–Muchnik lemma states: the number of stages at which R_m is injured is less than 2^m. [2]

[2] For a more detailed account of the finite injury priority method see Lachlan [9] or Sacks [10, p. 43].

It is proved by induction on m. It implies that R_m can be satisfied if there are infinitely many stages at which fresh opportunities to satisfy R_m present themselves.

No difficulty arises when lifting the above argument to α if L_α satisfies the Σ_2 replacement axiom schema of **ZF**. Recursive and finite are lifted to α-recursive and α-finite, and the proof of the Friedberg–Muchnik lemma remains valid, largely because the function

$$f\delta = \text{first stage after which } R_\delta \text{ is not injured}$$

is Σ_2.[3] If L_α does not satisfy Σ_2 replacement, then the Friedberg–Muchnik reasoning can fail totally. Stillwell [11] conjectured that some modification of the finite injury argument would show that the ordertype of the set of stages at which R_δ is injured is less than 2^δ, where 2^δ denotes ordinal exponentiation. But R. Shore found a counterexample that created a vacuum into which the present paper rushed.

The details of the α-finite injury method for a given α depend on the Σ_1 substructures of L_α. L_β is a Σ_1 substructure of L_α if $\beta \leq \alpha$ and each Σ_1 sentence (in the language of **ZF** with parameters from L_β) is true in L_β if and only if it is true in L_α; in symbols, $L_\beta \prec_1 L_\alpha$. Let

$$S_\alpha = \{\beta \mid \beta < \alpha \;\&\; L_\beta \prec_1 L_\alpha\} .$$

The ordertype of S_α determines the outcome of the method when applied to α.

Section 1 reviews α-recursion theory and draws attention to those parameters that occur in the Σ_1 definitions of certain key α-recursive functions. Such parameters are important ingredients of Σ_1 substructure arguments. Section 2 develops the model theoretic properties of L_α needed in Sections 4 and 5. Section 3 touches on some intuitive aspects of the α-finite injury method, and Sections 4 and 5 apply the method to Post's problem. Section 6 adds a notion related to forcing to obtain the strongest possible incomparability. And Section 7 discusses further results and open questions.

[3] Similarly no difficulty arises when lifting the infinite injury method ([10, p. 86]) to α if L_α satisfies the Σ_3 replacement axiom schema of ZF.

§1. α-recursion

The details missing from what follows can be extracted from [4, 12, 13].

The set theoretic definition of α-recursive given in section 0 must be rephrased in terms of computations before familiar recursion theoretic ideas can be applied. The Kleene–Kripke equation calculus for α includes numerals that name the ordinals less than α, substitution rules, and an infinitary rule needed to evaluate supremums as in the definition of addition of ordinals by transfinite induction. S^E_σ is the set of all equations computable from E (a finite initial set of equations) in less than σ steps. An important convention requires that every ordinal mentioned in S^E_σ be less than σ. The natural enumeration of S^E_α is Σ_1 over L_α. It follows that α is Σ_1 admissible (as defined in section 0) if and only if $S^E_{\alpha+1} = S^E_\alpha$ for all E. In addition a partial function $f \subset \alpha \times \alpha$ is α-recursive if and only if there is an E such that

$$f\gamma = \delta \leftrightarrow \underline{f}\underline{\gamma} = \underline{\delta} \in S^E_\alpha$$

for all $\gamma, \delta < \alpha$.

From now on α is Σ_1 admissible. A subset of α is α-recursive if its representing function is α-recursive. A set is α-finite if it is α-recursive and bounded by some ordinal less than α. If f is partial α-recursive and defined on an α-finite K, then $f[K]$ is α-finite.

There exists an α-recursive function $k(\gamma, \eta)$ such that

(i) if $k(\gamma, \eta) = 0$, then $\gamma < \eta$;

(ii) if K is α-finite, then $K = \{\gamma | k(\gamma, \eta) = 0\}$ for just one $\eta < \alpha$;

(iii) the E that defines k mentions no infinite ordinals. (Thus the same E works for every α.)

If $K = \{\gamma | k(\gamma, \eta) = 0\}$, then η is said to be the *canonical index* of K, in symbols $K = K_\eta$. A property of α-finite sets is α-recursive if the equivalent property of canonical indices is α-recursive.

If an equation can be computed from E in less than α steps, then it owes its existence to some α-finite, wellfounded tree called a computation. The set of all computations is α-recursively enumerable, so there exists a simultaneous enumeration of all α-recursively enumerable sets. To be more precise, there is an α-recursive function $r(\sigma, \epsilon)$ such that

(iv) $K_{r(\sigma,\epsilon)} \subset K_{r(\tau,\epsilon)}$ if $\sigma \leq \tau$;

(v) if B is recursively enumerable, then $B = \bigcup \{K_{r(\sigma,\epsilon)} \mid \sigma < \alpha\}$ for some $\epsilon < \alpha$;

(vi) the E that defines r mentions no infinite ordinals.

Let $R^\sigma_\epsilon = K_{r(\sigma,\epsilon)}$ and $R_\epsilon = \bigcup \{R^\sigma_\epsilon \mid \sigma < \alpha\}$. Then $\{R_\epsilon \mid \epsilon < \alpha\}$ is a standard enumeration of the α-recursively enumerable sets. Similarly there is a standard enumeration $\{\{\epsilon\} \mid \epsilon < \alpha\}$ of the partial α-recursive functions. There is no harm in blurring the difference between ordinals and ordered n-tuples $\langle \delta_1, \ldots, \delta_n \rangle$ of ordinals. With that in mind let $C \subset \alpha$ and define $[\epsilon]^C_\sigma(\gamma) = \delta$ by

$$(\mathrm{E}\rho)(\mathrm{E}\eta)[\langle \gamma, \delta, \rho, \eta \rangle \in R^\sigma_\epsilon \;\&\; K_\rho \subset C \;\&\; K_\eta \subset \alpha - C].$$

A useful convention is: the computation of $[\epsilon]^C_\sigma(\gamma) = \delta$ mentions no ordinal $\geq \sigma$. A partial function $f \subset \alpha \times \alpha$ is weakly α-recursive in C ($f \leq_{w\alpha} C$) if there is an $\epsilon < \alpha$ such that for all $\gamma, \delta < \alpha$,

$$f(\gamma) = \delta \leftrightarrow (\mathrm{E}\sigma)[[\epsilon]^C_\sigma(\gamma) = \delta] \,.$$

It is possible for $[\epsilon]^C_\sigma(\gamma)$ and $[\epsilon]^C_\tau(\gamma)$ to differ in value; if no such difference occurs and $[\epsilon]^C_\sigma(\gamma)$ is defined for some σ, then $[\epsilon]^C(\gamma)$ is defined. Thus $f \leq_{w\alpha} C$ if and only if $f = [\epsilon]^C$ for some ϵ. $B \leq_{w\alpha} C$ if the representing function of B is weakly α-recursive in C. Driscoll [14] showed $\leq_{w\alpha}$ is not transitive on the α-recursively enumerable sets for many α's. The notion of "α-recursive in" ($\leq_\alpha$) is transitive. $A \leq_\alpha B$ if there exists a partial α-recursive function ϕ such that:

$$J \subset A \leftrightarrow (\mathrm{E}H)(\mathrm{E}K)[\phi(J, H, K) = 0 \;\&\; H \subset B \;\&\; K \subset \alpha - B]$$

$$J \subset \alpha - A \leftrightarrow (\mathrm{E}H)(\mathrm{E}K)[\phi(J, H, K) = 1 \;\&\; H \subset B \;\&\; K \subset \alpha - B],$$

for all α-finite J; H and K are α-finite set variables. A and B have the same α-degree if each is α-recursive in the other.

§2. α-stability

Again α is Σ_1 admissible. The projectum of α, denoted by α^*, is the least β such that β is the supremum of the range of some one-one (into) α-recursive function. Clearly $\alpha^* \leq \alpha$.

Proposition 2.1. *If $\eta < \alpha^*$ and B is an α-recursively enumerable subset of η, then B is α-finite.*

Proof. Let f be a partial α-recursive function whose domain is an initial segment of α and which recursively enumerates B without repetitions. Since $\eta < \alpha^*$, the domain of f must be bounded by an ordinal less than α. Consequently the range of f is α-finite.

Suppose $\beta < \alpha$. β is an *α-cardinal* if there is no α-finite one-to-one correspondence between β and any $\gamma < \beta$. An α-cardinal β is regular if there is no α-finite $j : \gamma \to \beta$ such that $\gamma < \beta$ and $\beta = \bigcup \{j\delta \mid \delta < \gamma\}$. The α-cardinality of δ is denoted by α-card δ.

Proposition 2.2. (1). *If $\alpha^* < \alpha$, then α^* is the largest α-cardinal.*
(2). *If an α-cardinal is not regular, then it is the limit of regular α-cardinals.*

Lemma 2.3. *Suppose β is an infinite regular α-cardinal, $\nu < \beta$ and $\{I_\rho \mid \rho < \nu\}$ is a sequence of simultaneously α-recursively enumerable sets. If for each $\rho < \nu$, I_ρ is α-finite and* α-card $I_\rho < \beta$, *then* $\bigcup \{I_\rho \mid \rho < \nu\}$ *is α-finite and* α-card $\bigcup \{I_\rho \mid \rho < \nu\} < \beta$.

Proof. Let f be a partial α-recursive function whose domain is $\gamma \leq \alpha$ and which enumerates $\bigcup \{I_\rho \mid \rho < \nu\}$ without repetitions. Suppose for the sake of a reductio ad absurdum that $\gamma \geq \beta$. Then $f[\beta]$ is an α-finite subset of $\bigcup \{I_\rho \mid \rho < \nu\}$. Thus

$$f[\beta] = \bigcup \{J_\rho \mid \rho < \nu\},$$

where $\{J_\rho \mid \rho < \nu\}$ is an α-finite sequence of α-finite sets such that $J_\rho \subset I_\rho$ for all $\rho < \nu$. Clearly α-card $J_\rho < \beta$, since α-card $I_\rho < \beta$ and J_ρ is α-finite. The regularity of β implies

$$\alpha\text{-card} \bigcup \{J_\rho \times \{\rho\} \mid \rho < \nu\} < \beta .$$

On the other hand f is one-one, so α-card $\bigcup \{J_\rho \mid \rho < \nu\} = \beta$.

Let $\mathcal{E}_\beta$ be the set of all equations in the equation calculus for β as sketched in section 1. Assume $\gamma > \beta \geq \omega$. β is said to be γ-stable if $S_\beta^E = S_\gamma^E \cap \mathcal{E}_\beta$ for all finite $E \subset \mathcal{E}_\beta$. (Neither γ nor β are assumed to be Σ_1 admissible.) It is readily verified that β is γ-stable if and only if $L_\beta \prec_1 L_\gamma$ (cf. Section 0) and $\beta < \gamma$.

Jensen and Kripke have refined Gödel's downward Skolem argument in L to show: every α-cardinal $> \omega$ is α-stable; $\alpha = \alpha^* > \omega$ if and only if α is the limit of α-stable cardinals. The next lemma is the only instance of the above two facts needed in this paper.

Lemma 2.4. *If $\alpha^* = \alpha > \omega$ and there is a largest α-cardinal, then α is the limit of α-stable ordinals.*

Proof. Let $\aleph$ be the largest α-cardinal, and assume $\alpha > \gamma > \aleph$ with the intent of locating some α-stable $\beta \geq \gamma$. Define H to be the set of ordinals of the form $f(\gamma_1, ..., \gamma_n)$, where: $n < \omega$; $\gamma_1, ..., \gamma_n < \gamma$; and f is a partial α-recursive function whose defining set E of equations mentions no infinite ordinals. (H is the Σ_1 Skolem hull of γ in L_α.) Fix $\sigma \in H$ to see that H is an initial segment of α. Each member of H has a Σ_1 definition with parameters $< \gamma$. Let ρ be the least canonical index δ (cf. Section 1) such that K_δ is an α-finite function from $\aleph$ onto σ. Then $\rho \in H$. It follows that the partial α-recursive function K_ρ can be defined without mentioning any ordinals $\geq \gamma$, and so $K_\rho[\aleph] \subset H$.

Thus H equals some $\beta \leq \alpha$. $\beta < \alpha$ because $\alpha^* = \alpha$ and H can be mapped by a one-one partial α-recursive function into the α-finite set of all finite sequences of ordinals less than γ.

Fix $E \subset \mathcal{E}_\beta$ and $z \in S_\alpha^E \cap \mathcal{E}_\beta$ to see that β is α-stable. z is the last line of some α-finite computation whose roots are in E. Let g be an α-recursive function that enumerates (the Gödel numbers of) all α-finite computations of z from E. Choose g, as in clause (vi) of Section 1, so that the only ordinals occurring in the definition of g are those in z. Every such ordinal is less than β, and so g can be defined without mentioning any ordinals $\geq \gamma$. Hence $g0 < \beta$ and $z \in S_\beta^E$.

§3. Towards Post's problem

Assume α is Σ_1 admissible. Two α-recursively enumerable sets A and B, and witness functions f and g, are sought such that for each $\epsilon < \alpha$

(i) $f\epsilon \in A \leftrightarrow (\mathrm{E}\sigma)([\epsilon]^B_\sigma(f\epsilon) = 1)$;

(ii) $g\epsilon \in B \leftrightarrow (\mathrm{E}\sigma)([\epsilon]^A_\sigma(g\epsilon) = 1)$.

If A and B meet the ϵth instance of requirement (i), denoted by $f\epsilon$, then $f\epsilon$ is a witness to the fact that A is not weakly α-recursive in B via Gödel number ϵ, as defined in section 1. Conflicts are inevitable because requirement $f\epsilon$ (respectively $g\epsilon$) is met by adding $f\epsilon$ to A (respectively $g\epsilon$ to B) and excluding some α-finite set from B (respectively A) in order to preserve the value of $[\epsilon]^B_\sigma$ (respectively $[\epsilon]^A_\sigma$). They are resolved by appealing to priorities: it is permitted to add $f\epsilon$ to A – even if it means injuring requirement $g\delta$ which at some earlier stage excluded $f\epsilon$ from A – when $f\epsilon$ has higher priority than $g\delta$. The obvious wellordering of priorities $f0 < g0 < f1 < g1 < \ldots$ ($f0$ has highest priority, $g0$ next highest etc.) fails if $\alpha^* < \alpha$, but the remedy is simple enough. Let $t : \alpha \to \alpha^*$ be a one-one α-recursive function. Then say $f\epsilon$ has higher priority than $g\delta$ if $t\epsilon \leq t\delta$. Thus when $\alpha^* < \alpha$ the wellordering of priorities has order type less than α. All goes well, as will be seen in section 4, largely because of Lemmas 2.2 and 2.3. (If $\alpha^* = \omega$, the combinatorics of Friedberg suffice.)

If $\alpha^* = \alpha$ and the set $\{\beta \mid \beta < \alpha \;\&\; \beta \text{ is } \alpha\text{-stable}\}$ has order type α, then the obvious wellordering of priorities succeeds – thanks again to 2.2 and 2.3.

The most interesting case is when $\alpha^* = \alpha$ and the set of α-stable ordinals has order type less than α. In that event the predicate "$f\epsilon$ has higher priority than $g\delta$" is not α-recursive, and it becomes necessary to guess at what the priorities are. Thus in addition to the expected convergence problems associated with the finite injury method, namely the limits (if any) of $[\epsilon]^B_\sigma$ and $[\epsilon]^A_\sigma$ as σ increases towards α, there are convergence problems associated with the transfinite sequences of guesses of priorities.

§4. When $\alpha^* < \alpha$

The following theorem is now proved under the assumption $\alpha^* < \alpha$.

Theorem 4.1. *There exist α-recursively enumerable sets A and B such that neither is weakly α-recursive in the other.*

Let $C \subset \alpha$ and define $\{\epsilon\}_\sigma^C$ for each $\epsilon < \alpha^*$ as follows. Fix $t : \alpha \to \alpha^*$ so that t is one-one and α-recursive. If $t^{-1}\epsilon$ is undefined via computations whose Gödel numbers are less than σ, then so is $\{\epsilon\}_\sigma^C$. If $t^{-1}\epsilon$ is so defined, then $\{\epsilon\}_\sigma^C$ is $[t^{-1}\epsilon]_\sigma^C$ (as defined in section 1). Define $\{\epsilon\}^C$ similarly.

α-recursive functions A^σ, B^σ, $f(\sigma, \epsilon)$ and $g(\sigma, \epsilon)$ $(\sigma < \alpha, \epsilon < \alpha^*)$ are defined by recursion on $\sigma < \alpha$. $\{A^\sigma \mid \sigma < \alpha\}$ and $\{B^\sigma \mid \sigma < \alpha\}$ will be non-decreasing sequences of α-finite sets. A will be $\bigcup \{A^\sigma \mid \sigma < \alpha\}$ and B will be $\bigcup \{B^\sigma \mid \sigma < \alpha\}$.

For each $\epsilon < \alpha^*$; $f(\sigma, \epsilon)$ and $g(\sigma, \epsilon)$ will be *nondecreasing* functions of σ. It will turn out that $f\epsilon = \lim_\sigma f(\sigma, \epsilon)$ and $g\epsilon = \lim_\sigma g(\sigma, \epsilon)$ exist and are less than α, and that:

$$f\epsilon \in A \leftrightarrow (\mathrm{E}\sigma)(\{\epsilon\}_\sigma^B (f\epsilon) = 1) ;$$

$$g\epsilon \in B \leftrightarrow (\mathrm{E}\sigma)(\{\epsilon\}_\sigma^A (g\epsilon) = 1) .$$

Stage σ of the construction is ushered in by setting $A^{<\sigma} = \bigcup \{A^\tau \mid \tau < \sigma\}$, $B^{<\sigma} = \bigcup \{B^\tau \mid \tau < \sigma\}$, $f(<\sigma, \epsilon)$ = the least member of Z_ϵ (defined below) $\geqslant$ every member of $\{f(\tau, \epsilon) \mid \tau < \sigma\}$, and $g(<\sigma, \epsilon)$ similarly.

The witness functions $f(\sigma, \epsilon)$ and $g(\sigma, \epsilon)$ must satisfy the following proviso (WFP):

$$(\epsilon)(\epsilon')(\sigma)(\tau)[\epsilon \neq \epsilon' \to f(\sigma, \epsilon) \neq f(\tau, \epsilon')] ,$$

and similarly for g. So let $\{Z_\epsilon \mid \epsilon < \alpha^*\}$ be an α-recursive sequence of disjoint, unbounded α-recursive subsets of α, and insist $f(\sigma, \epsilon), g(\sigma, \epsilon) \in Z_\epsilon$.

Case 0. $\sigma = (\alpha^* \cdot \mu) + 2\epsilon$ for some $\mu < \alpha$ and $\epsilon < \alpha^*$. Set $B^\sigma = B^{<\sigma}$, and $f(\sigma, \epsilon') = f(<\sigma, \epsilon')$ for all $\epsilon' < \alpha^*$. If $f(\sigma, \epsilon) \notin A^{<\sigma}$ and

$$\{\epsilon\}_\sigma^{B^\sigma}(f(\sigma, \epsilon)) = 1 ,$$

then set

$$A^\sigma = A^{<\sigma} \cup \{f(\sigma, \epsilon)\}$$

$$g(\sigma, \epsilon') = g(<\sigma, \epsilon') \quad \text{for} \quad \epsilon' < \epsilon ,$$

and choose $g(\sigma, \epsilon')$ $(\epsilon \leq \epsilon' < \alpha^*)$ so that $g(\sigma, \epsilon') > \sigma + g(<\sigma, \epsilon')$ and the witness function proviso (WFP) for g is met. Computations are preserved by increasing the witness function. Thus if $\epsilon \leq \epsilon'$ and $\tau \geq \sigma$, then $g(\tau, \epsilon') \geq \sigma$ and the addition of $g(\tau, \epsilon')$ to B cannot injure the computation of $\{\epsilon\}_\sigma^{B^\sigma}(f(\sigma, \epsilon)) = 1$, since that computation depends only on $B^\sigma \cap \sigma$. Otherwise set $A^\sigma = A^{<\sigma}$ and $g(\sigma, \epsilon') = g(<\sigma, \epsilon')$ for all $\epsilon' < \alpha$.

Case 1. $\sigma = (\alpha^* \cdot \mu) + 2\epsilon + 1$ for some $\mu < \alpha$ and $\epsilon < \alpha^*$. Set $A^\sigma = A^{<\sigma}$ and $g(\sigma, \epsilon') = g(<\sigma, \epsilon')$ for all $\epsilon' < \alpha^*$. If $g(\sigma, \epsilon) \notin B^{<\sigma}$ and

$$\{\epsilon\}_\sigma^{A^\sigma}(g(\sigma, \epsilon)) = 1 ,$$

then set

$$B^\sigma = B^{<\sigma} \cup \{g(\sigma, \epsilon)\}$$

$$f(\sigma, \epsilon') = f(<\sigma, \epsilon') \quad \text{for} \quad \epsilon' \leq \epsilon ,$$

and choose $f(\sigma, \epsilon')$ $(\epsilon < \epsilon' < \alpha^*)$ so that $f(\sigma, \epsilon') > \sigma + f(<\sigma, \epsilon')$ and the witness function proviso for f is met. Otherwise set $B^\sigma = B^{<\sigma}$ and $f(\sigma, \epsilon') = f(<\sigma, \epsilon')$ for all $\epsilon' < \alpha^*$.

That concludes stage σ. Define

$$I_{2\epsilon} = \{\sigma \mid \sigma = (\alpha^* \cdot \mu) + 2\epsilon \ \&\ A^\sigma \neq A^{<\sigma}\}$$

$$I_{2\epsilon+1} = \{\sigma \mid \sigma = (\alpha^* \cdot \mu) + 2\epsilon + 1 \ \&\ B^\sigma \neq B^{<\sigma}\} .$$

The next lemma corresponds to Friedberg's result that card $I_n \leq 2^n$ when $\alpha = \omega$.

Lemma 4.2. *If* $\nu < \beta$ *and* β *is an infinite regular* α*-cardinal, then* I_ν *is* α*-finite and its* α*-cardinality is less than* β.

Proof. By induction on ν. Fix ν and assume for each $\rho < \nu$ that I_ρ is α-finite and its α-cardinality is less than β. By 2.3, $\bigcup\{I_\rho \mid \rho < \nu\}$ is α-finite and its α-cardinality is less than β. Consequently its ordertype is less than β. The construction of $f(\sigma, \epsilon)$ and $g(\sigma, \epsilon)$ guarantees that any two distinct members of I_ρ are separated by some member of $\bigcup\{I_\nu \mid \nu < \rho\}$; e.g. if $\rho = 2\epsilon'$, $\tau < \sigma$ and $\tau, \sigma \in I_\rho$, then $f(\tau, \epsilon') < f(\sigma, \epsilon')$ and there is an $\epsilon < \epsilon'$ and a γ such that $\tau < \gamma < \sigma$,

$$f(\tau, \epsilon') < f(\gamma, \epsilon') \leq f(\sigma, \epsilon'),$$

$\gamma = (\alpha^* \cdot \mu) + 2\epsilon + 1$ and $\gamma \in I_{2\epsilon+1}$. The interlacing of I_ρ and $\bigcup\{I_\nu \mid \nu < \rho\}$ implies the ordertype of I_ρ is less than β. Since I_ρ is bounded and recursively enumerated in increasing order, it must be α-finite, and so its α-cardinality must be less than β.

Lemma 4.3. *For each* $\nu < \alpha^*$, $\bigcup\{I_\rho \mid \rho < \nu\}$ *is* α*-finite.*

Proof. By 2.2, α^* is either a regular α-cardinal or the limit of regular α-cardinals. Choose an infinite regular α-cardinal $\beta \leq \alpha^*$ so that $\nu < \beta$. Then apply 4.2 and 2.3.

Clearly $f(< \sigma, \epsilon) = f(\sigma, \epsilon)$ whenever $\sigma \notin \bigcup\{I_\nu \mid \nu < 2\epsilon\}$. So by 4.3

$$f\epsilon = \lim_\sigma f(\sigma, \epsilon) < \alpha .$$

The same holds for $g\epsilon = \lim_\sigma g(\sigma, \epsilon)$.

Lemma 4.4. *For each* $\epsilon < \alpha^*$:

$$f\epsilon \in A \leftrightarrow (\mathrm{E}\sigma)(\{\epsilon\}^B_\sigma(f\epsilon) = 1);$$

$$g\epsilon \in A \leftrightarrow (\mathrm{E}\sigma)(\{\epsilon\}^A_\sigma(g\epsilon) = 1).$$

Proof. Suppose $f\epsilon \in A$. Let σ be the first stage such that $f\epsilon \in A^\sigma$. Then $\sigma \in I_{2\epsilon}$ and $f\epsilon = f(\tau, \epsilon)$ for all $\tau \geq \sigma$. Also

$$\{\epsilon\}_\sigma^{B\sigma}\ (f\epsilon) = 1\ .$$

Suppose for a reducio ad absurdum it is not the case that $\{\epsilon\}_\sigma^B\ (f\epsilon) = 1$. Then for some $\tau > \sigma$, it is not the case that $\{\epsilon\}_\tau^{B\tau}(f\epsilon) = 1$; let τ' be the least such τ. Then $\tau' \in I_{2\epsilon'+1}$ for some $\epsilon' \geq \epsilon$, since $f(\sigma, \epsilon) = f\epsilon$. At stage τ', $g(\tau', \epsilon')$ is added to $B^{<\tau}$ and $g(\tau', \epsilon') < \sigma$, since (by convention) the computation of $\{\epsilon\}_\sigma^{B\sigma}\ (f\epsilon)$ mentions no ordinal $\geq \sigma$. But $g(\tau', \epsilon') \geq \sigma$ by definition of g, since $\epsilon \leq \epsilon'$ and $\tau' > \sigma$.

Now suppose $(\mathrm{E}\sigma)(\{\epsilon\}_\sigma^B\ (f\epsilon) = 1)$. Choose τ large enough so that

$$\{\epsilon\}_\sigma^{B\sigma}\ (f\epsilon) = 1$$

and $f(\sigma, \epsilon) = f\epsilon$ for all $\sigma \geq \tau$. Then for some $\sigma \geq \tau$, either $f\epsilon \in A^{<\sigma}$ or case 0 applies and $f\epsilon \in A^\sigma$.

Theorem 4.1 is an immediate consequence of Lemma 4.4.

§5. When $\alpha = \alpha^*$

The proof of Theorem 4.1 is now completed under the assumption $\alpha^* = \alpha$.

First suppose there is no largest α-cardinal. Then α is the limit of regular α-cardinals. We now proceed as we did in Section 4 when α^* was the limit of regular α-cardinals. Let $t : \alpha \to \alpha$ be an α-recursive function that enumerates every ordinal less than α unboundedly often.

Stage σ. If $t\sigma = 2\epsilon$ (respectively $2\epsilon + 1$), proceed as in Section 4 when $\sigma = \alpha^* \cdot \mu + 2\epsilon$ (respectively $\alpha^* \cdot \mu + 2\epsilon + 1$).

The proofs of 4.2, 4.3 and 4.4 remain valid: 4.2 and 4.4 needed no assumption on α^*; 4.3 did not require that α^* be less than α, only that α^* be a regular α-cardinal or the limit of regular α-cardinals.

Now suppose for the remainder of this section that there is a largest α-cardinal; call it $\aleph$. Let

$$S = \{\beta \mid \aleph < \beta < \alpha \ \ \& \ \ \beta \text{ is } \alpha\text{-stable}\} ,$$

and let λ be the ordertype of S. Arrange S as $\beta_0 < \beta_1 < \ldots < \beta_\xi < \ldots$ $\ldots$ $(\xi < \lambda)$. Let K_ξ be the α-finite function from $\aleph$ onto β_ξ of least possible canonical index (cf. Section 1). Define

$$k(\aleph \cdot \xi + \gamma) = K_\xi(\gamma)$$

for all $\xi < \lambda$ and $\gamma < \aleph$. Thus k is a function from $\aleph \cdot \lambda$ onto α. Note that k cannot be α-recursive if $\lambda < \alpha$. k will be the source of a strange wellordering of priorities of length $\aleph \cdot \lambda$, as predicted in Section 3. It will be seen that the set of stages at which requirement ϵ ($\epsilon < \aleph \cdot \xi$ and $\xi < \lambda$) is injured is bounded by β_ξ.

From now on let ϵ be a variable that ranges over the ordinals less than $\aleph \cdot \lambda$.

It will be shown in a moment that k is a Σ_2 function, i.e. its graph has a Σ_2 definition over (with parameters in) L_α. The strength of k is its tameness (as the Good Book predicts), a term suggested by M. Lerman [15]. Not every Σ_2 function is tame. Let j be any function from $\aleph \cdot \lambda$ onto α. It is not hard to see that j is Σ_2 if and only if there is an α-recursive $j(\sigma, \epsilon)$ such that

$$(\epsilon)(\mathrm{E}\tau)(\sigma)_{\sigma\geq\tau}\ [j(\sigma,\epsilon)=j\epsilon]\ .$$

j is said to be *tame* Σ_2 if there is an α-recursive $j(\sigma, \epsilon)$ such that for each $\rho < \aleph\cdot\lambda$,

$$(\mathrm{E}\tau)(\epsilon)_{\epsilon<\rho}\ (\sigma)_{\sigma\geq\tau}\ [j(\sigma,\epsilon)=j\epsilon]\ .$$

Lemma 5.1. *k is tame Σ_2.*

Proof. Let S^σ be $\{\beta \mid \aleph < \beta < \sigma \ \&\ \beta \text{ is } \sigma\text{-stable}\}$. Arrange S^σ as $\beta_0^\sigma < \beta_1^\sigma < ... < \beta_\xi^\sigma < ...$ $(\xi < \lambda^\sigma)$, where λ^σ is the ordertype of S^σ. Let $\beta_\xi^\sigma = \sigma$ when $\lambda^\sigma \leq \xi < \lambda$. Let K_ξ^σ be the α-finite function from $\aleph$ onto β_ξ^σ of least possible canonical index. Define

$$k(\sigma, \aleph\cdot\xi+\gamma) = K_\xi^\sigma(\gamma)$$

for all $\xi < \lambda$ and $\gamma < \aleph$.

Assume $\sigma \geq \beta_\xi$ to see that $\beta_\delta^\sigma = \beta_\delta$ for all $\delta < \xi$. Clearly β_δ is σ-stable, since $\beta_\delta < \sigma < \alpha$ and β_δ is α-stable. It follows that $\beta_\delta^\sigma \leq \beta_\delta$. But then β_δ^σ is α-stable, because β_δ^σ is β_δ-stable (or is β_δ) and β_δ is α-stable.

If $\sigma > \beta_\xi$, then $K_\xi^\sigma = K_\xi$ since $\beta_\xi^\sigma = \beta_\xi$ and $\aleph < \beta_\xi$. Note that the α-stability of $\beta_{\xi+1}$ puts the canonical index of K_ξ below $\beta_{\xi+1}$. (This last point is a consequence of the fact that the α-recursive assignment of canonical indices is defined without any infinite parameters.) Thus $k(\sigma, \aleph\cdot\xi+\gamma) = k(\aleph\cdot\xi+\gamma)$ for all $\sigma > \beta_\xi$.

Now α-recursive functions A^σ, B^σ, $f(\sigma,\epsilon)$ and $g(\sigma,\epsilon)$ are defined as in Section 4. Let $t: \alpha \to \aleph\cdot\lambda$ be an α-recursive function that enumerates every $\epsilon < \aleph\cdot\lambda$ unboundedly often.

Case 0. $t\sigma = 2\epsilon$. Let ϵ^* be of the form $\aleph\cdot\xi^* + \gamma^*$ $(\xi^* < \lambda$ and $\gamma^* < \aleph)$. If $\beta_{\xi^*}^\sigma = \beta_{\xi^*}^{<\sigma} = \sup\{\beta_{\xi^*}^\tau \mid \tau < \sigma\}$, set $f(\sigma,\epsilon^*) = f(<\sigma,\epsilon^*)$. If $\beta_{\xi^*}^\sigma \neq \beta_{\xi^*}^{<\sigma}$, choose $f(\sigma,\epsilon^*) > \sigma + f(<\sigma,\epsilon^*)$ and in accord with the witness function proviso (WFP) of Section 4. Set $B^\sigma = B^{<\sigma}$.

If $f(\sigma,\epsilon) \notin A^{<\sigma}$ and

$$[k(\sigma,\epsilon)]_\sigma^{B^\sigma}\ (f(\sigma,\epsilon)) = 1\ ,$$

then set

$$A^\sigma = A^{<\sigma} \cup \{f(\sigma, \epsilon)\}$$

$$g(\sigma, \epsilon') = g(<\sigma, \epsilon') \text{ for } \epsilon' < \epsilon ,$$

and choose $g(\sigma, \epsilon')$ $(\epsilon \leq \epsilon' < \aleph \cdot \lambda)$ so that $g(\sigma, \epsilon') > \sigma + g(<\sigma, \epsilon')$ and the witness function proviso (WFP) for g is met.

Case 1. $t\sigma = 2\epsilon + 1$. Let ϵ^* be of the form $\aleph \cdot \xi^* + \gamma^*$ ($\xi^* < \lambda$ and $\gamma^* < \aleph$). If $\beta^\sigma_{\xi^*} = \beta^{<\sigma}_{\xi^*}$, set $g(\sigma, \epsilon^*) = g(<\sigma, \epsilon^*)$. If $\beta^\sigma_{\xi^*} \neq \beta^{<\sigma}_{\xi^*}$, choose $g(\sigma, \epsilon^*) > \sigma + g(<\sigma, \epsilon^*)$ and in accord with the witness function proviso. Set $A^\sigma = A^{<\sigma}$.

If $g(\sigma, \epsilon) \notin B^{<\sigma}$ and

$$[k(\sigma, \epsilon)]^{A^\sigma}_\sigma (g(\sigma, \epsilon)) = 1 ,$$

then set

$$B^\sigma = B^{<\sigma} \cup \{g(\sigma, \epsilon)\}$$

$$f(\sigma, \epsilon') = f(<\sigma, \epsilon') \text{ for } \epsilon' \leq \epsilon ,$$

and choose $f(\sigma, \epsilon')$ $(\epsilon < \epsilon' < \aleph \cdot \lambda)$ so that $f(\sigma, \epsilon') \geq \sigma$ and the witness function proviso for f is met.

Now define the injury sets:

$$I_{2\epsilon} = \{\sigma \mid t\sigma = 2\epsilon \;\&\; A^\sigma \neq A^{<\sigma}\}$$

$$I_{2\epsilon+1} = \{\sigma \mid t\sigma = 2\epsilon + 1 \;\&\; B^\sigma \neq B^{<\sigma}\} .$$

Lemma 5.2. *If* $\xi < \lambda$, *then*

$$\bigcup \{I_{\aleph \cdot \xi + \gamma} \mid \gamma < \aleph\} \subset \beta_{\xi+1} .$$

Proof. By induction on ξ. Fix ξ and assume $\bigcup \{I_\nu \mid \nu < \aleph \cdot \xi\} \subset \beta_\xi$. Clearly $\aleph$ is the largest $\beta_{\xi+1}$-cardinal, since $\beta_{\xi+1}$ is α-stable. Fix $\gamma < \aleph$

and assume for each $\delta < \gamma$, $I_{\aleph\cdot\xi+\delta}\backslash\beta_\xi$ is a $\beta_{\xi+1}$-finite set whose $\beta_{\xi+1}$-cardinality is less than $\aleph$. As in the proof of 4.2, any two consecutive members of $I_{\aleph\cdot\xi+\gamma}\backslash(\beta_\xi + 1)$ are separated by some $\sigma \in \bigcup\{I_\nu \mid \nu < \aleph\cdot\xi+\gamma\}$. By the induction hypothesis on ξ, $\sigma \notin \bigcup\{I_\nu \mid \nu < \aleph\cdot\xi\}$. So $\sigma \in \bigcup\{I_{\aleph\cdot\xi+\delta}\backslash\beta_\xi \mid \delta < \gamma\}$. The existence of σ is termed an interlacement.

By induction on γ, $\{I_{\aleph\cdot\xi+\delta}\backslash\beta_\xi \mid \delta < \gamma\}$ is a sequence of simultaneously $\beta_{\xi+1}$-recursively enumerable sets. Assume $\aleph$ is regular. Then 2.3 implies $\bigcup\{I_{\aleph\cdot\xi+\delta}\backslash\beta_\xi \mid \delta < \gamma\}$ is $\beta_{\xi+1}$-finite and its $\beta_{\xi+1}$-cardinality is less than $\aleph$. It follows from the interlacement described above that $I_{\aleph\cdot\xi+\gamma}\backslash\beta_\xi$ is a $\beta_{\xi+1}$-finite set whose $\beta_{\xi+1}$-cardinality is less than $\aleph$.

If $\aleph$ is not regular then it is the limit of regular α-cardinals. Let κ be a regular α-cardinal below $\aleph$. Fix $\gamma < \kappa$ and assume for each $\delta < \gamma$, $I_{\aleph\cdot\xi+\delta}\backslash\beta_\xi$ is a $\beta_{\xi+1}$-finite set whose $\beta_{\xi+1}$-cardinality is less than κ. Then apply 2.3 as above.

It follows from 5.1, 5.2 and the definition of $f(\sigma, \epsilon)$ that $f(\sigma, \epsilon) = f(<\sigma, \epsilon)$ whenever $\sigma > \beta_{\xi+1}$, where $\epsilon = \aleph\cdot\xi + \gamma$. Define

$$f\epsilon = \lim_\sigma f(\sigma, \epsilon) < \alpha\ .$$

Define $g\epsilon$ similarly.

Lemma 5.3. *For each* $\epsilon < \alpha^*$:

$$f\epsilon \in A \leftrightarrow (\mathsf{E}\sigma)([k\epsilon]^B_\sigma (f\epsilon) = 1)\ ;$$

$$g\epsilon \in B \leftrightarrow (\mathsf{E}\sigma)([k\epsilon]^A_\sigma (f\epsilon) = 1)\ .$$

Proof. Suppose $f\epsilon \in A$. Let σ be the first stage such that $f\epsilon \in A^\sigma$. The witness function proviso (WFP) implies

$$[k(\epsilon, \sigma)]^{B^\sigma}_\sigma (f\epsilon) = 1\ .$$

Since $f(\tau, \epsilon) = f\epsilon$ for all $\tau \geq \sigma$, it follows that $\beta^\sigma_\xi = \beta_\xi$ and $k(\epsilon, \sigma) = k\epsilon$. Suppose for a reductio ad absurdum it is not the case that $[k\epsilon]^B_\sigma (f\epsilon) = 1$. Then, as in 4.4, there is a $\tau' > \sigma$ such that $g(\tau', \epsilon')$ is added to $B^{<\tau'}$ and

$g(\tau', \epsilon') < \sigma$. Now $\epsilon' \geq \epsilon$ since $f(\sigma, \epsilon) = f\epsilon$. But then $g(\tau', \epsilon') \geq \sigma$ by definition of g, since $\epsilon' \geq \epsilon$.

Now suppose $(\mathrm{E}\sigma)([k\epsilon]_\sigma^B (f\epsilon) = 1)$. Choose τ large enough so that

$$[k(\sigma, \epsilon)]_\sigma^{B^\sigma} (f(\sigma, \epsilon)) = 1 ,$$

$f(\sigma, \epsilon) = f\epsilon$, and $k(\sigma, \epsilon) = k\epsilon$ for all $\sigma \geq \tau$. Then for some $\sigma \geq \tau$, either $f\epsilon \in A^{<\sigma}$ or case 0 applies and $f\epsilon \in A^\sigma$.

Theorem 4.1 is an immediate consequence of Lemma 5.3.

§6. Hyperregularity

G. Kreisel, a hard man to satisfy, pointed out that Theorem 4.1 is not a totally satisfying generalization of the Friedberg–Muchnik solution of Post's problem because $\leq_{w\alpha}$ (weakly α-recursive in) is only one of several possible generalizations of Turing reducibility. So in this section we develop what appears to be the strongest possible incomparability result (Theorem 6.2) for α-recursively enumerable sets.

If $B \subset \alpha$, then the diagram of B, denoted by $\triangle_B$, is

$$\{\underline{g}(\underline{\gamma}) = \underline{0} \mid \gamma \in B\} \cup \{\underline{g}(\underline{\gamma}) = \underline{1} \mid \gamma \notin B\} .$$

If E is a finite set of equations whose parameters are ordinals less than α (i.e. $E \subset \mathcal{E}_\alpha$), then $S^{E,B}$ is the set of all equations in $\mathcal{E}_\alpha$ deducible from $E \cup \triangle_B$ in *any* number of steps. A partial function $f \subset \alpha \times \alpha$ is *α-calculable* ($\leq_{c\alpha}$) from B if for some E,

$$f(\gamma) = \delta \longleftrightarrow \underline{f}(\underline{\gamma}) = \underline{\delta} \in S^{E,B}$$

for all $\gamma, \delta < \alpha$. $A \leq_{c\alpha} B$ if the representing function of A is α-calculable from B. Clearly $A \leq_{w\alpha} B$ implies $A \leq_{c\alpha} B$, but not conversely. The difference is clarified by the next lemma.

$R^{E,B}$ is the set of all equations in $S^{E,B}$ whose computation trees are α-finite. f is *α-finitely calculable* from B if for some E, $f(\gamma) = \delta \longleftrightarrow \underline{f}(\underline{\gamma}) = \underline{\delta} \in R^{E,B}$ for all $\gamma, \delta < \alpha$.

Lemma 6.1. *$f \leq_{w\alpha} B$ if and only if f is α-finitely calculable from B.*

Proof. Same as that of [12].

Kreisel has proposed a model theoretic notion of reducibility based on that of invariant implicit definability (cf. Kunen [16] for the definition); by Barwise completeness the notion coincides with $\leq_{c\alpha}$ when α is countable but rarely otherwise. Consequently Theorem 6.2 satisfies Kreisel when α is countable but rarely otherwise.

Theorem 6.2. *Suppose α is Σ_1 admissible. Then there exist α-recursively enumerable sets A and B such that $A \not\leq_{c\alpha} B$ and $B \not\leq_{c\alpha} A$.*

A relatively simple case of 6.2, $\alpha = \omega_1^{CK}$, was proved in [17]. There as here the key notion is hyperregularity. $B \subset \alpha$ is *hyperregular* if $f[\gamma]$ is bounded whenever $\gamma \subset$ domain f, $f \leq_{w\alpha} B$ and $\gamma < \alpha$. B is *regular* if $B \cap \gamma$ is α-finite for every $\gamma < \alpha$.

Lemma 6.3 ([12], [17]). *B is regular and hyperregular if and only if all computations from B are α-finite (i.e. $S^{E,B} = R^{E,B}$ for all finite $E \subset \mathcal{E}_\alpha$).*

Lemma 6.4 ([12], [17]). *If B is α-recursively enumerable and hyperregular, then B is regular.*

It follows from 6.1, 6.3 and 6.4 that one way of proving 6.2 is to find hyperregular, α-recursively enumerable A and B such that $A \not\leq_{w\alpha} B$ and $B \not\leq_{w\alpha} A$. The next lemma implies there is no other way.

Lemma 6.5. *If B is α-recursively enumerable but not hyperregular, then every α-recursively enumerable set is α-calculable from B.*

Proof. Similar to that of Spector's classical result that every Π_1^1 subset of ω is hyperarithmetic in every nonhyperarithmetic Π_1^1 subset of ω. Since B is not hyperregular, there is an $f \leq_{w\alpha} B$ and a $\gamma < \alpha$ such that $\gamma \subset$ domain f and $f[\gamma]$ is not bounded below α. Let g be an α-recursive function whose range is A. Thus

$$\delta \in A \longleftrightarrow (\mathrm{E}\beta)_{\beta<\alpha}\,[g\beta = \delta]$$

But then

$$\delta \in A \longleftrightarrow (\mathrm{E}\rho)_{\rho<\gamma}(\mathrm{E}\beta)_{\beta<f\rho}\,[g\beta = \delta]\,.$$

The last formula implies each membership fact about A can be calculated from f, hence B, by a computation of height at most $\alpha + 1$.

In order to prove 6.2, something a shade stronger will be proved.

Theorem 6.6. *If* α *is* Σ_1 *admissible, then there exist* α*-recursively enumerable sets* A *and* B *such that* $A \not\leq_{w\alpha} B$, $B \not\leq_{w\alpha}$ *and*

$$C = \{2\nu \mid \nu \in A\} \cup \{2\nu + 1 \mid \nu \in B\}$$

is hyperregular.

The proof of 6.6 resembles that of 4.1 save for some additional preservation requirements designed to make C hyperregular. Since the preservation requiremants of 4.1 were met by increasing the values of the witness functions $f(\sigma, \epsilon)$ and $g(\sigma, \epsilon)$ at appropriate σ's, the additional requirements will be met by increasing them still further. An auxiliary function $p(\sigma, \epsilon)$ will embody the additional requirements. $f(< \sigma, \epsilon)$ and $g(< \sigma, \epsilon)$ must now satisfy:

$$f(<\sigma, \epsilon) \geqslant p(\sigma, \epsilon) + \lim_{\tau<\sigma} f(\tau, \epsilon) \text{ if } \lim_{\tau<\sigma} f(\tau, \epsilon) \notin A^{<\sigma}$$

$$g(< \sigma, \epsilon) \geqslant p(\sigma, \epsilon) + \lim_{\tau<\sigma} g(\tau, \epsilon) \text{ if } \lim_{\tau<\sigma} g(\tau, \epsilon) \notin B^{<\sigma} .$$

Essentially $p(\sigma, \epsilon)$ is the supremum of all ordinals mentioned in certain computations worthy of preservation at stage σ. The definition of $p(\sigma, \epsilon)$ splits into five cases, the first two being the most instructive.

Case 1. $\alpha^* < \alpha$ and α^* is a regular cardinal. ϵ ranges over the ordinals less than α. Define

$$C^{<\sigma} = \{2\nu \mid \nu \in A^{<\sigma}\} \cup \{2\nu + 1 \mid \nu \in B^{<\sigma}\}$$

$$y(\sigma, \gamma, \epsilon) = \mu\tau_{\tau<\sigma} (\{\epsilon\}_\tau^{C^{<\sigma}} (\gamma) \text{ is defined})$$

$$m(\sigma, \epsilon) = \mu\gamma_{\gamma<\alpha^*} (y(\sigma, \gamma, \epsilon) = \sigma)$$

$$p(\sigma, \epsilon) = \sup \{y(\sigma, \gamma, \epsilon') \mid \epsilon' \leq \epsilon \ \& \ \gamma < m(\sigma, \epsilon')\} .$$

The usual convention for the bounded least number operator is observed: if $\sim (\mathrm{E}\tau)_{\tau<\sigma} \ldots$, then $\mu\tau_{\tau<\sigma} \ldots$ is equal to σ. Thus $y(\sigma, \gamma, \epsilon) = \sigma$ if and only if $\{\epsilon\}_\tau^{C^{<\sigma}}(\gamma)$ is undefined for all $\tau < \sigma$. The value of $p(\sigma, \epsilon)$ has been chosen with an eye to preserving $\{\epsilon\}^C(\gamma)$ $(\gamma < \alpha^*)$ as it develops and thereby forcing it to be an α-finite partial function.

Define $f(<\sigma, \epsilon)$ and $g(<\sigma, \epsilon)$ in accord with the inequalities preceding case 1 if they apply, otherwise as in Section 4. Then define A^σ, B^σ, $f(\sigma, \epsilon)$ and $g(\sigma, \epsilon)$ as in Section 4. Do not forget to honor the witness function proviso (WFP).

$I_{2\epsilon}$ and $I_{2\epsilon+1}$ have the same meaning they did in Section 4. So the proofs of 4.2 and 4.3 remain valid. Thus $\bigcup \{I_\rho \mid \rho < \nu\}$ is α-finite for each $\nu < \alpha^*$. But now a further argument is needed to show $\lim_\sigma f(\sigma, \epsilon)$ exists and is less than α.

Lemma 6.7. *Suppose* $\bigcup \{I_\nu \mid \nu < 2\epsilon\} \subset \sigma$ *and* $\gamma < m(\sigma, \epsilon)$. *Then for all* $\tau \geq \sigma$: $y(\sigma, \gamma, \epsilon) = y(\tau, \gamma, \epsilon)$ *and* $\gamma < m(\tau, \epsilon)$.

Proof. Fix $\tau \geq \sigma$. The only way $y(\tau, \gamma, \epsilon)$ can fail to be $y(\sigma, \gamma, \epsilon)$ is if some ordinal less than $y(\sigma, \gamma, \epsilon)$ is added to C at stage τ. But any such ordinal is of the form $2f(\tau, \epsilon^*)$ or $2g(\tau, \epsilon^*) + 1$ for some $\epsilon^* \geq \epsilon$, hence $\geq p(\sigma, \epsilon^*) \geq y(\sigma, \gamma, \epsilon)$.

Lemma 6.8. *For each* $\epsilon < \alpha^*$,

$$\{y(\sigma, \gamma, \epsilon') \mid \sigma < \alpha \;\;\&\;\; \epsilon' < \epsilon \;\;\&\;\; \gamma < m(\sigma, \epsilon')\}$$

is α-*finite.*

Proof. Fix $\epsilon < \epsilon^*$. By 6.7 there is a σ_0 such that $m(\sigma, \epsilon')$ is (for each $\epsilon' < \epsilon$) a nondecreasing function of σ for all $\sigma \geq \sigma_0$. Let $m\epsilon' = \lim_\sigma m(\sigma, \epsilon')$. Clearly $m\epsilon' \leq \alpha^*$. By 2.1 $\{\epsilon' < \epsilon \mid m\epsilon' = \alpha^*\}$ is α-finite, and so $\{\epsilon' < \epsilon \mid m\epsilon' < \alpha^*\}$ is α-finite. Now apply 2.3 and remember that α^* is (by assumption) a regular α-cardinal to see that the partial function

$$\{\langle m\epsilon', \epsilon'\rangle \mid \epsilon' < \epsilon\}$$

is α-finite. Then 6.8 follows from 6.7.

The existence of $\lim_\sigma f(\sigma, \epsilon)$ $(= f\epsilon < \alpha)$ is a consequence of 6.8 and the α-finiteness of $\bigcup \{I_\nu \mid \nu < 2\epsilon\}$. Similarly $\lim_\sigma g(\sigma, \epsilon)$ $(= g\epsilon < \alpha)$ exists. Then as in 4.4, A and B are α-recursively enumerable sets such that neither is weakly α-recursive in the other. It remains only to see that C is hyperregular.

Lemma 6.9. *C is hyperregular.*

Proof. Suppose $\{\epsilon\}^C(\gamma)$ is defined for all $\gamma < \alpha^*$. By 6.7 and 6.8 $\{\epsilon\}_\sigma^{C\sigma}(\gamma) = \{\epsilon\}^C(\gamma)$ for all $\gamma < \alpha^*$ and all sufficiently large σ. Consequently $\{\{\epsilon\}^C(\gamma) \mid \gamma < \alpha^*\}$ is bounded.

Case 2. $\alpha^* < \alpha$ and α^* is not a regular α-cardinal. Same as case 1 save that the definition of $m(\sigma, \epsilon)$ now reads:

$$m(\sigma, \epsilon) = (\mu\gamma)_{\gamma<\epsilon}(y(\sigma, \gamma, \epsilon) = \sigma) .$$

Thus the value of $\{\epsilon\}^C(\gamma)$ is preserved for $\gamma < \epsilon$ instead of $\gamma < \alpha^*$. The proof of 6.7 remains valid. The proof of 6.8 can no longer apply 2.3, but all is well by 2.1 since $m\epsilon' \leq \epsilon'$.

The proof of hyperregularity of C undergoes a slight change. Suppose $\{\epsilon\}^C(\gamma)$ is defined for all $\gamma < \alpha^*$ and $\{\{\epsilon\}^C(\gamma) \mid \gamma < \alpha^*\}$ is unbounded. Let ρ be a regular α-cardinal less than α^*. There is an $\epsilon_\rho \geq \rho$ such that $\{\epsilon\}^C(\gamma) = \{\epsilon_\rho\}^C(\gamma)$ for all $\gamma < \rho$. The construction of case 2 tends to preserve $\{\epsilon_\rho\}^C(\gamma)$ for all $\gamma < \epsilon_\rho$, hence the argument of 6.9 shows $\{\langle\gamma, \{\epsilon\}^C(\gamma)\rangle \mid \gamma < \rho\}$ is an α-finite partial function equal to

$$\{\langle\gamma, \{\epsilon\}_\sigma^{C\sigma}(\gamma)\rangle \mid \gamma < \rho\}$$

for all sufficiently large σ. By 2.2 (2), there is an α-finite sequence $\{\rho_\delta \mid \delta < \delta_0\}$ $(\delta_0 < \alpha^*)$ of regular α-cardinals whose limit is α^*. For each $\delta < \delta_0$ let $h\delta$ be the least σ such that for all $\tau \geq \sigma$,

$$\{\epsilon\}_\tau^{C\tau}(\gamma) = \{\epsilon\}^C(\gamma)$$

for all $\gamma < \rho_\delta$. Clearly $h \leq_{w\alpha} C$.

The argument of 6.9 shows the range of h restricted to δ_0 is bounded. It follows that the range of $\{\epsilon\}^C$ restricted to α^* is bounded.

Case 3. $\alpha^* = \alpha$, there is a largest α-cardinal $\aleph$, and $\aleph$ is regular. ϵ ranges over the ordinals less than $\aleph \cdot \lambda$, where λ is the ordertype of the set of α-stable ordinals beyond $\aleph$. Define $k(\sigma, \epsilon)$ as in Section 5. Let

$$y(\sigma, \gamma, \epsilon) = \mu\tau_{\tau<\sigma}\,([k(\sigma, \epsilon)]_{\tau}^{C<\sigma}(\gamma) \text{ is defined})\ ,$$

$$m(\sigma, \epsilon) = \mu\gamma_{\gamma<\aleph}\,(y(\sigma, \gamma, \epsilon) = \sigma)\ .$$

The proof of 5.2 remains valid. The statement of 6.7 is altered somewhat since $y(\sigma, \gamma, \epsilon)$ may change when σ increases because of a change in $k(\sigma, \epsilon)$. In the statement of 6.8, "α-finite" is replaced by "$\beta_{\xi+1}$-finite" where $\epsilon = \aleph \cdot \xi + \delta$ for some $\delta < \aleph$. In short the argument of case 3 parallels that of case 1 with $\beta_{\xi+1}$ corresponding to α, and $\aleph$ to α^*, the parallel being sound by virtue of the α-stability of $\beta_{\xi+1}$.

Case 4. $\alpha^* = \alpha$, there is a largest α-cardinal $\aleph$, and $\aleph$ is not regular. Define $y(\sigma, \gamma, \epsilon)$ as in case 3, but set

$$m(\sigma, \epsilon) = \mu y_{y<\delta}\,(y(\sigma, \gamma, \epsilon) = \sigma)\ ,$$

where $\epsilon = \aleph \cdot \xi + \delta$, $\delta < \aleph$, $\xi < \lambda$ and λ is defined as in case 3. Proceed as in Section 5 and in case 2 to show all the functions of interest associated with ϵ converge below $\beta_{\xi+1}$. Then proceed as in case 2 to show C is hyperregular.

Case 5. There is no largest α-cardinal. Define $y(\sigma, \gamma, \epsilon)$ and $m(\sigma, \epsilon)$ as in case 2. (ϵ now ranges over all the ordinals less than α.) Proceed as in Section 5 when there was no largest α-cardinal. The arguments of case 2 still work since α is the limit of regular α-cardinals.

§7. Open questions

1. α denotes a Σ_1-admissible ordinal. Do there exist (asks Kreisel) semi-invariantly implicitly definable subsets of α such that neither is invariantly implicitly definable from the other? (See [16] for the necessary definitions.) The answer is yes when α is countable, thanks to Theorem 6.2.

2. Suppose A is a non-α-recursive, regular, α-recursively enumerable set. Do there exist disjoint α-recursively enumerable sets B and C such that $A = B \cup C$, $B \not\leq_{w\alpha} C$ and $C \not\leq_{w\alpha} B$? The answer is yes when $\alpha^* = \omega$.

3. Do there exist finite Gödel numbers m and n such that for all α, R_m and R_n are α-recursively enumerable sets whose α-degrees are incomparable? The arguments of Sections 4 and 5 fail to produce such an m and n, because of the nonconstructive split into cases therein. An affirmative answer is likely.

4. Does there exist a generalization of the infinite injury method ([10, p. 86]) to every Σ_1 admissible α powerful enough to show the α-recursively enumerable degrees are dense? It is likely that a powerful generalization exists for every Σ_2 admissible α.

5. It is obvious that Sections 4 and 5 can be adapted to show: for each n and each Σ_n admissible α, there exist two Σ_n subsets of α such that neither is Δ_n in the other (over L_α)?

References

[1] G. Takeuti, On the recursive functions of ordinals, J. Math. Soc. Japan 12 (1960) 119–128.
[2] S. Kripke, Transfinite recursions on admissible ordinals, J. Symbolic Logic (Abstracts) 29 (1964) 161–162.
[3] R. Platek, Foundations of recursion theory, Ph. D. Thesis, Stanford (1966).
[4] G. Kreisel and G.E. Sacks, Metarecursive sets, J. Symbolic Logic 30 (1965) 318–338.
[5] K. Gödel, The consistency of the continuum hypothesis, Ann. Math. Study 3 (Princeton, 1940).
[6] R. Friedberg, Two recursively enumerable sets of incomparable degrees of unsolvability, Proc. Natl. Acad. Sci. U.S.A. 43 (1957) 236–238.
[7] A.A. Muchnik, Negative answer to the problem of reducibility of the theory of algorithms, Dokl. Akad. Nauk SSSR 108 (1956) 194–197.
[8] G. Kreisel, Some reasons for generalizing recursion theory, in: Logic Colloq. '69, eds. R. Gandy and C.E.M. Yates (North-Holland, Amsterdam, 1971) 139–198.
[9] A.H. Lachlan, The priority method I, Z. Math. Logic 13 (1967) 1–10.
[10] G.E. Sacks, Degrees of unsolvability, Ann. Math. Study 55 (Princeton, 1966).
[11] J. Stillwell, Ph. D. Thesis, M.I.T. (1970).
[12] G.E. Sacks, Metarecursion theory, in: Sets, models and recursion theory, ed. J.N. Crossley, (North-Holland, Amsterdam, 1967).
[13] G.E. Sacks, Higher recursion theory (Springer, Berlin, to appear).
[14] G. Driscoll, Metarecursively enumerable sets and their metadegrees, J. Symbolic Logic 33 (1968) 389–411.
[15] M. Lerman and G.E. Sacks, Some minimal pairs of α-recursively enumerable degrees, Ann. Math. Logic 4 (1972) 415–442 (this issue).
[16] K. Kunen, Implicit definability and infinitary languages, J. Symbolic Logic 33 (1968) 446–451.
[17] G.E. Sacks, Post's problem, admissible ordinals and regularity, Trans. Amer. Math. Soc. 124 (1966) 1–24.

J.E.Fenstad, P.G.Hinman (eds.), Generalized Recursion Theory

THE 1-SECTION OF A TYPE *n* OBJECT

Gerald E. SACKS [1]
Harvard University, Massachusetts Institute of Technology

1. Introduction

This paper is a puffing up of the proof of the plus-one theorem for the case $k = 1$. Let ${}^nE$ be the representing function of the equality predicate for all objects X, Y of type less than n : ${}^nE(X, Y) = 0$ if $X = Y$, and $= 1$ if $X \neq Y$.

Plus-One Theorem. *Let U be of type n. Suppose ${}^nE$ is recursive in U and $k < n$. Then there exists a V of type k+1 such that the objects of type k recursive in V are the same as those recursive in U. Furthermore ${}^{k+1}E$ is recursive in V.*

Recursion in objects of finite type was discovered by Kleene [2]. An equivalent formulation, needed for the forcing argument of section 4, is given in section 2. The proof of the plus-one theorem for the case $k > 1$ will be given in [3]. It is largely a consequence of a stability lemma described in section 5.

${}_k\mathrm{sc}U$ is the set of all objects of type k recursive in U, and is called the k-section of U. The plus-one theorem states that all k-sections generated by finite type objects (in which the appropriate equality predicates are recursive) are generated by type k+1 objects. The first result on k-sections was Kleene's [2]: ${}_1\mathrm{sc}^2E$ is the set of hyperarithmetic reals. He asked if the Δ^1_2 reals constituted the 1-section of some type 2 object. They do by virtue of the complete characterization of 1-sections of type 2 objects (in which ${}^2E$ is recursive) developed in section 4.

[1] The preparation of this paper was partially supported by NSF contract GP-29079. Its principal results were announced in [1]. The author is grateful to T. Grilliot for steering him towards k-section problems.

Section 2 redoes some of the elements of recursion in objects of finite type. Section 3 introduces the notion of abstract 1-section and shows that many familiar collections of reals, among then the 1-section of ^nE $(n \geq 2)$ and the set of lightface Δ^i_j reals $(\min(i,j) \geq 1)$, are countable abstract 1-sections. Section 4 proves every countable abstract 1-section is the 1-section of some type 2 object in which ^2E is recursive by means of a forcing argument of the sort associated with generic classes rather than sets. Section 5 describes some further results based on the technique of section 4, and speculates on the nature of abstract k-sections when $k > 1$.

2. Recursion in higher types

The objects of type 0 are the nonnegative integers. An object of type $n > 0$ is a total function whose arguments range over all objects of type $< n$ and whose values are objects of type $< n$. Any object of type i can be inflated to an equivalent object of type $j > i$ by adding dummy arguments. Any object of type $n > 0$ is equivalent to one of type n whose values are either 0 or 1; e.g. the function $F(x)$ is equivalent to the representing function of the predicate $F(x) = y$. Any finite sequence of objects is equivalent to a single object; e.g. the pair $F_0(x), F_1(x)$ is equivalent to $H(n,x)$, where $H(n,x) = F_n(x)$ for $n < 2$, and $= 0$ otherwise. The previous three sentences should make the ambiguities in what follows tolerable.

Fix $n \geq 0$, and let $F, G, \ldots$ be objects of type $n+2$ called *functionals*. The definition of $G \leq F, {}^{n+2}E$ (read G is recursive in $F, {}^{n+2}E$) is given by means of a hierarchy inspired by those of Shoenfield [4] and Grilliot [5]. D. MacQueen (cf. [6]) has shown that $G \leq F, {}^{n+2}E$ if and only if G is recursive in F, ^{n+2}E in the sense of Kleene [2]. MacQueen's argument is not unlike that of Grilliot [5]. The hierarchical approach via total objects is equivalent to Kleene's schematic approach via partial objects because of the presence of ^{n+2}E.

Functions $f, g, h, \ldots$ and *sets* $R, S, T, \ldots$ are objects of type $n+1$. *Individuals* $a, b, c, \ldots$ are objects of type at most n. *Subindividuals* $r, s, t, \ldots$ are objects of type less than n when $n > 0$ and are nonnegative integers when $n = 0$.

Fix F and g. The hierarchy $\{S_\sigma\}$ of sets, defined by induction on σ, is designed to define "G is recursive in $F, {}^{n+2}E, g$ via e" by induction on n.

Stage 0. S_0 contains $\langle 1, a\rangle$ for every a, and $\langle 1, a, m\rangle$ whenever $ga = m$. $\langle 1, a\rangle$ is an index for S_0 for every a.

Stage $\sigma+1$. $\langle 2^e, a\rangle$ is an index for $S_{\sigma+1}$ if (i) $\langle 2^e, a\rangle \notin S_\sigma$, (ii) $\langle m, a\rangle$ is an index for S_σ for some integer m, and (iii) there is a function f recursive in S_σ, ${}^{n+1}E$, a via e.

Clause (iii) makes sense by induction on n. If $n = 0$ the clause states f is Turing reducible to S via e. The f of clause (iii) is denoted by $\lambda b\{e\}^{S_\sigma, {}^{n+1}E, a}(b)$.

$S_{\sigma+1}$ is S_σ augmented by: (1) all indices for $S_{\sigma+1}$; (2) all quadruples $\langle 3^e, a, b, m\rangle$ if $\langle 2^e, a\rangle$ is an index for $S_{\sigma+1}$ and

$$\{e\}^{S_\sigma, {}^{n+1}E, a}(b) = m\ ;$$

and (3) all triples $\langle 5^e, a, m\rangle$ if $\langle 2^e, a\rangle$ is an index for S_σ and

$$F(\lambda b\{e\}^{S_\sigma, {}^{n+1}E, a}(b)) = m\ .$$

$|\langle e, a\rangle| = \sigma$ means $\langle e, a\rangle$ is an index for S_σ.

Stage λ, *where* λ *is a limit*. $\langle 7^e, a\rangle$ is an index for S_λ if $\langle 2^e, a\rangle$ is an index for some $S_{\delta+1} < \lambda$ and $\lambda b\{e\}^{S_\sigma, {}^{n+1}E, a}(b)$ is the characteristic function of a set T of indices such that

$$\lambda = \sup\{|b|\,|\,b \in T\}\ .$$

S_λ is $\bigcup\{S_\delta | \delta < \lambda\}$ augmented by all indices for S_λ.

f is said to be recursive in F, ${}^{n+2}E, g, a$ via e, written

$$f = \{e\}^{F, {}^{n+2}E, g, a}$$

if $\langle 2^e, a\rangle$ is an index for some $S_{\sigma+1}$ and f is $\lambda b\{e\}^{S_\sigma, {}^{n+1}E, a}(b)$. G is said to be recursive in F, ${}^{n+2}E$, g via e, written

$$G = \{e\}^{F, {}^{n+2}E, g}$$

if $G(h)$ is $\{e\}^{F, {}^{n+2}E, g, h}(0)$ for all h.

R is recursive in F, ${}^{n+2}E$ if its characteristic function is. ${}_k\mathrm{sc}(F, {}^{n+2}E)$ is the set of all objects of type k recursive in F, ${}^{n+2}E$. If ${}^{n+2}E$ is recursive in F in the sense of Kleene [2] then ${}_k\mathrm{sc}(F, {}^{n+2}E) = {}_k\mathrm{sc}F$.

R is recursively enumerable in F, ${}^{n+2}E$ via e if $R = \{a | \langle 2^e, a\rangle$ is an index$\}$. This notion of recursive enumerability coincides with Kleene's [2]. Thus R is recursive in $F, {}^{n+2}E$ if and only if both R and its complement are recursively enumerable in $F, {}^{n+2}E$.

The principal fact needed in the next section is the existence of a selection operator discovered by R. Gandy [7], who had to wrestle with Kleene's definition. His result follows more readily from the hierarchial definition of this section. [2]

Gandy's Selection Operator.
There exists a recursive function $\lambda e | e^*$ such that if T is a nonempty subset of ω recursively enumerable in $F, {}^{n+2}E$ via e, then $\{e^*\}^{F,{}^{n+2}E}(0)$ is defined and belongs to T.

3. Abstract 1-sections

Let A be a nonempty transitive set. A is said to be an abstract 1-section if it is closed under the operations of pairing and union, and satisfies axiom (1) and schemas (2)–(3), where $a \in A$ and $\mathcal{G}(y)$ and $\mathcal{G}(x,y)$ are Δ_0 formulas of **ZF** (i.e. formulas whose quantifiers are restricted, cf. Levy [10]) with parameters in A.

(1) Local countability: (x) (x is countable).

(2) Δ_0 separation:

$$(\mathsf{E}x)(y)[y \in x \leftrightarrow y \in a \ \& \ \mathcal{G}(y)].$$

(3) Δ_0 dependent choice:

$$(x)(\mathsf{E}y)\mathcal{G}(x,y) \rightarrow (\mathsf{E}h)(n)\mathcal{G}(hn, h(n+1)),$$

where $h: \omega \rightarrow A$.

If A is an abstract 1-section, then A is an admissible set as defined by Platek [8], and every member of A is hereditarily countable.

Each hereditarily countable set x can be encoded by a real number Y. Let m be the function that takes a code Y to the set mY encoded by Y. The relation

$$Y \text{ is a code } \& \ mY = x$$

[2] Not surprising, since his argument is based on a shadowy hierarchy of computations arising from the Kleene schemas of recursion. The existence of Gandy's selection operator was proved by him when $n = 0$, by Moschovakis [16] when $n = 1$, and by Platek [8] when $n > 1$.

is defined by induction on the rank of x: Y is a code for $\{m(Y_n) | n<\omega\}$, where $Y_n = \{a | 2^n \cdot 3^a \in Y\}$. The set of all codes is $\mathbf{\Pi}_1^1$, since a code is little more than a wellfounded relation on ω.

Proposition 3.1. *Let d be a code and* $\mathcal{G}(x)$ *a* $\mathbf{\Delta}_0$ *formula of* **ZF**. *The predicate* $P(e)$, *defined by*

$$(\mathrm{E}n)[e=d_n] \ \& \ \mathcal{G}(me)\,,$$

is recursive in $d, {}^2E$.

Proof. If sets correspond to codes, then restricted set quantifiers correspond to number quantifiers.

For each $K \subset 2^\omega$ let $\mathcal{M}K$ be the set of all sets with codes in K. It is convenient to ignore the distinction between K and $\mathcal{M}K$ (e.g. to say K is an abstract 1-section instead of $\mathcal{M}K$ is an abstract 1-section) when every member of K is coded by some member of K. The next proposition extends a result of Platek [8] to the effect that the least ordinal not in $\mathcal{M}({}_1\mathrm{sc}^n E)$ is $\mathbf{\Sigma}_1$ admissible.

Proposition 3.2. *Suppose* $n > 1$ *and U is a type n object in which* ${}^nE$ *is recursive. Then the* 1*-section of U is a countable abstract* 1*-section.*

Proof. It is immediate that $\mathcal{M}({}_1\mathrm{sc}\,U)$ is closed under the operations of pairing and union. To check $\mathbf{\Delta}_0$ separation let $\mathcal{G}(y)$ be a $\mathbf{\Delta}_0$ formula and $d \in {}_1\mathrm{sc}\,U$ be a code. It suffices to find a code $c \in {}_1\mathrm{sc}\,U$ such that
$(y)[y \in mc \leftrightarrow y \in md \ \& \ \mathcal{G}(y)]$ holds in $\mathcal{M}({}_1\mathrm{sc}\,U)$. The predicate $P(e)$, defined by

$$(\mathrm{E}n)[e = d_n] \ \& \ \mathcal{G}(me)\,,$$

is recursive in U by Proposition 3.1. The desired c is such that $\{c_n | n<\omega\}$ is an enumeration of all the e's that satisfy $P(e)$.

To check $\mathbf{\Delta}_0$ dependent choice let $\mathcal{G}(x,y)$ be a $\mathbf{\Delta}_0$ formula such that

$$(x)(\mathrm{E}y)\mathcal{G}(x,y)$$

holds in $\mathcal{M}({}_1\text{sc}\, U)$. Suppose $\{p\}^U$ is a code. Let Q_p be the set of all n such that

$$\{n\}^U \text{ is a code } \& \ \mathcal{G}(m\{p\}^U, m\{n\}^U).$$

Q_p is recursively enumerable in U (uniformly in p). Gandy's selection operator yields a partial function t recursive in[3] U such that $tp \in Q_p$ whenever $\{p\}^U$ is a code. Define g recursively in U by:

$$g0 = e_0 \quad (\{e_0\}^U \text{ is a code for } 0),$$

$$g(n+1) = t(gn).$$

Then $(n)\mathcal{G}(m\{gn\}^U, m\{g(n+1)\}^U)$.

For each ordinal α let L_α be the set of sets constructible in the sense of Gödel [11] via ordinals less than α. α is said to be Σ_1 admissible if L_α satisfies the Σ_1 replacement axiom schema of **ZF**.

Proposition 3.3. *If α is a Σ_1 admissible ordinal, then $L_\alpha \cap 2^\omega$ is an abstract 1-section.*

Proof. Gödel [11] shows $L \cap 2^\omega = L_{\omega_1} \cap 2^\omega$, where ω_1 is the least ordinal not countable in L. His argument restricted to L_α shows

$$L_\alpha \cap 2^\omega = L_{\omega_1^\alpha} \cap 2^\omega,$$

where ω_1^α is the least ordinal not countable in L_α. It follows $L_{\omega_1^\alpha}$ satisfies local countability. Gödel's wellordering of L restricted to $L_{\omega_1^\alpha}$ is Σ_1 over $L_{\omega_1^\alpha}$.

Let Δ_j^i be the set of all lightface Δ_j^i reals.

Proposition 3.4. *If* $\min(i,j) \geq 1$, *then Δ_j^i is a countable abstract 1-section.*

Proof. If $i = j = 1$, then the proposition follows from Spector's boundedness theorem [9] for Σ_1^1 subsets of Kleene's O and Kreisel's selection operator [12] for Π_1^1 predicates of numbers.

[3] I.e. the graph of t is recursively enumerable in U.

Assume $\max(i,j) > 1$. Let **HC** be the set of hereditarily countable sets. The Kondo–Addison uniformization of Π_1^1 predicates of reals implies **HC** satisfies Δ_0 dependent choice. Consequently **HC** is an abstract 1-section. It suffices to show $\mathcal{M}(\Delta_j^i)$ is a Σ_1 substructure of **HC**. Let $\mathcal{G}(x)$ be a Δ_0 formula with parameters in $\mathcal{M}(\Delta_j^i)$ such that

$$\mathbf{HC} \models (\mathrm{E}x)\mathcal{G}(x)\,.$$

There exists an arithmetic predicate $A(Y)$ such that

$$[Y \text{ is a code } \ \& \ A(Y)] \leftrightarrow \mathbf{HC} \models \mathcal{G}(mY)\,.$$

The set parameters occurring in $\mathcal{G}(x)$ correspond to Δ_j^i codes occurring in $A(Y)$. The Kondo–Addison uniformization supplies a code Z such that Z satisfies $A(Y)$ and is Δ_2^1 in the Δ_j^i codes occurring in $A(Y)$. Since $\max(i,j) > 1$, $Z \in \Delta_j^i$. Hence $\mathcal{M}(\Delta_j^i) \models (\mathrm{E}x)\mathcal{G}(x)$.

4. Generic type 2 objects

Let K be a countable abstract 1-section. Suppose F maps ω^ω into ω and is 0 off K. If F is generic in the sense of the following forcing relation, then the 1-section of $(F, {}^2E)$ is K.

Let p be a partial function from ω^ω into ω. p generates a hierarchy $\{T_\sigma\}$ of reals as defined below. If p is total, then the T_σ's are equivalent to the S_σ's of section 2 when $n = 0$ and $F = p$. If p is not total, then there may be a σ such that $T_{\sigma+1}$ has an index but is not total.

Stage 0. $T_0 = \{1\}$. 1 is an index for T_0. T_0 is total.

Stage $\sigma+1$. 2^e is an index for $T_{\sigma+1}$ if T_σ has an index, T_σ is total, $2^e \notin T_\sigma$, and $\{e\}^{T_\sigma}(m)$ is defined for all m. ($\{e\}^{T_\sigma}$ is the unique partial function from ω into ω recursive in T_σ via Gödel number e.)

$T_{\sigma+1}$ is total if it has an index and $p(\lambda m\{e\}^{T_\sigma}(m))$ is defined whenever 2^e is an index for $T_{\sigma+1}$.

$T_{\sigma+1}$ is T_σ augmented by: all indices for $T_{\sigma+1}$; all triples $\langle 3^e, m, n\rangle$ such that $\{e\}^{T_\sigma}(m) = n$ and 2^e is an index for $T_{\sigma+1}$; and all pairs $\langle 5^e, n\rangle$ such that $p(\lambda m\{e\}^{T_\sigma}(m)) = n$ and 2^e is an index for $T_{\sigma+1}$.

$$|m| = \sigma \text{ means } m \text{ is an index for } T_\sigma\,.$$

Stage λ (limits). 7^e is an index for T_λ if 2^e is an index for $T_{\delta+1}$ for some $\delta < \lambda$ and $\lambda m\{e\}^{T_\delta}(m)$ is the characteristic function of a set R of indices such that

$$\lambda = \sup\{|m| \mid m \in R\}.$$

T_λ is total if it has an index. T_λ is $\bigcup\{T_\delta \mid \delta < \lambda\}$ augmented by all indices for T_λ.

p is said to *generate* T_σ if T_σ has an index and is total. Fact H is easily proved by induction on σ.

Fact H. If $\sigma < \gamma$ and p generates T_σ and T_γ, then T_σ has lower Turing degree than T_γ.

Suppose p is total and S is a real. The arguments of Shoenfield [4] show S is Turing reducible to some T_σ generated by p if and only if S is recursive in p, ${}^2E$ in the sense of Kleene [2].

If $T_{\sigma+1}$ has an index but is not total, then $T_{\sigma+1}$ is said to be the *maximum* of p. p is a *forcing condition* if it meets two requirements.

(1) $p \in \mathcal{M}K$ and has a maximum.

(2) X is in the domain of p if and only if X is Turing reducible to T_δ for some $\delta < \sigma$, where $T_{\sigma+1}$ is the maximum of p.

Requirement (2) is not as limiting as it may appear, because Fact H implies that the generation of T_σ by p utilizes the value of $p(X)$ only if X is Turing reducible to T_δ for some $\delta < \sigma$.

From this point on $p, q, r, \ldots$ denote forcing conditions. p is extended by q (in symbols $p \supset q$) if the graph of p is contained in the graph of q.

The language $\mathcal{L}(K)$ will be used to define the desired generic F's. The individual constants of $\mathcal{L}(K)$ are: $\underline{m}$ for each $m \in \omega$; $\underline{f}$ for each $f \in \omega^\omega \cap \mathcal{M}K$; $\underline{\sigma}$ and $T_{\underline{\sigma}}$ for each ordinal $\sigma \in \mathcal{M}K$; and $\underline{S}$ for each $S \in 2^\omega \cap \mathcal{M}K$. The variables of $\mathcal{L}(K)$ are: $x, y, \ldots$ (numbers); $\mu, \nu, \ldots$ (ordinals); and $T_\mu, T_\nu, \ldots$ (sets). The atomic formulas of $\mathcal{L}(K)$ are of the form: $|x| = \mu$, $|fx| = \mu$, $\mu < \nu$, and $\underline{S} \leq T_\mu$. The sentences of $\mathcal{L}(K)$ are built up from the atomic formulas by substitution of appropriate individual constants for variables and by application of propositional connectives (& and $\sim$) and existential number and ordinal quantifiers.

$\mathcal{G}$ is a *ranked* sentence of rank σ if $\mathcal{G}$ contains no ordinal quantifiers and σ is the least ordinal greater than every ordinal occurring in $\mathcal{G}$. Such an $\mathcal{G}$ is *true* in $\{T_\delta \mid \delta < \sigma\}$ if it is true when $\underline{\delta} < \underline{\gamma}$ is interpreted as $\delta < \gamma$, $|\underline{m}| = \underline{\delta}$ as

m is an index for T_δ, and $\underline{S} \leq \underline{T_\delta}$ as S is Turing reducible to T_δ. The forcing relation $\Vdash$ is defined inductively.

(i) $p \Vdash \mathcal{G}$ if $\mathcal{G}$ is of rank σ, p generates T_δ for all $\delta < \sigma$, and $\mathcal{G}$ is true in $\{T_\delta | \delta < \sigma\}$.

(ii) $p \Vdash (\mathsf{E}x)\mathcal{G}(x)$ if $p \Vdash \mathcal{G}(\underline{m})$ for some m.

(iii) $p \Vdash (\mathsf{E}\mu)\mathcal{G}(\mu)$ if $p \Vdash \mathcal{G}(\underline{\sigma})$ for some σ.

(iv) $p \Vdash \mathcal{G}$ & $\mathcal{G}$ if $p \Vdash \mathcal{G}$ and $p \Vdash \mathcal{G}$.

(v) $p \Vdash \sim \mathcal{G}$ if $(q)_{p \supset q} \sim [q \Vdash \mathcal{G}]$ and $\mathcal{G}$ is not ranked.

A sentence $\mathcal{G}$ is said to be $\mathbf{\Sigma}_1$ if it is in prenex normal form and contains no universal quantifiers.

Proposition 4.1. *The relation* $p \Vdash \mathcal{G}$, *restricted to* $\mathbf{\Sigma}_1$ $\mathcal{G}$'s, *is* $\mathbf{\Sigma}_1$ *over* $\mathcal{M}K$.

Proof. Suppose $w \in \mathcal{M}K$ and is a partial function from ω^ω into ω. The set of T_σ's generated by w is $\mathbf{\Sigma}_1$ over $\mathcal{M}K$ uniformly in w. (This last is a consequence of the $\mathbf{\Sigma}_1$ admissibility of $\mathcal{M}K$ and the autonomous fashion in which indices are assigned to T_λ when λ is a limit ordinal.) Thus if w has a maximum, then that maximum belongs to $\mathcal{M}K$. It follows that the set of forcing conditions is $\mathbf{\Sigma}_1$ over $\mathcal{M}K$.

Let F map ω^ω into ω and be 0 off K. F satisfies p (in symbols $F \in p$) if the graph of p is contained in the graph of F. F is *generic* if for each sentence $\mathcal{G}$ of the language $\mathcal{L}(K)$ there is a p such that $F \in p$ and either $p \Vdash \mathcal{G}$ or $p \Vdash \sim \mathcal{G}$. Generic F's exist because there are only countably many sentences to be forced. Standard arguments [15] show: if F is generic, then each true statement about F (expressible in the language $\mathcal{L}(K)$) is forced by some p satisfied by F.

Lemma 4.2. *If F is generic, then* $K \subset {}_1\mathrm{sc}(F, {}^2E)$.

Proof. Suppose S: $\omega \to \omega$ belongs to K. Fix p; since F is generic it suffices to find a $q \subset p$ such that

$$q \Vdash \underline{S} \leq \underline{T_\sigma}$$

for some σ. Since p has a maximum there is an e and a γ such that p generates T_γ, 2^e is an index for $T_{\gamma+1}$ and

$$p(\lambda m\{e\}^{T_\gamma}(m))$$

is undefined. Let $\lambda n | e_n$ be a recursive function such that

$$\{e_n\}^X \simeq n + \{e\}^X$$

for all $X \subset \omega$ and $n \in \omega$. Clearly $\{e_n\}^X$ is total if and only if $\{e\}^X$ is. It follows 2^{e_n} is an index for $T_{\gamma+1}$ because 2^e is. In addition

$$p(\lambda m\{e_n\}^{T_\gamma}(m))$$

is undefined, because the domain of p is an initial segment of Turing degrees.

Choose $q \subset p$ so that the domain of q consists of all functions Turing reducible to T_γ, and so that

$$q(\lambda m\{e_n\}^{T_\gamma}(m)) = Sn$$

for all $n \in \omega$. Thus q generates $T_{\gamma+1}$ but not $T_{\gamma+2}$, since $q(T_{\gamma+1})$ is undefined. And S is Turing reducible to $T_{\gamma+1}$ since

$$Sn = r \leftrightarrow \langle 5^{e_n}, r\rangle \in T_{\gamma+1}$$

for all n and r.

Lemma 4.3. *If F is generic, then ${}_1\mathrm{sc}(F, {}^2E) \subset K$.*

Proof. Suppose $S \in {}_1\mathrm{sc}(F, {}^2E) - K$ for the sake of a reductio ad absurdum. Then S is Turing reducible to some T_σ generated by F but not in $\mathcal{M}K$; $\sigma \notin \mathcal{M}K$ since F is generic. Let α be the least ordinal not in $\mathcal{M}K$. Then F generates some T_α with index 7^e. Thus 2^e is an index for some $T_{\delta+1}$ generated by F in $\mathcal{M}K$, and $\{e\}^{T_\delta}$ is the characteristic function of a set R such that

$$\alpha = \sup\{|n| \,|\, n \in R\}.$$

Let $f \in \mathcal{M}K$ enumerate R. Since F is generic there is a p satisfied by F such that:

(a) $p \Vdash (x)(\mathbf{E}\mu)[|\underline{f}x| = \mu]$;
(b) $p \Vdash (\mu)(\mathbf{E}x)(\mathbf{E}\nu)[\mu < \nu \ \& \ |\underline{f}x| = \nu]$;
(c) $p \Vdash \underline{R} \leq T_{\underline{\delta}}$;
(d) $p \Vdash \sim(\mathbf{E}\mu)[|\underline{7}^e| = \mu]$.

(a) is equivalent to

(a*) $(m)(q)_{p\supset q}(\mathbf{E}r)_{q\supset r}(\mathbf{E}\sigma)[r \Vdash |\underline{f}\underline{m}| = \underline{\sigma}]$,

and (b) is equivalent to

(b*) $(\sigma)(q)_{p\supset q}(\mathbf{E}r)_{q\supset r}(\mathbf{E}m)(\mathbf{E}\gamma)_{\sigma<\gamma}[r \Vdash |\underline{f}\underline{m}| = \underline{\gamma}]$.

It follows from (a*), (b*), Proposition 4.1 and the validity of the Σ_1 dependent choice schema in $\mathcal{M}K$ that there exist functions $\lambda m | p_m$ and $\lambda m | \sigma_m$ in $\mathcal{M}K$ such that

$$p \supset p_m \supset p_{m+1}, \qquad p_m \Vdash |\underline{f}\underline{m}| = \underline{\sigma}_m$$

and $(m)(\mathbf{E}n)[\delta < \sigma_m < \sigma_n]$. Let $\lambda = \sup\{\sigma_m | m \in \omega\}$ and $w =$ = $\mathbf{U}\{p_m | m \in \omega\}$. Clearly $\lambda, w \in \mathcal{M}K$. w generates $T_{\underline{\delta+1}}$ with index 2^e thanks to (c) and the fact that $\delta < \lambda$. Consequently w generates some T_λ with index 7^e. None of the p_m's generate T_λ by (d). It follows from requirement (2) of the definition of forcing condition and Fact H that $w(T_\lambda)$ is undefined. Thus w has a maximum (namely $T_{\lambda+1}$) and so is a forcing condition. But then $w \Vdash |7^e| = \underline{\lambda}$, an impossibility according to (d).

Theorem 4.4. *K is a countable abstract* 1-*section if and only if K is the* 1-*section of some type* 2 *object in which* ^2E *is recursive.*

Proof. If G is a type 2 object in which ^2E is recursive, then $_1\mathrm{sc}\,G$ is a countable abstract 1-section by Lemma 3.2.

Suppose K is a countable abstract 1-section. Let F be generic in the sense of Lemma 4.2, and let G be the recursive join of F and ^2E. Then $K = {}_1\mathrm{sc}\,G$ by Lemmas 4.2 and 4.3.

The following three corollaries of Theorem 4.4 are consequences of Propositions 3.2–3.4.

Corollary 4.5. *Suppose* $n > 2$ *and U is a type n object in which* ^nE *is recursive. Then there exists a type* 2 *object V such that*

$$_1\mathrm{sc}\, U = {}_1\mathrm{sc}\, V$$

and ^2E *is recursive in* V.

Corollary 4.6. *Suppose* α *is a countable* Σ_1 *admissible ordinal. Then there exists a type* 2 *object* V *such that*

$$L_\alpha \cap 2^\omega = {}_1\mathrm{sc}\, V$$

and ^2E *is recursive in* V.

Corollary 4.7. *If* $\min(i,j) \geq 1$, *then the set of all lightface* Δ^i_j *reals is the* 1-*section of some type* 2 *object in which* ^2E *is recursive.*

5. Further results

The method of section 4 is applicable to the study of Gandy's superjump [7]. Theorems 5.1 and 5.2 are typical results of [14] and were inspired by some questions raised by P. Hinman at the 1969 Manchester Logic Colloquium. Let F and G be objects of type 2, G' the superjump of G, and E_1 the superjump of ^2E. ${}_1\mathrm{sc}\, G$ is said to be closed under hyperjump if

$${}_1\mathrm{sc}(E_1, X) \subset {}_1\mathrm{sc}\, G$$

for every $X \in {}_1\mathrm{sc}\, G$.

Theorem 5.1. *Suppose* ${}_1\mathrm{sc}\, G$ *is closed under hyperjump. Then there exists an* F *such that*

$${}_1\mathrm{sc}\, G = {}_1\mathrm{sc}(F').$$

Theorem 5.2. (Assume $2^\omega = \omega_1$.) *There exists an* H *such that* $(G)(\mathbf{E}F)[H \leq G \rightarrow F' \equiv G]$.

The method of section 4 does not appear to suffice for the proof of the plus-one theorem when $k > 1$. A stability result is needed to overcome problems caused by gaps in the hierarchy of section 2, gaps that fall between objects recursive in $F, {}^{n+2}E$ when $n > 0$. Call R subrecursive in $F, {}^{n+2}E$ if R is recursive in some S_σ (as defined in section 2) with an index of the form $\langle 2^e, r \rangle$, where r is a subindividual. The stability result in question says: each nonempty recursively enumerable (in $F, {}^{n+2}E$) collection of subrecursive (in $F, {}^{n+2}E$) sets must have a recursive (in $F, {}^{n+2}E$) set among its members.

At this writing it is not known if there exists a decent notion of abstract k-section when $k > 1$. Decency requires that Theorem 4.4 remain true when "1-section" is replaced by "k-section" and "2" by "$k+1$".

References

[1] G.E. Sacks, Recursion in objects of finite type, Proceedings of the International Congress of Mathematicians 1 (1970) 251–254.

[2] S.C. Kleene, Recursive functionals and quantifiers of finite type, Trans. Amer. Math. Soc. 91 (1959) 1–52; 108 (1963) 106–142.

[3] G.E. Sacks, The k-section of a type n object, to appear.

[4] J. Shoenfield, A hierarchy based on a type 2 object, Trans. Amer. Math. Soc. 134 (1968) 103–108.

[5] T. Grilliot, Hierarchies based on objects of finite type, Jour. Symb. Log. 34 (1969) 177–182.

[6] G.E. Sacks, Higher Recursion Theory, Springer Verlag, to appear.

[7] R. Gandy, General recursive functionals of finite type and hierarchies of functions, University of Clermont-Ferrand (1962).

[8] R. Platek, Foundations of Recursion Theory, Ph.D. Thesis, Stanford (1966).

[9] C. Spector, Recursive well-orderings, Jour. Symb. Log. 20 (1955) 151–163.

[10] A. Levy, A hierarchy of formulas in set theory, Memoirs of the American Mathematical Society, Number 57 (1965).

[11] K. Gödel, The Consistency of the Axiom of Choice and of the Generalized Continuum Hypothesis (Princeton University Press, Princeton, 1966).

[12] G. Kreisel, Set theoretic problems suggested by the notion of potential totality, in: Infinitistic Methods (Pergamon Press, Oxford, and PWN, Warsaw, 1961) pp. 103–140.

[13] J. Shoenfield, The problem of predicativity, Essays on the Foundations of Mathematics (Magnes Press, Jerusalem, 1961 and North-Holland, Amsterdam, 1962).

[14] G.E. Sacks, Inverting the superjump, to appear.

[15] S. Feferman, Some applications of the notion of forcing and generic sets, Fund. Math. 56 (1965) 325–345.

[16] Y. Moschovakis, Hyperanalytic predicates, Trans. Amer. Math. Soc. 129 (1967) 249–282.

HISTORIA MATHEMATICA 2 (1975), 523-528

REMARKS AGAINST FOUNDATIONAL ACTIVITY

BY GERALD SACKS, HARVARD UNIVERSITY & M.I.T.

I want to discourage people -- mathematicians and others -- from trying to build foundations for *all* of mathematics. There is no need for such an enterprise. Moreover, any such attempt is bound to assume a negative puritanical flavor: "don't do this"; "you can't do that"; etc. Puritanism is not fruitful for mathematics.

However, foundational activity is of interest in some limited spheres. For example, when there were difficulties in set theory, a few small changes were needed to straighten things out. There were difficulties, but not paradoxes; there was no need to write *Principia Mathematica* to straighten them out. It was necessary to clarify ideas -- and that is when foundational activity is of interest. Another example is category theory in logic. (Reference is being made to the new theory of "topoi", developed by Lawvere and others. The role of category theory in logic is not to be confused with its role in mathematics, as described for example in S. MacLane, *Categories for the Working Mathematician*, Springer, 1971.) There are a number of paradoxical constructions which can probably be straightened out by a minimal effort. There is no need to write down axioms for category theory.

Foundational activities should be very narrow, very restrained. They should be geared to solve the difficulties at hand. Since there have not yet been any substantial paradoxes, there has not yet been any need for a wider sphere of foundational activity. For example, there were confusions, not paradoxes, in late nineteenth and early twentieth century mathematics, despite all claims to the contrary.

There are no substantial paradoxes in the history of mathematics. Sometimes results are called paradoxes, but not for long. For example, the Banach-Tarski paradox is actually just a theorem. [Or consider eighteenth century geometry and Kramer's paradox -- interjected by Dieudonné.] When one called these paradoxes, one simply meant that the conclusion was startling, e.g. Peano's paradox of the space-filling curve. After a while such results become accepted, and don't seem nearly so strange.

Most mathematical logicians today are not interested in foundations. They are more interested in seeing what ideas from mathematical logic can be applied to other areas of mathematics. These days mathematical logic is a peripheral field of mathematics, and its utility is in its applicability to the more central fields. However, that may change.

A good example of this is the work of Abraham Robinson. He was aware of the foundational problems and discussed them with various people. Yet if one peruses his publications, one will find few investigations of such problems. Instead one finds an interest in applications of logic to certain problems in algebra and analysis. And one of his final interests was applying his notion of nonstandard analysis to algebraic number theory.

You can think of Robinson's work in nonstandard analysis as foundational if you like, namely, as an alternative foundation of analysis. However I am convinced that this was not his motivation at all. His interest in nonstandard analysis lay in its applications to classical analysis. His idea was to get new theorems for classical analysis by looking at classical problems in a slightly different way. His approach has been fairly successful. For example, for quite a few years the most useful way of considering the invariant subspace problem was through nonstandard analysis, even though this approach has been recently superseded. Also, there are applications of nonstandard analysis to mathematical economics, unexpected as they may be.

There are many other examples of the sorts of services that logic can provide to other fields of mathematics. One of the most paradigmatic is due to a result of Gödel. Gödel defined a certain class of statements which, if provable using the Axiom of Choice, would be provable without the Axiom of Choice. Logic is needed to make precise just what this class of statements is. One consequence is that if you could prove the Riemann hypothesis using the Axiom of Choice (there is no reason to believe that you can), you could prove it without the Axiom of Choice. This method due to Gödel was actually used in a paper by Ax and Kochen [a], although a direct method was eventually found to reprove their result without the Axiom of Choice and without Gödel's method.

Also exemplary of the use of logic in other fields of mathematics is its use in making sense of Lefschetz's Principle and the idea of a universal domain. Essentially, Lefschetz's Principle [b] says that any algebraically closed field of characteristic zero and of infinite transcendence rank is as good as any other as far as algebraic geometry is concerned, i.e. if you can prove a statement over one of these fields, it is true over all others. Attempts have been made by Feferman and, more recently, by Barwise to make sense of this notion, by defining a broad class of statements (including all those relevant to algebraic geometry) which are based on Lefschetz's Principle.

Although there haven't been any substantial paradoxes in mathematics, there have been errors. It is important that histories of mathematics be written which discuss real, substantial errors made by mathematicians! A classic example is related to a mistake made by Lebesgue when he tried to show that the projection of any Borel set in the plane on the line is Borel. The motivation behind this was to show that the projection of a Borel set is measurable -- which, of course, it is, but by a more intricate argument. The error was in print for 10 or 12 years before it was discovered by Souslin [c]. This led to the definition of analytic set and to the development of descriptive set theory. Histories of mathematical logic should be wary of over-stressing foundational aspects, although the various attempts to erect foundational systems should be included.

The axiomatic aspects of mathematical logic has encouraged clarity and precision to a dangerous degree. The attempt to provide maximum precision has a killing effect upon mathematics. Clarity should come at the end, not at the beginning of a mathematical endeavor. Consider Euler's discovery of all the integral solutions to the equation $x^3 = y^2 + 2$. This was accomplished with an argument which must have been incomprehensible, if not meaningless, to most people of his time. As Dieudonné points out, Euler himself did not know what he was doing when he broke expressions into complex factors. Students today are discouraged from proceeding as Euler did. Too often, they are unwilling to move from one point to the next until they have painfully analyzed the minutiae of each argument. The effect is altogether too often, a long and uninspired thesis.

NOTES

a. *Annals of Math.* 83 (1966), 437-56.

b. As formulated in A. Weil, *Foundations of Algebraic Geometry*, Am. Math. Soc., 1946.

c. H. Lebesgue, M. Souslin, *C.R. Paris* 164 (1917), 88-; N. Lusin, *Fund. Math.* 10 (1927), 1-95.

DISCUSSION

Dreben: Sacks is right in saying that mathematical logic today is no longer concerned with foundations in the traditional sense, and has not been since World War II. Sacks might have misled some people, however, into thinking that the reason Russell and Whitehead wrote *Principia Mathematica* was primarily because of the paradoxes. That is not true: The reason that Russell and Whitehead, or Frege, or their immediate followers engaged in this type of enterprise was not in trying to do mathematics *per se*, but because they had a very general philosophical conception and enterprise (although I don't agree with it). The *form* that *Principia Mathematica* took was because of the paradoxes; this was the reason for the theory of types. But the reason behind *Principia Mathematica*, or much that is of interest in it was not the result of such paradoxes. This is important.

These are called paradoxes -- and not contradictions -- because most people see what Sacks was saying, i.e. they are genuine mathematical results in set theory. What is somewhat misleading is the view -- a view currently held by many logicians that there never was any real problem in Cantorian set theory, because set theorists are talking about the iterative notion of a set (the notion that you start at the bottom and build up) first made explicit by Von Neumann. This iterative notion underlies the contemporary view of set theory.

The truth is, as Gödel has pointed out, that there are really two basic notions of set. There is the logical notion of a set; not in the narrow sense of mathematics but in the broad sense that Frege and Russell intended it to be used. This is a continuation of the medieval idea of the extension of a predicate. Then there is the iterative notion of set, implicit in Cantor and used in mathematical work. This notion never had any troubles in it. Gödel thinks we still don't fully understand the logical notion of set.

Moore: (to Sacks) You had said there have been no real, substantial paradoxes in the history of mathematics. What would constitute a real paradox?

Dreben: That's a contradiction in terms.

Sacks: I think it would have to have the nature of an experience that every mathematician has at least once a year, that of proving a theorem one day and refuting it the next day. Like this -- except that nobody can find any mistakes in either the proof or the refutation. It might mean we would have to retreat in our mathematics to a level where the paradox does not appear.

Kline: I believe that historically you are incorrect. The word "paradox" does mean a seeming contradiction which has been resolved. But paradoxes like Russell's class of all classes or the Richard paradox were genuine contradictions as far as the mathematicians were then concerned. They used the "paradox" to ease their troubled souls because the word "paradox" implied that they were going to resolve the problems.

Dreben: We should make this clear. The Russell paradox, the Richard paradox, the Burali-Forti paradox -- these are all logical contradictions. Up until the turn of the century, up until the statement of the paradoxes, people talked of wanting to have a general logic. Even mathematicians of the first rank like Zermelo, those building axiom systems, were under the illusion that their work was connected with general logical investigations. Zermelo and even Von Neumann in his 1925 paper say explicitly that the only justification for his setting of the axioms is that they make the paradoxes appear not to exist. Von Neumann, and probably Zermelo, knew that you could have a working set theory based on the iterative notion of set (cumulative type theory) in which there is no contradiction. This is what underlies Von Neumann's proof that the axiom of foundation is consistent with the other axioms of set theory. This is such a powerful notion that all young people in mathematical logic today take it for granted that this is what set theory is about.

These two notions of set come together in working set theory although they should be kept separate. Yet, as Gödel pointed out in his 1947 paper[C10], there never have been any contradictions in the mathematical notion of set.

Dieudonné: What caused people to think of the set of all sets?

Dreben: About 1899, Cantor probably had two notions of set theory: *absolutist* set theory, which is what we have in mind when we think of sets as extensions of predicates, and *cumulative* set theory which could be done just for mathematical purposes. It was the absolutist notion that would lead to the paradoxes.

Von Neumann (1925) says that some omission is essential to avoid the paradoxes; what that omission is, is arbitrary. He was the first to formulate the cumulative notion of set explicitly [C31, vol. 1, 35-57; C11, pp. 393-413].

...Mr. Bishop is engaged in foundational work, a reconstruction of mathematics with some picture in mind. Proof theory is not this.

Kline: What was the purpose of Hilbert's work if not to prove consistency? Reconstruction would have been useless to Hilbert if he had not had his *Beweistheorie* (proof theory).

Dreben: The original motivation for proof theory was to prove the consistency of a formalization of mathematics, but after Gödel that kind of motivation was pre-empted. But this does not stop proof theory from being interesting.

Kline: Thus, we must distinguish between proof theory before and after Gödel.

Dreben: Mathematical logic grew out of philosophical purposes, but it is absolutely essential to realize that even though people use the same words, e.g. "foundations", they may mean entirely different things by them today and have entirely different purposes.

Putnam: There are still conceptual problems with the mathematical, cumulative notion of a set. The problem of a working set theorist is this: he wants to have things both ways. At that point, you are right back in the Russell paradox. There was a good discussion of this at the May meeting of the American Philosophical Society, to be published in *Nous*. There is still a vestige of the paradox hanging over the mathematical notion of a set, although it does not threaten contradiction.

Dieudonné: For example, when one is talking about an arbitrary group, he tends to think of the set of all groups. But it is just an illusion that you must talk of this "large" set (of all groups).

Birkhoff: Weierstrass provides a good example of the good influence of precision coming at the end of a mathematical development. Here precision seemed to crystallize analysis.

Also, constructive proof is better than non-constructive proof, other considerations aside. It would be of service, and not much effort, to indicate whether proofs are constructive or need be non-constructive. I still feel uneasy about the Banach-

Tarski decomposition; the "components" defined (non-constructively) just do not seem very meaningful.

In response to Birkhoff's comments, Sacks said that in closely following Gödel he represents the majority view. Birkhoff agreed.

Dieudonné: Since Cohen's work, we have as many different mathematics as we want, for we have as many choices of alephs for the continuum as we want... [Ed.: His implication seems to be that this freedom of choice makes the resulting theories not very meaningful.]

Kahane: Hausdorff mixed his set theory with measure theory and topology. For his predecessors, set theory was the study of linear sets. When did the change come so that the study of abstract *sets* became part of mathematics?

Sacks: Cantor speaks of quite arbitrary sets (absolute, general sets). Dreben and Browder added that, in a general working sense, it was in Fréchet's Thesis (1906) that abstract *spaces* were first studied, and that he had applications to functional analysis in mind. [See Part E of the Proceedings.]

May: To clarify the discussion, we should distinguish the question as to when certain individuals thought of sets and spaces abstractly from the question as to when the idea of considering abstract topological spaces and measure spaces became generally diffused. Widespread diffusion of the abstract approach may not have come until the 1920's.

The discussion concluded with an exchange of views between Profs. May and Dieudonné concerning a point made by Sacks: that axiomatization often comes late in the historical development of a subject. The following is a statement of consensus written by the editor.

Sacks' assertion was notably true of Euclidean geometry and group theory. Likewise, Boolean algebra was decades old before it became axiomatized, first incompletely by C.S. Peirce and then almost too completely by E. Schröder. (In a fruitful compromise, Dedkind used variants of the axioms of Boolean algebra to define lattices, modular lattices, and distributive lattices.) But by itself, axiomatization usually contributes more to the precise formulation of a theory than to its development. Carried to excess, it may even tend to make a subject seem moribund, though it certainly did not do this to the theory of finite groups! What seems to be true is that, for an axiom system to have continuing interest, it must have many variants, each with important realizations. (Bourbaki sometimes refers to the mathematical structures which, dually, satisfy several axiom systems, as "mother structures".) Axiomatization also makes it possible to *synthesize* whole theories of mathematics from a few basic ingredients. Once such a synthesis has taken place, however, its fundamental concepts tend to become static.

Reprinted from ADVANCES IN MATHEMATICS Vol. 20, No. 2, May 1976

Countable Admissible Ordinals and Hyperdegrees

GERALD E. SACKS*

Department of Mathematics, Harvard University, and Massachusetts Institute of Technology, Cambridge, Massachusetts 02139

1. INTRODUCTION

This paper began sometime ago with the question: Does every countable set of hyperdegrees have a minimal upper bound? That question is still open, however the following partial answer is provided by Section 3.

COROLLARY 3.14. *If A is a countable admissible set that satisfies Σ_1 dependent choice, then the set of all hyperdegrees contained in A has a minimal upper bound.*

The search for minimal upper bounds uncovered the answer to a question raised by R. Platek: Is every countable admissible ordinal beyond ω of the form $\omega_1{}^X$ for some $X \subset \omega$? ($\omega_1{}^X$ is the least ordinal not recursive in X. Platek's question was prompted by his observation that $\omega_1{}^X$ is admissible for every X.) Theorem 4.26 was proved by techniques originally developed to handle some special cases of Corollary 3.14.

THEOREM 4.26. *If α is a countable admissible ordinal greater than ω, then there exists an $X \subset \omega$ such that*

(i) $\omega_1{}^X = \alpha$, *and*

(ii) $\omega_1{}^Y < \alpha$ *for every Y of lower hyperdegree than X.*

Subsequently a model theoretic proof of 4.26(i) based on Barwise's compactness theorem was found by Friedman and Jensen [1]. Their argument was simplified by Grilliot [2] and reduced to its essentials by Keisler [3]. The model theoretic approach, primarily that of omitting

* The preparation of this paper was partially supported by NSF Grant P 29079.

a type, does not yield the minimality property expressed by 4.26(ii)[1]. That sort of minimality is a natural side effect of forcing with pointed perfect conditions, the principal technique of this paper.

Let P be a perfect set of reals. P is said to be hyperarithmetically pointed if the standard encoding of P by a real is hyperarithmetic in every member of P. By 2.2 the hyperdegrees of the members of P are the same as those greater than or equal to that of P. Thus P can be thought of as a condition that forces the hyperdegree of some generic real to lie in the cone of hyperdegrees greater than or equal to that of P without ruling out any member of the cone. The key technical problem associated with hyperarithmetically pointed P's arises in the proofs of splitting lemmas as indicated in 3.8. The problem is to find precautions to take in the course of defining a contracting sequence of hyperarithmetically pointed perfect P's so that the intersection of the elements of the sequence will be hyperarithmetically pointed and perfect. To make the intersection perfect never calls for much effort, but to make it pointed requires machinations whose minutia vary widely in the several situations encountered in Sections 3 and 4.

Recursively pointed P's were applied in [4, p. 353] to show certain sequences of Turing degrees have 2-least upper bounds.[2] They were also used with very little understanding of their nature in [5] to prove every countable set of Turing degrees has a minimal upper bound. Recently, Friedman [6] made use of Δ_2^1 pointed P's to show that every countable set of Δ_2^1 degrees has a minimal upper bound. One of the reasons that the minimal upper bound question has been answered completely for Turing degrees and Δ_2^1 degrees, but not for hyperdegrees, is that the problem of showing certain intersections of hyperarithmetically pointed P's are hyperarithmetically pointed has no counterpart in any of the existing applications of recursively pointed or Δ_2^1 pointed P's. Sequences of hyperdegrees are surprisingly less tractable than sequences of Turing or Δ_2^1 degrees.

Section 2 is largely groundwork for Sections 3 and 4; it covers: perfect sets, a brief review of the definitions of admissible set and hyperdegree and of the expressibility of a Π_1^1 relation on reals restricted to an admissible set A as a Σ_1 relation on A, the use of local Cohen forcing to con-

[1] Simpson [22] needs 4.26(ii) to show: If V is a generic extension of L, then the upper semilattice of hyperdegrees of V does not include a cone of minimal covers.

[2] Deeper applications of recursive pointedness to show certain sequences of Turing degrees have definable upper bounds have been announced by Jockusch and Simpson [23].

struct uniformly hyperarithmetically pointed perfect sets, and an application of unbounded Levy forcing.

Section 3 is mainly concerned with the existence of minimal upper bounds via hyperarithmetically pointed forcing; it deals with: The forcing relation $\Vdash^A$ for an admissible set A that satisfies Σ_1 dependent choice but is otherwise arbitrary, the nature of $\Vdash^A$ when A is an $L(\alpha)$, the forcing relation $\Vvdash^A$, and the location of least upper bounds. The principal technical lemma of Section 3 is 3.11, which establishes the equivalence of genericity in the sense of $\Vdash^A$ with that of $\Vvdash^A$.

Section 4 is singlemindedly devoted to the construction of a minimal solution of $\alpha = \omega_1{}^X$; it defines: perfect conditions on K (where K is a certain type of collapsing map), K-uniform pointedness, and forcing with complex conditions of the form $(H, \mathscr{P})$, where H is a perfect condition on K and $\mathscr{P}$ is a term forced by H to denote a hyperarithmetically pointed perfect set. The principal technical lemmas of Section 4 are 4.13 and 4.21, each of which surmounts some of the obstacles encountered in iterated forcing with pointed perfect sets.

Section 5 hints at some further results and mentions some open questions; it touches on Σ_n admissibility, the use of constructibly pointed conditions to collapse uncountable regular cardinals to ω_1, and the relation between forcing and compactness arguments.

2. Machinery

2.1. *Pointed Perfect Sets*

Let $p, q, r, \ldots$ be sequence numbers that encode finite initial segments of characteristic functions of subsets of ω. $p, q, r, \ldots$ are called finite conditions. A set T satisfies p (in symbols $T \in p$) if p encodes a finite initial segment of the characteristic function of T. p is extended by q (in symbols $p \supset q$) if every T that satisfies q also satisfies p. p and q are incompatible (in symbols $p \mid q$) if neither extends the other. Suppose $\lambda ij \mid p_j{}^i$ is a function such that $p_j{}^i$ is a finite condition for all $i < \omega$ and $j < 2^i$. $\lambda ij \mid p_i{}^j$ encodes a perfect subset of 2^ω if p_{2j}^{i+1} and p_{2j+1}^{i+1} are incompatible extensions of $p_j{}^i$ for all $i < \omega$ and $j < 2^i$. The perfect set encoded by $\lambda ij \mid p_j{}^i$ is

$$\{T \mid (i)(Ej)(j < 2^i \;\&\; T \in p_j{}^i)\}.$$

Each perfect set P has a standard encoding readily defined by thinking of P as a tree with binary branching: The branches of the tree correspond to the members of P; the branch points correspond to the finite conditions in the range of the standard encoding of P. Note that the standard encoding of P is recursive (uniformly) in every encoding of P.

Let P, Q, R,... ambiguously denote perfect sets and their standard encodings. Thus "$T \in P$" means T belongs to the perfect set P, and "$P \leqslant_h T$" means the standard encoding of P is hyperarithmetic in T. $\leqslant_r$ is a reducibility relation on the reals if it is reflexive, transitive and satisfies (2.1.1) and (2.1.2):

(2.1.1) X recursive in $Y \to X \leqslant_r Y$,

(2.1.2) $X \leqslant_r Z \;\&\; Y \leqslant_r Z \to X \oplus Y \leqslant_r Z$.

($X \oplus Y$ is $\{2n \mid n \in X\} \cup \{2n+1 \mid n \in Y\}$, the recursive join of X and Y.)

Examples of reducibility relations are "recursive in," "hyperarithmetic in," and "constructible from." X and Y have the same r-degree (in symbols $X \equiv_r Y$) if $X \leqslant_r Y$ and $Y \leqslant_r X$.

P is said to be r-pointed if

$$(T)[T \in P \to P \leqslant_r T];$$

i.e., the standard encoding of P is r-reducible to every member of P.

The notion of pointedness was inspired by the Kleene–Post construction [7] of a Turing degree above those of the arithmetic sets and below that of the truth set for arithmetic. Their construction is a forerunner of forcing with recursively pointed P's whose standard encodings are arithmetic. It will follow from Propositions 2.2 and 2.3 that forcing with r-pointed perfect conditions imposes no upper bound on the r-degrees of generic reals.

Proposition 2.2. *If P is r-pointed, then the r-degrees of $\{T \mid T \in P\}$ are the same as those of $\{X \mid P \leqslant_r X\}$.*

Proof. Let $\lambda ij \mid p_j{}^i$ be the standard encoding of P. Fix an X such that $P \leqslant_r X$. Define h by

$$\begin{aligned} h0 &= 0, \\ h(i+1) &= 2(hi) \qquad \text{if } i \in X, \\ &= 1 + 2(hi) \qquad \text{if } i \notin X. \end{aligned}$$

Let T be such that $T \in p^i_{hi}$ for all i. Clearly, $T \leqslant_r P \oplus h$. Hence, $T \leqslant_r X$, since $P \oplus h \leqslant_r X$. On the other hand, $X \leqslant_r T$, since

$X \leqslant_r P \oplus T$ and $P \leqslant_r T$. Thus $X \equiv_r T$. In short, the natural homeomorphism of 2^ω onto P preserves all r-degrees greater than or equal to the r-degree of P.

PROPOSITION 2.3. *If P is r-pointed and $P \leqslant_r Y$, then there exists an r-pointed $Q \subset P$ such that $Q \equiv_r Y$.*

Proof. There exist functions f and g recursive in P such that for each $p \in P$: fp and gp are incompatible extensions of p in P. ($p \in P$ means p encodes an initial segment of some $T \in P$.) $q_j^{\,i}$ ($i < \omega$ & $j < 2^i$) is defined by induction on i.

$$
\begin{aligned}
& q_0^{\,0} \in P, \\
& q_{2j}^{i+1} = ffq_j^{\,i} && \text{if } \ i \in Y, \\
& \phantom{q_{2j}^{i+1}} = gfq_j^{\,i} && \text{if } \ i \notin Y, \\
& q_{2j+1}^{i+1} = fgq_j^{\,i} && \text{if } \ i \in Y, \\
& \phantom{q_{2j+1}^{i+1}} = ggq_j^{\,i} && \text{if } \ i \notin Y.
\end{aligned}
$$

Let Q be the perfect set encoded by $\lambda ij \mid q_j^{\,i}$. Clearly, $Q \subset P$ and $Q \leqslant_r Y$. Fix $T \in Q$ to see Q is r-pointed. Let hi be the unique j such that $T \in q_j^{\,i}$. A simple recursion on i defines $\lambda i \mid q_{hi}^{i}$ from T, f, g. Consequently Y is recursive in T, f, g because

$$i \in Y \leftrightarrow q_{h(i+1)}^{i+1} \in \{ffq_{hi}^{i}, fgq_{hi}^{i}\}.$$

$P \leqslant_r T$ since $T \in P$. Hence $Y \leqslant_r T$ and so $Q \leqslant_r T$. $Y \leqslant_r Q$ because T_Q, the leftmost path through Q, is recursive in Q. ($T_Q \in q_0^{\,i}$ for all i.)

2.4. *Admissible Sets and Hyperdegrees*

A transitive set A is said to be admissible (Platek [8], Barwise [9]) if it is closed under the operations of pairing and set union, and satisfies all instances of the Δ_0 separation and bounding schemes of *ZF*. (A formula of *ZF* is Δ_0 if all its quantifiers are restricted.)

A typical instance of Δ_0 separation is

$$(Eb)(x)[x \in b \leftrightarrow \mathscr{F}(x) \ \& \ x \in a],$$

where $a \in A$ and $\mathscr{F}(x)$ is a Δ_0 formula with parameters in A.

A typical instance of Δ_0 bounding is

$$(x)_{x \in a}(Ey)\, \mathscr{F}(x, y) \rightarrow (Eb)(x)_{x \in a}(Ey)_{y \in b}\, \mathscr{F}(x, y),$$

where $a \in A$ and $\mathscr{F}(x, y)$ is a Δ_0 formula with parameters in A.

The economy of the above definition sometimes obscures the fact that admissible sets satisfy Δ_1 separation and Σ_1 bounding. Consequently many familiar constructions can be carried out within an admissible set A. For example each wellordering in A is order isomorphic to an ordinal in A.

For the remainder of this paper every admissible A satisfies the axiom of infinity, i.e., $\omega \in A$. It follows that if X and Y are reals such that X is hyperarithmetic in Y and $Y \in A$, then $X \in A$. Thus it makes sense to speak of hyperdegrees contained in A.

A typical instance of Σ_1 dependent choice is

$$(x)(Ey)\,\mathscr{F}(x, y) \rightarrow (Ef)(n)\,\mathscr{F}(fn, f(n+1)),$$

where f denotes a function with domain ω and $\mathscr{F}(x, y)$ is a Δ_0 formula of ZF with parameters. Friedman has shown there exists an admissible A that does not satisfy Σ_1 dependent choice.

PROPOSITION 2.5. *Let A be an admissible set and let D be a $\Pi_1{}^1$ set of reals. Then $A \cap D$ is Σ_1 over A.*

Proof. Since $D(X)$ is $\Pi_1{}^1$ there must be an e that satisfies (2.5.1) and (2.5.2) for all X:

(2.5.1) $\{e\}^X$ (the eth binary relation partial recursive in X) is a linear ordering of ω.

(2.5.2) $D(X) \leftrightarrow \{e\}^X$ is a wellordering.

Consequently $D(X)$ & $X \in A$ holds if and only if

(2.5.3) $(E\delta)(Ef)[f$ is an order preserving map of $\{e\}^X$ onto the ordinal $\delta]$ holds in A.

2.6. *Admissible Ordinals*

For each ordinal α let $L(\alpha)$ be the set of all sets constructible in the sense of Gödel via ordinals less than α. α is admissible (Kripke [10], Platek [8]) if $L(\alpha)$ is an admissible set, or equivalently, if $L(\alpha)$ satisfies the Σ_1 replacement axiom of ZF (with parameters in $L(\alpha)$). From now on α always denotes an admissible ordinal. b is α-finite means $b \in L(\alpha)$.

2.7. $L(\alpha, T)$ *and* $\mathscr{L}(\alpha, \mathscr{T})$

As in 2.1, $T \subset \omega$. $L(\alpha, T)$ is the set of all sets constructible from T via ordinals less than α. The language $\mathscr{L}(a, \mathscr{T})$ is the syntactical counterpart of the structure $L(\alpha, T)$. The primitive symbols of $\mathscr{L}(\alpha, \mathscr{T})$ are: ϵ; unranked set variables $x, y, z, \ldots$; ranked set variables x^β, y^β, $z^\beta, \ldots$, for each $\beta < \alpha$; existential quantifiers for all variables; and the propositional connectives & (conjunction) and $\sim$ (negation). A set $\mathscr{C}(\alpha)$ of constants intended to name the members of $L(\alpha, T)$ is defined by recursion on $\beta < \alpha$.

$$\mathscr{C}(0) \text{ is } \{\bar{n} \mid n < \omega\}.$$

$\mathscr{T}$ belongs to $\mathscr{C}(\beta + 1)$. Every other member of $\mathscr{C}(\beta + 1)$ is of the form $\hat{x}^\beta \mathscr{F}(x^\beta)$, where $\mathscr{F}(x^\beta)$ is a formula built up from the primitive symbols $\mathscr{T}$, and the constants in $\bigcup \{\mathscr{C}(\gamma) \mid \gamma \leqslant \beta\}$, and such that all quantified variables in $\mathscr{F}(x^\beta)$ are of rank at most β.

$\mathscr{C}(\lambda)$ is $\bigcup \{\mathscr{C}(\beta) \mid \beta < \lambda\}$ when λ is a limit ordinal less than or equal to α.

A sentence $\mathscr{G}$ is ranked if all its variables are ranked. The *ordinal rank* of a ranked $\mathscr{G}$ is the greatest member of $\{\beta \mid (Ex^\beta) \text{ occurs in } \mathscr{G}\} \cup \{\beta + 1 \mid \hat{x}^\beta$ occurs in $\mathscr{G}\}$. If $t \in \mathscr{C}(\alpha)$, then the ordinal rank of t is the ordinal rank of $\bar{0} \in t$.

The atomic formulas are of the form $t_1 \in t_2$, where t_1 and t_2 are variables or members of $\mathscr{C}(\alpha)$.

The following predicates can be defined simultaneously by recursion on $\beta < \alpha$: $c \in \mathscr{C}(\beta)$ and $b \in L(\beta, T)$ and c denotes b; the ordinal rank of $\mathscr{G}$ is at most β and $\mathscr{G}$ is true in $L(\beta, T)$. Simply keep in mind that x^β ranges over $L(\beta, T)$ and that the members of $L(\beta, T)$ are named by the constants of $\mathscr{C}(\beta)$. Of course x ranges over $L(\alpha, T)$ and $\mathscr{T}$ is interpreted as T.

2.8. *Local Cohen Forcing*

Assume $P \in L(\alpha)$. $\Vdash_\alpha^P$ debotes the Cohen forcing relation over $L(\alpha)$ localized at P. It is useful for constructung perfect subsets of a given subset as in 2.10 below and [4, p. 350], and it has roots in Spector [11, p. 588]. Let $p, q, r, \ldots$ be finite conditions on T and $\mathscr{F}$ a sentence of $\mathscr{L}(\alpha, \mathscr{T})$. $p \Vdash_\alpha^P \mathscr{F}$ is defined as is the ordinary Cohen forcing relation over $L(\alpha)$ except that all finite conditions must belong to P. For example:

$$p \Vdash_\alpha^P \bar{n} \in \mathscr{T} \quad \text{iff} \quad p \in P \text{ and } (T)(T \in p \rightarrow n \in T).$$

$$p \Vdash_\alpha^P \sim \mathscr{F} \quad \text{iff} \quad p \in P \text{ and } (q)_{p \supset q \in P} \sim [q \Vdash_\alpha^P \mathscr{F}].$$

It follows that any T generic in the sense of $\Vdash_\alpha^P$ must belong to P.

2.9. *Uniform Hyperarithmetic Pointedness*

P is hyperarithmetically pointed if

$$(T)[T \in P \to P \leqslant_h T],$$

where $\leqslant_h$ is the relation "hyperarithmetic in." A more tractable notion, needed for the forcing arguments of Section 3, is obtained by insisting on some uniformity. Recall that the reals hyperarithmetic in T are the same as the reals in $L(\omega_1^T, T)$, where ω_1^T is the least ordinal not recursive in T. For each $\delta < \omega_1^T$ let $\{\delta\}^T$ be the δth set in Gödel's standard wellordering of $L(\omega_1^T, T)$. His wellordering of $L(\omega_1^T, T)$ is uniform in T, since $\{\delta\}$ can be thought of as an instruction given by a ranked formula of $\mathscr{L}(\omega_1^T, \mathscr{T})$ to be applied to an arbitrary T. Thus $\{\delta\}^T$ might be

$$\{x \mid L(\beta, T) \models \mathscr{F}(x)\},$$

where $\mathscr{F}(x)$ is a formula of $\mathscr{L}(\omega_1^T, \mathscr{T})$ of rank $\beta < \omega_1^T$.

P is uniformly hyperarithmetically pointed if

$$(E\delta)(T)[T \in P \to P = \{\delta\}^T \;\&\; \delta < \omega_1^T]. \tag{2.9.1}$$

In other words, P is hyperarithmetically recoverable from every T via a procedure independent of T. Note that δ must be less than ω_1^P, since T_P (the leftmost branch of P) is recursive in P; it follows that (2.9.1) expresses a Π_1^1 property of P. The next lemmas shows that the notion of uniform pointedness is not as restrictive as it first appears to be.

LEMMA 2.10. *Suppose P is hyperarithmetically pointed. Then there exists a uniformly hyperarithmetically pointed $Q \subset P$ such that $Q \equiv_h P$.*

Proof. Let $K(b, T)$ be

$$T \in P \to b \in O^T \;\&\; P \text{ is recursive in } H_b^T.$$

Clearly $(T)(Eb)K(b, T)$ since P is hyperarithmetically pointed. $K(b, T)$ is Π_1^1 with parameter P, hence by Kreisel [12] there is a function b_T, hyperarithmetic in P, such that

$$(T)[b_T \in O^T \;\&\; K(b_T, T)]. \tag{2.10.1}$$

Since O^P is not Σ_1^1 in P, it follows from (2.10.1) that there is a $\gamma_0 < \omega_1^P$ such that $|\, b_T \,| < \gamma_0$ for all T. In other words,

$$(T)[T \in P \to P \in L(\gamma_0, T)]. \tag{2.10.2}$$

Let $\mathscr{L}(\omega_1^P, \mathscr{T}, \mathscr{P})$ be the language that results from adding a new set constant $\mathscr{P}$ to the language $\mathscr{L}(\omega_1^P, \mathscr{T})$. The constants of $\mathscr{L}(\omega_1^P, \mathscr{T}, \mathscr{P})$ name the sets constructible from $T \oplus P$ via ordinals less than ω_1^P. Let $\alpha = \omega_1^P$. The local Cohen forcing relation of 2.8, $\Vdash_\alpha^P$, is easily extended to all sentences of $\mathscr{L}(\alpha, \mathscr{T}, \mathscr{P})$ by interpreting $\mathscr{P}$ as P. Thus, for all $q \in P$,

$$q \Vdash_\alpha^P \bar{n} \in \mathscr{P} \quad \text{iff} \quad n \in P,$$

where "$n \in P$" means n belongs to the standard encoding of P by a real. Let T_0 be generic in the sense of $\Vdash_\alpha^P$ with respect to all sentences of $\mathscr{L}(\alpha, \mathscr{T}, \mathscr{P})$. Since $T_0 \in P$, it follows from (2.10.2) that $P = \{\delta\}^{T_0}$ for some $\delta < \gamma_0$. Since T_0 is generic, there is a p_0 satisfied by T_0 such that

$$p_0 \Vdash_\alpha^P \mathscr{P} = \{\delta\}^{\mathscr{T}}. \tag{2.10.3}$$

The definition of $\lambda ij \mid q_j^i$, an encoding of the desired Q, is by induction on i.

$$q_0^0 = p_0 .$$

q_{2j}^{i+1} and q_{2j+1}^{i+1} are incompatible extensions of q_j^i, and for each $m < 2$, either $q_{2j+m}^{i+1} \Vdash_\alpha^P \mathscr{F}_i$ or $q_{2j+m}^{i+1} \Vdash_\alpha^P \sim \mathscr{F}_i$, where $\{\mathscr{F}_i \mid i < \omega\}$ is an enumeration of all sentences of $\mathscr{L}(\omega_1^P, \mathscr{T}, \mathscr{P})$ of rank $\leqslant \delta + 1$.

Since $\delta < \omega_1^P$, the restriction of $\Vdash_\alpha^P$ to $\{\mathscr{F}_i \mid i < \omega\}$ is hyperarithmetic in P. It follows that Q can be defined hyperarithmetically from P. Every q_j^i belongs to P, so $Q \subset P$. Every $T \in Q$ satisfies p_0 and is generic with respect to all sentences of rank $\leqslant \delta + 1$, hence (2.10.3) implies $P = \{\delta\}^T$ for all $T \in Q$. Consequently Q is uniformly hyperarithmetically pointed. P is hyperarithmetic in the leftmost branch of Q, and hence in Q.

LEMMA 2.11. *If P is hyperarithmetically pointed and $P \leqslant_h Y$, then there exists a uniformly hyperarithmetically pointed $Q \subset P$ such that $Q \equiv_h Y$.*

Proof. First 2.3 then 2.10.

PROPOSITION 2.12. *For each $X \in L(\alpha) \cap 2^\omega$ there is a $Y \in L(\alpha) \cap 2^\omega$ such that X is recursive in Y and $Y \in L(\omega_1^Y)$.*

Proof. Suppose $X \in L(\delta + 1) - L(\delta)$. According to Boolos and Putnam [13] there is a $Z \in L(\delta + 1) - L(\delta)$ such that $\delta < \omega_1^Z$. Let $Y = X \oplus Z$.

2.13. $L[\alpha, K]$ *and* $\mathscr{L}[\alpha, \mathscr{K}]$

From now to 4.1, K is always a function from $(\alpha - \omega) \times \omega$ into α such that for each $\beta \in (\alpha - \omega)$, $\lambda m \mid K(\beta, m)$ is a one–one map of ω onto β. $L[\alpha, K]$ is defined by recursion on $\beta \leqslant \alpha$

$$L[0, K] = \omega, \qquad K^{\beta} = K \cap ((\beta - \omega) \times \omega \times \beta).$$

$L[\beta + 1, K]$ is the set of all subsets of $L[\beta, K]$ first order definable over $L[\beta, K]$ with parameters in $L[\beta, K] \cup \{K^{\beta}\}$.

$$L[\lambda, K] = \bigcup \{L[\beta, K] \mid \beta < \lambda\}, \qquad \text{for every limit } \lambda \leqslant \alpha.$$

$L[\alpha, K]$ will be useful in Sections 3 and 4 solely because every member of $L[\alpha, K]$ is countable in $L[\alpha, K]$. If α is admissible and K is generic in some reasonable sense, then $L[\alpha, K]$ is an admissible set. One reasonable sense is defined in the next subsection and another, perhaps less reasonable, in Section 4.

The language $\mathscr{L}[\alpha, \mathscr{K}]$ is developed by adding constants to the language of ZF (as in Subsection 2.7) so that all the members of $L[\alpha, K]$ have names; in particular K^{β} is named by $\mathscr{K}^{\beta}$. $\mathscr{L}[\alpha, \mathscr{K}]$ also includes a two-place function symbol $\mathscr{K}$.

2.14. *Unbounded Levy Forcing*

Let $s, t, u \ldots$ denote finite conditions on K. s is a finite condition with domain η and of length l if:

(2.14.1) η is a finite subset of $\alpha - \omega$.

(2.14.2) l is a finite initial segment of ω.

(2.14.3) s maps $\eta \times l$ into α.

(2.14.4) For each $\beta \in \eta$, $\lambda m \mid s(\beta, m)$ is a one–one map of l into β.

s is extended by t (in symbols $s \geqslant t$) if the graph of s is contained in the graph of t. K satisfies s (in symbols $K \in s$) if the graph of s is contained in the graph of K. The Levyesque forcing relation $s \Vdash_{\alpha} \mathscr{F}$, where s is a finite condition on K and $\mathscr{F}$ is a sentence of $\mathscr{L}[\alpha, K]$, is defined in the standard manner. For example:

$$s \Vdash_{\alpha} \mathscr{K}(\bar{\beta}, \bar{m}) = \bar{\delta} \qquad \text{iff} \quad s(\beta, m) = \delta,$$

$$s \Vdash_{\alpha} \sim \mathscr{G} \qquad \text{iff} \quad (t)_{s \geqslant t} \sim [t \Vdash_{\alpha} \mathscr{G}].$$

K is *generic* if for each $\mathscr{F}$ there is an s such that $K \in s$ and either $s \Vdash_\alpha \mathscr{F}$ or $s \Vdash_\alpha \sim\mathscr{F}$.

Suppose s has domain η and length l. Then $s^{<\beta}$ denotes the restriction of s to $\eta \cap \beta$. Clearly $s^{<\beta} \geqslant s$.

The notion of ordinal rank for sentences of $\mathscr{L}[\alpha, \mathscr{K}]$ is essentially the same as that for $\mathscr{L}(\alpha, \mathscr{T})$ in Subsection 2.7. Let it suffice to say that a sentence of ordinal rank at most β is interpretable in $L[\beta, K]$, and that its interpretation requires values of $K(\gamma, m)$ only for $\gamma < \beta$.

PROPOSITION 2.15. *Suppose the ordinal rank of* $\mathscr{F}$ *is at most* β. *If* $s \Vdash_\alpha \mathscr{F}$, *then* $s^{<\beta} \Vdash_\alpha \mathscr{F}$.

Proof. By induction on the ordinal rank and logical complexity of $\mathscr{F}$. The most interesting case occurs when $\mathscr{F}$ is of the form $\sim\mathscr{G}$. It is enough to check there is no $t \leqslant s^{<\beta}$ such that $t \Vdash_\alpha \mathscr{G}$. If there were such a t, then it would follow by induction that $t^{<\beta} \Vdash_\alpha \mathscr{G}$, and consequently that $t^{<\beta} \cup (s - s^{<\beta}) \Vdash_\alpha \mathscr{G}$, an impossibility since $s \geqslant t^{<\beta} \cup (s - s^{<\beta})$.

A formula of $\mathscr{L}[\alpha, \mathscr{K}]$ is Σ_1 if it is ranked or of the form $(Ex)\mathscr{G}$, where $\mathscr{G}$ is ranked.

PROPOSITION 2.16. *The forcing relation* $s \Vdash_\alpha \mathscr{F}$, *restricted to* Σ_1 $\mathscr{F}$'s, *is* Σ_1 *over* L_α.

Proof. It suffices to consider ranked $\mathscr{F}$'s only. Then, the definition of $\Vdash_\alpha$ proceeds by transfinite recursion on the ordinal rank and logical complexity of $\mathscr{F}$. The recursion establishes that the forcing relation $s \Vdash_\alpha \mathscr{F}$, restricted to $\mathscr{F}$'s of rank at most β, is Δ_1 over $\mathscr{L}(\alpha)$ uniformly in β. Proposition 2.15 is needed to reduce unbounded universal quantification over finite conditions to bounded quantification. For example, suppose $\mathscr{F}$ is $\sim\mathscr{G}$ and is of ordinal rank β. Then, by 2.15, $s \Vdash_\alpha \mathscr{F}$ iff

$$(t)[t = t^{<\beta} \;\&\; s^{<\beta} \geqslant t^{<\beta} \rightarrow \sim(t \Vdash_\alpha \mathscr{G})].$$

LEMMA 2.17. *Suppose* $\mathscr{F}(x, y)$ *is a formula of* $\mathscr{L}[\alpha, \mathscr{K}]$ *with no unranked quantifiers. If* $s \Vdash_\alpha (x)(Ey)\mathscr{F}(x, y)$, *then* $s \Vdash_\alpha (x^\rho)(Ey^\rho)\,\mathscr{F}(x^\rho, y^\rho)$ *for some* $\rho < \alpha$.

Proof. By 2.16, there exist functions h and γ Σ_1 over $L(\alpha)$ such that for each constant $c \in \mathscr{C}(\alpha)$ and $t \leqslant s$:

$$h(c, t) \leqslant t \;\&\; h(c, t) \Vdash_\alpha (Ey^{\gamma(c,t)})\,\mathscr{F}(c, y^{\gamma(c,t)}).$$

Since α is admissible there must be a $\rho < \alpha$ such that:

$$s = s^{<\rho}, \tag{2.16.1}$$

$$\text{if } c \in \mathscr{C}(\rho) \text{ and } t^{<\rho} = t \leqslant s, \quad \text{then} \quad h(c, t) = h(c, t)^{<\rho} \text{ and } \gamma(c, t) < \rho. \tag{2.16.2}$$

Fix $c \in \mathscr{C}(\rho)$ and $t \leqslant s$ to see that there is a $w \leqslant t$ such that $w \Vdash_\alpha (Ey^\rho)\mathscr{F}(c, y^\rho)$. By (2.16.1), $t^{<\rho} \leqslant s$, hence by (2.16.2)

$$h(c, t^{<\rho}) \Vdash_\alpha (Ey^\rho)\,\mathscr{F}(c, y^\rho).$$

Take w to be $h(c, t^{<\rho}) \cup (t - t^{<\rho})$.

LEMMA 2.18. *Suppose K is generic in the sense of $\Vdash_\alpha$ (2.14). Then, $L[\alpha, K]$ is an admissible set that satisfies Σ_1 dependent choice, and α is the least ordinal not countable in $L[\alpha, K]$.*

Proof. Suppose $L[\alpha, K]$ satisfies $(x)(Ey)\mathscr{F}(x, y)$, where $\mathscr{F}(x, y)$ is a formula of $\mathscr{L}[\alpha, \mathscr{K}]$ with no unranked quantifiers. It follows from 2.17 that $L[\rho, K]$ satisfies $(x^\rho)(Ey^\rho)\mathscr{F}(x^\rho, y^\rho)$ for some $\rho < \alpha$. Now $L[\rho, K]$ is a member of $L[\alpha, K]$, hence countable in $L[\alpha, K]$. Thus there is a function $f: \omega \to L[\rho, K]$ such that $f \in L[\alpha, K]$ and $L[\rho, K]$ satisfies $(n)\mathscr{F}(fn, f(n+1))$.

3. MINIMAL UPPER BOUNDS

3.1 *Hyperarithmetically Pointed Forcing*

Throughout Section 3, A is a countable admissible set that satisfies the Σ_1 dependent choice schema of 2.4. The ostensible purpose of Section 3 is to show that the set of all hyperdegrees in A has a minimal upper bound. Its real purpose, perhaps confessed too soon, is to demonstrate that certain perfect set constructions, readily performed when $A = L(\alpha)$ with the aid of a natural forcing relation $\Vdash^A$, can still be performed, but less gracefully, when $A \neq L(\alpha)$ by means of an unnatural forcing relation $\Vvdash^A$ defined in Subsection 3.9.

Let P, Q, R,... be uniformly hyperarithmetically pointed perfect sets as defined in 2.9. Throughout Section 3, all perfect sets belong to A; i.e., their standard encodings are reals in A. Let α be the least ordinal not countable in A. Clearly α is admissible. The forcing relation

$P \Vdash^A \mathscr{F}$, where $\mathscr{F}$ is a sentence of the language $\mathscr{L}(\alpha, \mathscr{T})$ defined in 2.7, is defined by recursion on the logical complexity of $\mathscr{F}$.

(3.1.1) $\mathscr{F}$ is ranked. $P \Vdash^A \mathscr{F}$ iff the ordinal rank of $\mathscr{F}$ is less than $\omega_1{}^P$ and

$$(T)[T \in P \rightarrow L(\alpha, T) \models \mathscr{F}].$$

(3.1.2) $\mathscr{F}$ & $\mathscr{G}$ is not ranked. $P \Vdash^A \mathscr{F}$ & $\mathscr{G}$ iff $P \Vdash^A \mathscr{F}$ and $P \Vdash^A \mathscr{G}$.

(3.1.3) $(Ex^\beta)\mathscr{F}(x^\beta)$ is not ranked. $P \Vdash^A (Ex^\beta)\mathscr{F}(x^\beta)$ iff $P \Vdash^A \mathscr{F}(c)$ for some $c \in \mathscr{C}(\beta)$.

(3.1.4) $P \Vdash^A (Ex)\mathscr{F}(x)$ iff $P \Vdash^A (Ex^\beta)\mathscr{F}(x^\beta)$ for some $\beta < \alpha$.

(3.1.5) $\mathscr{G}$ is not ranked. $P \Vdash^A \sim\mathscr{G}$ iff $(Q)_{P \supset Q} \sim [Q \Vdash^A \mathscr{G}]$.

PROPOSITION 3.2. *The forcing relation $P \Vdash^A \mathscr{F}$, restricted to Σ_1 $\mathscr{F}$'s, is Σ_1 over A.*

Proof. Clause (3.1.4) implies that only ranked $\mathscr{F}$'s need be considered. Let $\mathscr{F}$ be any ranked sentence, and let $D_{\mathscr{F}}$ be the set of *all* uniformly hyperarithmetically pointed Z's (not just those in A) such that the ordinal rank of $\mathscr{F}$ is less than $\omega_1{}^Z$ and

$$(T)[T \in Z \rightarrow L(\alpha, T) \models \mathscr{F}].$$

$D_{\mathscr{F}}$ is Π_1^1 uniformly in any Y such that $\omega_1{}^Y$ exceeds the ordinal rank of $\mathscr{F}$. Now apply 2.5 relativized to a suitable $Y \in A$. Then $D_{\mathscr{F}}$ is Σ_1 over A uniformly in $\mathscr{F}$ and Y.

LEMMA 3.3. *For each $\mathscr{F}$ and P, there is a $Q \subset P$ such that $Q \Vdash^A \mathscr{F}$* or *$Q \Vdash^A \sim\mathscr{F}$.*

Proof. Clause (3.1.5) makes it safe to assume $\mathscr{F}$ is ranked; suppose its rank is δ. Since $\delta < \alpha$, and α is the least ordinal not countable in A, there is an $X \in A \cap 2^\omega$ such that both P and δ are recursive in X. By 2.11, there is an $R \subset P$ such that $R \equiv_h X$. The virtue of R resides in the fact that $\delta < \omega_1{}^R$. It follows that the local forcing relation of 2.8,

$$p \Vdash_\alpha^R \mathscr{G},$$

restricted to $\mathscr{G}$'s of rank at most δ, is hyperarithmetic in R.

Define $\lambda ij \mid q_j^i$, an encoding of the desired Q, so that

$$q_0^0 \Vdash_\alpha^R \mathscr{F}, \qquad \text{or} \qquad q_0^0 \Vdash_\alpha^R \sim\mathscr{F}.$$

q_{2j}^{i+1} and q_{2j+1}^{i+1} are incompatible extensions of q_j^i.

For each $m < 2$, either $q_{2j+m}^{i+1} \Vdash_\alpha^R \mathscr{G}_i$ or $q_{2j+m}^{i+1} \Vdash_\alpha^R \sim\mathscr{G}_i$, where $\{\mathscr{G}_i \mid i < \omega\}$ is an enumeration of all sentences of $\mathscr{L}(\alpha, \mathscr{T})$ of rank at most δ.

Since $\delta < \omega_1^R$, the enumeration $\{\mathscr{G}_i\}$ can be taken to be hyperarithmetic in R, and consequently Q can be defined hyperarithmetically from R. $Q \subset R$ since every q_j^i belongs to R. It follows that Q is uniformly hyperarithmetically pointed because R is. Every $T \in Q$ satisfies q_0^0 and is generic in the sense of $\Vdash_\alpha^R$ with respect to all sentences of rank at most δ. Consequently $Q \Vdash^A \mathscr{F}$ or $Q \Vdash^A \sim\mathscr{F}$.

Let $\{\delta\}^T$ be defined as in Subsection 2.9.

Lemma 3.4. *Suppose $\delta < \alpha$. For each P there is a $Q \subset P$ satisfying* (3.4.1) *or* (3.4.2):

$$(EX)_{X \leqslant_h Q}(T)_{T \in Q}[\{\delta\}^T = X]. \tag{3.4.1}$$

$$(T)_{T \in Q}[T \leqslant_h \{\delta\}^T, Q]. \tag{3.4.2}$$

Proof. The idea, as in Spector [11], is to find a Q such that the restriction of $\{\delta\}$ to Q is continuous and either constant or one-one. As in the proof of 3.3, P can be contracted to an R such that $\delta \leqslant \omega_1^R$ and the local forcing relation $p \Vdash_\alpha^R \mathscr{G}$, restricted to $\mathscr{G}$'s of rank at most δ, is hyperarithmetic in R.

Case 1. Assume there is a $p \in R$ such that

$$(q)_{p \supset q}(r)_{p \supset r}(n) \sim [q \Vdash_\alpha^R \bar{n} \in \{\delta\}^{\mathscr{T}} \text{ and } r \Vdash_\alpha^R \bar{n} \notin \{\delta\}^{\mathscr{T}}].$$

Let $q_0^0 = p$ and define $\lambda ij \mid q_j^i$, an encoding of Q, as in the proof of 3.3. Let X be

$$\{n \mid (Eq)[q \in Q \;\&\; q \Vdash_\alpha^R \bar{n} \in \{\delta\}^{\mathscr{T}}]\}.$$

Then $\{\delta\}^T = X$ for all $T \in Q$, since otherwise, there would be a q and an r in Q, and an n, such that $p \supset q$, $p \supset r$,

$$q \Vdash_\alpha^R \bar{n} \in \{\delta\}^{\mathscr{T}} \qquad \text{and} \qquad r \Vdash_\alpha^R \bar{n} \notin \{\delta\}^{\mathscr{T}}.$$

As in 3.3, $Q \subset R$, $Q \leqslant_h R$, and Q is uniformly hyperarithmetically pointed. $R \leqslant_h Q$ because the leftmost branch of Q is recursive in Q and belongs to R, and R is hyperarithmetically pointed. It follows that $X \leqslant_h Q$.

Case 2. There is no p that satisfies the assumption of Case 1. Thus, for each $p \in R$, there exist q, r and n such that

$$p \supset q, \quad p \supset r, \quad q \Vdash_{\alpha}^{R} \bar{n} \in \{\delta\}^{\mathscr{T}}, \quad \text{and} \quad r \Vdash_{\alpha}^{R} \bar{n} \notin \{\delta\}^{\mathscr{T}}.$$

Define $\lambda ij \mid q_j{}^i$, an encoding of Q, and $\lambda ij \mid n_j{}^i$ hyperarithmetically from R so that: For each $m < 2$, $q_{2j+m}^{i+1} \subset q_j{}^i$, and either $q_{2j+m}^{i+1} \Vdash_{\alpha}^{R} \mathscr{G}_i$ or $q_{2j+m}^{i} \Vdash_{\alpha}^{R} \sim\mathscr{G}_i$, where $\{\mathscr{G}_i\}$ is an enumeration of all sentences of rank at most δ; also

$$q_{2j}^{i+1} \Vdash_{\alpha}^{R} \bar{n}_j{}^i \in \{\delta\}^{\mathscr{T}} \qquad \text{and} \qquad q_{2j+1}^{i+1} \Vdash \bar{n}_j{}^i \notin \{\delta\}^{\mathscr{T}}.$$

It follows that $\{\delta\}$ is one–one on Q since the results of applying $\{\delta\}$ to any two distinct members of Q will differ on some $n_j{}^i$. Fix $T \in Q$ to see that the one–one-ness of $\{\delta\}$ on Q implies $T \leqslant_h \{\delta\}^T, Q$. There is a unique $t\colon \omega \to \omega$ such that $t0 = 0$ and:

$$t(i+1) = 2(ti) \qquad \text{if} \quad n_{ti}^{i} \in \{\delta\}^T,$$

$$t(i+1) = 2(ti) + 1 \qquad \text{if} \quad n_{ti}^{i} \notin \{\delta\}^T.$$

Clearly $T \in q_j{}^i$ iff $j = ti$, hence $T \leqslant_h t, Q$. Now $t \leqslant_h \lambda ij \mid n_j{}^i, \{\delta\}^T$; and $\lambda_{ij} \mid n_j{}^i \leqslant_h R$. Thus $t \leqslant_h Q, \{\delta\}^T$, since $R \leqslant_h Q$ as was shown in Case 1.

3.5. $\Vdash^A$ *Genericity*

A set D is said to be A-definable if there is a formula $\mathscr{F}(x)$ of ZF (with parameters in A) such that

$$D = \{b \mid b \in A \text{ and } A \models \mathscr{F}(\underset{\sim}{b})\},$$

where $\underset{\sim}{b}$ is a constant that denotes b. Suppose D is a set of P's; D is said to be *dense* if

$$(\underset{\sim}{Q})(EP)[P \in D \;\&\; \underset{\sim}{Q} \supset P].$$

T is $\Vdash^A$ generic if $T \in \bigcup \{P \mid P \in D\}$ for every dense A-definable D.

Proposition 3.6. *If T is $\Vdash^A$ generic, then*:

(3.6.1) *$X \leqslant_h T$ for every $X \in A \cap 2^\omega$;*

(3.6.2) *for each $\delta < \alpha$, there exists an $X \in A \cap 2^\omega$ such that either $\{\delta\}^T = X$ or $T \leqslant_h \{\delta\}^T, X$.*

Proof. Fix $X \in A \cap 2^\omega$. The set of all Q such that $X \leqslant_h Q$ is dense by 2.11 and A-definable by 2.5. Hence, there is a Q such that $T \in Q$ and $X \leqslant_h Q$. Then $Q \leqslant_h T$ since Q is hyperarithmetically pointed.

Let D be the set of all Q such that Q satisfies either (3.4.1) or (3.4.2). D is dense by 3.4 and A-definable by 2.5. Hence there is a Q such that either $\{\delta\}^T \in A \cap 2^\omega$ or $T \leqslant_h \{\delta\}^T, Q$.

3.7. *Δ_0 Bounding*

Assume T is $\Vdash^A$ generic. Theorem 3.13, the principal result of Section 3, will follow from 3.6 after it is seen that $\omega_1{}^T \leqslant \alpha$. To limit $\omega_1{}^T$ by α it is more than enough to show $L(\alpha, T)$ satisfies the Δ_0 bounding principle (BP):

$$(n)(Ex)\,\mathscr{F}(\bar{n}, x) \rightarrow (E\delta)_{\delta<\alpha}(n)(Ex^\delta)\,\mathscr{F}(\bar{n}, x^\delta), \tag{3.7.1}$$

where $\mathscr{F}(\bar{n}, x)$ is a formula of $\mathscr{L}(\alpha, \mathscr{T})$ with no unranked quantified variables. (BP) is an immediate consequence of its generic counterpart (GBP): If $P \Vdash^A (n)(Ex)\mathscr{F}(\bar{n}, x)$, then there is a $Q \subset P$ such that $Q \Vdash^A (n)(Ex^\delta)\mathscr{F}(\bar{n}, x^\delta)$ for some $\delta < \alpha$. The proof of GBP is based on some rather obscure manipulations in Subsection 3.11, which become considerably simpler when $A = L(\alpha)$. The problem of proving GBP reduces to the problem mentioned in Section 1 of what steps to take in the construction of a contracting sequence of pointed perfect sets so that its intersection will be pointed.

3.8 *When $A = L(\alpha)$.*

Assume $P \Vdash^A (n)(Ex)\mathscr{F}(\bar{n}, x)$ to find a Q that satisfies GBP. Thus,

$$(n)(Q)_{P\supset Q}(ER)_{Q\supset R}(E\delta)_{\delta<\alpha}[R \Vdash^A (Ex^\delta)\,\mathscr{F}(\bar{n}, x^\delta)].$$

It follows from 3.2 (and the fact that Σ_1 dependent choice holds in $L(\alpha)$) that there exist functions $\lambda n \mid P_n$ and $\lambda n \mid \delta^n$ Σ_1 over $L(\alpha)$ such that:

$$P_0 = P \quad \text{and} \quad P_n \supset P_{n+1}, \tag{3.8.1}$$

$$P_{n+1} \Vdash^A (Ex^{\delta^n})\,\mathscr{F}(\bar{n}, x^{\delta^n}). \tag{3.8.2}$$

Let $Q = \bigcap \{P_n \mid n < \omega\}$ and $\delta = \sup\{\delta^n \mid n < \omega\}$. Clearly Q is a closed subset of P. If Q were perfect and uniformly hyperarithmetically pointed, then it would be the case that $Q \Vdash^A (n)(Ex^\delta)\mathscr{F}(\bar{n}, x^\delta)$.

Now it is routine to arrange that Q be perfect by means of the *splitting trick* of [4, 16]. For each n, $P_n = \bigcup\{P_n^{\,j} \mid j < 2^n\}$, and P_{n+1}^{2j} and P_{n+1}^{2j+1} are disjoint perfect subsets of $P_n^{\,j}$. The splitting trick raises no dust since any perfect set can be split effectively into two disjoint perfect sets by a finite condition.

Thus the only substantial problem that arises in the construction of Q is how to ensure that Q be hyperarithmetically pointed. (2.10 eliminates any concern about whether or not Q is uniformly pointed.) The solution consists of requiring that Q be *absolutely hyperarithmetically pointed*

$$(X)[X \in Q \rightarrow Q \in L(\omega_1^{X})].$$

Thus the hyperarithmetic recovery of Q from $X \in Q$ does not utilize X itself but only some ordinal recursive in X. It follows from 2.11 and 2.12 that every hyperarithmetically pointed $P \in L(\alpha)$ can be contracted to an absolutely pointed $Q \in L(\alpha)$. The local forcing proof of 3.3 is easily adapted to show: If $\mathscr{F}$ is a sentence of $\mathscr{L}(\alpha, \mathscr{T})$ and P is absolutely pointed, then there is an absolutely pointed $Q \subset P$ such that $Q \Vdash^A \mathscr{F}$ or $Q \Vdash^A \sim\mathscr{F}$.

Now the functions $\lambda n \mid P_n$ and $\lambda n \mid \delta^n$ can be made to satisfy some additional requirements:

(3.8.3) P_n is absolutely hyperarithmetically pointed;

(3.8.4) $\delta^n < \omega_1^{P_{n+1}}$ (satisfiable by 2.11 and 2.12);

(3.8.5) the choices of P_n and δ^n are always the least possible with respect to Gödel's standard Δ_1 wellordering of $L(\alpha)$.

Fix $X \in Q$ $(= \bigcap \{P_n \mid n < \omega\})$ to see that Q is absolutely hyperarithmetically pointed. Let

$$\gamma_0 = \sup\{\omega_1^{P_n} \mid n < \omega\}.$$

$\gamma_0 \leqslant \omega_1^{X}$, thanks to (3.8.3). The definition by recursion of $\lambda n \mid P_n$ and $\lambda n \mid \delta^n$ took place over $L(\alpha)$ but by virtue of (3.8.4) did not involve any ordinals greater than γ_0. Hence, the recursion can be visualized as taking place over $L(\omega_1^{X})$ rather than $L(\alpha)$. By (3.8.5) the results of the recursion over $L(\alpha)$ and $L(\omega_1^{X})$ are identical, since the standard Δ_1 well-orderings of $L(\alpha)$ and $L(\omega_1{}^X)$ are compatible. Thus $\lambda n \mid P_n$ and $\lambda n \mid \delta^n$ are Σ_1 over $L(\omega_1{}^X)$ as well as $L(\alpha)$. Since $\omega_1{}^X$ is admissible, $Q \in L(\omega_1{}^X)$.

3.9. *The Relation* $\Vdash\!\vdash^A$

The line of argument pursued in 3.8 succeeded only because the assumption $A = L(\alpha)$ made it possible to force with absolutely pointed P's. Then the recovery of Q $(= \bigcap \{P_n \mid n < \omega\})$ from $X \in Q$ could be accomplished by a manipulation of ordinals rather than reals. The recovery of Q when A is arbitrary is based on a uniqueness feature of $\Vdash\!\vdash^A$ expressed by 3.10, a feature that makes it possible for ordinals to do all the work in the recovery of Q even when $A \neq L(\alpha)$.

According to 3.2, there is a Δ_0 formula $\mathscr{B}(u, v, y)$ such that

$$P \Vdash^A \mathscr{F} \quad \text{iff} \quad A \models (Ey)\, \mathscr{B}(P, \mathscr{F}, y), \tag{3.9.1}$$

whenever $\mathscr{F}$ is a Σ_1 sentence of $\mathscr{L}(\alpha, \mathscr{T})$. The proof of 2.5 shows $\mathscr{B}(u, v, y)$ can be chosen so that it has no infinite parameters; thus it is absolute in nature. A useful interpretation of $\mathscr{B}(P, \mathscr{F}, y)$ is: y is a proof that $P \Vdash^A \mathscr{F}$. A more extended notion of proof will be needed in (3.9.8) below. The Δ_0 predicate, y proves $(P \Vdash^A \mathscr{F})$, is generated by four rules.

(3.9.2) If $\mathscr{B}(P, \mathscr{F}, y)$, then y proves $(P \Vdash^A \mathscr{F})$.

(3.9.3) If y proves $(P \Vdash^A \mathscr{F})$ and $P \supset Q$, then y proves $(Q \Vdash^A \mathscr{F})$.

(3.9.4) Suppose $Q = \bigcap \{Q_i \mid i < \omega\}$ and $\{\mathscr{F}_i \mid i < \omega\}$ is an α-finite sequence of Σ_1 sentences. If y_i proves $Q_i \Vdash^A \mathscr{F}_i$ for each $i < \omega$, then $\lambda i \mid y_i$ proves $(Q \Vdash^A \&_{i<\omega} \mathscr{F}_i)$.

(3.9.5) Suppose $Q \subset P$, $Q \Vdash^A \mathscr{F}$, and Q is constructed from P via the local forcing method of 3.3. If y encodes the construction, then y proves $(Q \Vdash^A \mathscr{F})$.

Define

$$(P, \delta) \Vdash^A \mathscr{F},$$

where $\mathscr{F}$ is Σ_1 and $\delta < \alpha$ as follows:

$$(X)[X \in P \rightarrow P = \{\delta\}^X] \text{ (uniform hyperarithmetic pointedness as in 2.9)}, \tag{3.9.6}$$

$$P \Vdash^A \mathscr{F}, \tag{3.9.7}$$

$$(Ey)[y \text{ proves } (P \Vdash^A \mathscr{F}) \,\&\, (X)(X \in P \rightarrow y \in L(\delta + 1, X))]. \tag{3.9.8}$$

Clause (3.9.8) will turn out to be less restrictive than it appears for two reasons: The construction of a P intended to force some Σ_1 $\mathscr{F}$ will always be encodable by a y that is hyperarithmetic in P, hence hyperarithmetic in $X \in P$; and the hierarchy $L(\omega_1^X, X)$ has an inherent repetiveness that can be exploited (as in 3.11) to juggle the relative locations of y and P.

The y of clause (3.9.8) includes a proof that P is uniformly hyperarithmetically pointed via some ordinal; that ordinal is always assumed to be the δ of clause (3.9.6).

It follows from (3.9.1)–(3.9.5) that there is a Δ_0 formula $\mathscr{B}^*(u, v, y)$ such that for all X and δ

$$\begin{aligned} &X \in \{\delta\}^X \,\&\, (\{\delta\}^X, \delta) \Vdash^A \mathscr{F} \\ &\qquad \text{iff} \quad L(\delta + 1, X) \models (Ey^{\delta+1})\, \mathscr{B}^*(\{\delta\}^X, \mathscr{F}, y^{\delta+1}). \end{aligned} \tag{3.9.9}$$

The formulas $\mathscr{B}_\gamma(u, v, y)$ and the minimal forcing relation

$$(P, \gamma) \Vdash\!\Vdash^A \mathscr{F}$$

are defined simultaneously by recursion on $\gamma < \alpha$ for all Σ_1 $\mathscr{F}$'s.

Fix γ. Assume for all X and $\delta < \gamma$:

$$\begin{aligned} &X \in \{\delta\}^X \,\&\, (\{\delta\}^X, \delta) \Vdash\!\Vdash^A \mathscr{F}, \\ &\qquad \text{iff} \quad L(\delta + 1, X) \models (Ey^{\delta+1})\, \mathscr{B}_\delta(\{\delta\}^X, \mathscr{F}, y^{\delta+1}), \end{aligned} \tag{3.9.10}$$

where $\mathscr{B}_\delta(u, v, y)$ is a formula all of whose parameters are ordinals less than δ, and all of whose quantifiers are of rank at most δ. Define $(P, \gamma) \Vdash\!\Vdash\!\Vdash^A \mathscr{F}$ by:

$$(P, \gamma) \Vdash^A \mathscr{F}, \tag{3.9.11}$$

$$(P, \gamma) \Vdash^A (\delta)_{\delta<\gamma} \sim (Ey^{\delta+1})\, \mathscr{B}_\delta(\{\delta\}^{\mathscr{T}}, \mathscr{F}, y^{\delta+1}). \tag{3.9.12}$$

It follows from (3.9.9) that there exists a formula $\mathscr{B}_\gamma(u, v, y)$ such that (3.9.10) remains true when δ is replaced by γ.

The proof of 3.11 will clarify the definition of $\Vdash\!\Vdash^A$. Proposition 3.10 singles out the property of $\Vdash\!\Vdash^A$ needed for the proof of 3.14.

PROPOSITION 3.10. *For each X and Σ_1 $\mathscr{F}$, there is at most one (P, γ) such that*

$$X \in P \,\&\, (P, \gamma) \Vdash\!\Vdash^A \mathscr{F}.$$

Proof. By (3.9.12) and (3.9.10).

LEMMA 3.11. *Suppose* $P \Vdash^{A} \mathscr{F}$, *where* $\mathscr{F}$ *is a* Σ_1 *sentence of* $\mathscr{L}(\alpha, \mathscr{T})$. *Then there exists a* $Q \subset P$ *and a* $\lambda_0 < \alpha$ *such that* $Q \equiv_h P$ *and* $(Q, \lambda_0) \Vdash^{A} \mathscr{F}$.

Proof. By (3.1.1) and (3.1.3) it is safe to assume that $\mathscr{F}$ is ranked. The requirement that Q be contained in P is of no real concern, since P can be thought of as forcing the ranked sentence $\mathscr{F}$ & $\mathscr{T} \in P$ and consequently Q can be required to force $\mathscr{F}$ & $\mathscr{T} \in P$ instead of $\mathscr{F}$.

The search for λ_0 is simplified by perturbing the standard wellordering of $L(\omega_1^P, X)$. Recall the description of $\{\delta\}^X$ given in 2.9. The standard wellordering is derived from an onto function $f: \omega_1^P \to L(\omega_1^P, X)$ whose graph is Σ_1 over $L(\omega_1^P, X)$ uniformly in X. $f\delta$ coincides with $\{\delta\}^X$. $L(\delta, X)$ is $\{\{\gamma\}^X \mid \gamma < \delta\}$, and $\{\delta\}^X$ is first-order definable over $L(\delta, X)$. It is not difficult to perturb f so that it retains the above properties and yet every $\{\delta\}$ is repeated unboundedly often at limit ordinal stages. To be precise: For each $\delta < \omega_1^P$, there is a limit ordinal β such that

$$\delta < \beta < \omega_1^P \,\&\, (X)[\{\delta\}^X = \{\beta\}^X]. \tag{3.11.1}$$

It follows from (3.11.1) and some further mild perturbations of f, that there exists a limit ordinal $\lambda < \omega_1^P$ that satisfies (3.11.2)–(3.11.5).

(3.11.2) If $\delta < \lambda$, then there exists an onto map $w: \omega \to \delta$ and a $\rho < \lambda$ such that $w = \{\rho\}^X$ for all $X \in P$. w and ρ are easy to find since every $\delta < \omega_1^P$ is recursive in P and P is uniformly hyperarithmetically pointed.

(3.11.3) If $\delta < \lambda$ and Z is first-order definable over $L(\delta, X)$, then there exists a $\rho < \lambda$ such that $Z = \{\rho\}^X$. Furthermore, ρ is independent of X, i.e., ρ depends only on δ and the definition of Z.

(3.11.4) There is a $\beta < \lambda$ such that $(P, \beta) \Vdash^{A} \mathscr{F}$. To find β, note that the rank of $\mathscr{F}$ is less than ω_1^P by (3.1.1). Consequently a proof y (in the sense of (3.9.2)) of $P \Vdash^{A} \mathscr{F}$ can be found in $L(\omega_1^P, P)$ by paying attention to the details of 2.5. Since P is uniformly hyperarithmetically pointed, there is a $\gamma < \omega_1^P$ such that $y \in L(\beta, X)$ for all $\beta \geqslant \gamma$ and all $X \in P$. By (3.11.1), there is a $\beta \geqslant \gamma$ such that $P = \{\beta\}^X$ for all $X \in P$.

(3.11.5) $\{\lambda\}^X$ is defined in two stages. The first stage yields the same onto function $h: \omega \to \lambda$ for all $X \in P$ as in (3.11.2) above. The second stage yields $\{\lambda\}^X$ itself. If $X \in P$, then $\{\lambda\}^X$ is the perfect set $R =$

$\bigcap \{R_n \mid n < \omega\}$. The functions R_n and γ_n are first-order definable over $L(\lambda, X)$ uniformly in $X \in P$. The dependence of R_n and γ_n on X is minor, since X is used only at the outset to define $R_0 = P$; subsequently R_{n+1} and γ_n are defined from R_n via the ordinals less than λ:

$$R_0 = P \qquad \text{and} \qquad R_n \supset R_{n+1}, \tag{3.11.5a}$$

$$(R_{n+1}, \gamma_n) \in L(\lambda, X), \tag{3.11.5b}$$

$$(R_{n+1}, \gamma_n) \Vdash^A \sim\mathscr{B}^n \qquad \text{or} \qquad (R_{n+1}, \gamma_n) \Vdash^A \mathscr{B}^n, \tag{3.11.5c}$$

where $\mathscr{B}^n$ is $(Ey^{hn+1})\mathscr{B}_{hn}(\{hn\}^{\mathscr{T}}, \mathscr{F}, y^{hn+1})$, h is as in (3.11.5), and $\mathscr{B}_\delta$ is as in (3.9.10).

The local forcing argument of 3.3 is applied to satisfy (3.11.5abc). Suppose R_n is given. Then, as in the proof of 3.3, there is an $R_{n+1} \subset R_n$ such that

$$R_{n+1} \Vdash^A \sim\mathscr{B}^n \qquad \text{or} \qquad R_{n+1} \Vdash^A \mathscr{B}^n. \tag{3.11.6}$$

The nature of local forcing allows R_{n+1} to be first-order definable over any $L(\delta, X)$ that contains R_n, the ordinal hn, and a counting of hn, since the rank of $\mathscr{B}^n$ is $hn + 1$. In addition, the construction of R_{n+1} can be construed as a proof (in the sense of (3.9.5)) of (3.11.6). Hence, by (3.11.1)–(3.11.3), there is a (R_{n+1}, γ_n) that satisfies (3.11.5abc).

The splitting trick of 3.8 can be added to the definition of R_n to ensure that $R = \bigcap \{R_n \mid n < \omega\}$ is perfect.

To obtain the desired $\underset{\sim}{Q}$ and λ_0 first suppose there is an n such that

$$(R_{n+1}, \gamma_n) \Vdash^A \mathscr{B}^n. \tag{3.11.7}$$

Let Z be the leftmost path through R_{n+1}. Then Z is recursive in R_{n+1}, and

$$L(\lambda, Z) \models \mathscr{B}^n,$$

since the rank of $\mathscr{B}^n$ is less than λ. But then

$$(\{hn\}^Z, hn) \Vdash\!\vdash^A \mathscr{F}$$

by (3.9.10). Thus, (Q, λ_0) is $(\{hn\}^Z, hn)$.

Now, suppose there is no n that satisfies (3.11.7). Then, (3.11.5c) implies

$$(R, \lambda) \Vdash^A (\delta)_{\delta<\lambda} \sim (Ey^{\delta+1})\, \mathscr{B}_\delta(\{\delta\}^{\mathscr{T}}, \mathscr{F}, y^{\delta+1}),$$

and consequently (Q, λ_0) is (R, λ).

LEMMA 3.12. *Let $\mathscr{F}(\bar{n}, x)$ be a formula of $\mathscr{L}(\alpha, \mathscr{T})$ with no unranked quantifiers. Suppose $P \Vdash^A (n)(Ex)\mathscr{F}(\bar{n}, x)$. Then there exists a $Q \subset P$ and a $\delta < \alpha$ such that $Q \Vdash^A (n)(Ex^\delta)\mathscr{F}(\bar{n}, x^\delta)$.*

Proof. Clearly

$$(n)(Q)_{P \supset Q}(ER)_{Q \supset R}[R \Vdash^A (Ex)\, \mathscr{F}(\bar{n}, x)].$$

Hence, by 3.11, for each n and $Q \subset P$

$$(ER)(E\gamma)[(R, \gamma) \Vdash\!\!\Vdash^A (Ex)\, \mathscr{F}(\bar{n}, x)\ \&\ \mathscr{T} \in Q]. \tag{3.12.1}$$

The effect of $R \Vdash^A \mathscr{T} \in Q$ is intended to be the same as that of $Q \supset R$. Strictly speaking $\mathscr{T} \in Q$ is not a ranked sentence of $\mathscr{L}(\alpha, \mathscr{T})$, however there is no difficulty in treating it as such.

The argument of 3.2 shows that the relation $R \Vdash\!\!\Vdash^A \mathscr{G}$, restricted to $\Sigma_1\mathscr{G}$'s, is Σ_1 over A. Since A satisfies Σ_1 dependent choice, it follows from (3.12.1) that there exist functions $\lambda n \mid Q_n$ and $\lambda n \mid \gamma_n$ in A such that $Q_0 = P$, Q_0 is uniformly hyperarithmetically pointed via γ_0, and for all n,

$$(Q_{n+1}, \gamma_{n+1}) \Vdash\!\!\Vdash^A (Ex)\, \mathscr{F}(\bar{n}, x)\ \&\ \mathscr{T} \in Q_n\,.$$

Clearly, $Q_n \supset Q_{n+1}$ for all n. Let $Q = \bigcap \{Q_n \mid n < \omega\}$. The splitting trick of 3.8 can be used to ensure that Q is perfect.

Let $X \in Q$ to see that Q is uniformly hyperarithmetically pointed. For each n, $\gamma_n < \omega_1{}^X$ and $Q_n = \{\gamma_n\}^X$ since $X \in Q_n$. Proposition 3.10 makes it possible to define $\lambda n \mid \gamma_n$ and $\lambda n \mid Q_n$ by a recursion Σ_1 over $L(\omega_1{}^X, X)$: γ_{n+1} is the least β such that $X \in \{\beta\}^X$ and

$$(\{\beta\}^X, \beta) \Vdash\!\!\Vdash^A (Ex)\, \mathscr{F}(\bar{n}, x)\ \&\ \mathscr{T} \in Q_n\,.$$

Thus there is a ρ such that for all $X \in Q$, $\lambda n \mid \gamma_n = \{\rho\}^X$ and $\rho < \omega_1{}^X$, and consequently Q possesses the desired uniform pointedness.

For each n, let δ_n be the least δ such that

$$Q \Vdash^A (Ex^\delta)\, \mathscr{F}(\bar{n}, \delta),$$

δ_n is well defined since $Q \subset Q_n$. $\lambda n \mid \delta_n$ belongs to A by 3.2. Then $\delta = \bigcup \{\delta_n \mid n < \omega\} < \alpha$.

THEOREM 3.13. *Suppose T is generic in the sense of $\Vdash^A$ (3.5). Then $\omega_1{}^T = \alpha$ and every real in A is hyperarithmetic in T. Furthermore there exists no hyperdegree less than that of T and greater than every one of A.*

Proof. By (3.6.1) every real in A is hyperarithmetic in T and so $\omega_1^T \geqslant \alpha$. For each $\delta < \alpha$, (3.6.2) implies $\{\delta\}^T$ has the same hyperdegree as T or some real in A. Hence, it suffices to show $\omega_1^T \leqslant \alpha$. As noted in 3.7, it is enough to see that $L(\alpha, T)$ satisfies the Δ_0 bounding principle (3.7.1). Let $\mathscr{F}(\bar{n}, x)$ be a formula of $\mathscr{L}(\alpha, \mathscr{T})$ with no unranked quantifiers. Let D be the set of all P such that either $P \Vdash^A \sim(n)(Ex)\mathscr{F}(\bar{n}, x)$, or

$$(E_\delta)_{\delta<\alpha}[P \Vdash^A (n)(Ex^\delta)\, \mathscr{F}(\bar{n}, x^\delta)].$$

Lemma 3.12 implies D is dense (as in 3.5). By 3.2, D is A-definable. Consequently, there is a $P \in D$ such that $T \in P$, and so $L(\alpha, T)$ satisfies (3.7.1).

COROLLARY 3.14. *If A is a countable admissible set that satisfies Σ_1 dependent choice, then the set of all hyperdegrees contained in A has a minimal upper bound.*

Proof. If the set H of hyperdegrees of all reals in A has no least upper bound, then 3.13 furnishes a minimal upper bound for H.

COROLLARY 3.15. *If β is a countable ordinal, then the set of hyperdegrees of reals in $L(\beta)$ has a minimal upper bound.*

Proof. Boolos and Putnam [13] call δ an index ordinal if $(L(\delta + 1) - L(\delta)) \cap 2^\omega \neq 0$. They show $L(\delta + 1)$ contains a one–one correspondence between δ and ω whenever δ is an index ordinal. Thus it is safe to assume that either β is the limit of index ordinals or β is the successor of an index ordinal. It follows that the hyperdegree of Y is an upper bound for the hyperdegrees of $L(\beta)$ if and only if $\omega_1^Y \geqslant \beta$.

Let γ be the least admissible ordinal $\geqslant \beta$. Let α be the least ρ such that

$$L(\rho) \cap 2^\omega = L(\gamma) \cap 2^\omega.$$

α is admissible because if $\alpha < \gamma$, then α sits inside $L(\gamma)$ in very much the same way constructible aleph–one sits inside L. If $\alpha \leqslant \beta$, then the sought after minimal upper bound exists by 3.14. Suppose $\alpha > \beta$. Then $\alpha = \gamma$ and there is an index ordinal δ such that $\alpha > \delta \geqslant \beta$. It follows there is an $X \in L(\alpha)$ such that $\omega_1^X > \beta$. Clearly the hyperdegree of X is an upper bound for the hyperdegrees of $L(\beta)$. X is in fact a least upper bound. Suppose every real in $L(\beta)$ is hyperarithmetic in Y. Then, $\omega_1^Y > \beta$ and $X \in L(\alpha) \subset L(\omega_1^Y, Y)$, since ω_1^Y is admissible.

COROLLARY 3.16 (cf. 4.26). *α is a countable admissible ordinal greater than ω if and only if there is a $T \subset \omega$ such that $\alpha = \omega_1^T$.*

Proof. Let K be generic in the sense of $\Vdash_\alpha$ (2.14). By 2.18, $L[\alpha, K]$ is a countable admissible set satisfying Σ_1 dependent choice. In addition, the least ordinal not countable in $L[\alpha, K]$ is α. Let T be generic in the sense of $\Vdash^A$ with $A = L[\alpha, K]$. By 3.13 $\omega_1^T = \alpha$.

Let F be a type n object ($n \geqslant 2$). F is said to be *normal* if ${}^nE$ (the equality predicate for objects of type $< n$) is recursive in F in the sense of Kleene [17]. ${}_1\mathrm{sc}\, F$ is the set of all reals recursive in F and was christened the 1-section of F in [17].

COROLLARY 3.17. *If F is a normal type n object, then the hyperdegrees of the reals in the 1-section of F have a minimal upper bound.*

Proof. It was observed in [18] that ${}_1\mathrm{sc}\, F = 2^\omega \cap A$ for some countable admissible A that satisfies Σ_1 dependent choice.

COROLLARY 3.18. *Let A be a countable admissible set such that the hyperjump of each real in A also belongs to A. Then the hyperdegrees in A have a minimal upper bound.*

Proof. Let HC_A be the set of all members of A hereditarily countable in the sense of A. The standard arguments that show that HC (the set of all hereditarily countable sets) is an admissible set satisfying Σ_1 dependent choice also show that HC_A has the same properties. The point to remember is that the closure of HC_A under hyperjump implies that every linear ordering of ω in HC_A is a wellordering in the sense of HC_A if and only if it is an actual wellordering.

THEOREM 3.19. *Let A be a countable admissible set. Then there exist reals S and T such that for all real X: $X \in A$ if and only if $X \leqslant_h S$ and $X \leqslant_h T$.*[3]

Proof. By a type omitting argument similar to the one given in Keisler [3, p. 58]. Let α be the least ordinal not countable in A. Define A_1 to be HC_A if $\alpha \in A$, and A if $a \notin A$. A_1 is admissible as in the proof of 3.18, because HC_A is closed under hyperjump when $\alpha \in A$. (It is not

[3] The original version of 3.19 assumed A satisfied Σ_1 dependent choice and was proved by pointed forcing (cf. 5.2 and (5.4.1)).

known if HC_A is admissible for every admissible A.) Also α is the least ordinal not in A_1. Let $\mathscr{S}$ be the following collection of sentences:

$$S \subset \omega, \qquad \underset{\sim}{X} \text{ recursive in } S \; (X \in A \cap 2^\omega),$$

$$(y)\left[y \in \underset{\sim}{\delta} \leftrightarrow \bigvee_{\beta<\delta} y = \underset{\sim}{\beta}\right], \qquad (\delta < \alpha).$$

Clearly $\mathscr{S}$ is consistent and Σ_1 over A_1. Since A_1 is admissible, the argument of Keisler cited above shows $\mathscr{S}$ has a model that "omits" α. Thus there is a real S such that every real X in A is recursive in S and $\omega_1^S = \alpha$.

Let $\mathscr{T}$ be the set of sentences obtained by substituting T for S throughout $\mathscr{S}$.

The S and T of 3.19 are developed by constructing models of $\mathscr{S}$ and $\mathscr{T}$ simultaneously, models that "omit" α, and by interweaving some additional steps to ensure

$$\{\delta\}^S = \{\rho\}^T \rightarrow \{\delta\}^S \in A, \tag{3.19.1}$$

for all δ and ρ less than α. Suppose that only finitely many steps in the Henkinesque constructions of models of $\mathscr{S}$ and $\mathscr{T}$ have occurred. Then, (3.19.1) can be satisfied for a fixed δ and ρ as follows. Assume δ is such that $\{\delta\}^S$ must be the characteristic function of a real.

Case (i). There exists an n such that *each* of the sentences, $\{\delta\}^S(n) = 0$ and $\{\delta\}^S(n) = 1$, is consistent with the decisions made up to now in the construction of the $\mathscr{S}$ model. Choose an m so that $\{\rho\}^T(n) = m$ is consistent with the construction up to now of the $\mathscr{T}$ model. Then $\{\delta\}^S(n) \neq m$ can be added consistently to the construction of the $\mathscr{S}$ model, and so $\{\delta\}^S \neq \{\rho\}^T$ is guaranteed.

Case (ii). There is no n that fits the needs of Case (i). Thus for each n, only one of the statements, $\{\delta\}^S(n) = 0$ and $\{\delta\}^S(n) = 1$, can be added consistently to the construction of S. Define a real B by: $n \in B$ if and only if $\{\delta\}^S(n) \neq 0$ is a logical consequence of all the decisions made up to now in the construction of the $\mathscr{S}$ model. That set of decisions is a finitely generated extension of $\mathscr{S}$, and hence, is Σ_1 over HC_A. Consequently, the hypothesis of Case (ii) implies B is Δ_1 over HC_A, hence a member of HC_A, and so $\{\delta\}^S \in A$ is assured.

Corollary 3.20 (cf. Theorem 5.2). *Let A be a countable admissible set. If the hyperdegrees contained in A have a least upper bound, then, that least upper bound is contained in A.*

Proof. The S and T of 3.19 force the least upper bound, if it exists, into A.

Richter [19] defined and studied the following hierarchy of hyperdegrees associated with iteration of the hyperjump.

$h(0)$ is the hyperdegree of the empty set.

$h(\gamma + 1)$ is the hyperjump of $h(\gamma)$.

$h(\lambda)$ is the least upper bound of $\{h(\gamma) \mid \gamma < \lambda\}$, when that least upper bound exists; otherwise, $h(\lambda)$ is undefined.

Richter defined δ to be the least λ such that $h(\lambda)$ is undefined. He showed $\aleph_1 > \delta \geqslant \omega_1^{E_1}$, where E_1 is the Souslin operator and $\omega_1^{E_1}$ is the least ordinal not recursive in E_1. A more current characterization of $\omega_1^{E_1}$ is: the least admissible ordinal that is a limit of admissible ordinals.

COROLLARY 3.21. $\delta = \omega_1^{E_1}$.

Proof. Let α be $\omega_1^{E_1}$. Some of Richter's results [19] are equivalent to: (1) for each $\gamma < \alpha$, $h(\gamma)$ is defined and belongs to $L(\alpha)$; (2) each hyperdegree in $L(\alpha)$ is less than $h(\gamma)$ for some $\gamma < \alpha$. It follows from 3.20 that $\{h(\gamma) \mid \gamma < \alpha\}$ has no least upper bound, since it has none in $L(\alpha)$.

4. MINIMAL SOLUTIONS OF $\alpha = \omega_1{}^X$

α is a countable admissible ordinal. In Section 3 a solution X of $\alpha = \omega_1{}^X$ was developed in two steps. First a K was chosen generically using the finite conditions of 2.14 so that $L[\alpha, K]$ would be an admissible set in which every ordinal less than α was countable. Then it was seen in Section 3 that every T generic in the sense of $\Vdash^A$, with $A = L[\alpha, K]$, was a solution of $\alpha = \omega_1{}^X$. The purpose of Section 4 is to find a minimal solution, an X such that no real of lower hyperdegree than X is a solution. Again the argument is in two steps. The second step resembles that of Section 3, but the first differs markedly. K is now constructed generically via perfect conditions that lend themselves to the splitting arguments needed for the desired minimality. The principal technical lemmas are 4.13 and 4.21. Both of them deal with some of the thorny problems encountered in splitting arguments when the notions of pointed and iterated perfect forcing are combined.

4.1. *Perfect Conditions on K*

From now on K is always a map from $(\alpha - \omega) \times \omega$ into α such that for each $\beta \in (\alpha - \omega)$, $\lambda m \mid K(\beta, m)$ is a one–one map of ω into β with the following property: if

$$L(\alpha) \models [\beta \text{ is a cardinal \& cofinality } \beta > \omega], \tag{4.1.1}$$

then the range of $\lambda m \mid K(\beta, m)$ is an unbounded subset of β. Clearly, every ordinal less than α is countable in $L[\alpha, K]$.

Let $s, t,\ldots$ be finite conditions on K as in 2.14. Let $H, I, J\ldots$ denote α-finite perfect conditions on K. A typical H is given by an α-finite partial function $\lambda i\sigma \mid s_\sigma{}^i$ ($i < \omega$ and $\sigma \in \alpha^i$) that meets requirements (4.1.2)–(4.1.7).

(4.1.2) If $s_\sigma{}^i$ is defined, then $s_\sigma{}^i$ is a finite condition on K; $s_\sigma{}^i$ is said to belong to H (in symbols $s_\sigma{}^i \in H$).

(4.1.3) $s_0{}^0$ is defined.

(4.1.4) If s_τ^{i+1} is defined and $(j)_{j<i}(\tau j = \sigma j)$, then $s_\sigma{}^i$ is defined and $s_\sigma{}^i$ is extended by s_τ^{i+1} ($s_\sigma{}^i \geqslant s_\tau^{i+1}$).

(4.1.5) Suppose $s_\sigma{}^i$ is defined. Let $b_\sigma{}^i$ be the set of all $\tau \in \alpha^{i+1}$ such that s_τ^{i+1} is defined and $(j)_{j<i}(\tau j = \sigma j)$. $\{s_\tau^{i+1} \mid \tau \in b_\sigma{}^i\}$ is called the set of *immediate successors* of $s_\sigma{}^i$ in H. All the immediate successors of $s_\sigma{}^i$ have the same length $l_\sigma{}^i$ and the same domain $\eta_\sigma{}^i$ as defined in 2.14. The *relevant domain* $d_\sigma{}^i$ is the set of all $\beta \in \eta_\sigma{}^i$ such that β has property (4.1.1) above.

(4.1.6) Let $d_\sigma{}^i$ be $\{\beta_k \mid k < z\}$. For each $\gamma \in \Pi\{\beta_k \mid k < z\}$, there is an $\langle m_i{}^\gamma\rangle_{i<z}$ and a $\tau \in b_\sigma{}^i$ such that

$$(i)_{i<z}[s_\tau^{i+1}(\beta_i\,, m_i{}^\gamma) > \gamma_i]. \tag{4.1.7}$$

The requirement imposed by (4.1.6) and defined by (4.1.7) is called *diversification* and is needed for the proof of complex minimization (4.21). $\{s_\tau^{i+1} \mid \tau \in b_\sigma{}^i\}$ is said to be diversified.

A path through H is a sequence $\lambda i \mid s_{\sigma i}^i$ such that $s_{\sigma i}^i$ is defined and $s_{\sigma i}^i \geqslant s_{\sigma(i+1)}^{i+1}$ for all i. K satisfies H (in symbols $K \in H$) if there is a path $\lambda i \mid s_{\sigma i}^i$ through H such that K satisfies $s_{\sigma i}^i$ for all i.

4.2. *The Forcing Relation* $H \Vdash_\alpha \mathscr{F}$

Let $\mathscr{F}$ be a sentence of the language $\mathscr{L}[\alpha, \mathscr{K}]$ as in 2.13, and let H be an α-finite perfect condition on K. The forcing relation $H \Vdash_\alpha \mathscr{F}$ is similar to that of 3.1.

(4.2.1) $\mathscr{F}$ is ranked. In this case, $H \Vdash_\alpha \mathscr{F}$ iff $L[\alpha, K] \models \mathscr{F}$ true for every $K \in H$. (Keep in mind the convention concerning K established at the beginning of 4.1. The definition of $H \Vdash_\alpha \mathscr{F}$ for ranked $\mathscr{F}$ refers only to those K's in H that obey the convention. The existence of such K's in H follows from (4.1.7).)

(4.2.2) $\mathscr{F}$ & $\mathscr{G}$ is not ranked. $H \Vdash_\alpha \mathscr{F}$ & $\mathscr{G}$ iff $H \Vdash_\alpha \mathscr{F}$ and $H \Vdash_\alpha \mathscr{G}$.

(4.2.3) $(Ex^\beta)\mathscr{F}(x^\beta)$ is not ranked. $H \Vdash_\alpha (Ex^\beta)\mathscr{F}(x^\beta)$ iff $H \Vdash_\alpha \mathscr{F}(c)$ for some $c \in \mathscr{C}(\beta)$.

(4.2.4) $H \Vdash_\alpha (Ex)\mathscr{F}(x)$ iff $H \Vdash_\alpha \mathscr{F}(c)$ for some $c \in \mathscr{C}(\alpha)$.

(4.2.5) $\mathscr{G}$ is not ranked. $H \Vdash_\alpha \sim\mathscr{G}$ iff it is not the case that $I \Vdash_\alpha \mathscr{G}$ for any $I \subset H$. ($I \subset H$ means I extends H, i.e., every K that satisfies I also satisfies H.)

K is said to be *generic* if for each sentence $\mathscr{F}$ of $\mathscr{L}[\alpha, \mathscr{K}]$ there exists an H such that $K \in H$ and either $H \Vdash_\alpha \mathscr{F}$ or $H \Vdash_\alpha \sim\mathscr{F}$. A splitting lemma (4.4) is needed to establish the existence of generic K's as well as the validity of Δ_0 bounding (with $x \in K$ as an additional atomic predicate) in $L[\alpha, K]$ for all generic K's.

LEMMA 4.3. *The forcing relation* $H \Vdash_\alpha \mathscr{F}$, *restricted to* Σ_1 $\mathscr{F}$'*s, is* Σ_1 *over* $L(\alpha)$.

Proof. Suppose $\delta < \alpha$ and X is any real such that $\delta \leqslant \omega_1{}^X$. The set of all H's that are perfect conditions on K, and that belong to $L(\delta)$, can be construed via indices as a set of numbers $\Pi_1{}^1$ in X. The indexing of the H's is a slight modification of the indexing of the ordinals less than $\omega_1{}^X$ by elements of O^X. Similarly the set of all $\mathscr{F}$'s of rank less than δ is $\Pi_1{}^1$ in X. Thus, the predicate $P_\delta(H, \mathscr{F})$, defined by

$$H \in L(\delta) \text{ \& rank } \mathscr{F} < \delta \text{ \& } H \Vdash_\alpha \mathscr{F},$$

is $\Pi_1{}^1$ uniformly in any X such that $\delta \leqslant \omega_1{}^X$, since the universal quantifier on K in clause (4.2.1) of the definition of $\Vdash_\alpha$ can be regarded as a universal quantifier on reals thanks to the countability in $L(\omega_1^X, X)$ of the set of all finite conditions in H.

Let X_δ encode a Levyesque collapse of δ to ω generic over $L(\alpha)$. For example X_δ could be

$$\{\langle m, n\rangle \mid K(\delta, m) < K(\delta, n)\},$$

where $\lambda m \mid K(\delta, m)$ is a one–one map of ω onto δ generic over $L(\alpha)$ in the manner of 2.14. Then, $P_\delta(H, \mathscr{F})$ is Π_1^1 in X_δ, and consequently is Σ_1 over $L(\alpha, X_\delta)$ by 2.5. To be precise, there is a Σ_1 formula $Z(\delta, H, \mathscr{F})$ such that

$$P_\delta(H, \mathscr{F}) \quad \text{iff} \quad L(\alpha, X_\delta) \models Z(\delta, H, \mathscr{F}). \tag{4.3.1}$$

But (4.3.1) holds for all generic X_δ's, hence

$$P_\delta(H, \mathscr{F}) \quad \text{iff} \quad \varnothing \Vdash_\delta \sim\sim Z(\delta, H, \mathscr{F}), \tag{4.3.2}$$

where $\varnothing$ is the null forcing condition, and $\Vdash_\delta$ is the Levyesque forcing relation designed to collapse δ to ω generically over $L(\alpha)$. Since Z is Σ_1 and the collection of forcing conditions needed to collapse δ is a set in $L(\alpha)$, it follows that the right side of (4.3.2) is Σ_1 over $L(\alpha)$ uniformly in δ.

Hence $H \Vdash \mathscr{F}$, restricted to ranked $\mathscr{F}$'s, is Σ_1 over $L(\alpha)$, since it is equivalent to $(E\delta)P_\delta(H, \mathscr{F})$.

Lemma 4.4. *Let $\{\mathscr{F}_i \mid i < \omega\}$ be an α-finite sequence of Σ_1 sentences of $\mathscr{L}[\alpha, \mathscr{K}]$. Suppose H is such that*

$$(i)(I)_{H\supset I}(EJ)_{I\supset J}[J \Vdash_\alpha \mathscr{F}_i].$$

Then there exists a $J \subset H$ such that $(i)[J \Vdash_\alpha \mathscr{F}_i]$.

Proof. Partial functions I_σ^i, t_σ^i, and c_σ^i are defined by recursion on i for all $i < \omega$ and some $\sigma \in \alpha^i$. Their definitions require certain choices to be made. All such choices are the least possible with respect to the standard Δ_1 wellordering of $L(\alpha)$ as in (3.8.5). It will follow from 4.3 that all three partial functions are α-finite.

Let t_0^0 be the null condition on K, and let I_0^0 be H.

Fix $i < \omega$ and $\sigma \in \alpha^i$. Assume t_σ^i and I_σ^i have been defined so that $t_\sigma^i \in I_\sigma^i$ and I_σ^i is a perfect condition on K contained in H. Choose b_σ^i, t_τ^{i+1} and I_τ^{i+1} in accord with the following requirements.

(4.4.1) $\{t_\tau^{i+1} \mid \tau \in b_\sigma{}^i\}$ is a diversified set of immediate successors of $t_\sigma{}^i$ inside $I_\sigma{}^i$.

$$t_\tau^{i+1} \in I_\tau^{i+1} \subset I_\sigma{}^i, \tag{4.4.2}$$

$$(t)[t \in I_\tau^{i+1} \to t_\tau^{i+1} \geqslant t], \tag{4.4.3}$$

$$I_\tau^{i+1} \Vdash_\alpha \mathscr{F}_i, \qquad \text{for all } \tau \in b_\sigma{}^i. \tag{4.4.4}$$

The choice of I_τ^{i+1} is the least possible in the following sense. By 4.3 there is a Σ_1 formula $R(I, \mathscr{F}_i)$ such that

$$L(\alpha) \models R(I, \mathscr{F}_i) \qquad \text{iff} \quad I \Vdash_\alpha \mathscr{F}_i.$$

Suppose $R(I, \mathscr{F}_i)$ is $(Ex)P(x, I, \mathscr{F}_i)$, where P is a Δ_0 formula. Let (x^0, I^0) be the least pair (x, I) that satisfies $P(x, I, \mathscr{F}_i)$. Then I^0 is said to be least among those I's that force $\mathscr{F}_i$.

Let J be the perfect condition given by the α-finite partial function $\lambda i\sigma \mid t_\sigma{}^i$.

Suppose $\mathscr{F}_i$ is $(Ex)\mathscr{G}_i(x)$, where x is the only unranked variable in $\mathscr{G}_i(x)$, to see $J \Vdash_\alpha \mathscr{F}_i$. Let b be the α-finite set of all $\tau \in \alpha^{i+1}$ such that I_τ^{i+1} is defined. Then for each $\tau \in b$,

$$I_\tau^{i+1} \Vdash_\alpha \mathscr{G}_i(c_\tau)$$

for some $c_\tau \in \mathscr{C}(\alpha)$. By 4.3 c_τ can be construed as an α-finite function of τ. Hence, there is a $\beta < \alpha$ such that $\{c_\tau \mid \tau \in b\} \subset \mathscr{C}(\beta)$. Consequently

$$I_\tau^{i+1} \Vdash_\alpha (Ex^\beta)\, \mathscr{G}_i(x^\beta)$$

for all $\tau \in b$, and so $J \Vdash_\alpha (Ex)\mathscr{G}_i(x)$, since $J \subset \bigcup \{I_\tau^{i+1} \mid \tau \in b\}$.

LEMMA 4.5. *$(\mathscr{F})(H)(EJ)_{H\supset J}[J \Vdash_\alpha \mathscr{F}$ or $J \Vdash_\alpha \sim\mathscr{F}]$.*

Proof. By induction on the complexity of $\mathscr{F}$. Clause (4.2.5) makes it safe to restrict attention to ranked $\mathscr{F}$'s. Let $(Ex^\beta)\mathscr{G}(x^\beta)$ be a ranked sentence. If there is an $I \subset H$ such that $I \Vdash_\alpha \mathscr{G}(c)$ for some $c \in \mathscr{C}(\beta)$, then any such I can serve as the desired J. Suppose no such I exists. Then by induction

$$(c)_{c\in\mathscr{C}(\beta)}(I)_{H\supset I}(EJ)_{I\supset J}[J \Vdash_\alpha \sim\mathscr{G}(c)], \tag{4.5.1}$$

since $\mathscr{G}(c)$ is less complex than $(Ex)\mathscr{G}(x)$ for all $c \in \mathscr{C}(\beta)$.

Let f be a constant of $\mathscr{L}[\alpha, \mathscr{K}]$ that denotes a map of ω onto $\mathscr{C}(\beta)$ in $L[\alpha, K]$ for every K that obeys the convention established at the beginning of 4.1. Thus

$$I \Vdash_\alpha f\colon \omega \xrightarrow{\text{onto}} \mathscr{C}(\beta), \tag{4.5.2}$$

for every I. Consequently for each i and I, there is $I_i \subset I$ and $c_i \in \mathscr{C}(\beta)$ such that

$$I_i \Vdash_\alpha fi = c_i\,. \tag{4.5.3}$$

By (4.5.1)–(4.5.3)

$$(i)(I)_{H\supset I}(EJ)_{I\supset J}[J \Vdash_\alpha \sim\mathscr{G}(fi)]. \tag{4.5.4}$$

It follows from 4.4 that there is a $J \subset H$ such that

$$(i)[J \Vdash_\alpha \sim\mathscr{G}(fi)].$$

By (4.5.2) $J \Vdash_\alpha \sim(Ex^\beta)\mathscr{G}(x^\beta)$.

It is an immediate consequence of Lemma 4.5 that there exist K's generic in the sense of 4.2.

LEMMA 4.6. *Let $\mathscr{F}(x, y)$ be a Σ_1 formula of $\mathscr{L}[\alpha, \mathscr{K}]$. Suppose $H \Vdash_\alpha (x^\beta)(Ey)\mathscr{F}(x^\beta, y)$ for some $\beta < \alpha$. Then, there exists a $J \subset H$ such that*

$$J \Vdash_\alpha (x^\beta)(Ey^\gamma)\,\mathscr{F}(x^\beta, y^\gamma)$$

for some $\gamma < \alpha$.

Proof. Clearly

$$(I)_{H\supset I}(c)_{c\in\mathscr{C}(\beta)}(EJ)_{I\supset J}[J \Vdash_\alpha (Ey)\,\mathscr{F}(c, y)].$$

Let f be a constant of $\mathscr{L}[\alpha, \mathscr{K}]$ with the property claimed in (4.5.2). Then, as in the proof of 4.5, there is a $J \subset H$ such that

$$(i)[J \Vdash_\alpha (Ey)\,\mathscr{F}(fi, y)].$$

It follows from 4.3 that there is an α-finite $h\colon \omega \to \alpha$ such that

$$(i)[J \Vdash_\alpha (Ey^{hi})\,\mathscr{F}(fi, y^{hi})].$$

Let γ be $\sup\{hi \mid i < \omega\}$. Then, $J \Vdash_\alpha (x^\beta)(Ey^\gamma)\mathscr{F}(x^\beta, y^\gamma)$.

THEOREM 4.7. *Suppose K is generic in the sense of* 4.2. *Then* $L[\alpha, K]$ *satisfies* (x) *(x is countable) and* Δ_0 *bounding (with* $x \in K$ *as an additional atomic predicate). Furthermore* $L[\alpha, K]$ *has a* Δ_1 *(in K) wellordering.*

Proof. Lemma 4.6 implies Δ_0 bounding holds in $L[\alpha, K]$, just as 3.12 does in the proof of 3.13. The Δ_1 welloıdering of $L[\alpha, K]$ is derived from the standard order in which the elements of $L[\alpha, K]$ are constructed from K via ordinals less than α.

For the remainder of Section 4 assume K is generic in the sense of 4.2. By 4.7 $L[\alpha, K]$ is a countable admissible set that satisfies Σ_1 dependent choice, and α is the least ordinal not countable in $L[\alpha, K]$. Hence the pointed perfect forcing argument of 3.13 is applicable to $L[\alpha, K]$. Thus if T is generic (in the sense of $\Vdash^A$ with $A = L[\alpha, K]$), then $\omega_1{}^T = \alpha$. The primary objective of Section 4 is to prove that T has the following additional property:

$$(X)[X <_h T \to \omega_1{}^X < \alpha].$$

Of course the proof will draw heavily upon the fact that K was constructed from the perfect conditions of 4.1 rather than the finite Levy conditions of 2.14. On the other hand it is quite possible that finite conditions suffice.

4.8. *K-Uniform Pointedness*

Let $P \in L[\alpha, K]$ be perfect as in 2.1, and let $\mathscr{P}$ be a term of $\mathscr{L}[\alpha, \mathscr{K}]$ that denotes P. $\mathscr{P}$ can be viewed as a function that takes K's as arguments and perfect P's as values; thus $P = \mathscr{P}(K)$. Only a bounded part of K is needed to evaluate $\mathscr{P}(K)$; thus $P = \mathscr{P}(K^\beta)$, where β is the rank of $\mathscr{P}$ as a term of $\mathscr{L}[\alpha, \mathscr{K}]$.

P is said to be K-uniformly hyperarithmetically pointed via $\mathscr{P}$ and δ if for all $T \in P$:

$$\beta < \omega_1{}^T \quad \text{and} \quad P = \mathscr{P}(K^\beta), \tag{4.8.1}$$

$$\delta < \omega_1{}^T \quad \text{and} \quad K^\beta = \{\delta\}^T. \tag{4.8.2}$$

Clearly K-uniform pointedness implies uniform pointedness in the sense of 2.9, but the converse is false. However forcing with K-uniform P's is the same as forcing with uniform P's, since each one of the latter is readily extendible to one of the former with the help of 2.3.

The forcing relation $P \Vdash^K \mathscr{F}$, where P is K-uniformly hyperarithmetically pointed and $\mathscr{F}$ is a sentence of $\mathscr{L}(\alpha, \mathscr{T})$, is defined in the same fashion as the forcing relation $\Vdash^A$ of 3.1. The only difference between the two is in the notion of P each employs.

LEMMA 4.9. *The relation* $P \Vdash^K \mathscr{F}$, *restricted to* Σ_1 $\mathscr{F}$*'s, is* Σ_1 *over* $L[\alpha, K]$.

Proof. Similar to that of 3.2. Again, most of the work is done by 2.5.

4.10. *Complex Forcing*

By 4.9 there exists a formula $\mathscr{P}^\delta \Vdash^{\mathscr{K}} \mathscr{F}$ that expresses $P \Vdash^K \mathscr{F}$ as follows:

(4.10.1) $\mathscr{P}^\delta \Vdash^{\mathscr{K}} \mathscr{F}$ is a Σ_1 formula of $\mathscr{L}[\alpha, \mathscr{K}]$ with $\mathscr{P}$, δ, and $\mathscr{F}$ as free variables.

(4.10.2) If $L[\alpha, K] \models [\mathscr{P}^\delta \Vdash^{\mathscr{K}} \mathscr{F}]$, then $\mathscr{P}$ is a term of $\mathscr{L}[\alpha, \mathscr{K}]$ that denotes a P that is K-uniformly hyperarithmetically pointed via $\mathscr{P}$ and δ, $\mathscr{F}$ is a Σ_1 sentence of $\mathscr{L}(\alpha, \mathscr{T})$ and $P \Vdash^K \mathscr{F}$.

(4.10.3) If $\mathscr{F}$ is a Σ_1 sentence of $\mathscr{L}(\alpha, \mathscr{T})$ and $P \Vdash^K \mathscr{F}$, then $L[\alpha, K] \models [\mathscr{P}^\delta \Vdash^{\mathscr{K}} \mathscr{F}]$ for some $\mathscr{P}$ and δ such that P is K-uniformly hyperarithmetically pointed via $\mathscr{P}$ and δ.

Formulas such as $\mathscr{P}^\delta \Vdash^{\mathscr{K}} \mathscr{F}$ are characteristic of iterated forcing situations. Define the complex forcing relation

$$(H, \mathscr{P}^\delta) \Vdash_\alpha \mathscr{F} \qquad \text{by} \quad H \Vdash_\alpha [\mathscr{P}^\delta \Vdash^{\mathscr{K}} \mathscr{F}] \tag{4.10.4}$$

for every Σ_1 sentence $\mathscr{F}$ of $\mathscr{L}(\alpha, \mathscr{T})$. (The relation $H \Vdash_\alpha \mathscr{G}$ was defined in 4.2.) Since $\mathscr{F}$ is Σ_1, the meaning of $(H, \mathscr{P}^\delta) \Vdash_\alpha \mathscr{F}$ is: for every $K \in H$ it is the case that $\mathscr{P}(K)$ is K-uniformly hyperarithmetically pointed via $\mathscr{P}$ and δ, and that $L(\alpha, T) \models \mathscr{F}$ for every $T \in \mathscr{P}(K)$.

A pair $(H, \mathscr{P}^\delta)$ as in (4.10.4) is called a complex forcing condition. $\mathscr{P}^\delta$ is called the virtual part of $(H, \mathscr{P}^\delta)$. The advantage held by a virtual $\mathscr{P}^\delta$ over an actual $\mathscr{P}(K)$ is membership in $L(\alpha)$, an advantage used heavily in the splitting arguments of 4.13 and 4.21.

From now on the ordinal superscript in (H, P^δ) will be omitted for the sake of notational simplicity. Thus $(H, \mathscr{P})$ will denote a typical complex condition.

LEMMA 4.11. *The complex forcing relation* $(H, \mathscr{P}) \Vdash_\alpha \mathscr{F}$, *restricted to* Σ_1 $\mathscr{F}$*'s, is* Σ_1 *over* $L(\alpha)$.

Proof. By 4.3 and (4.10.4).

PROPOSITION 4.12a. *Suppose P is K-uniformly hyperarithmetically pointed and $X \in L[\alpha, K]$. Then there exists a K-uniformly hyperarithmetically pointed $Q \subset P$ such that X is hyperarithmetic in Q.*

Proof. Assume P is K-uniformly hyperarithmetically pointed via $\mathscr{P}$ and δ. Thus $P = \mathscr{P}(K^\beta)$, where β is the rank of $\mathscr{P}$. Choose $\gamma \geqslant \beta$ so that X is hyperarithmetic in K^γ. (Strictly speaking K^γ was defined to be that part of K that collapses all infinite ordinals $< \gamma$ to ω, but there is no harm in thinking of K^γ as a relation on ω.) Construct Q from P and K^γ as Q was constructed from P and Y in 2.3. Let $\mathscr{Q}$ be a term of $\mathscr{L}[\alpha, \mathscr{K}]$ that composes $\mathscr{P}$ with a description of the construction of Q from P and K^γ. Then $Q = \mathscr{Q}(K^\gamma)$, since P is hyperarithmetic in K^γ. Fix $T \in Q$ to see Q is suitably pointed. As in 2.3, $Q \subset P$ and K^γ is recursive in P, T, hence hyperarithmetic in T since $T \in P$. The rank of $\mathscr{Q}$ $(= \gamma)$ is less than $\omega_1{}^T$, and $K^\gamma = \{\rho\}^T$ for some $\rho < \omega_1{}^T$ determined by γ and δ. $X \leqslant_h Q$ by the concluding argument of 2.3.

Let U, V, W,... be complex forcing conditions. U is extended by V (in symbols $U \supset V$) if $U = (H, \mathscr{P})$, $V = (I, \mathscr{Q})$, $H \supset I$ and $I \Vdash_\alpha \mathscr{P} \supset \mathscr{Q}$. The relation $U \supset V$ is Σ_1 over $L(\alpha)$ by 4.3.

PROPOSITION 4.12b. *Suppose $(H, \mathscr{P})$ is a complex forcing condition and $\beta < \alpha$. Then there exists a complex forcing condition $(H, \mathscr{Q}) \subset (H, \mathscr{P})$ such that $\beta \leqslant$ rank $\mathscr{Q}$.*

Proof. Define $\mathscr{Q}$ as in the proof of 4.12a. Thus rank $\mathscr{Q} = \gamma$; and for each $K \in H$, $\mathscr{Q}(K^\gamma)$ is K-uniformly hyperarithmetically pointed.

LEMMA 4.13 (Complex Splitting). *Let $\{\mathscr{F}_i \mid i < \omega\}$ be an α-finite sequence of Σ_1 sentences of $\mathscr{L}(\alpha, \mathscr{T})$. Suppose U is a complex forcing condition such that*

$$(i)(V)_{U \supset V}(EW)_{V \supset W}[W \Vdash_\alpha \mathscr{F}_i].$$

Then there exists a $W \subset U$ such that $(i)[W \Vdash_\alpha \mathscr{F}_i]$.

Proof. The splitting of 3.8 and 4.4 are combined. α-finite partial functions $t_\sigma{}^i$, $I_\sigma{}^i$, $\mathscr{Q}_\sigma^{i,j}$, and $r_\sigma^{i,j}$ are defined by recursion on i for all $i < \omega$, some $\sigma \in \alpha^i$ and all $j < 2^i$.

Suppose U is $(H, \mathscr{P})$. Let $I_0{}^0$ be H, $\mathscr{Q}_0^{0,0}$ be $\mathscr{P}$, let $t_0{}^0$ be the null finite condition on K, and let $r_0^{0,0}$ be the null finite condition on T.

Fix i and σ. Assume $t_\sigma{}^i$, $I_\sigma{}^i$, $\mathscr{Q}_\sigma^{i,j}$ and $r_\sigma^{i,j}$ have been defined for all $j < 2^i$. Choose the t_τ^{i+1}'s as in 4.4. (All choices are the least possible with respect to the standard Δ_1 wellordering of $L(\alpha)$ as in (3.8.5).) Thus the t_τ^{i+1}'s constitute a diversified set of immediate successors to $t_\sigma{}^i$ inside $I_\sigma{}^i$. Choose I_τ^{i+1}, $\mathscr{Q}_\tau^{i+1,j}$ and $r_\tau^{i+1,j}$ $(j < 2^{i+1})$ to meet the following requirements.

(4.13.1) $t_\tau^{i+1} \in I_\tau^{i+1} \subset I_\sigma{}^i$.

(4.13.2) $(t)[t \in I_\tau^{i+1} \rightarrow t_\tau^{i+1} \geqslant t]$.

(4.13.3) $I_\tau^{i+1} \Vdash_\alpha [r_\tau^{i+1,2j}$ and $r_\tau^{i+1,2j+1}$ are incomparable extensions of $r_\sigma^{i,j}$ inside $\mathscr{Q}_\sigma^{i,j}]$.

(4.13.4) $I_\tau^{i+1} \Vdash_\alpha [r_\tau^{i+1,2j} \in \mathscr{Q}_\tau^{i+1,2j} \subset \mathscr{Q}_\sigma^{i,j}]$.

(4.13.5) Same as (4) with $2j$ replaced by $2j + 1$.

(4.13.6) $(I_\tau^{i+1}, \mathscr{Q}_\tau^{i+1,j}) \Vdash_\alpha \mathscr{F}_i$ for all $j < 2^{i+1}$.

(4.13.7) Rank $\mathscr{Q}_\tau^{i+1,j} > \beta_i$ for all $j < 2^{i+1}$, where β_i is the supremum of all ordinals occurring in the definitions of $t_\rho{}^m$, $I_\rho{}^m$, $\mathscr{Q}_\rho^{m,j}$ and $r_\rho^{m,j}$ for all $m \leqslant i$ and $\rho \in \alpha^i$ for which they are defined. Requirement (4.13.7) is satisfied with the help of 4.12.

The desired W is $(J, \mathscr{R})$. J is given by $\lambda i\sigma \mid t_\sigma{}^i$ as in 4.4. For each $K \in J$, $\mathscr{R}(K)$ is given by $\lambda ij \mid r_{\sigma i}^{i,j}$, where σi is the unique τ such that K satisfies $t_\tau{}^i$.

By (4.13.6) $\mathscr{R}(K) \Vdash_\alpha \mathscr{F}_i$ for all i and all $K \in J$. It remains only to show that $(J, \mathscr{R})$ is a complex forcing condition, i.e., to find a δ such that for all $K \in J$, $\mathscr{R}(K)$ is K-uniformly hyperarithmetically pointed via $\mathscr{R}$ and δ. Fix $K \in J$ and $T \in \mathscr{R}(K)$ to clarify the definition of δ. (The value of δ, of course, will not depend on the choices of K and T.) For each τ, let

$$\mathscr{Q}_\tau{}^i = \bigcup \{\mathscr{Q}_\tau^{i,j} \mid j < 2^i\}.$$

$(I_\tau{}^i, \mathscr{Q}_\tau{}^i)$ is a complex forcing condition since it is a finite union of complex conditions that are pairwise disjoint by (4.13.3) and (4.13.4). Again, let σi be the unique τ such that K satisfies $t_\tau{}^i$. Since $(I_{\sigma i}^i, \mathscr{Q}_{\sigma i}^i)$ is a complex condition and $T \in \mathscr{Q}_{\sigma i}^i(K)$, it follows that

$$\omega_1{}^T > \text{rank } \mathscr{Q}_{\sigma i}^i .$$

But then by (4.13.7), ω_1^T is greater than every ordinal occurring in the definitions of t_σ^i, I_σ^i, $\mathcal{Q}_\sigma^{i,j}$, and $r_\sigma^{i,j}$. Consequently these four α-finite partial functions are in fact ω_1^T-finite, because all choices made in the course of their definitions were actually made within $L(\omega_1^T)$, (Keep in mind that the standard Δ_1 wellorderings of $L(\alpha)$ and $L(\omega_1^T)$ are compatible; see the last paragraph of Subsection 3.8.) Thus the entire construction outlined in (4.13.1)–(4.13.7) is an ω_1^T-finite object. Hence

$$\text{rank}\, \mathcal{R} < \omega_1^T.$$

In addition there is an ω_1^T-finite partial function δ_τ^i such that

$$K_0^{\text{rank}\mathcal{Q}_\tau^i} = \{\delta_\tau^i\}^{T_0} \tag{4.13.8}$$

for all $K_0 \in I_\tau^i$ and $T_0 \in \mathcal{Q}_\tau^i(K_0)$, i.e., $\mathcal{Q}_\tau^i(K_0)$ is uniformly K-pointed via $\mathcal{Q}_\tau^i$ and δ_τ^i for all $K_0 \in I_\tau^i$.

Equation (4.13.8) suggests the following ω_1^T-finite definition of $\lambda i \mid \sigma i$ by induction on i

$$\sigma 0 = 0$$

$$K^{\text{rank}\mathcal{Q}_{\sigma i}^i} = \{\delta_{\sigma i}^i\}^T$$

$$\sigma(i+1) = \text{unique } \tau \text{ such that } K^{\text{rank}\mathcal{Q}_{\sigma i}^i} \in t_\tau^{i+1} \in I_{\sigma i}^i\,. \tag{4.13.9}$$

The correctness of (4.13.9) needs to be checked. Clearly $\sigma(i+1)$ is the unique τ such that $K \in t_\tau^{i+1} \in I_{\sigma i}^i$. By 4.12b it is safe to assume rank $\mathcal{Q}_{\sigma i}^i$ is greater than any ordinal mentioned by any member of $I_{\sigma i}^i$. Hence $K^{\text{rank}\,\mathcal{Q}_{\sigma i}^i}$ contains all the information about K needed to single out the unique t_τ^{i+1} (in $I_{\sigma i}^i$) satisfied by K.

4.14. *Generic* (K, T)'s

A pair (K, T) is said to be generic if K is generic with respect to the forcing relation $\Vdash_\alpha$ of 4.2 and T is generic with respect to the forcing relation $\Vdash^K$ of 4.8. Suppose (K, T) is generic and $\mathcal{F}$ is true in $L[\alpha, T]$. Then there is a $P \in L[\alpha, K]$ such that $T \in P$ and $P \Vdash^K \mathcal{F}$. As in 4.10 there is a formula $\mathcal{P} \Vdash^{\mathcal{K}} \mathcal{F}$ of $\mathcal{L}[\alpha, \mathcal{K}]$ that expresses $P \Vdash^K \mathcal{F}$. Thus $P = \mathcal{P}(K)$ and

$$L[\alpha, K] \models [\mathcal{P} \Vdash^{\mathcal{K}} \mathcal{F}].$$

Since K is generic, there is an H such that $K \in H$ and

$$H \Vdash_\alpha [\mathcal{P} \Vdash^{\mathcal{K}} \mathcal{F}]. \tag{4.14.1}$$

Formula (4.14.1) makes it possible to extend the complex forcing relation of 4.10 to all sentences of $\mathscr{L}(\alpha, \mathscr{T})$. Let U be a complex forcing condition. Then $U \Vdash_\alpha \mathscr{F}$ if $U = (H, \mathscr{P})$ and (4.14.1) holds. To make the definition of complex forcing precise, observe that for each $n \geqslant 1$ there is a Σ_n formula of $\mathscr{L}[\alpha, \mathscr{K}]$ that expresses $P \Vdash^K \mathscr{F}$ for all $\Sigma_n \mathscr{F}$'s of $\mathscr{L}(\alpha, \mathscr{T})$.

Proposition 4.15. *Suppose* $\delta < \alpha$ *and*

$$(H, \mathscr{P}) \Vdash_\alpha \{\delta\}^{\mathscr{T}} \notin L[\alpha, \mathscr{K}].$$

Then there exist m, $(I, {}^0\mathscr{Q})$, *and* $(I, {}^1\mathscr{Q})$ such that $(H, \mathscr{P}) \supset (I, {}^v\mathscr{Q})$ *and*

$$(I, {}^v\mathscr{Q}) \Vdash_\alpha \{\delta\}^{\mathscr{T}}(m) = v$$

for all $v < 2$.

Proof. There is no harm in assuming $\{\delta\}^{\mathscr{T}}$ denotes the characteristic function of a subset of ω. Fix a generic $K \in H$ and let $P = \mathscr{P}(K)$. Then

$$P \Vdash_\alpha \{\delta\}^{\mathscr{T}} \notin L[\alpha, K].$$

As in the proof of 3.4 there exist Q, R, and m such that

$$P \supset Q, \qquad P \supset R$$

$$Q \Vdash_\alpha \{\delta\}^{\mathscr{T}}(m) = 0,$$

and

$$R \Vdash_\alpha \{\delta\}^{\mathscr{T}}(m) = 1;$$

otherwise $\{\delta\}^T$ would be hyperarithmetic in P and hence belong to $L[\alpha, K]$ for every $T \in P$. Thus

$$H \Vdash_\alpha (Em)(E^0\mathscr{Q})(E^1\mathscr{Q})(v)_{v<2}(\{\delta\}^{\mathscr{T}}(m) = v \;\&\; \mathscr{P} \supset {}^v\mathscr{Q}).$$

Hence there exist $I \subset H$ and m, ${}^0\mathscr{Q}$, ${}^1\mathscr{Q}$ such that $(H, \mathscr{P}) \supset (I, {}^v\mathscr{Q})$ and

$$(I, {}^v\mathscr{Q}) \Vdash_\alpha \{\delta\}^{\mathscr{T}}(m) = v$$

for all $v < 2$.

Lemma 4.16. *Let* $\mathscr{F}(x, y)$ *be a* Σ_1 *formula of* $\mathscr{L}(\alpha, \mathscr{T})$. *Suppose* $\beta < \alpha$ *and* $U \Vdash_\alpha (x^\beta)(Ey)\mathscr{F}(x^\beta, y)$. *Then there exists a* $V \subset U$ *and a* $\gamma < \alpha$ *such that*

$$V \Vdash_\alpha (x^\beta)(Ey^\gamma)\, \mathscr{F}(x^\beta, y^\gamma).$$

Proof 4.16 follows from 4.13 and 4.11 in the same manner that 4.6 follows from 4.4 and 4.3.

4.17. *Unbounded δ's and Inconsistent Formulas*

Assume $\delta < \alpha$. δ is said to be unbounded via the complex forcing condition $(H, \mathscr{P})$ if

$$(H, \mathscr{P}) \Vdash_\alpha \omega_1^{\{\delta\}^{\mathscr{T}}} = \alpha.$$

Suppose δ is unbounded. Let $\mathscr{L}(\alpha, \{\delta\}^{\mathscr{T}})$ be the language obtained by replacing every occurrence of $\mathscr{T}$ in the language $\mathscr{L}(\alpha, \mathscr{T})$ by $\{\delta\}^{\mathscr{T}}$. There exists a Σ_1 formula $\mathscr{K}_\delta(\gamma, n) = \rho$ of $\mathscr{L}(\alpha, \{\delta\}^{\mathscr{T}})$ such that for each $\gamma \in \alpha - \omega$,

$$(H, \mathscr{P}) \Vdash_\alpha \sim\sim[\lambda n \mid \mathscr{K}_\delta(\gamma, n)\colon \omega \xrightarrow[\text{onto}]{1-1} \gamma]. \tag{4.17.1}$$

$\lambda n \mid \mathscr{K}_\delta(\gamma, n)$ is simply the first counting of γ constructible from $\{\delta\}^{\mathscr{T}}$ via an ordinal less than α.

A set S of sentences of $\mathscr{L}(\alpha, \{\delta\}^{\mathscr{T}})$ is said to be pairwise inconsistent if the conjunction of any two distinct members of S is false in $L(\alpha, \{\delta\}^T)$ for every T.

The next three propositions isolate the splitting of finite conditions on K needed for the proof of complex minimization (4.21).

PROPOSITION 4.18. *Suppose δ is unbounded via $(H, \mathscr{P})$, $t \in H$ and $\mathscr{S}$ is an α-finite diversified set of immediate successors of t in H. Then there exist α-finite sets $\{s_\rho \mid \rho \in b\}$, $\{\mathscr{F}_\rho \mid \rho \in b\}$ and $\{(H_\rho, \mathscr{P}_\rho) \mid \rho \in b\}$ such that: $\{s_\rho \mid \rho \in b\}$ is a diversified subset of $\mathscr{S}$; $\{\mathscr{F}_\rho \mid \rho \in b\}$ is a set of pairwise inconsistent ranked sentences of $\mathscr{L}(\alpha, \{\delta\}^{\mathscr{T}})$; and $s_\rho \in H_\rho$, $(H, \mathscr{P}) \supset (H_\rho, \mathscr{P}_\rho)$ and $(H_\rho, \mathscr{P}_\rho) \Vdash_\alpha \mathscr{F}_\rho$ for all $\rho \in b$.*

Proof. Recall from 4.1 that all immediate successors of t in H have the same relevant domain d. If d is empty, then the conclusion of 4.18 can be satisfied trivially by setting $\{s_\rho \mid \rho \in b\}$ equal to a singleton $\{s\}$ for some $s \in \mathscr{S}$. Thus suppose d is $\{\beta_i \mid i < z\}$, where $z > 0$. As in 4.1

$$L(\alpha) \models [\beta_i \text{ is a cardinal \& cofinality } \beta_i > \omega], \tag{4.18.1}$$

for all $i < z$. $\mathscr{S}$ is diversified, hence for each $\gamma \in \Pi\{\beta_i \mid i < z\}$, there is an $\langle n_i^\gamma\rangle_{i<z}$ and an $s \in \mathscr{S}$ such that

$$(i)_{i<z}[s(\gamma_i, n_i^\gamma) > \gamma_i]. \tag{4.18.2}$$

It follows from (4.18.1) that $\langle n_i^\gamma \rangle_{i<z}$ can be chosen independently of γ; i.e., there exists an $\langle n_i \rangle_{i<z}$ such that for each $\gamma \in \Pi\{\beta_i \mid i < z\}$, there is an $s \in \mathscr{S}$ such that

$$(i)_{i<z}[s(\gamma_i, n_i) > \gamma_i]. \tag{4.18.3}$$

Call two members of $\mathscr{S}$, say s_1 and s_2, equivalent if $s_1(\beta_i, n_i) = s_2(\beta_i, n_i)$ for all $i < z$. Form $\mathscr{S}_0 \subset \mathscr{S}$ by choosing just one member from each equivalence class of $\mathscr{S}$. The advantage of $\mathscr{S}_0$ over $\mathscr{S}$ is that distinct members of $\mathscr{S}_0$ correspond to distinct members of $\Pi\{\beta_i \mid i < z\}$. ($s$ corresponds to $\langle s(\beta_0, n_0),..., s(\beta_{z-1}, n_{z-1})\rangle$.) Thus, $\mathscr{S}_0 = \{s_\rho \mid \rho \in c\}$, where $c \subset \Pi\{\beta_i \mid i < z\}$. $\mathscr{S}_0$ is diversified thanks to (4.18.3).

For each $\rho \in c$ there is an $\langle m_i^\rho \rangle_{i<z}$ and an $(H_\rho, \mathscr{P}_\rho) \subset (H, \mathscr{P})$ such that $s_\rho \in H_\rho$ and

$$(H_\rho, \mathscr{P}_\rho) \Vdash_\alpha (i)_{i<z}[\mathscr{K}_\delta(\beta_i, m_i^\rho) = \rho_i], \tag{4.18.4}$$

where $\mathscr{K}_\delta$ is the Σ_1 formula of (4.17.1). By 4.11 m^ρ can be regarded as an α-finite function of ρ. Hence the same reasoning that established the existence of $\langle n_i \rangle$ in (4.18.3) also establishes the existence of an $\langle m_i \rangle$ such that for each $\gamma \in c$, there is a $\rho \in c$ such that

$$(i)_{i<z}[\rho_i > \gamma_i \;\&\; m_i^\rho = m_i]. \tag{4.18.5}$$

Let b be the set of all ρ's that satisfy (4.18.4). For each $\rho \in b$, let $\mathscr{F}_\rho$ be $(i)_{i<z}[\mathscr{K}_\delta(\beta_i, m_i) = \rho_i]$.

PROPOSITION 4.19. *Suppose $t \in H$ and*

$$(H, \mathscr{P}^j) \Vdash_\alpha [\omega_1^{\{\delta\}^{\mathscr{T}}} = \alpha]$$

for all $j < i$. Then, there exist α-finite sets $\{s_\rho \mid \rho \in c\}$, $\{\mathscr{F}_\rho^j \mid \rho \in c\}$ and $\{(H_\rho, \mathscr{P}_\rho^j) \mid \rho \in c\}$ $(j < i)$ such that: $\{s_\rho \mid \rho \in c\}$ is a diversified set of immediate successors of t in H; for each $j < i$, $\{\mathscr{F}_\rho^j \mid \rho \in c\}$ is a set of pairwise inconsistent ranked sentences of $\mathscr{L}(\alpha, \{\delta\}^{\mathscr{T}})$; and $s_\rho \in H_\rho$, $(H, \mathscr{P}^j) \supset (H_\rho, \mathscr{P}_\rho^j)$ and $(H_\rho, \mathscr{P}_\rho^j) \Vdash_\alpha \mathscr{F}_\rho^j$ for all $j < i$ and $\rho \in c$.

Proof (By induction on i). If $i = 1$, apply 4.18 with P as P^0 and $\mathscr{S}$ as the set of all immediate successors of t in H. Assume that $i > 1$, and that the α-finite sets $\{s_\rho \mid \rho \in c\}$, $\{\mathscr{F}_\rho^j \mid \rho \in c\}$ and $\{(H_\rho, \mathscr{P}_\rho^j) \mid \rho \in c\}$ $(j < i)$ satisfy the conclusions of 4.19. It is safe to assume $\{s_\rho \mid \rho \in c\}$ has the same form as $\mathscr{S}_0$ in the proof of 4.18. Thus $c \subset \Pi\{\beta_i \mid i < z\}$ and s_ρ is the unique member of $\{s_\rho \mid \rho \in c\}$ with the property that $s_\rho(\beta_i, n_i) = \rho_i$ for all $i < z$.

For each $\rho \in c$ there is an $\langle m_i^\rho \rangle_{i<z}$ and an $(I_\rho, \mathscr{P}_\rho^i) \subset (H_\rho, \mathscr{P}^i)$ such that $s_\rho \in I_\rho$ and

$$(I_\rho, \mathscr{P}_\rho^i) \Vdash_\alpha (i)_{i<z}[\mathscr{K}_\delta(\beta_i, m_i^\rho) = \rho_i],$$

where $\mathscr{K}_\delta$ is the Σ_1 formula of (4.17.1). As in the second half of the proof of 4.18, there is an $\langle m_i \rangle$ such that for each $\gamma \in c$, there is a $\rho \in c$ such that

$$(i)_{i<z}[\rho_i > \gamma_i \,\&\, m_i^\rho = m_i]. \tag{4.19.1}$$

Let b be the set of all ρ's that satisfy (4.19.1). For each $\rho \in b$, let $\mathscr{F}_\rho^i$ be $(i)_{i<z}[\mathscr{K}_\delta(\beta_i, m_i) = \rho_i]$. Then the α-finite sets $\{s_\rho \mid \rho \in b\}$, $\{\mathscr{F}_\rho^j \mid \rho \in b\}$ and $\{(I_\rho, \mathscr{P}_\rho^j) \mid \rho \in b\}$ $(j < i + 1)$ satisfy the conclusions of 4.19.

PROPOSITION 4.20. *Suppose* $t \in H$ *and*

$$(H, \mathscr{P}^j) \Vdash_\alpha [\omega_1^{\{\delta\}^{\mathscr{T}}} = \alpha \,\&\, \{\delta\}^{\mathscr{T}} \notin L[\alpha, \mathscr{K}]],$$

for all $j < i$. *Then there exist* α-*finite sets* $\{s_\rho \mid \rho \in c\}$, $\{\mathscr{F}_\rho^j \mid j < i \,\&\, \rho \in c\}$ *and* $\{(H_\rho, \mathscr{P}_\rho^j) \mid j < i \,\&\, \rho \in C\}$ *such that*: $\{s_\rho \mid \rho \in c\}$ *is a diversified set of immediate successors of* t *in* H; $\{\mathscr{F}_\rho^j \mid j < i \,\&\, \rho \in c\}$ *is a set of pairwise inconsistent ranked sentences of* $\mathscr{L}(\alpha, \{\delta\}^{\mathscr{T}})$; *and* $s_\rho \in H_\rho$, $(H, \mathscr{P}^j) \supset (H_\rho, \mathscr{P}_\rho^j)$ *and* $(H_\rho, \mathscr{P}_\rho^j) \Vdash_\alpha \mathscr{F}_\rho^j$ *for all* $j < i$ *and* $\rho \in c$.

Proof. Consider the α-finite sets supplied by Proposition 4.19. They satisfy all the claims of 4.20 save one: pairwise inconsistency of the members of $\{\mathscr{F}_\rho^i \mid j < i \,\&\, \rho \in c\}$. But 4.19 does provide inconsistency for the members of $\{\mathscr{F}_\rho^j \mid \rho \in c\}$ for each $j < i$. Hence the missing inconsistency can be realized by developing a set $\{\mathscr{G}^j \mid j < i\}$ of pairwise inconsistent sentences of $\mathscr{L}(\alpha, \{\delta\}^{\mathscr{T}})$, and then replacing each $\mathscr{F}_\rho^j$ by $\mathscr{F}_\rho^j \,\&\, \mathscr{G}^j$. It also will be necessarily to contract c to b and $(H_\rho, \mathscr{P}_\rho^j)$ to $(I_\rho, \mathscr{Q}_\rho^j)$ in such a way that: $\{s_\rho \mid \rho \in b\}$ is a diversified set of immediate successors of t in H; and $s_\rho \in I_\rho$ and

$$(I_\rho, \mathscr{Q}_\rho) \Vdash_\alpha \mathscr{F}_\rho^j \,\&\, \mathscr{G}^j,$$

for all $j < i$ and $\rho \in b$.

Each $\mathscr{G}^j$ is of the form $\&\{\mathscr{G}_k^j \mid k \neq j \,\&\, k < i\}$. The $\mathscr{G}_k^j$'s are defined by means of a process with $\binom{i}{2}$ stages. A typical stage is devoted to choosing a single pair $(\mathscr{G}_k^j, \mathscr{G}_j^k)$ of inconsistent ranked sentences of $\mathscr{L}(\alpha, \{\delta\}^{\mathscr{T}})$.

For the sake of notational simplicity, only the first stage is described; it has all the features of the later stages.

The first stage is a search for $\mathscr{G}_1{}^0$ and $\mathscr{G}_0{}^1$; their existence will follow from the premise that $(H, \mathscr{P}^0) \Vdash_\alpha [\{\delta\}^{\mathscr{T}} \notin L(\alpha, K)]$. By 4.15, for each $\rho \in c$, there exist m^ρ and $(J_\rho, {}^v\mathscr{Q}_\rho) \subset (H_\rho, \mathscr{P}_\rho{}^0)$ such that

$$(J_\rho, {}^v\mathscr{Q}_\rho) \Vdash_\alpha \{\delta\}^{\mathscr{T}}(m^\rho) = v$$

for all $v < 2$. As in the second half of the proof of 4.18 there is an m and a $b_0 \subset c$ such that $\{s_\rho \mid \rho \in b_0\}$ is a diversified set of immediate successors of t in H and $m^\rho = m$ for all $\rho \in b_0$. (4.18 relied on the fact that if a diversified set is divided into countably many parts, then one of those parts is diversified.)

For each $\rho \in b_0$ there is a $v^\rho < 2$ and an $(I_\rho, \mathscr{R}_\rho) \subset (J_\rho, \mathscr{P}_\rho{}^1)$ such that

$$(I_\rho, \mathscr{R}_\rho) \Vdash_\alpha \{\delta\}^{\mathscr{T}}(m) = v^\rho.$$

As in the previous paragraph there is a $w < 2$ and a $b_1 \subset b_0$ such that $\{s_\rho \mid \rho \in b_1\}$ is a diversified set of immediate successors of t in H and $v^\rho = w$ for all $\rho \in b_1$. Let $\mathscr{G}_1{}^0$ be $[\{\delta\}^{\mathscr{T}}(m) = 1 - w]$ and let $\mathscr{G}_0{}^1$ be $[\{\delta\}^{\mathscr{T}}(m) = w]$. Then each $\rho \in b_1$ has the following properties: $H_\rho \supset I_\rho$, $P_\rho{}^0 \supset {}^{(1-w)}\mathscr{Q}_\rho$, $\mathscr{P}_\rho{}^1 \supset \mathscr{R}_\rho$,

$$(I_\rho, {}^{(1-w)}\mathscr{Q}_\rho) \Vdash_\alpha \mathscr{F}_\rho^0 \,\&\, \mathscr{G}_1{}^0,$$

and

$$(I_\rho, \mathscr{R}_\rho) \Vdash_\alpha \mathscr{F}_\rho^1 \,\&\, \mathscr{G}_0^1.$$

Each of the remaining $\binom{i}{2} - 1$ stages has the same pattern of contractions of forcing conditions and diversified sets as the first stage.

LEMMA 4.21 (Complex Minimization). *Suppose*

$$(H, \mathscr{P}) \Vdash_\alpha [\omega_1^{\{\delta\}^{\mathscr{T}}} = \alpha \,\&\, \{\delta\}^{\mathscr{T}} \notin L[\alpha, \mathscr{K}]].$$

Then there exists a $(J, \mathscr{R}) \subset (H, \mathscr{P})$ *such that*

$$(J, \mathscr{R}) \Vdash_\alpha [\mathscr{T} \in L(\alpha, \{\delta\}^{\mathscr{T}})].$$

Proof. The stratagems of 3.4 (Case 2) and 4.13 are combined. $(J, \mathscr{R})$ will be defined so that $\{\delta\}$ is one–one in the following sense: suppose

$K_i \in J$ and $T_i \in \mathscr{R}(K_i)$ $(i < 2)$; if $(K_0, T_0) \neq (K_1, T_1)$, then $\{\delta\}^{T_0} \neq \{\delta\}^{T_1}$. The one–one-ness of $\{\delta\}$ will make it possible to recover $(K^{\text{rank}\mathscr{R}}, T)$ from $\{\delta\}^T$ whenever $K \in J$ and $T \in \mathscr{R}(K)$.

α-finite partial functions $t_\sigma{}^i$, $I_\sigma{}^i$, $\mathscr{Q}_\sigma^{i,j}$, $r_\sigma^{i,i}$, and $\mathscr{F}_\sigma^{i,j}$ are defined by recursion on i for all $i < \omega$, some $\sigma \in \alpha^i$ and all $j < 2^i$. The definition proceeds as in 4.13 with no change in (4.13.1)–(4.13.5) and (4.13.7), but with (4.13.6) replaced by (4.13.6*) below.

$$(I_\tau^{i+1}, \mathscr{Q}_\tau^{i+1,j}) \Vdash_\alpha \mathscr{G}_\tau^{i+1,j}, \tag{4.13.6*}$$

where $\{\mathscr{G}_\tau^{i+1,j} \mid j < 2^{i+1} \;\&\; \tau \in b_\sigma{}^i\}$ is a set of pairwise inconsistent ranked sentences of $\mathscr{L}(\alpha, \{\delta\}^{\mathscr{T}})$. Requirement (6*) is satisfied with the help of 4.20 and 4.15. By 4.20 there exist $(J_\tau^{i+1}, \mathscr{P}_\tau^{i,j})$ and $\mathscr{F}_\tau^{i,j}$ $(j < 2^i$ and $\tau \in b_\sigma{}^i)$ such that: $t_\tau^{i+1} \in J_\tau^{i+1}$,

$$(I_\sigma{}^i, \mathscr{Q}_\sigma^{i,j}) \supset (J_\tau^{i+1}, \mathscr{P}_\tau^{i,j}),$$

$$(J_\tau^{i+1}, \mathscr{P}_\tau^{i,j}) \Vdash_\alpha \mathscr{F}_\tau^{i,j},$$

and $\mathscr{F}_\tau^{i,j}$'s constitute a set of pairwise inconsistent ranked sentences of $\mathscr{L}(\alpha, \{\delta\}^{\mathscr{T}})$. By 4.15, there exist I_τ^{i+1}, $\mathscr{Q}_\tau^{i+1,2j}$, $\mathscr{Q}_\tau^{i+1,2j+1}$, and $m_\tau^{i,j}$ $(j < 2^i$ and $\tau \in b_\sigma{}^i)$ such that: $t_\tau^{i+1} \in I_\tau^{i+1}$,

$$(J_\tau^{i+1}, \mathscr{P}_\tau^{i,j}) \supset (I_\tau^{i+1}, \mathscr{Q}_\tau^{i+1,2j}), \tag{4.21.1}$$

$$(I_\tau^{i+1}, \mathscr{Q}_\tau^{i+1,2j}) \Vdash_\alpha \{\delta\}^{\mathscr{T}}(m_\tau^{i,j}) = 0, \tag{4.21.2}$$

and (4.21.1) and (4.21.2) with $2j$ replaced by $2j + 1$ and 0 by 1. Then, $\mathscr{G}_\tau^{i+1,2j}$ is $[\mathscr{F}_\tau^{i,j} \;\&\; \{\delta\}^{\mathscr{T}}(m_\tau^{i,j}) = 0]$ and $\mathscr{G}_\tau^{i+1,2j+1}$ is $[\mathscr{F}_\tau^{i,j} \;\&\; \{\delta\}^{\mathscr{T}}(m_\tau^{i,j}) = 1]$.

Define $(J, \mathscr{R})$ as in 4.13. The argument given in 4.13 to show $(J, \mathscr{R})$ is a complex forcing condition still holds. Fix $K \in J$ and $T \in \mathscr{R}(K)$ to see how to recover $(K^{\text{rank}\mathscr{R}}, T)$ from $\{\delta\}^T$ by a uniform method encoded by an ordinal less than α. $K^{\text{rank}\mathscr{R}}$ corresponds to a unique path through J. Thus there is a function σi such that

$$K^{\text{rank}\mathscr{R}} = \bigcup \{t_{\sigma i}^i \mid i < \omega\}.$$

Similarly there is a function hi such that

$$\{T\} = \bigcap \{\mathscr{Q}_{\sigma i}^{i,hi}(K^{\text{rank}\mathscr{R}}) \mid i < \omega\}.$$

Thanks to (4.13.6*) a simple recursion on i yields σi and hi: $\sigma 0 =$

$h0 = 0$; $(\sigma(i+1), h(i+1))$ is the unique (τ, j) such that $\tau \in b^i_{\sigma i}$, $j < 2^{i+1}$ and

$$L(\alpha, \{\delta\}^T) \models \mathscr{G}^{i+1,j}_\tau.$$

The functions σi and hi belong to $L(\alpha, \{\delta\}^T)$, because the entire construction of $(J, \mathscr{R})$ can be viewed as an element of $L(\alpha)$, and because just one of the ranked sentences $\mathscr{G}^{i+1,j}_\tau$ ($\tau \in b_\sigma{}^i$ and $j < 2^{i+1}$) is true in $L(\alpha, \{\delta\}^T)$.

LEMMA 4.22. *Let K be generic in the sense of* 4.2. *Suppose that the set of all $\beta < \alpha$ such that*

$$L(\alpha) \models [\beta \textit{ is a cardinal}]$$

is not bounded below α. Then $\omega_1{}^X < \alpha$ for every $X \in L[\alpha, K]$.

Proof. Suppose not. Then there is a term c of rank $\gamma < \alpha$ such that

$$\omega_1^{c(K^\gamma)} = \alpha.$$

In addition there is an H such that $K \in H$ and

$$H \Vdash_\alpha \omega_1{}^c = \alpha. \tag{4.22.1}$$

Let β be a cardinal in the sense of $L(\alpha)$ such that c, $H \in L(\beta)$. A sequence $\{I_n \mid n < \omega\}$ of perfect conditions on K is defined by recursion on n. I_0 is H. I_{n+1} is the "least" I such that

$$I_n \supset I \qquad \text{and} \qquad I \Vdash_\alpha [n \in O^c]$$

if there is such an I; if not, I_{n+1} is I_n. $[n \in O^c]$ is a Σ_1 sentence of $\mathscr{L}[\alpha, \mathscr{K}]$ that says n is a notation for an ordinal recursive in $c(\mathscr{K}^\gamma)$. Least means least with respect to the standard Δ_1 wellordering of $L(\alpha)$. Since β is an $L(\alpha)$-cardinal, $L(\beta)$ is a Σ_1 substructure of $L(\alpha)$. It follows from 4.3 that each $I_n \in L(\beta)$. Also, if $I_n \Vdash_\alpha [n \in O^c]$, then I_n forces n to be a notation for an ordinal $< \beta$.

Let J be $\bigcap \{I_n \mid n < \omega\}$. The splitting argument of 4.4 can be used to ensure J is a perfect condition on K. Then $J \Vdash_\alpha \omega_1{}^c \leqslant \beta$, an impossibility according to (4.22.1).

4.23. *K as a Bounded Set*

Let Z^* be the set of all $\beta < \alpha$ such that

$$L(\alpha) \models [\beta \text{ is a cardinal \& cofinality } \beta > \omega].$$

When Z^* is not bounded below α, Theorem 4.26 follows fairly directly from Lemmas 4.21 and 4.22, but when Z^* is bounded, a modification of 4.21, namely 4.25, is required.

Suppose the supremum of Z^* is $\beta_\infty < \alpha$. Clearly β_∞ is an $L(\alpha)$-cardinal. If $\beta_\infty \in Z^*$, define $Z_0^* = \{\beta_\infty\}$. If $\beta_\infty \notin Z^*$, define $Z_0^* = \{\beta_i \mid i < \omega\}$, where $\lambda i \mid \beta i$ is an α-finite sequence of members of Z^* whose supremum is β_∞. Z_0^* is used to simplify the convention on K stated in 4.1 as follows: K is a map from $(Z_0^* - \omega) \times \omega$ into β_∞ such that if $\beta \in Z_0^*$, then the range of $\lambda m \mid K(\beta, m)$ is an unbounded subset of β. The new convention on K implies $K \in L[\alpha, K]$ as well as

$$L[\alpha, K] \models (x)(x \text{ is countable}).$$

When working with the new convention, it is convenient to reduce the collection of forcing conditions $H, I, J,\ldots$, of 4.1 to those that mention only those β's in Z_0^*. Since Z_0^* is countable, no difficulty is caused by requiring that each forcing condition H mention *every* member of Z_0^*. Thus each path through H is the complete graph of a map from $(Z_0^* - \omega) \times \omega$ into β_∞.

Let $\{n\}^{\mathscr{K}}$ be a term of $\mathscr{L}[\alpha, \mathscr{K}]$ that denotes a map from $(Z_0^* - \omega) \times \omega$ into β_∞. The n is said to be *unbounded via* H if

$$H \Vdash_\alpha [\lambda m \mid \{n\}^{\mathscr{K}}(\beta, m) \colon \omega \xrightarrow[\text{onto}]{1-1} \beta],$$

for all $\beta \in Z_0^*$.

Let $\mathscr{L}[\alpha, \{n\}^{\mathscr{K}}]$ be the language obtained by replacing every occurrence of $\mathscr{K}$ in the language $\mathscr{L}[\alpha, \mathscr{K}]$ by $\{n\}^{\mathscr{K}}$. A set S of sentences of $\mathscr{L}[\alpha, \{n\}^{\mathscr{K}}]$ is said to be pairwise inconsistent if the conjunction of any two distinct members of S is false in $L[\alpha, \{n\}^{\mathscr{K}}]$ for every K.

LEMMA 4.24. *Suppose n is unbounded via H as in 4.23, $t \in H$ and $\mathscr{S}$ is an α-finite diversified set of immediate successors of $t \in H$. Then there exist α-finite sets $\{s_\rho \mid \rho \in b\}$, $\{\mathscr{F}_\rho \mid \rho \in b\}$ and $\{H_\rho \mid \rho \in b\}$ such that: $\{s_\rho \mid \rho \in b\}$ is a diversified subset of $\mathscr{S}$; $\{\mathscr{F}_\rho \mid \rho \in b\}$ is a set of pairwise inconsistent ranked sentences of $\mathscr{L}[\alpha, \{n\}^{\mathscr{K}}]$; and $s_\rho \in H_\rho$, $H \supset H_\rho$, and $H_\rho \Vdash_\alpha \mathscr{F}_\rho$, for all $\rho \in b$.*

Proof. Same as that of 4.18 with $\mathscr{K}_\delta$ replaced by $\{n\}^{\mathscr{K}}$, *and* $(H, \mathscr{P})$ by H. Thus (4.18.4) becomes

$$H \Vdash_\alpha (i)_{i<z}[\{n\}^{\mathscr{K}}(\beta_i, m_i^\rho) = \rho_i],$$

and $\mathscr{F}_\rho$ becomes $(i)_{i<z}[\{n\}^{\mathscr{K}}(\beta_i, m_i) = \rho_i]$.

LEMMA 4.25. *Suppose n is unbounded via H as in 4.23. Then there exists a $J \subset H$ such that*

$$J \Vdash_{\alpha} (\mathscr{K} \in L[\alpha, \{n\}^{\mathscr{K}}]).$$

Proof. Similar to, but much simpler than, the proof of 4.21, with 4.24 taking the place of 4.20. J is defined so that n is 1–1 on J, thereby making it possible to recover K from $\{n\}^{\mathscr{K}}$ when $K \in J$.

As in 4.21, α-finite partial functions $I_\sigma{}^i$, $t_\sigma{}^i$, and $\mathscr{F}_\sigma{}^i$ are defined by recursion on i for all i and some $\sigma \in \alpha^i$. $I_0{}^0 = H$ and $t_0{}^0 \in I_0{}^0$. The t_τ^{i+1}'s constitute a diversified set of immediate successors to $t_\sigma{}^i$ inside $I_\sigma{}^i$. In addition the following requirements are met with the help of 4.24.

$$t_\tau^{i+1} \in I_\tau^{i+1} \subset I_\sigma{}^i, \tag{4.25.1}$$

$$(t)[t \in I_\tau^{i+1} \rightarrow t_\tau^{i+1} \geqslant t], \tag{4.25.2}$$

$$I_\tau^{i+1} \Vdash_\alpha \mathscr{F}_\tau{}^i, \tag{4.25.3}$$

where the $\mathscr{F}_\tau{}^i$'s in (4.25.3) form a set of pairwise inconsistent ranked sentences of $\mathscr{L}[\alpha, \{n\}^{\mathscr{K}}]$.

Let J be the perfect condition given by $\lambda i\sigma \mid t_\sigma{}^i$. Fix $K \in J$ to see how to recover K from $\{n\}^{\mathscr{K}}$ by a uniform method encoded by an ordinal less than α. K corresponds to a unique path through J. (Keep in mind that the conventions on K and J established in 4.23 hold throughout 4.24 and 4.25.) Thus there is a function σi such that

$$K = \bigcup \{t_{\sigma i}^i \mid i < \omega\}.$$

A simple recursion on i yields σi. $\sigma(i+1)$ is the unique τ such that

$$L[\alpha, \{n\}^{\mathscr{K}}] \models \mathscr{F}_\tau{}^i.$$

THEOREM 4.26. *Let α be a countable admissible ordinal greater than ω. Then there exists an $X \subset \omega$ such that*

(i) $\omega_1{}^X = \alpha$;

(ii) $(Y)[Y <_h X \rightarrow \omega_1{}^Y < \alpha]$.

Proof. There are four cases of which the third is the most troublesome.

Case 1.

There is an $X \in L(\alpha)$ such that $\omega_1^X = \alpha$. Every such X satisfies (ii), since every real in $L(\alpha)$ is hyperarithmetic in every Y with the property that $\omega_1^Y \geqslant \alpha$.

Case 2.

Z^* (defined in 4.23) is empty and the hypothesis of Case 1 does not hold. Thus $L(\alpha)$ satisfies (x) (x is countable). Let T be generic in the sense of 3.5 with A equal to $L(\alpha)$. The argument of 3.8 shows $\omega_1^T \leqslant \alpha$. By (3.6.1), $\omega_1^T \geqslant \alpha$. Suppose $Y \leqslant_h T$ and $\omega_1^Y = \alpha$. Then $Y \notin L(\alpha)$ since Case 1 does not hold. Consequently $T \leqslant_h Y$ by (3.6.2).

Case 3.

Z^* is an unbounded subset of α. Let (K, T) be generic in the sense of 4.14. It follows from 4.12a that $\omega_1^T \geqslant \alpha$, and from 4.16 that $\omega_1^T \leqslant \alpha$. Suppose $Y \leqslant_h T$ and $\omega_1^Y = \alpha$. Then, $Y \notin L[\alpha, K]$ by 4.22 since Z^* is unbounded. Consequently $T \leqslant_h Y$ by 4.21.

Case 4.

Z^* is bounded and nonempty. Let K be generic as in 4.2, but with the stipulation that all forcing conditions $H, I, J,\dots$ obey the conventions of 4.23. Thus $\Vdash_\alpha$ is defined as in 4.2 save for the limitations on K and $H, I, J,\dots$ imposed by 4.23.

Subcase 4a. There is an $X \in L[\alpha, K]$ such that $\omega_1^X = \alpha$. Suppose $Y \leqslant_h X$ and $\omega_1^Y = \alpha$. Then there is an $n < \alpha$ and an H such that n is unbounded via H (as defined in 4.23), and such that $\{n\}^K$ is encoded by a subset of ω hyperarithmetic in Y. But then $K \in L[\alpha, \{n\}^K]$ by 4.25, and so $X \leqslant_h Y$.

Subcase 4b. The hypothesis of Subcase 4a is false. It is safe to assume there is a T such that (K, T) is generic in the sense of 4.14, but with the stipulation that all forcing conditions $H, I, J,\dots$ obey the conventions of 4.23. Now proceed as in Case 3 with 4.22 replaced by the hypothesis of Subcase 4b.

5. Further Results and Open Questions

Let $n \geqslant 1$. An ordinal α is said to be Σ_n admissible if $L(\alpha)$ satisfies Rep(n), the replacement axiom schema of ZF restricted to Σ_n formulas.

For each real T, $\omega_{\Sigma_n}^T$ is the least α such that $L(\alpha, T)$ satisfies Rep(n). A real S is Σ_n recursive in T if $S \in L(\omega_{\Sigma_n}^T, T)$. Two reals have the same Σ_n degree if each is Σ_n recursive in the other.

THEOREM 5.1. *Let α be a countable Σ_n admissible ordinal greater than ω. Then, there exists a T such that:*

(i) $\omega_{\Sigma_n}^T = \alpha$;

(ii) $\omega_{\Sigma_n}^S < \alpha$ *for every S of lesser Σ_n degree than T.*

For $n > 1$ a T with property (i) was first discovered by Jensen. A sketch of the proof of 5.1 was given in [20]. That sketch should not be difficult to complete after a reading of Section 4 of the present paper. The appropriate notions of pointedness are as follows. P is Σ_n pointed if

$$(T)[T \in P \rightarrow P \in L(\omega_{\Sigma_n}^T; T)].$$

P is uniformly Σ_n pointed via δ if

$$(T)[T \in P \rightarrow P = \{\delta\}^T \;\&\; \delta < \omega_{\Sigma_n}^T].$$

The proof of 2.10 can be modified to show: Each Σ_n pointed P contains a uniformly Σ_n pointed Q of the same Σ_n degree as P. Note that 2.2 and 2.3 hold for Σ_n pointedness.

P is K-uniformly Σ_n pointed via $\mathscr{P}$ and δ if for all $T \in P$:

$$\beta < \omega_{\Sigma_n}^T \quad \text{and} \quad P = \mathscr{P}(K^\beta),$$

$$\delta < \omega_{\Sigma_n}^T \quad \text{and} \quad K^\beta = \{\delta\}^T.$$

When $n > 1$, the proof of Lemma 4.13 (complex splitting) needs only one substantial change to make it work for Σ_n $\mathscr{F}_i$'s and complex conditions whose virtual components are Σ_n pointed. Requirement (4.13.7) is augmented by

$$L(\beta_i) \prec_{n-1} L(\beta_{i+1}) \prec_{n-1} L(\alpha).$$

($\mathfrak{A} \prec_{n-1} \mathfrak{B}$ means every Σ_n sentence with parameters in $\mathfrak{A}$ is true in $\mathfrak{B}$ if it is true in $\mathfrak{A}$.) It then follows from the elementary chain principle that

$$L(\sup_i \beta_i) \prec_{n-1} L(\alpha). \tag{5.1.1}$$

The forcing relation $\Vdash_\alpha$ restricted to Σ_n formulas is Σ_n, hence (5.1.1) implies that the $(I_\tau^{i+1}, \mathcal{Q}_\tau^{i+1,j})$ chosen in $L(\alpha)$ to force $\mathcal{F}_i$ could just as well have been chosen in $L(\sup_i \beta_i)$. Thus, as in 4.13, the entire construction of $(J, \mathcal{R})$ is an $\omega_{\Sigma_n}^T$-finite object when $T \in \mathcal{R}(K)$ for some $K \in J$, since in that event $\omega_{\Sigma_n}^T \geqslant \sup_i \beta_i$. Note that the predicate $L(\beta) <_{n-1} L(\alpha)$ is Σ_n over $L(\alpha)$.

Call a transitive set A Σ_n admissible if it is closed under the operations of pairing and set union, and satisfies the Δ_n separation and bounding schemas of ZF (cf. 2.4).

THEOREM 5.2. *Let A be a countable Σ_n admissible set that satisfies Σ_n dependent choice. Then, the Σ_n degrees contained in A have a minimal upper bound; if they have a least upper bound, then, that least upper bound is contained in A.*

The proof of 5.2 is similar to that of 3.14. The only difference is the use of an elementary chain argument as in 5.1.

THEOREM 5.3. *Let M be a countable standard model of ZF and $V = L$. Suppose $\mathcal{K}$ is a regular uncountable cardinal of M. Then there exists a $T \subset \omega$ such that:*

(i) *$M(T)$ is a model of ZF in which $\mathcal{K}$ is the least uncountable cardinal;*

(ii) *if $S \in M(T)$ and $M(S)$ has the same properties as $M(T)$ in* (i), *then $M(S) = M(T)$.*

The proof of 5.3 for $\mathcal{K} = \omega_2$ is due to Prikry. A T with property (i) was given by Jensen and Solovay in [21]. The proof of 5.3 is similar to that of 4.26. The appropriate notions of pointedness are as follows. P is constructibly pointed if

$$(T)[T \in P \to P \in L(T)].$$

P is uniformly constructibly pointed via δ if

$$(T)[T \in P \to P = \{\delta\}^T \mathbin{\&} \delta < \omega_1^{L(T)}],$$

where $\omega_1^{L(T)}$ is the least uncountable cardinal of $L(T)$. P is K-uniformly constructibly pointed via $\mathcal{P}$ and δ if for all $T \in P$:

$$P = \mathcal{P}(K^\beta) \quad \text{and} \quad \beta < \omega_1^{L(T)},$$

$$K^\beta = \{\delta\}^T \quad \text{and} \quad \delta < \omega_1^{L(T)}.$$

Of course, P, β, and δ belong to M.

5.4 *Questions*

(5.4.1) Suppose A is a countable Σ_n admissible set such that the Σ_n degrees contained in A have a least upper bound. Does that least upper bound belong to A? If it is assumed that A satisfies Σ_n dependent choice, then the forcing argument of 5.2 provides an affirmative answer. The compactness argument of 3.20 does not require Σ_n dependent choice but appears to succeed only when $n = 1$. It is tempting to think that there exist compactness theorems for Σ_n admissible sets ($n > 1$) substantially different from Barwise's for the Σ_1 case.

(5.4.2) Does every countable set of hyperdegrees have a minimal upper bound?

ACKNOWLEDGMENT

The author wishes to thank B. Dreben, G. Kreisel, A. Nerode, and H. Rogers for their unfailing encouragement.

REFERENCES

1. H. FRIEDMAN AND R. JENSEN, A note on admissible ordinals, *in* "Lecture Notes in Mathematics 72," pp. 77–78, Springer–Verlag, 1968.
2. T. GRILLIOT, Omitting types; applications to recursion theory, *J. Symbolic Logic* **37** (1972), 81–89.
3. H. J. KEISLER, "The Model Theory of Infinitary Languages," North-Holland, 1971.
4. G. E. SACKS, Forcing with perfect closed sets, *in* "Proc. Symposia Pure Math. XIII," pp. 331–355, Amer. Math. Soc., 1971.
5. G. E. SACKS, "Degrees of Unsolvability," Princeton Univ. Press, 1963; 2nd ed., 1966.
6. H. FRIEDMAN, Minimality in the Δ_2^1 degrees, *Fund. Math.*, to appear.
7. S. C. KLEENE AND E. L. POST, The upper semilattice of degrees of recursive unsolvability, *Ann. Math.* **59** (1954), 379–407.
8. R. PLATEK, Foundations of recursion theory, Ph.D. Thesis, Stanford Univ., 1965.
9. J. BARWISE, Infinitary logic and admissible sets, *J. Symbolic Logic* **34** (1969), 226–252.
10. S. KRIPKE, Transfinite recursions on admissible ordinals (Abstracts), *J. Symbolic Logic* **29** (1964), 161–162.
11. C. SPECTOR, On degrees of recursive unsolvability, *Ann. Math.* **64** (1956), 581–592.
12. G. KREISEL, The axiom of choice and the class of hyperarithmetic functions, *Indag. Math.* **24** (1962), 307–319.
13. G. BOOLOS AND H. PUTNAM, Degrees of unsolvability of constructible sets of integers, *J. Symbolic Logic* **33** (1968), 497–513.
14. C. SPECTOR, Hyperarithmetical quantifiers, *Fund. Math.* **48** (1959), 313–320.
15. S. FEFERMAN, Some applications of the notion of forcing and generic sets, *Fund. Math.* **56** (1965), 325–345.

16. R. GANDY AND G. E. SACKS, A minimal hyperdegree, *Fund. Math.* **61** (1967), 215–223.
17. S. C. KLEENE, Recursive functionals and quantifiers of finite type, *Trans. Amer. Math. Soc.* **91** (1959), 1–52; **108** (1963), 106–142.
18. G. E. SACKS, The 1-section of a type *n* object, *in* "Generalized Recursion Theory," pp. 81–96, North-Holland, 1974.
19. W. RICHTER, Constructive transfinite number classes, *Bull. Amer. Math. Soc.* **73** (1967), 261–265.
20. G. E. SACKS, *F*-recursiveness, *in* "Logic Colloquium '69," pp. 289–303, North-Holland, 1971.
21. R. B. JENSEN AND R. M. SOLOVAY, Some applications of almost disjoint sets, *in* "Mathematical Logic and Foundations of Set Theory," pp. 84–104, North-Holland, 1970.
22. S. SIMPSON, Minimal covers and hyperdegrees, *Trans. Amer. Math. Soc.*, to appear.
23. C. JOCKUSCH AND S. SIMPSON, A degreetheoretic hierarchy for ramified analysis, *Ann. Math. Logic*, to appear.

THE K-SECTION OF A TYPE n OBJECT.*

By Gerald E. Sacks.

1. Introduction. This paper includes a proof of the plus-one theorem for the case $k>1$. The case $k=1$ was covered in [1].

Plus-one Theorem. *Suppose $n>k\geqslant 1$ and F is a normal type n object. Then there exists a normal type $k+1$ object G whose k-section equals that of F.*

The k-section of a finite type object H consists of the type k objects recursive in H; it was first defined and studied by Kleene [2]. H is said to be normal if the equality predicate for objects of type less than that of H is recursive in H. The plus-one theorem states that k-sections generated by normal finite type objects are in fact generated by normal type $k+1$ objects. Thus for each $k\geqslant 1$, very little about the type of a normal object can be inferred from its k-section. The role of normality in the plus-one theorem is uncertain, because there exist substantially nonnormal objects whose k-sections are the same as those of normal objects. But at the least it may be noted that the normality of H imposes a hierarchic structure on the computations associated with recursion in H, and that hierarchies are often amenable to forcing and stability arguments.

The proof of the plus-one theorem appears to split into two irreconcilable cases: $k=1$ and $k>1$. The difference between the two merits more reflection than a brief introduction allows. Perhaps it will suffice to draw attention to the most immediate difference between the 1- and 2-sections of a normal type n object ${}^nH$. Each member of the 1-section of ${}^2H$ owes its existence to a computation with the property that it and all its subcomputations are encoded by members of the 1-section. However more than one member of the 2-section of ${}^3H$ derives from a computation such that it but not all its subcomputations are encoded by members of the 2-section. It is loose yet not misleading to say the 2-section of ${}^3H$ exhibits gaps. The presence of gaps is to be expected in the study of effective computations of uncountable length.

Manuscript received November 22, 1974.

*The preparation of this paper was partially supported by NSF contract GP-29079 and by the Institute for Advanced Study, Princeton, New Jersey. The author is indebted to A. Kechris and L. Harrington for many discriminating suggestions.

Section 2 reviews the elements of recursion in a normal object via the hierarchic approach, introduces the notion of admissible set with gaps and shows it to be exemplified by the k-section of any normal object of type greater than k, defines a notion of reflection for recursion in a normal object, and proves the lemmas of simple and further reflection, the latter lemma a central result due to Harrington [3].

Section 3 introduces the notion of countable abstract k-section ($k>1$), a combination of admissibility and reflection, and illustrates it by means of the k-section of any normal object of type greater than k, devises a version of forcing to show each countable abstract k-section is the k-section of some normal type $k+1$ object, and ends with the concept of stable counting, equivalent to but simpler than the concept of countable abstract k-section.

Section 4 sketches further results, some actual, others potential, by stating Harrington's plus-two theorem [3] for k-envelopes (defined by Moschovakis [4]), commenting on an extended plus-one hypothesis, and formulizing a plus-three problem.

2. Recursion and Reflection.

2.1. *Finite Types.* Let m be a nonnegative integer. $Tp(m)$ is the set of all objects of type m. $Tp(0)=\omega$, and $Tp(m+1)$ is the set of all functions from $Tp(m)$ into ω. Standard, effective coding devices make it safe to view an object of type m as an object of type q for any $q>m$, to construe a finite sequence $a_0,\ldots,a_r$ of objects of types at most m as a single object $\langle a_0,\ldots,a_r\rangle$ of type m, to regard a function from $Tp(m)$ into $Tp(m+1)$ as an object of type $m+1$, and to think of an arbitrary object of type $m+1$ as a set of objects of type m. Sometimes it will prove convenient to equate $Tp(m)$ with its set-theoretic counterpart $R(\omega+m)$. ($R(0)=0$, $R(\delta+1)=2^{R(\delta)}$, and $R(\lambda)=\cup\{R(\delta)|\delta<\lambda\}$ when λ is a limit ordinal.)

Fix $n\geqslant 0$ in order to establish some conventions which will on occasion be violated. The objects of type $n+2$ are denoted by $F,G,\ldots,$ and are called functionals. The objects of type $n+1$ are denoted by $f,g,\ldots,$ and are called functions. The objects of type at most n are denoted by $a,b,\ldots,$ and are called individuals. The set of all individuals is denoted by I. A typical function is a map of I into ω. Subsets of I are denoted by $R,S,\ldots,$ and are called sets. The objects of type less than n, together with the nonnegative integers, are denoted by $r,s,\ldots,$ and are called subindividuals; the set of all such is denoted by SI. These conventions make it possible to state certain results about objects of finite type without making any explicit references to the type involved.

E is the equality functional for sets:

$$E(R,S)= \begin{matrix} 0 & \text{if} & R=S \\ 1 & \text{if} & R\neq S. \end{matrix}$$

(E is equivalent to Kleene's ${}^{n+2}E$.)

F is said to be normal if F is of the form $\langle F',E\rangle$ where F' is a functional of the same type as F. More generally F is said to be normal if E is recursive in F, since in that event F and $\langle F,E\rangle$ have the same degree; i.e. each is recursive in the other. The assumption of normality renders the study of sets recursive in F amenable to methods associated with the hyperarithmetic and constructible hierarchies. Every F in this paper is normal unless otherwise indicated.

2.2. *The $L(F)$ Hierarchy.* Let F and I be as in 2.1. Harrington [3] found the following hierarchy of sets constructible relative to F to be useful in the study of recursion in F.

> $L(0,F)=I=R(\omega+n)=Tp(n)$.
> $X\in L(\sigma+1,F)$ iff $X\subset L(\sigma,F)$ and X is first order definable with parameters over $M(\sigma,F)$.
> $M(\sigma,F)=\langle L(\sigma,F),\varepsilon,F\restriction(Tp(n+1)\cap L(\sigma,F))\rangle$.
> $L(\lambda,F)=\cup\{L(\delta,F)\mid\delta<\lambda\}$ when λ is a limit ordinal.
> $L(F)=\cup\{L(\sigma,F)\mid\sigma\}$.

Note that if $n=0$ and $F={}^2E$, then $L(F)=L$. Let $\mathcal{L}(\mathfrak{F})$ be a first order language suitable for dissecting $L(F)$. $\mathcal{L}(\mathfrak{F})$ is essentially the language of ZF augmented by ranked variables $x^\sigma,y^\sigma,\ldots$ for each σ, set constants to name the elements of $L(F)$, and a function symbol $\mathfrak{F}$.

$L(\sigma,F)$ will be of primary interest only if σ is recursive in F in the sense of 2.3. If F is ${}^2E$, then the ordinals recursive in F turn out to be the familiar recursive ordinals, but if $n>0$ then certain complexities occur, the simplest of which is the existence of gaps between some of the ordinals recursive in F.

2.3. *Set and Ordinals Recursive in F.* The use of hierarchies to develop Kleene's notion of recursion in a normal F goes back to [5], [6] and [7]. The following version, devised by Harrington [3], stays close to the idea of defining the hyperarithmetic reals by iterating the Turing jump. Let $X\subset I$. The F jump of X, denoted by jX, is the effective disjoint union of certain definable subsets of I. Let $\{e\}^X_\pi$ be the e-th function (e-th according to some standard Gödel numbering of formulas) from I into ω whose graph is definable over $\langle I, \varepsilon, X\rangle$ by means of a first order formula whose only parameters are nonnegative integers, and whose atomic parts are of the form $y \in z$ and $y \in \mathcal{X}$, where y and z range over I and $\mathcal{X}$ is a set constant corresponding to X. (Here I is of course equated with $R(\omega+n)$.)

Let W_e^X be $\{a \mid a \in I \;\&\; \{e\}_\pi^X(a)=0\}$.

Then jX is the set of all triples $\langle e,a,0\rangle$ such that $a \in W_e^X$, and all triples $\langle e,a,y+1\rangle$ such that $F(\{e\}_\pi^{\langle X,a\rangle})=y$.

The first part of jX encodes subsets of I first order definable using only integer parameters; the second part encodes the F-images of subsets of I definable using arbitrary individuals as parameters.

If F is ${}^2E$, then $j0$ is equivalent to the effective disjoint union of the arithmetic reals.

O^F, $|\;|^F : O^F \longrightarrow On$, and H_σ^F ($\sigma \in \text{range} \mid\;|^F$) are defined by a simultaneous recursion.

(1) For all $a \in I$, $\langle 1,a\rangle \in O^F$, $|\langle 1,a\rangle|^F=0$, and $H_0^F=0$.

(2) If $\langle e,a\rangle \in O^F$ and $|\langle e,a\rangle|^F=\sigma$, then $|\langle 2^e,a\rangle| \in O^F$, $|\langle 2^e,a\rangle|=\sigma+1$, and $H_{\sigma+1}^F = jH_\sigma^F$.

(3) Suppose $\langle m,a\rangle \in O^F$, $|\langle m,a\rangle|^F=\sigma$, and $W_e^{\langle H_\sigma^F,a\rangle} \subset O^F$. Then $\langle 3^m \cdot 5^e, a\rangle \in O^F$ and $|\langle 3^m\cdot 5^e,a\rangle|^F=\lambda$, where λ is the least limit ordinal greater than $|b|^F$ for all $b \in W_e^{\langle H_\sigma^F,a\rangle} \cup \{\langle m,a\rangle\}$. In addition H_λ^F is the set of pairs $\langle b,c\rangle$ such that

$$b \in O^F \quad \& \quad |b|^F=\sigma<\lambda \quad \& \quad c \in H_\sigma^F.$$

Let κ^F be the least ordinal that does not have a notation in O^F. The hierarchy $\{H_\sigma^F \mid \sigma<\kappa^F\}$ is in essence a hierarchy of all the computations that arise in the computing of all functions (from I into ω) recursive in $\langle F,a\rangle$ for all $a \in I$.

$\{e\}^F(a)$ is said to be defined and equal to x if $e=\langle e_0,e_1\rangle$, $\langle e_0,a\rangle \in O^F$, $|\langle e_0,a\rangle|^F=\sigma$, and $\{e_1\}_\pi^{H_\sigma^F}(a)=x$.

f is recursive in F (in symbols $f \leqslant F$) if there is an e such that $f(a)=\{e\}^F(a)$ for all a.

A set R of individuals is recursive in F (in symbols $R \leqslant F$) if the characteristic function of R is recursive in F.

An ordinal σ is constructive in F if $\sigma=|\langle e,0\rangle|^F$ for some $\langle e,0\rangle \in O^F$. (In short σ has an integer among its notations.) σ is recursive in F if σ is the height of a prewellordering of I recursive in F. Every σ constructive in F is recursive in F, and each ordinal recursive in F is less than some ordinal constructive in F.

The hierarchy $\{H_\sigma^F \mid \sigma < \kappa^F\}$ obeys what might be termed a *bounding principle*: if $f: I \longrightarrow O^F$ is recursive in F, then there exists an ordinal σ constructive in F such that $|fa|^F < \sigma$ for all a. It follows from the bounding principle that a set R is recursive in F iff there is an ordinal σ recursive in F such that R is one-one reducible to $H_{\sigma+1}^F$ as follows. The last clause means there is an e such that

$$a \in R \leftrightarrow \langle e, a, 0 \rangle \in H_{\sigma+1}^F \tag{4}$$

for all a.

A set R is recursively enumerable in F if there is an e such that for all a,

$$a \in R \leftrightarrow \{e\}^F(a) \quad \text{is defined.} \tag{5}$$

Further properties of the hierarchy are based on Grilliot's selection principle proved by Harrington and MacQueen [8]: there exists a uniform method for selecting a nonempty recursive (in F) subset of a nonempty recursively enumerable (in F) set of *subindividuals*. To be more precise, there exists a recursive number-theoretic function t such that for all e:

$$\begin{aligned} &\text{if} \quad (Er)\big[\{e\}^F(r) \quad \text{is defined}\big], \\ &\text{then} \quad (r)\big[\{te\}^F(r) \quad \text{is defined}\big] \\ &\text{and} \quad (r)\big[\{te\}^F(r) = 0 \to \{e\}^F(r) \quad \text{is defined}\big]. \end{aligned} \tag{6}$$

Moschovakis [9] has shown that when $n > 0$ there exists a set of *individuals* recursively enumerable in F with no nonempty subset recursive in F. *Gandy's selection principle*, which antedates Grilliot's, is also useful. It provides a uniform method for selecting an integer from a nonempty recursively enumerable (in F) set of integers.

For each $\sigma < \kappa^F$, H_σ^F and $M(\sigma, F)$ convey roughly the same information, so it is not surprising that they are equivalent in a sense made precise in Lowenthal [10]. The H_σ^F's are needed only to see which ordinals are constructive in F. Once those ordinals are known they can be used to read off the sets recursive in F from the $M(\sigma, F)$'s. The equivalence of Kleene's notion of recursion in F via schemes with the above is proved in detail in [10]. It must not be forgotten that the equivalence holds only for normal F's, and consequently that nothing in this paper casts any light on Kleene recursion in nonnormal objects.

2.4. *Admissible Sets with Gaps.* It is not hard to verify that a set of integers is recursively enumerable in F iff it is Σ_1 definable over $M(\kappa_0^F, F)$ by a formula of $\mathcal{L}(\mathcal{F})$ with integer parameters only. When $n=0$, $M(\kappa_0^F, F)$ is an admissible structure, but when $n>0$ it fails to satisfy Δ_0 collection for reasons associated with the existence of gaps between some of the ordinals recursive in F. The failure can be circumvented by invoking the notion of admissible set with gaps, a slight modification of Barwise's notion of admissible set with urelements [11]. A set A is admissible with gaps if A is closed under the operations of pairing and union, and satisfies the principles of Δ_0 separation and collection when the quantifiers occurring in those principles are suitably interpreted. An unrestricted quantifier ranges over A, a restricted one over the transitive closure of A. Consider a typical instance of Δ_0 separation:

$$(Ex)(y)[\, y \in x \leftrightarrow y \in b \quad \& \quad \mathcal{G}(y)].$$

b and all parameters of the Δ_0 formula $\mathcal{G}(y)$ belong to A. The proposed interpretation requires x to be a member of A and y to range over *all* elements of x, not merely those in A.

Δ_0 collection is treated similarly. A typical instance is:

$$(x)_{x \in b}(Ey)\mathcal{G}(x,y) \rightarrow (Ez)(x)_{x \in b}(Ew)_{w \in z}\mathcal{G}(x,w),$$

where b and all parameters of the Δ_0 formula $\mathcal{G}(x,y)$ belong to A; y and z must be members of A, while x can be any member of b.

$\langle A, F\rangle$ is admissible with gaps means A remains admissible with gaps when the Δ_0 formulas are augmented by $\mathcal{F}(x)=y$ and the latter is interpreted as $F(x)=y$.

The k-section of F, denoted by ${}_k sc\,F$, is the set of all objects of type k recursive in F. It can be viewed as a set theoretic object, in fact an admissible set with gaps when $k \leqslant n+1$, by interpreting objects of type k as codes for sets hereditarily of cardinality at most $\beth_{k-1}$. (Recall $\beth_0=\omega$ and $\beth_{p+1}=2^{\beth_p}$.) Let m be the map that takes a code b to the set mb coded by b. The relation

$$b \text{ is a code } \;\&\; mb = x, \tag{1}$$

where $b: T_p(k-1) \rightarrow T_p(k)$ and x is hereditarily of cardinality at most $\beth_{k-1}$, is defined by induction on the rank of x: b is a code for $\{m(br) | r \in T_p(k-1)\}$. The set of all codes of type k is recursively enumerable in ${}^{k+1}E$ when $k=1$, and recursive in ${}^{k+1}E$ when $k>1$, since a code of type k is essentially a reflexive wellfounded binary relation on $T_p(k-1)$. It is important to note that a coding relation need not be antisymmetric. Thus an ordinal σ recursive in ${}^3E$ is in general coded by a prewellordering, rather than a wellordering, of the reals recursive in ${}^3E$, since there may not exist a wellordering recursive in ${}^3E$ of height σ.

PROPOSITION 2.5. *Let b and c be codes and $\mathcal{G}(y)$ a Δ_0 formula of ZF whose parameters are coded by c. Then the predicate $P(d)$, defined by*

$$(Er)\big[d=br \quad \& \quad \mathcal{G}(md)\big],$$

is recursive in ${}^{k+1}E, b, c$.

Proof. The restricted set quantifiers of $\mathcal{G}(y)$ correspond to quantifiers ranging over $Tp(k-1)$, and the latter are equivalent to ${}^{k+1}E$.

For each $Z \subset Tp(k)$ let $\mathfrak{M}Z$ be the set of all sets with codes in Z. There is no harm in dropping the distinction between Z and $\mathfrak{M}Z$ (e.g. to say Z, rather than $\mathfrak{M}Z$, is an admissible set with gaps) when each member of $\mathfrak{M}Z$ is coded by some member of $\mathfrak{M}Z$.

PROPOSITION 2.6. *Assume $k \leqslant n+1$. Then $\langle \mathfrak{M}({}_k sc\,F), F\rangle$ is admissible with gaps (cf. 3.2).*

Proof. To verify Δ_0 separation, let $b, c \in {}_k sc\,F$ be codes and $\mathcal{G}(y)$ a Δ_0 formula whose parameters are coded by c. It suffices to find a code $b^* \in {}_k sc\,F$ such that

$$(y)\big[\,y \in mb^* \leftrightarrow y \in mb \quad \& \quad \mathcal{G}(y)\big]$$

holds in $\langle \mathfrak{M}({}_k sc\,F), F\rangle$. By 2.5 the desired b^* exists and is a map from $Tp(k-1)$ into $Tp(k)$ whose range is the set of all d that satisfy the predicate $P(d)$ of 2.5. Note that the presence of F in $\mathcal{G}(y)$ causes no difficulty, because only part of the graph of F is relevant to

$$y \in mb \quad \& \quad \mathcal{G}(y),$$

and that part is coded by an element of ${}_k sc\,F$ determined by b, c.

To verify Δ_0 collection, let $b, c \in {}_k sc\,F$ be codes and $\mathcal{G}(x,y)$ a Δ_0 formula whose parameters are coded by c and such that

$$(x)_{x \in mb}(Ey)\mathcal{G}(x,y)$$

holds in $\mathfrak{M}({}_k sc\,F)$. For each $r \in Tp(k-1)$ let Q_r be the set of all e such that $\{e\}^F$ is a code and

$$\mathcal{G}\big(m(br), m\{e\}^F\big)$$

holds in $\mathfrak{M}({}_k sc\,F)$. Since Q_r is recursively enumerable in $\langle F,r\rangle$, Gandy's selection principle (2.3) yields a function t such that t is recursive in F and $tr \in Q_r$ for all r. It follows from the bounding principle (2.3) that there is an ordinal σ recursive in F such that $\{tr\}^F$ is one-one reducible to $H_{\sigma+1}$ for all r. Since $H_{\sigma+1}$ is recursive in F, a code $d \in {}_k sc\,F$ can be extracted from $H_{\sigma+1}$ with the property that

$$(x)_{x \in mb}(Er)\mathcal{G}(x, m(dr)).$$

2.7. *Ordinals that Reflect to* κ_0^F. The notion of reflection (or stability) is needed to describe some further properties of ${}_k sc\,F$, properties that stem from Grilliot's selection principle rather than Gandy's. In section 3 it will be shown that the notions of reflection and admissibility with gaps suffice to characterize k-sections of normal objects of type greater than k for all $k > 1$.

For each $i \leqslant n$ let κ_i^F be the supremum of all ordinals with notations in O^F of type i; i.e. $\sigma < \kappa_i^F$ if $\sigma = |\langle e,a\rangle|^F$ for some $\langle e,a\rangle \in O^F$ such that $a \in Tp(i)$. Thus $\kappa^F = \kappa_n^F$.

An ordinal τ is said to be reflecting if for every Σ_1 sentence $\mathcal{G}$ of $\mathfrak{L}(\mathfrak{F})$ with integer parameters only, it is the case that

$$M(\tau, F) \models \mathcal{G} \qquad \text{implies} \qquad M(\kappa_0^F, F) \models \mathcal{G}.$$

The Σ_1 sentence $(E\sigma)[e \in H_\sigma^F]$ is typical, since each $H_{\gamma+1}^F$ encodes all first order facts about $\{H_\sigma^F | \sigma < \gamma\}$ in a positive way. Thus to say τ is reflecting is to say: an H_σ^F that stands in a Δ_0 relation to some integer is developed before stage τ only if one is developed before stage γ for some γ recursive in F. Note that if γ is recursive in F and $e \in H_\sigma^F$ holds for some $\sigma < \gamma$, then the least such σ is recursive in F because it is first order definable over H_γ^F using only integer parameters.

From now on n is assumed to be positive in any statement about reflecting ordinals. Let κ_r^F be the greatest reflecting ordinal*. It follows from Moschovakis [9] that $\kappa^F > \kappa_r^F$ (otherwise κ_r^F would be of no interest); on the other hand Harrington [3] has shown κ_r^F is much greater than κ_{n-1}^F in a sense whose precise meaning is not relevant here. In order to prove the plus-one theorem it is enough to know $\kappa_r^F \geqslant \kappa_{n-1}^F$, an inequality equivalent to the stability result stated at the end of [1].

*Kechris and Harrington [3] have found an intrinsic definition of κ_r^F, i.e. a definition in terms of the notion of recursive enumerability in F rather than in terms of a hierarchy generated by F in a somewhat arbitrary fashion.

LEMMA 2.8. (Simple Reflection). $\kappa_r^F \geqslant \kappa_{n-1}^F$.

Proof. Fix e and assume $e \in H_\sigma^F$ is satisfied by some $\sigma < \kappa_{n-1}^F$ with the intent of showing it is satisfied by some $\sigma < \kappa_0^F$. Thus there is a subindividual s such that

$$s \in O^F \quad \& \quad (E\sigma)\left[\sigma < |s|^F \quad \& \quad e \in H_\sigma^F\right]. \tag{1}$$

Let S be the set of all s that satisfy (1). By Grilliot's selection principle (2.3 (6)) there is a nonempty recursive (in F) $T \subset S$. But then the bounding principle (2.3) yields a γ recursive in F such that $|s|^F < \gamma$ for all $s \in T$. Clearly $e \in H_\sigma^F$ for some $\sigma < \gamma$.

The next lemma will be needed in section 3 to prove the converse of the result that implies the plus-one theorem when $k=2$; i.e. to show the 2-section of a normal type 3 object is an abstract 2-section.

LEMMA 2.9. (Further Reflection, Harrington [3]). *If R is a recursively enumerable (in F) set of subindividuals, then* $\kappa_r^F \geqslant \kappa_{n-1}^{\langle F,R\rangle}$.

Proof. Fix e and assume $e \in H_\sigma^F$ is satisfied by some $\sigma < \kappa_{n-1}^{\langle F,R\rangle}$ with the intent of showing it is satisfied by some $\sigma < \kappa_0^F$. Thus there is a subindividual s such that

$$s \in O^{\langle F,R\rangle} \quad \& \quad (E\delta)\left[\delta < |s|^{\langle F,R\rangle} \quad \& \quad e \in H_\delta^F\right].$$

Let W be the set of all individuals c such that c is a wellordering of all subindividuals. Note that W is recursive in F because $n>0$ and F is normal. For each $c \in W$, let $|c|$ be the ordertype of c. Call an ordinal γ *subconstructive* (in F) if $\gamma = |s|^F$ for some subindividual $s \in O^F$. Let λ be the ordertype of the subconstructive (in F) ordinals. λ splits W in a curious fashion:

(A) $\{c \mid |c| < \lambda\}$ is recursively enumerable in F.
(B) If $|c| \geqslant \lambda$, then $e \in H_\sigma^F$ for some $\sigma < \kappa_0^{\langle F,c\rangle}$.

For each $s \in O^F$, let $|s|$ be the ordertype of the set of all subconstructive (in F) ordinals less than $|s|^F$. To prove (A) observe that

$$(Es)\left[s \in O^F \quad \& \quad |c| < |s|\right]$$

is recursively enumerable in F, since Grilliot's selection principle implies the recursively enumerable (in F) predicates are closed under existential quantification over subindividuals ([12]).

To prove (B) assume $|c| \geqslant \lambda$. Then there is an initial segment of c, call it c_s, such that $|c_s| = \lambda$, where c_s is either c or that part of c below some s in the field of c. R is recursive in $\langle F,c,s\rangle$, since the recursive (in F) enumeration of R comes to an end as soon as $\{H^F_\sigma | \sigma<\kappa^F_{n-1}\}$ is developed, and the latter occurs as soon as O^F has been enumerated to the point where the subconstructive ordinals have ordertype equal to $|c_s|$. Consequently $\kappa^{\langle F,R\rangle}_{n-1} \leqslant \kappa^{\langle F,c\rangle}_{n-1}$, and so there is a $\sigma<\kappa^{\langle F,c\rangle}_{n-1}$ such that $e \in H^F_\sigma$. It follows from 2.8 relativized to c that

$$(E\sigma)\left[\sigma<\kappa_0^{\langle F,c\rangle} \quad \& \quad e\in H^F_\sigma\right]. \tag{1}$$

The set of all c that satisfy (1) is recursively enumerable in F, since an ordinal is constructible from $\langle F,c\rangle$ iff it has a notation in O^F of the form $\langle m,c\rangle$ for some integer m. (A) and (B) make it possible to enumerate all of W recursively in F as follows: enumerate c if $|c|<\lambda$ or if c satisfies (1). (Gandy's selection principle implies that the union of two sets recursively enumerable in F is recursively enumerable in F.) Since W is recursive in F, the bounding principle implies the enumeration of W is complete by some $\sigma_0<\kappa_0^F$. The enumeration of W includes the development of all σ's that arise in (B), so $e\in H^F_\sigma$ for some $\sigma<\sigma_0$.

3. *k*-Sections.

3.1. *Abstract k-Sections.* Assume $k>1$. A set A is said to be locally of type k if $Tp(k-1)\in A$ and each member of A is coded as in 2.4 (1) by some element of $Tp(k)\cap A$. A pair (A,B) of sets is called stable if $A\in B$, and $A \prec_1 B$ (i.e. $A\subset B$ and each Σ_1 sentence of ZF (with parameters in A) is true in $\langle B,\varepsilon\rangle$ only if it is true in $\langle A,\varepsilon\rangle$). A is closed under recursion in ${}^{k+1}E$ if $y\in A$ whenever $x,y\in Tp(k)$, $x\in A$ and y is recursive in ${}^{k+1}E,x$.

A is an abstract k-section $(k>1)$ if there is a B such that:

(1) A and B are locally of type k.
(2) A and B are admissible sets with gaps.
(3) A and B are closed under recursion in ${}^{k+1}E$.
(4) (A,B) is stable.

The above definition has not been chosen with much respect for economy (e.g. in the presence of (4), (3) is equivalent to something much weaker), because in all practical cases the problem of showing A is an abstract k-section quickly reduces to finding a B such that (A,B) is stable. As a rule if there exists a B satisfying (4), then there exists one which can be developed by some inductive definition over A.

Clause (3) is not superfluous: there does exist an admissible set with gaps which is locally of type 2 but not closed under recursion in ${}^3E$. Thus the gaps do make a difference, since every admissible set (without gaps) which is locally of type k is closed under recursion in ${}^{k+1}E$ for all $k \geqslant 1$. But the only substantial difference between an abstract k-section ($k>1$) and an abstract 1-section ([1], pg. 84) is the stability imposed by (4).

An abstract k-section A is said to be countable if the B of (1)–(4) has the additional property that

$$\langle B, \varepsilon \rangle \models (A \text{ is countable}). \tag{5}$$

Suppose $Z \subset Tp(k)$. Z is said to be a countable abstract k-section if $\mathfrak{M}Z$ satisfies (1)–(5) (cf. 3.12).

LEMMA 3.2. *Assume* $k \leqslant n+1$. *Then* ${}_k sc\, F$ *is a countable abstract* k-*section.*

Proof. Fix $k>1$. Let P be the complete recursively enumerable (in F) set of integers, and let B be $\mathfrak{M}({}_k sc\langle F, P\rangle)$. Clearly B is closed under recursion in ${}^{k+1}E$ since $k \leqslant n+1$ and F is normal. By 2.6 B is an admissible set with gaps. The presence of P in B implies

$$\langle B, \varepsilon \rangle \models (\mathfrak{M}({}_k sc\, F) \text{ is countable}).$$

The stability of $\mathfrak{M}({}_k sc\, F)$ in B follows from the reflection of $\kappa_0^{\langle F,P\rangle}$ to κ_0^F. Simple reflection (2.8) suffices if $n>1$, since in that case $P \in Tp(n-1)$. But further reflection (2.9) is needed if $n=1$.

3.3. *The Forcing Relation* $\Vdash^Z$. Assume Z is a countable abstract k-section for some $k>1$. The construction of a type $k+1$ G whose k-sections is Z is given in forcing theoretic terms chiefly for the sake of future developments, since the construction utilizes little of the power of forcing.

$p, q, r\ldots$ denote forcing conditions. p is a forcing condition if:

(1) $p: \tau+1 \to Tp(k)$, where τ is an ordinal; and

(2) there is a normal $G \in Tp(k+1)$ such that τ is constructive in G and $(\sigma)[\sigma \leqslant \tau \to p\sigma = H_\sigma^G]$.

The G of clause (2) is unique in the following sense. Let G_p be the restriction of G to those $R \in Tp(k)$ such that

$$(Ee)(E\sigma)_{\sigma<\tau}(Ea)\left[a \in Tp(k-1) \quad \& \quad R = \{e\}_\pi^{\langle p\sigma, a\rangle}\right]. \tag{3}$$

Then G_p is uniquely determined by p, and moreover is equivalent to p, because the generation of $\{H_\sigma^G | \sigma \leqslant \tau\}$ requires the value of $G(R)$ if and only if R satisfies (3), and because all such values are coded within $\{H_\sigma^G | \sigma \leqslant \tau\}$.

The definition of $\Vdash$ has two clauses.

(4) $p \Vdash \bar{m} \in O^{\mathcal{G}}$ iff the τ and G of (2) are such that for some $\sigma \leqslant \tau, |m|^G = \sigma$. (Note that G_p decides $|m|^G = \sigma$ for all $\sigma \leqslant \tau$.)

(5) $p \Vdash \bar{R} \leqslant \mathcal{G}$ iff the τ and G of (2) are such that for some $\sigma < \tau$, σ is constructive in G, and R is one-one reducible to $H_{\sigma+1}^G$.

The definition of $\Vdash^Z$ has two clauses; in each of them $\mathcal{H}$ denotes a sentence of the sort forced in (4) or (5).

(6) $p \Vdash^Z \mathcal{H}$ iff $p \in \mathfrak{M}Z$ and $p \Vdash \mathcal{H}$.

(7) $p \Vdash^Z \sim \mathcal{H}$ iff $p \in \mathfrak{M}Z$ and $(q)_{p \supset q} \sim [q \Vdash^Z \mathcal{H}]$. ($p \supset q$ means p is extended by q.)

PROPOSITION 3.4. *Let A be an admissible set with gaps. Then the relation* $p \Vdash^Z \bar{m} \in O^{\mathcal{G}}$, *restricted to* $p \in A$, *is* Σ_1 *over A.*

Proof. An immediate consequence of the fact that G_p (defined by 3.3 (3)) decides whether or not p is a forcing condition, and if so, whether or not $p \Vdash^Z \bar{m} \in O^{\mathcal{G}}$. Since A may not be a transitive set, it is worth noting that if $p \in A$ and $p \Vdash^Z \bar{m} \in O^G$, then G_p forces m to be a notation for some $\sigma \in A$.

3.5. *Genericity for* $\Vdash^Z$. Assume $G = {}^{k+1}E$ on $Tp(k) - Z$. G is said to be generic with respect to $\Vdash^Z$ if for each sentence $\mathcal{H}$ of the form

$$\bar{m} \in O^{\mathcal{G}} \quad \text{or} \quad \bar{R} \leqslant \mathcal{G}, \tag{1}$$

where $R \in Z$, there is a p such that either

$$p \Vdash^Z \mathcal{H} \quad \text{or} \quad p \Vdash^Z \sim \mathcal{H},$$

and the graph of G_p is contained in the graph of G. Since Z is countable, generic G's exist. As usual a sentence about a generic G of the form occurring in (1) is true if and only if it is forced by some p such that $G_p \subset G$.

LEMMA 3.6. *If G is generic, then* $Z \subset_k sc\, G$.

Proof. Suppose $R: Tp(k-1) \longrightarrow \omega$ belongs to Z. Fix p. Since G is generic it suffices to find a q extending p such that

$$q \Vdash^Z \bar{R} \leqslant \mathcal{G}.$$

Let τ and G_p be as in 3.3 (1)–(3). For each $a \in Tp(k-1)$, $\langle H_\tau^G, a\rangle$ is not first order definable over

$$\langle Tp(k-1), \varepsilon, \langle H_\sigma^G, a\rangle\rangle$$

for any $\sigma < \tau$, because it encodes all such definitions. Consequently

$$G_p(\langle H_\tau^G, a\rangle)$$

is undefined for all a. Extend G_p to G_q by setting

$$G_q(\langle H_\tau^{G_p}, a\rangle) = Ra$$

for all a. Let e_0 be a trivial index; i.e. $\{e_0\}_\pi^X = X$ for all X. Then

$$Ra = y \leftrightarrow \langle e_0, a, y+1\rangle \in H_{\tau+1}^{G_q}$$

for all $a \in Tp(k-1)$.

3.7. *B-Genericity.* Since Z is a countable abstract k-section, there is a B such that $(\mathfrak{M}Z, B)$ satisfies 3.1 (1)–(5) with $\mathfrak{M}Z$ for A. Assume G is generic in the sense of 3.5. Note that the restriction of G to Z (denoted by $G \restriction Z$) is readily encoded by a type k object since Z is countable, and that G is equivalent to $\langle {}^{k+1}E, G\restriction Z\rangle$. The existence of a generic G such that $(G\restriction Z) \in B$ follows readily from the countability of $\mathfrak{M}Z$ in B and the admissibility (with gaps) of B.

It is not difficult to find a generic G whose k-section exceeds Z.

3.8. If G is generic and $(G\restriction Z) \in B$, then ${}_k sc\, G \subset Z$.

Proof. Suppose $R \in ({}_k sc\, G) - Z$ for the sake of a reductio ad absurdum. Thus for some τ and m, R is one-one reducible to H_τ^G and $\tau = |m|^G$. There is no p such that $G_p \subset G$ and

$$p \Vdash^Z \overline{m} \in O^{\mathcal{G}};$$

if there were such a p, then R would belong to $\mathfrak{M}Z$ because τ would. Consequently there is a q such that $G_q \subset G$ and

$$q \Vdash^Z \sim(\overline{m} \in O^{\mathcal{G}}). \tag{1}$$

Observe that $H_\tau^G \in B$, because $(G\restriction Z) \in B$, $G = \langle {}^{k+1}E, G\restriction Z\rangle$ and B is closed under recursion in ${}^{k+1}E$. Define r by $r\sigma = H_\sigma^G$ for all $\sigma \leqslant \tau$. Clearly $r \in B$,

$$q \subset r \quad \text{and} \quad r \Vdash^Z \overline{m} \in O^{\mathcal{G}}. \tag{2}$$

It follows from 3.4 and the stability of $\mathfrak{M}Z$ in B that some r in $\mathfrak{M}Z$ satisfies (2), an impossibility according to (1).

THEOREM 3.9. *Z is a countable abstract k-section iff Z is the k-section of some normal type k+1 object.*

Proof. Assume $k>1$, since the case of $k=1$ is covered by 4.4 of [1]. If Z is the k-section of some normal type $k+1$ object, then Z is a countable abstract k-section by 3.2.

Suppose Z is a countable abstract k-section. Let G be generic as in 3.7. Then $Z={}_k sc\, G$ by 3.6 and 3.8.

COROLLARY 3.10. (Plus-One Theorem). *Suppose $n>m>k\geqslant 1$ and F is a normal type n object. Then there exists a normal type m object G such that ${}_k sc\, F={}_k sc\, G$.*

Proof. Assume $k>1$, since $k=1$ is covered by [1]. Let $Z={}_{(m-1)}sc\, F$. By 3.2 and 3.9, there exists a normal type m G such that

$${}_{(m-1)}sc\, F={}_{(m-1)}sc\, G.$$

But $k<m$, so $Tp(k)\subset Tp(m-1)$, and consequently ${}_k sc\, F={}_k sc\, G$.

3.11. *Stable Countings.* Assume $k>1$ and $Z\subset Tp(k)$. $t(=\langle t_0,t_1\rangle)$ is said to be a counting of Z if:

$$t_0:\mathfrak{M}Z\to\omega \quad \text{is one-one;}$$

and

$$\mathfrak{M}Z\models \mathcal{G}_e(\bar{b}) \quad \text{iff} \quad t_1(e,t_0 b)=0, \tag{1}$$

where $\mathcal{G}_e(x)$ is the e-th Δ_0 formula of ZF and $b\in\mathfrak{M}Z$. Note that in (1) the restricted quantifiers of $\mathcal{G}_e$ are interpreted as in 2.4, since $\mathfrak{M}Z$ may not be transitive. Strictly speaking t is a type k object because t_0 is, but it is more to the point to regard t as a type 1 object, since most of the information carried by t is carried by t_1. Thus P, the complete recursively enumerable (in ${}^{k+1}E$) set of integers, can be regarded as the natural counting of ${}_k sc({}^{k+1}E)$ if the usual correlation between sets recursive in ${}^{k+1}E$ and their Gödel numbers is kept in mind.

A counting t of Z is said to be *stable* if

$$\mathfrak{M}Z\prec_1\mathfrak{M}\big({}_k sc(\langle {}^{k+1}E,t\rangle)\big).$$

Further reflection (2.9) implies P is a stable counting of ${}_k sc({}^{k+1}E)$. The next proposition is easily verified.

PROPOSITION 3.12. *Assume $k>1$ and $Z \subset Tp(k)$. Then Z is a countable abstract k-section iff Z has a stable counting.*

Proposition 3.12 clarifies what is needed for an application of Theorem 3.10. Let Z_1^2 be the collection of all lightface Δ_1^2 sets of reals. The natural counting of Z_1^2 is given by P_1^2, the complete Σ_1^2 set of integers. Whether or not P_1^2 is stable seems to depend on circumstances. If $V=L$, or if the axiom of determinateness holds, then P_1^2 is stable. At this writing it is within the realm of possibility that P_1^2 is not stable yet Z_1^2 possesses some unnatural stable counting. The stability of P_1^2 is equivalent to: for every π_1^1 predicate Q,

$$(E^2x)\left[Q({}^2x) \quad \& \quad {}^2x \leqslant {}^3E, P_1^2\right] \rightarrow (E^2x)\left[Q({}^2x) \quad \& \quad {}^2x \in Z_1^2\right],$$

where ${}^2x$ is a type 2 variable.

4. Further Results and Open Questions. Moschovakis [4] defined the k-envelope of F (denoted by ${}_k en F$) to be the collection of all $X \in Tp(k)$ such that X is recursively enumerable in F. He showed that if F is a normal object of type n and $n>k+1\geqslant 2$, then the k-envelope of F is not the k-envelope of any normal object of type $k+1$. Thus there is no plus-one theorem for k-envelopes, but there is a plus-two theorem for them discovered by Harrington [3].

PLUS-TWO THEOREM. *Suppose $n>k+1\geqslant 2$ and F is a normal type n object. Then there exists a normal type $k+2$ object G such that ${}_k en F = {}_k en G$.*

It is tempting to think that the plus-two theorem is a consequence of the plus-one theorem (3.10) by the following false argument: By 3.10 there is a type $k+2$ G such that

$$ {}_{k+1} sc\, F = {}_{k+1} sc\, G; \tag{1}$$

(2) But then ${}_k en F = {}_k en G$ because k-envelopes are determined by $(k+1)$-sections. Step (2) is plausible because ${}_k en\, G$ is Σ_1 over

$$\langle \mathfrak{M}({}_{(k+1)} sc\, G), G\rangle. \tag{3}$$

((3) is an admissible structure with gaps as in 2.6.) In short the k-envelope of G is determined by the restriction of G to its $(k+1)$-section, but not by its $(k+1)$-section alone. A slight modification of the arguments of section 2 yields a type $k+2$ G which satisfies (1) but whose k-envelope is different from that of F.

On the other hand Harrington [3] points out that the proof of the plus-two theorem is readily altered so as to produce a type $k+2$ G whose $(k+1)$-section and k-envelope are the same as that of F.

Moldestad [13] defines the (i,k)-section of F (denoted by ${}^{i}_{k}sc\,F$) to be the collection of all $X \in Tp(k)$ such that X is recursive in $\langle F, Y\rangle$ for some $Y \in Tp(i)$. (Thus ${}^{0}_{k}sc\,F = {}_{k}sc\,F$.) Harrington [3] remarks, and Moldestad [13] verifies, that the ideas occurring in the proof of the plus-two theorem suffice to prove a strengthened plus-one theorem: if $n > k \geqslant 2$ and F is a normal type n object, then there exists a normal type $k+1$ G such that

$$ {}^{k-2}{}_{k}F = {}^{k-2}{}_{k}G. $$

Nonetheless the following hypothesis remains unproved.

Extended Plus-One Hypothesis. *Suppose $n > k \geqslant 1$ and F is a normal type n object. Then there exists a normal type $k+1$ object G such that ${}^{k-1}{}_{k}F = {}^{k-1}{}_{k}G$.*

${}^{k-1}{}_{k}F$ is called the extended k-section of F. Note that the extended k-section of a normal type $(k+1)$ object G is equivalent to the hierarchy $\{H^G_\sigma \mid \sigma < \kappa^G\}$ of 2.3. Whether or not the extended plus-one hypothesis is true may depend on metaphysical circumstances. It is true if the generalized continuum hypothesis holds*. It is false (as Harrington has observed) when $k=2$ if the axiom of determinateness holds. Consequently an investigation of it for the case $k=2$ should begin by assuming the existence of a wellordering of the reals. Perhaps the wisest course would be to assume the wellordering is recursive in F, r for some real r.

Plus-Three Problem. Suppose $n > k+2 \geqslant 3$ and F is a normal type n object. Does there exist a normal type $k+3$ object G such that ${}_{k}sk\,G = {}_{k}sk\,F$?

${}_{k}sk\,F$ denotes the k-skeleton of F, a concept as yet undefined. Does there exist a concept of k-skeleton such that the answer to the plus-three problem is yes, and such that the k-skeleton of a normal type $k+3$ object cannot be the k-skeleton of any normal object of type $k+2$ or $k+1$? A suitable concept of k-skeleton might justify the study of recursion in normal type 5 objects. At the moment every phenomenon associated with recursion in a normal type 5 object can be exemplified by recursion in some normal object of lower type. But surely

*It is also true when $k=3$ if the axiom of determinateness is assumed.

the transcendence of type 5 over lower types can be made manifest by some concept accessible to recursion theorists.

HARVARD UNIVERSITY

AND

MASSACHUSETTS INSTITUTE OF TECHNOLOGY.

REFERENCES.

[1] G. E. Sacks, "The 1-section of a type n object," *Generalized Recursion Theory* (North-Holland, 1974), pp. 81–93.

[2] S. C. Kleene, "Recursive functionals and quantifiers of finite type," *Trans. Amer. Math. Soc.* **91** (1959), pp. 1–52; and **108** (1963), pp. 106–142.

[3] L. A. Harrington, "Contributions to recursion theory on higher types," Ph.D. thesis, Massachusetts Institute of Technology, 1973.

[4] Y. N. Moschovakis, "Structural characterizations of classes of relations," *Generalized Recursion Theory* (North-Holland, 1974), pp. 53–79.

[5] J. R. Shoenfield, "A hierarchy based on a type 2 object, *Trans. Amer. Math. Soc.* **134** (1968), pp. 103–108.

[6] T. Grilliot, "Hierarchies based on objects of finite type," *Jour. Symb. Log.* **34** (1969), pp. 177–182.

[7] G. E. Sacks, "Higher recursion theory," (Notes by D. B. MacQueen of lectures at the Massachusetts Institute of Technology, 1971–72).

[8] L. A. Harrington and D. B. MacQueen, "Selection in abstract recursion theory," *Jour. Symb. Log.*, to appear.

[9] Y. N. Moschovakis, "Hyperanalytic predicates," *Trans. Amer. Math. Soc.* **129** (1967), pp. 249–282.

[10] F. D. Lowenthal, "The Minimal Pair Problem for Higher Type Objects," Ph.D. thesis, Massachusetts Institute of Technology, 1974.

[11] K. J. Barwise, "Admissible sets over models of set theory," *Generalized Recursion Theory* (North-Holland, 1974), pp. 97–122.

[12] T. Grilliot, "Selection functions for recursive functionals," *Notre Dame Jour. Form. Log.* **10** (1969), pp. 225–234.

[13] J. Moldestad, *Report of a Recursion Theory Seminar* (Chapter IV), University of Oslo, 1973.

FUNDAMENTA
MATHEMATICAE
CIII (1979)

Effective bounds on Morley rank

by

Gerald E. Sacks * (Cambridge, Mass.)

To the Memory of Andrzej Mostowski

Abstract. Effective bounds are obtained on the Morley rank of a 1-type, the Morley rank α_T of a totally transcendental theory T, and the Blum density number d_T of a quasi-totally transcendental theory T by means of type-omitting and absoluteness arguments over admissible sets. The above restrictions on T imply that α_T and d_T are ordinals recursive in T. Every theory T is seen to have a universal domain hyperarithmetic in the hyperjump of T.

1. Introduction. This paper might better have been titled: On the Absolute Character of the Morley Derivative. For the bounds given below on Morley rank, and on Blum's density number for quasi-totally transcendental theories, are derived from some absoluteness properties of Morley's analysis of 1-types. Let T be a countable theory of first order logic. Assume T is complete and substructure complete ([1]) in order to smooth the application of Morley's rank-and-degree machine to T. (The details of his machine will be reviewed in Section 2. A full account was given in [12].) Suppose $\mathscr{A}$ is a substructure of a model of T, and p is a 1-type over $\mathscr{A}$ (in symbols $\mathscr{A} \in \mathscr{K}(T)$ and $p \in S\mathscr{A}$). If p has a Morley rank, then that rank is denoted by $r_M(p)$, and the existence of that rank is indicated by the inequality: $r_M(p) < \infty$.

Let N be an admissible set as defined in Section 2. Assume T and $\mathscr{A}$ belong to N. One aspect of the absoluteness of the Morley derivative is expressed by:

(1) $$r_M(p) = \beta < \infty \rightarrow p \in N \;\&\; \beta \in N\,.$$

Another aspect is the fact that the relation

$$p \in S\mathscr{A} \;\&\; r_M(p) \leqslant \beta$$

is Σ_1 over N. (1) implies a bound first obtained by Lachlan [8]:

(2) $$r_M(p) < \infty \rightarrow r_M(p) < \aleph_1\,,$$

---

* This article is based on a talk given at the 1972 Orléans Logic Conference. Its preparation was supported in part by NSF grant MCS 76-10430. The author is grateful to L. Harrington and S. Simpson for several helpful comments.

([1]) Substructure completeness is equivalent to admitting elimination of quantifiers.

since every Morley rank achieved by some p over some $\mathscr{A}$ is also achieved by some p over some countable $\mathscr{A}$, and since each countable $\mathscr{A}$ belongs to some countable N.

The notion of "next" admissible set, as in [1, p. 60], is needed for an effective version of (2). Let $\mathscr{A}^+$ be the least admissible set N such that $\mathscr{A} \in N$. Define $\omega_1^{\mathscr{A}}$ to be the least ordinal not in $\mathscr{A}^+$. An effective version of (2) is:

$$(3) \qquad p \in S\mathscr{A} \;\&\; r_{\mathrm{M}}(p) < \infty \rightarrow p \in \mathscr{A}^+ \;\&\; r_{\mathrm{M}}(p) < \omega_1^{\mathscr{A}}.$$

If $\mathscr{A}$ is countable, then there exists a real number C such that $\mathscr{A}$ is coded by C and $\omega_1^C = \omega_1^{\mathscr{A}}$. (One such C is in essence a counting of $\mathscr{A}$ generic over $\mathscr{A}^+$.) The conclusion of (3) now becomes: p is (i.e. coded by a real) hyperarithmetic in $\mathscr{A}$, and $r_{\mathrm{M}}(p)$ is an ordinal recursive in $\mathscr{A}$.

The proof of (3), given in Section 3, is in the main an omitting types argument over a countable admissible fragment of $L_{\omega_1,\omega}$. The argument succeeds thanks to some weak absoluteness properties of the Morley derivative extracted in Section 2 from the strong absoluteness of the Cantor–Bendixson derivative. The notion of universal domain also plays a part in the proof of (3). A universal domain $\mathscr{U}$ has the advantage that Morley rank equals Cantor–Bendixson rank for all $p \in S\mathscr{U}$. (3) is applied in Section 5 to prove the existence of a countable universal domain $\mathscr{U}$ for T with most of the effort directed towards holding down the ordinal needed to construct $\mathscr{U}$ from T.

As a further application of (3), a bound on d_T, Blum's density number, is obtained in Section 4. Blum [2] calls T quasi-totally transcendental (q.-t.t.) if the Morley ranked points of $S\mathscr{A}$ are dense in $S\mathscr{A}$ for every $\mathscr{A} \in \mathscr{K}(T)$. If T is q.-t.t., then d_T is defined to be the least β such that for all $\mathscr{A} \in \mathscr{K}(T)$, the points of $S\mathscr{A}$ of Morley rank less than β are dense in $S\mathscr{A}$. It will be shown that $d_T < \aleph_1$ by showing $d_T < \omega_1^T$. (If T is coded by a real, e.g. the set of Gödel numbers of axioms of T, then ω_1^T is the least ordinal not recursive in the code for T.) α_T, the Morley rank of T, defined for all T, is the least β such that for all $\mathscr{A} \in \mathscr{K}(T)$, every Morley ranked point of $S\mathscr{A}$ has rank less than β. L. Harrington has found a quasi-totally transcendental T such that $\alpha_T = \aleph_1$. In Section 4 it is also shown that $\alpha_T < \omega_1^T$ when T is totally transcendental.

A. Kechris [5] has pointed out to the author that some of the results of this paper can be derived from the work of Blass and Cenzer (cf. [3]) on monotonic Π_1^1 inductive definitions of sets of reals. Kechris's approach is a striking example of the power of generalized recursion theory. Yet another approach has been developed by Harnik and Makkai [4] via Vaught sentences.

Some open questions on the absoluteness of the Morley derivative are listed in Section 6.

2. Rank. Notation and definitions are taken from [12]. $\mathscr{K}(T)$ is the category of all substructures of all models of T. $f: \mathscr{A} \rightarrow \mathscr{B}$ is a monomorphism of $\mathscr{A}$ into $\mathscr{B}$. $S\mathscr{A}$ is the Stone space of 1-types over $\mathscr{A}$. A typical basic open subset of $S\mathscr{A}$ is

$$\{p \mid F(\underset{\sim}{a}, x) \in p\},$$

where $F(\underset{\sim}{a}, x)$ is a formula in the language of T with parameters $\underset{\sim}{a}$ in A, the universe of $\mathscr{A}$.

Sf: $S\mathscr{B} \to S\mathscr{A}$ is the continuous onto map induced by f. If $q \in S\mathscr{B}$, then

$$F(\underset{\sim}{a}, x) \in (Sf)q \leftrightarrow F(f\underset{\sim}{a}, x) \in q .$$

S is a contravariant, limit preserving functor from $\mathscr{K}(T)$ into the category of Stone spaces.

DS is the Morley derivative of S. $p \in DS\mathscr{A}$ iff $p \in S\mathscr{A}$ and there exists an f: $\mathscr{A} \to \mathscr{B}$ in $\mathscr{K}(T)$ such that $(Sf)^{-1}(p)$ contains a limit point of $S\mathscr{B}$. DSf is the restriction of Sf to $DS\mathscr{B}$, and maps $DS\mathscr{B}$ onto $DS\mathscr{A}$.

$D^{\beta}S$, the βth Morley derivative of S, is defined by transfinite recursion. $D^0S = S$, $D^{\beta+1}S = D(D^{\beta}S)$, and

$$D^{\lambda}S\mathscr{A} = \bigcap \{D^{\beta}S\mathscr{A} \mid \beta < \lambda\}$$

when λ is a limit ordinal. $D^{\beta}S$ is a contravariant, limit preserving functor from $\mathscr{K}(T)$ into the category of Stone spaces. If there is a β (necessarily unique) such that

$$p \in D^{\beta}S\mathscr{A} - D^{\beta+1}S\mathscr{A} ,$$

then p is said to be a *Morley ranked point of* $S\mathscr{A}$ of Morley β.

$dS\mathscr{A}$, the Cantor–Bendixson derivative of $S\mathscr{A}$, is the set of all limit points of $S\mathscr{A}$. $d^{\beta}S\mathscr{A}$ is defined by transfinite recursion. $d^0S\mathscr{A} = S\mathscr{A}$, $d^{\beta+1}S\mathscr{A} = dd^{\beta}S\mathscr{A}$, and

$$d^{\lambda}S\mathscr{A} = \bigcap \{d^{\beta}S\mathscr{A} \mid \beta < \lambda\} .$$

If there is a β such that $p \in d^{\beta}S\mathscr{A} - d^{\beta+1}S\mathscr{A}$, then p is said to be a *Cantor–Bendixson ranked point of* $S\mathscr{A}$ of Cantor–Bendixson rank β. The latter state of affairs is indicated by: $r_{\mathrm{CB}}(p) = \beta < \infty$.

PROPOSITION 2.1. $r_{\mathrm{CB}}(p) \leqslant r_{\mathrm{M}}(p)$.

Proof. It suffices to show $d^{\beta}S\mathscr{A} \subset D^{\beta}S\mathscr{A}$ for all β by transfinite induction. The latter is straightforward, because the identity monomorphism 1_A: $\mathscr{A} \to \mathscr{A}$ belongs to $\mathscr{K}(T)$. ■

Suppose $\mathscr{U}, \mathscr{V}, \mathscr{W} \in \mathscr{K}(T)$. $\mathscr{V}$ is finitely generated if there exists a finite $Y \subset V$ such that $\mathscr{V}$ is the least substructure of $\mathscr{V}$ whose universe contains Y. Suppose $p \in D^{\beta}S\mathscr{V}$ and i: $\mathscr{V} \subset \mathscr{W}$ is an inclusion. p is said to *split* in $D^{\beta}S\mathscr{W}$ if $(D^{\beta}Si)^{-1}(p)$ has at least two members.

$\mathscr{U}$ is an α-universal domain for T if for all $\beta < \alpha$, all finitely generated $\mathscr{V} \subset \mathscr{U}$, and all isolated $p \in D^{\beta}S\mathscr{V}$, the following holds: if p splits in $D^{\beta}S\mathscr{W}$ for some $\mathscr{W} \in \mathscr{K}(T)$, then p splits in $D^{\beta}S\mathscr{W}_0$ for some finitely generated $\mathscr{W}_0 \subset \mathscr{U}$.

$\mathscr{U}$ is a universal domain for T if $\mathscr{U}$ is an α-universal domain for T for all α.

LEMMA 2.2. *Suppose* $\mathscr{U}$ *is an* α*-universal domain for* T. *Then* $D^{\beta}S\mathscr{U} = d^{\beta}S\mathscr{U}$ *for all* $\beta < \alpha$.

Proof. By induction of β. For details, see the proof of Lemma 31.3 on page 190 of [12]. The point to remember is that $D^{\beta}S$ preserves limits. Let $\{\mathscr{V}_i\}$ be the direct system of all finitely generated substructures of $\mathcal{U}$. Then $D^\beta S\mathcal{U}$ is the inverse limit of the inverse system $\{D^\beta S\mathcal{V}_i\}$. ■

The existence of a countable universal domain for T is implied by Proposition 31.2 of [12]. Unfortunately there is a gap in the proof of 31.2. It will be filled in Section 5 by the type omitting argument of Section 3. The next proposition retains all that is correct in the proof of 31.2. It will be needed in Sections 3 and 4. card is an abbreviation for cardinality.

PROPOSITION 2.3. *Suppose* $\mathcal{A} \in \mathcal{K}(T)$ *is infinite and* $\operatorname{card}\alpha \leqslant \operatorname{card}\mathcal{A}$. *Then there exists a* $\mathcal{U} \supset \mathcal{A}$ *such that* $\operatorname{card}\mathcal{U} = \operatorname{card}\mathcal{A}$ *and* $\mathcal{U}$ *is an* α*-universal domain.*

Proof. A chain $\{\mathcal{U}_n |\ n<\omega\}$ is defined by recursion. $\mathcal{U}$ will be $\bigcup\{\mathcal{U}_n |\ n<\omega\}$.

1. $\mathcal{U}_0 = \mathcal{A}$.

2. For each $\beta<\alpha$, each finitely generated $\mathcal{V} \subset \mathcal{U}_n$, and each isolated $p \in D^\beta S\mathcal{V}$, choose a $\mathcal{W}^V_{\beta,p}$ finitely generated over $\mathcal{V}$ with the following property: if p splits in $D^\beta S\mathcal{W}$ for some $\mathcal{W} \in \mathcal{K}(T)$, then p splits in $D^\beta S\mathcal{W}^V_{\beta,p}$. The existence of $\mathcal{W}^V_{\beta,p}$ follows from the limit preserving property of $D^\beta S$ described in the proof of Lemma 2.2. Let $\mathcal{U}_{n+1}$ be a model of T which extends $\mathcal{U}_n$ and every $\mathcal{W}^V_{\beta,p}$, and which has the least possible cardinality.

$\mathcal{U}$ is α-universal, since each finitely generated substructure of $\mathcal{U}$ is contained in some $\mathcal{U}_n$. Assume $\operatorname{card}\mathcal{U}_n = \operatorname{card}\mathcal{A}$ in order to see that $\operatorname{card}\mathcal{U}_{n+1} = \operatorname{card}\mathcal{A}$. The set of all finitely generated $\mathcal{V} \subset \mathcal{A}$ has cardinality at most that of $\mathcal{A}$. For each such $\mathcal{V}$, $S\mathcal{V}$ has a countable base for its topology, hence only countably many isolated points. ■

A set N is said to be admissible (cf. Barwise [1]) if it is transitive, closed under the operations of pairing and set union, and satisfies the axiom schemes of Δ_0 separation and collection. Every admissible set N in this paper also satisfies the axiom of infinity: $\omega \in N$. Lemma 2.4 sums up the strong absoluteness properties of the Cantor–Bendixson derivative alluded to in Section 1. It is not known if the Morley derivative is equally absolute. 2.4 is inspired by an early result of Kreisel [7] to the effect that the recursive ordinals, and none less, suffice for the Cantor–Bendixson analysis of a closed, bounded Σ^1_1 set of reals.

LEMMA 2.4 (strong absoluteness of the Cantor–Bendixson derivative). *Let N be an admissible set and $\mathcal{A}$ a substructure of a model of T. If $\mathcal{A} \in N$, then* (i) *and* (ii) *are equivalent for all p and all $\beta<\infty$.*

(i) $p \in S\mathcal{A}\ \&\ r_{\mathrm{CB}}(p) = \beta$.

(ii) $p \in N\ \&\ \beta \in N\ \&\ N \models [p \in S\mathcal{A}\ \&\ r_{\mathrm{CB}}(p) = \beta]$.

In addition all values of the function g, defined by

$$g(\beta) = \{p |\ p \in S\mathcal{A}\ \&\ r_{\mathrm{CB}}(p) = \beta\},$$

belong to N, and g is Δ_1 over N.

Proof. Since $\mathcal{A}$ belongs to N, its complete first order theory also belongs to N. That theory is given by $T \cup \operatorname{Di}(\mathcal{A})$, where $\operatorname{Di}(\mathcal{A})$ is the diagram of $\mathcal{A}$, since it was assumed in Section 1 that T is substructure complete. A typical $p \in S\mathscr{A}$ is a consistent set of formulas of the form $F(x)$ with parameters in A. "Consistent" means consistent with respect to $T \cup \mathrm{Di}(\mathscr{A})$, and has the same meaning inside N that it has outside N $F(x)$ has a Cantor–Bendixson rank $\leqslant \infty$ defined by

$$r_{\mathrm{CB}}(F(x)) = \sup\{r_{\mathrm{CB}}(p) | \ F(x) \in p\} .$$

Part I of the proof is an induction on β intended to show the equivalence of (i) and (ii) for all $\beta \in N$. The induction is performed simultaneously with a recursion on β designed to show that the relation

(1) $$r_{\mathrm{CB}}(F(x)) < \beta$$

is Δ_1 over N. Part II of the proof will show that $r_{\mathrm{CB}}(p) < \infty$ implies $r_{\mathrm{CB}}(p) \in N$.

Part I. Fix $\beta \in N$ and suppose (i) holds. Let $G(x)$ isolate p in $d^{\beta}S\mathscr{A}$. Thus $G(x) \in p$, and for all q,

(2) $$G(x) \in q \ \rightarrow \ q = p \vee r_{\mathrm{CB}}(q) < \beta .$$

Then p is generated by the set $Z(G)$ defined by

(3) $$\{G(x)\} \cup \{\sim H(x) | \ r_{\mathrm{CB}}(H(x)) < \beta\} ;$$

that is p is the unique 1-type containing all of $Z(G)$. By induction (1) is Δ_1 over N, so $Z(G)$, and hence p, are members of N. Also by induction, $r_{\mathrm{CB}}(p) \geqslant \beta$ and (2) are true in N. Consequently (ii) holds.

Now suppose (ii) holds. Then p is generated in N by a set $Z(G)$ as in (3). Since the concept of first order logical consequence is Δ_0, $Z(G)$ generates p outside N. By induction $Z(G)$ insures that $r_{\mathrm{CB}}(p) = \beta$.

To see that $r_{\mathrm{CB}}(F(x)) \leqslant \beta$ is Δ_1, observe that $r_{\mathrm{CB}}(F(x)) = \beta$ if and only if there is at least one p, but not infinitely many, such that $F(x) \in p$ and p is generated by some $Z(G)$ as in (3). It is a Δ_0 matter to decide if $Z(G)$ is consistent and complete (that is generates a 1-type), and if two $Z(G)$'s generate the same 1-type.

Part II. Let α be the least ordinal not in N. Suppose $p \in S\mathscr{A}$ and $r_{\mathrm{CB}}(p) \leqslant \alpha$. Then there is a $G(x)$ such that

(4) $$\{G(x)\} \cup \{\sim H(x) | \ r_{\mathrm{CB}}(H(x)) < r_{\mathrm{CB}}(p)\}$$

generates p. Since (1) is Δ_1; it follows that (4) belongs to N, and is equal to (3) for some $\beta < \alpha$. But then $r_{\mathrm{CB}}(p) < \alpha$.

Each p of Cantor–Bendixson rank β is generated by some $Z(G)$ as in (3). The Δ_1-ness of (1) over N implies that $g(\beta) \in N$, and that g is Δ_1 over N. ■

Corollary 2.5. *If $\mathscr{A}$ is countable, $p \in S\mathscr{A}$ and $r_{\mathrm{CB}}(p) < \infty$, then p is hyperarithmetic in $\mathscr{A}$ and $r_{\mathrm{CB}}(p) < \omega_1^{\mathscr{A}}$* (cf. Kreisel [7]).

Proof. Let N be $\mathscr{A}^+$, the least admissible set with $\mathscr{A}$ as a member, and then apply 2.4. ■

3. Bounding Morley rank.

THEOREM 3.1 (weak absoluteness of the Morley derivative, cf. Lemma 5.2). *Let N be an admissible set, and $\mathscr{A}$ a substructure of a model of T. Suppose $\mathscr{A} \in N$. Then*

$$p \in S\mathscr{A} \;\&\; r_{\mathrm{M}}(p) = \beta \;\rightarrow\; p \in N \;\&\; \beta \in N\,.$$

Proof. By 2.1, $r_{\mathrm{CB}}(p) \leqslant \beta$, and so by 2.4, $p \in N$. Let ZFC_n be a finite fragment of set theory strong enough to carry out all routine arguments of model theory. The following collection $\mathscr{F}$ of axioms is Σ_1 over N, and talks about an end extension of N in which ZFC_n is true.

(i) ZFC_n.

(ii) The diagram of N. For each $a \in N$ there is an individual constant $\underline{a}$ and a sentence

$$(y)[y \in \underline{a} \leftrightarrow \bigvee_{b \in a} (y = \underline{b})]\,.$$

(iii) $\infty > r_{\mathrm{M}}(\underline{p}) \geqslant \underline{\delta}$. One such axiom for each ordinal $\delta \in N$.

(iv) $\underline{\mathscr{A}} \subset \dot{\mathscr{U}}$ and $\mathscr{U}$ is a countable $\underline{\delta}$-universal domain. One such axiom for each $\delta \in N$.

Assume $r_{\mathrm{M}}(p) \notin N$ in the hope of a contradiction. For the moment assume N is countable. Then $\mathscr{F}$ is countable, and by 2.3, $\mathscr{F}$ has a model, namely V, the real world. It follows that $\mathscr{F}$ has a model $\mathscr{N}$ that omits α, the least ordinal not in N. (For details concerning the construction of such an $\mathscr{N}$, see Keisler [6, p. 58]). (iii) implies $\mathscr{N}$ is a proper end extension, because $r_{\mathrm{M}}(\underline{p})$ must have a value in $\mathscr{N}$, and that value cannot belong to N.

For each $z \in \mathscr{N}$, let $\mathrm{tc}(z)$ be the transitive closure of z with respect to $\in_{\mathscr{N}}$, the $\in$ relation of $\mathscr{N}$. If the restriction of $\in_{\mathscr{N}}$ to $\mathrm{tc}(z)$ is well-founded, then z is said to be a *standard set of* $\mathscr{N}$. Let $\mathrm{St}(\mathscr{N})$ be the set of all standard sets of $\mathscr{N}$. $\mathrm{St}(\mathscr{N})$ is admissible, contains N, and its ordinals are just those less than α (cf. [1], p. 60). $\mathscr{N}$ can be regarded as an end extension of $\mathrm{St}(\mathscr{N})$.

The countability of $\mathscr{U}$ in $\mathscr{N}$ makes it possible to construe $\mathscr{U}$ as a member of $\mathrm{St}(\mathscr{N})$. The universe of $\mathscr{U}$ becomes ω, and the relations of $\mathscr{U}$ become sets of n-tuples of finite ordinals.

From now on superscripts will be used to indicate in which universe ranks are being computed. Since $\mathscr{A}$ can be injected into $\mathscr{U}$, there is a $q \in S\mathscr{U} \cap \mathscr{N}$ such that q is a pre-image of p and $r^{\mathscr{N}}_{\mathrm{M}}(q) = r^{\mathscr{N}}_{\mathrm{M}}(p)$. Since q is an actual pre-image of p in the real world and p has a Morley rank, q must also have a Morley rank. It follows from 2.1 and 2.4 that $r^{\mathrm{St}}_{\mathrm{CB}}(q) < \alpha$. A straightforward induction shows that any assignment of Cantor–Bendixson rank in $\mathrm{St}(\mathscr{N})$ persists in $\mathscr{N}$. In short an isolated point of $d^{\gamma}\mathscr{S}\mathscr{U}$ in $\mathrm{St}(\mathscr{N})$ remains isolated in $\mathscr{N}$. Thus $r^{\mathscr{N}}_{\mathrm{CB}}(q) < \alpha$; hence $r^{\mathscr{N}}_{\mathrm{M}}(q) < \alpha$ by 2.2. But this last contradicts axiom (iii).

Now suppose N is uncountable. It suffices to show β is countable. There exist a countable $\mathscr{A}_0 \subset \mathscr{A}$ and a $p \in S\mathscr{A}_0$ such that $r_{\mathrm{M}}(p_0) = \beta$. $\mathscr{A}_0^+$, the least admissible set having $\mathscr{A}_0$ as a member, is countable, hence susceptible to the above argument for countable N. ∎

Corollary 3.2. *Suppose $\mathscr{A}$ is a countable substructure of T and p is a 1-type over $\mathscr{A}$. If p has a Morley rank, then p is hyperarithmetic in $\mathscr{A}$ and its Morley rank is an ordinal recursive in $\mathscr{A}$.*

4. Bounding d_T and α_T. d_T, Blum's density number, was defined in Section 1.

Theorem 4.1. *If T is quasi-totally transcendental, then d_T is recursive in T.*

Proof. Similar to that of 3.1. Let $\mathscr{A}$ be a substructure of a model of T. Define $d_T^{\mathscr{A}}$ to be the least β such that the points of $S\mathscr{A}$ of Morley rank less than β are dense in β. As in the proof of 3.1, the following set $\mathscr{G}$ of axioms is Σ_1 over T^+, the least admissible set with T as a member.

(i) ZFC_n.

(ii) T is quasi-totally transcendental.

(iii) $\mathscr{A}$ is a countable substructure of a model of T.

(iv) $d_T^{\mathscr{A}} > \delta$. One such axiom for each $\delta < \omega_1^T$.

Assume $d_T \geq \omega_1^T$ in hope of a contradiction. Then for each $\delta < \omega_1^T$, there is a countable $\mathscr{A}$ such that $d_T^{\mathscr{A}} > \delta$, and so each T^+-finite ([2]) subset of $\mathscr{G}$ is modeled by the world. It follows that $\mathscr{G}$ has a model $\mathscr{N}$ that omits ω_1^T. (ii) requires $d_T^{\mathscr{A}}$ to have a value in $\mathscr{N}$, and (iv) implies that that value is nonstandard. As in 3.1, the countability of $\mathscr{A}$ implies $\mathscr{A} \in \mathrm{St}(\mathscr{N})$, the standard part of $\mathscr{N}$. Since T is quasi-totally transcendental, the isolated points of $S\mathscr{A}$ are dense in $S\mathscr{A}$ (cf. [12], page 200), and every isolated point of $S\mathscr{A}$ has a Morley rank. Let I be the set of isolated points of $S\mathscr{A}$. By 2.4, $I \in \mathrm{St}(\mathscr{N})$.

Fix $p \in I$ to show $r_{\mathrm{M}}(p) < \omega_1^T$. Suppose $r_{\mathrm{M}}^{\mathscr{N}}(p) = c$; by 3.1 c is a countable ordinal (possibly nonstandard) of $\mathscr{N}$. By (i) and 2.3, there is a $\mathscr{U} \in \mathscr{N}$ such that $\mathscr{A} \subset \mathscr{U}$ and

$$\mathscr{N} \models [\mathscr{U} \text{ is a countable, } (c+1)\text{-universal domain}].$$

Let $q \in S\mathscr{U}$ be a pre-image of p such that $r_{\mathrm{M}}^{\mathscr{N}}(p) = r_{\mathrm{M}}^{\mathscr{N}}(q)$. By 2.2, $r_{\mathrm{CB}}^{\mathscr{N}}(q) = r_{\mathrm{M}}^{\mathscr{N}}(q)$. It follows from 2.4 that $r_{\mathrm{CB}}^{\mathrm{St}(\mathscr{N})}(q) < \omega_1^T$. It was noted in the proof of 3.1 that any assignment of Cantor–Bendixson rank in $\mathrm{St}(\mathscr{N})$ persists in $\mathscr{N}$. Consequently $c < \omega_1^T$.

It now follows that the least upper bound (in $\mathscr{N}$) of

$$\{r_{\mathrm{M}}^{\mathscr{N}}(p) \mid p \in I\} \tag{1}$$

is less than ω_1^T, since $\omega_1^T \notin \mathscr{N}$. But (iv) implies that the least upper bound of (1) is nonstandard. ∎

Corollary 4.2. $d_T < \aleph_1$.

α_T, the Morley rank of T, was defined in Section 1. Morley [9] showed that $\alpha_T < \aleph_1$ when T is totally transcendental.

Theorem 4.3. *If T is totally transcendental, then α_T is recursive in T.*

([2]) I.e. member of T^+.

Proof. Same as that of 4.1. Let $\mathscr{A} \in \mathscr{K}(T)$. Define $\alpha_T^{\mathscr{A}}$ to be the least β such that every $p \in S\mathscr{A}$ has Morley rank less than β. $\mathscr{G}_0$ is the result of two changes in the axiom set $\mathscr{G}$ of 4.1: "quasi-totally" is replaced by "totally", and $d_T^{\mathscr{A}}$ by $\alpha_T^{\mathscr{A}}$. The part played by I in 4.1 is now played by all of $S\mathscr{A}$. The concluding contradiction concerns the least upper bound (in $\mathscr{N}$) of $\{r_M^{\mathscr{N}}(p) | \ p \in S\mathscr{A}\}$. ■

5. Countable universal domains. The purpose of this section is not to prove the existence of a countable universal domain $\mathscr{U}$ for T, for that fact follows easily from the Lachlan bound on Morley rank expressed by formula (2) of Section 1, but rather to estimate the least ordinal needed to construct such a $\mathscr{U}$ from T.

Let WO be the set of all countable wellorderings of integers, and for each $Y \in \mathrm{WO}$, let $|Y|$ be the ordinal represented by Y. Suppose $P(X, \beta)$ is a predicate with variable X ranging over the reals, and β ranging over the countable ordinals. $P(X, \beta)$ is said to be Σ_1^1 if $Y \in \mathrm{WO} \rightarrow P(X, |Y|)$ is Σ_1^1.

PROPOSITION 5.1. *The predicate* $p \in S\mathscr{A}$ & $r_M(p) \geqslant \beta$, *restricted to countable* $\mathscr{A}$*'s, is* Σ_1^1 (3).

Proof. The predicate in question, call it $P(p, \mathscr{A}, \beta)$, is equivalent to

$$(1) \qquad (E\mathscr{B})_{\mathscr{A} \subset \mathscr{B}}(Eq)_{p \subset q \in S\mathscr{B}}[r_{CB}(q) \geqslant \beta] .$$

(1) implies $P(p, \mathscr{A}, \beta]$ by 2.1. Suppose $P(p, \mathscr{A}, \beta)$ holds. Extend $\mathscr{A}$ to $\mathscr{B}$, a countable, $(\beta+1)$-universal domain provided by 2.3. Let $q \in S\mathscr{B}$ be a pre-image of p of the same Morley rank as p. Then $r_{CB}(q) \geqslant \beta$ by 2.2.

Since the predicate $q \in S\mathscr{B}$ is arithmetic, it need only be verified that $[r_{CB}(q) \geqslant \beta]$ is Σ_1^1. The proof of 2.4 shows that the restriction of

$$(2) \qquad Y \in \mathrm{WO} \ \rightarrow \ r_{CB}(q) \geqslant |Y|$$

to any admissible set N is Π_1 over N uniformly in N. The last phase refers to a Π_1 predicate independent of N; its only parameters are integers. Thus there is a Δ_0 formula R such that for all Y and q, (2) is equivalent to

$$(3) \qquad \mathrm{HYP}(Y, q) \models (v) R(v, Y, q) .$$

HYP(Y, q) is the least admissible set with Y and q as members; each element of HYP(Y, q) corresponds to a real hyperarithmetic in Y, q. A universal quantifier restricted to HYP(Y, q) is equivalent to an unrestricted existential quantifier (cf. [11], p. 418). Hence (3) is Σ_1^1. ■

An admissible set N is said to be *recursively inaccessible* if for each $x \in N$, there is a $y \in N$ such that $x \in y$ and y is admissible.

LEMMA 5.2. *Let N be recursively inaccessible and $\mathscr{A}$ a substructure of a model of T. Assume $\mathscr{A} \in N$, $T \in N$ and*

$N \models [T$ is countable & $(x)(x$ is wellorderable$)]$.

(3) Proposition 5.1 can be derived from Lemma 4.8 of [3] without any mention of universal domains.

Then (i) *and* (ii) *are equivalent for all* p *and all* $\beta<\infty$:

(i) $p \in S\mathscr{A}$ & $r_M(p) = \beta$,

(ii) $p \in N$ & $\beta \in N$ & $N \models [p \in S\mathscr{A}$ & $r_M(p) = \beta]$.

Proof. If p, $\mathscr{A} \in N$ and $p \in S\mathscr{A}$, let $r_M^N(p)$ be the Morley rank assigned to p inside N. By 3.1 it suffices to prove

$$r_M(p) = \beta \leftrightarrow r_M^N(p) = \beta \tag{1}$$

for all $\mathscr{A}$, p and β in N by induction on β. Since elementary arguments of model theory can be carried out in N, the value of $r_M^N(p)$ is determined, as in the real world V, by the countable structures. Thus it is enough to prove (1) for $\mathscr{A}$'s countable in N.

Fix β and assume $r_M(p) = \beta$. By induction $r_M^N(p) \geqslant \beta$. Suppose $r_M^N(p) > \beta$. Then in N there is a $\mathscr{B} \supset \mathscr{A}$ and a $q \supset p$ such that q is a limit point of $D^\beta S\mathscr{B}$. In V there is a neighborhood F of q that isolates q from all points of $S\mathscr{B}$ save those of Morley rank less than β. But then by induction F isolates q in $D^\beta S\mathscr{B}$ in N.

Now assume $r_M^N(p) = \beta$. By induction $r_M(p) \geqslant \beta$. Then in V the following is true for each n. There exist a countable $\mathscr{B}_n$ and distinct $q_0, \ldots, q_n$ such that

$$\mathscr{A} \subset \mathscr{B}_n \;\&\; p \subset q_i \in S\mathscr{B}_n \;\&\; r_M(q_i) \geqslant \beta \quad (i<n)\,. \tag{2}$$

Since $r_{CB}^N \leqslant r_M^N(p) < \infty$, it follows from 2.4 that p is hyperarithmetic in $\mathscr{A}$. Suppose $\beta < \omega_1^{\mathscr{A}}$. Then 5.1 implies (2) is Σ_1^1 with parameter $\mathscr{A}$. By Kleene's basis theorem ([11], p. 420), $\mathscr{B}_n$ can be assumed to be recursive in $\mathcal{O}^{\mathscr{A}}$, the hyperjump of the real that codes $\mathscr{A}$. Since N is recursively inaccessible, $\mathcal{O}^{\mathscr{A}} \in N$. Let $\mathscr{B}_\infty$ be the amalgamation in N of all substructures (of models of T) coded by reals recursive in $\mathcal{O}^{\mathscr{A}}$. In V, p has infinitely many pre-images in $\mathscr{B}_\infty$ of Morley rank $\geqslant \beta$, and by induction the same is true in N. But then $r_M^N(p) > \beta$.

Finally suppose $\beta \geqslant \omega_1^{\mathscr{A}}$. Since $r_M(p) \geqslant \beta$, it follows as in the proof of 5.1 that there exist $\mathscr{B}$ and q such that

$$p \subset q \in S\mathscr{B} \;\&\; r_{CB}(q) \geqslant \omega_1^{\mathscr{A}}\,. \tag{3}$$

The second half of (3) is equivalent to

$$(m)[m \in \mathcal{O}^{\mathscr{A}} \rightarrow r_{CB}(q) > |m|]\,.$$

($|m|$ is the ordinal less than $\omega_1^{\mathscr{A}}$ represented by m.) It follows from 5.1 that (3) is Σ_1^1 with parameter $\mathscr{A}$. As in the previous paragraph, (3) has a solution $\langle \mathscr{B}, q\rangle$ in N. By 2.4 and 2.5,

$$r_{CB}(q) = r_{CB}^N(q) = \infty\,.$$

But then $r_M^N(p) = \infty > \beta$. ■

THEOREM 5.3. *Let* $\mathscr{A}$ *be a countable substructure of a model of* T. *Then* $\mathscr{A}$ *can be extended to a countable universal domain* $\mathscr{U}$ *hyperarithmetic in the hyperjump of* $\mathscr{A}$.

Proof. A compound of 2.3 and the last part of 5.2. $\mathscr{U}$ is the limit of a chain $\{\mathscr{U}_n |\; n<\omega\}$ defined by recursion on n. $\mathscr{U}_n$ is recursive in $\mathscr{U}_{n+1}$ uniformly in n. (This can be taken to mean that the even part of the code for $\mathscr{U}_{n+1}$ is the code for $\mathscr{U}_n$.) Each $\mathscr{U}_n$ has lower hyperdegree than $\mathcal{O}^{\mathscr{A}}$ (in symbols $\mathscr{U}_n <_h \mathcal{O}^{\mathscr{A}}$) to insure that the ordinals $\leqslant \omega_1^{\mathscr{A}}$ suffice for the construction.

Let $\mathscr{U}_0$ be $\mathscr{A}$.

Choose some effective enumeration E of ω^4 such that for all n, the nth member of E, denoted by (i, j, k, l), has the property that $i \leqslant n$. Let $\mathscr{V}$ be the jth finitely generated substructure of $\mathscr{U}_i$. If $k \in \mathcal{O}^{\mathscr{A}}$, let β be $|k|$, the ordinal represented by k; otherwise let β be $\omega_1^{\mathscr{A}}$.

The second half of the proof of 5.1 shows

$$\{p \mid p \in S\mathscr{V} \ \& \ r_{\mathrm{CB}}(p) < \infty\} \tag{1}$$

is Π_1^1 with parameter $\mathscr{V}$. By 2.5 each member of (1) is hyperarithmetic in $\mathscr{V}$. Consequently the members of (1) can be arranged in a sequence hyperarithmetic in $\mathcal{O}^{\mathscr{V}}$. Let p be the lth element of that sequence.

Assume $\mathscr{U}_n < \mathcal{O}^{\mathscr{A}}$. Since $\mathscr{A}$ is recursive in $\mathscr{U}_n$, it follows that

$$\omega_1^{\mathscr{U}_n} = \omega_1^{\mathscr{A}} \quad \text{and} \quad \mathcal{O}^{\mathscr{U}_n} \equiv_h \mathcal{O}^{\mathscr{A}} . \tag{2}$$

Consequently $\omega_1^{\mathscr{V}} \leqslant \omega_1^{\mathscr{A}}$ and $\mathcal{O}^{\mathscr{V}} \leqslant_h \mathcal{O}^{\mathscr{A}}$.

Assume $p \in D^{\beta} S\mathscr{V}$ and that p splits in $D^{\beta} S\mathscr{W}$ for some $\mathscr{W} \in \mathscr{K}(T)$. (If not, let $\mathscr{U}_{n+1}$ be $\mathscr{U}_n$.) The definition of $\mathscr{U}_{n+1}$ has two cases.

Case I. $\beta < \omega_1^{\mathscr{V}}$. Thus $\beta < \omega_1^{\mathscr{U}_n}$, and so 5.1 implies that the predicate

$$q_0 \neq q_1 \ \& \ \mathscr{U}_n \subset \mathscr{W} \ \& \ (i)_{i<2}[p \subset q_i \in D^{\beta} S\mathscr{W}] \tag{3}$$

is Σ_1^1 with parameter $\mathscr{U}_n$. By Gandy's basis theorem, (3) has a solution $\langle q_0, q_1, \mathscr{W}\rangle <_h \mathcal{O}^{\mathscr{U}_n}$. Let $\mathscr{U}_{n+1}$ be $\mathscr{W}$. Then $\mathscr{U}_{n+1} <_h \mathcal{O}^{\mathscr{A}}$ by (2).

Case II. $\beta \geqslant \omega_1^{\mathscr{V}}$. By 3.2, $r_{\mathrm{M}}(p) = \infty$. Hence there exist q_0, q_1 and $\mathscr{W}$ such that $q_0, q_1 \in S\mathscr{W}$ and

$$q_0 \neq q_1 \ \& \ \mathscr{U}_n \subset \mathscr{W} \ \& \ (i)_{i<2}[p \subset q_i \ \& \ r_{\mathrm{M}}(q_i) = \infty] . \tag{4}$$

(4) is Σ_1^1 with parameter $\mathscr{U}_n$, because $[r_{\mathrm{M}}(q_i) = \infty]$ is equivalent to

$$(m)[m \in \mathcal{O}^{\mathscr{W}} \rightarrow q_i \in D^{|m|} S\mathscr{W}] , \tag{5}$$

and (5) is Σ_1^1 by 5.1. Now proceed as in case I.

$\mathscr{U}$ ($= \bigcup \{\mathscr{U}_n \mid n < \omega\}$) is hyperarithmetic in $\mathcal{O}^{\mathscr{A}}$, because $\mathscr{U}_n \leqslant \mathcal{O}^{\mathscr{A}}$ uniformly in n. The last assertion is a consequence of several uniformities, two of which are worth mentioning. (i) The set of indices of solvable Σ_1^1 predicates with parameter P is recursive in $\mathcal{O}^P$ uniformly in P. (ii) An index for computing $\mathcal{O}^Y$ hyperarithmetically from $\mathcal{O}^X$ can be obtained effectively from an index for computing Y hyperarithmetically from $\mathcal{O}^X$, if it is assumed that $Y <_h \mathcal{O}^X$.

To see that $\mathscr{U}$ is a universal domain for T, let $\mathscr{V}$ be a finitely generated substructure of $\mathscr{U}$. $\mathscr{V}$ is the jth finitely generated substructure of some $\mathscr{U}_i$. Fix β and p. If $\beta < \omega_1^{\mathscr{V}}$, let $k \in \mathcal{O}^{\mathscr{V}}$ be a notation for β; otherwise let k be some non-member of $\mathcal{O}^{\mathscr{V}}$. Suppose p is an isolated point of $D^{\beta}S\mathscr{V}$. Then $r_{\text{CB}}(p) < \infty$, since $d^{\gamma}S\mathscr{V} \subset D^{\gamma}S\mathscr{V}$ for all γ. Let p be the lth element of (1). Suppose p splits in $D^{\beta}S\mathscr{W}$ for some $\mathscr{W} \in \mathscr{K}(T)$. If (i, j, k, l) is the nth member of E, then p splits in $D^{\beta}\mathscr{U}_{n+1}$, hence in some finitely generated substructure of $\mathscr{U}$. ■

COROLLARY 5.4. *T has a countable universal domain that is hyperarithmetic in the hyperjump of T.*

6. Open questions.

1. To what extent is the assignment of Morley rank absolute? Does Lemma 2.4 remain true when the Cantor–Bendixson derivative is replaced by the Morley derivative? Can the assumption of recursive inaccessibility in Lemma 5.2 be substantially weakened?

2. Does every countable theory T have a countable universal domain $\mathscr{U}$ such that $\mathscr{U}$ is hyperarithmetic in T? If not, can it at least be arranged that $\omega_1^{\mathscr{U}} = \omega_1^{T}$?

Added in proof. C. Calmer (Stanford Ph. D Thesis, 1978) has answered all of the above questions. In particular, Lemma 2.4 holds for the Morley derivative.

References

[1] J. Barwise, *Admissible Sets and Structures*, Springer-Verlag, Berlin–Heidelberg–New York 1975.

[2] L. Blum, *Generalized Algebraic Structures: A Model Theoretic Approach*, Ph. D. Thesis, Massachusetts Institute of Technology 1968.

[3] D. Cenzer, *Monotone inductive definitions over the continuum*, J. Symb. Logic 41 (1976) pp. 188–198.

[4] V. Harnik and M. Makkai, *Applications of Vaught sentences and the covering theorem*, J. Symb. Logic 41 (1976), pp. 171–187.

[5] A. Kechris, *Σ_1^1 derivatives with applications*, MIT Logic Seminar Notes 1974.

[6] J. Keisler, *Model Theory for Infinitary Logic*, North-Holland 1971.

[7] G. Kreisel, *Analysis of the Cantor–Bendixson theorem by means of the analytic hierarchy*, Bull. Acad. Polon. Sci. 7 (1959), pp. 621–626.

[8] A. Lachlan, *The transcendental rank of a theory*, Pacific J. Math. 37 (1971), pp. 119–122.

[9] M. Morley, *Categoricity in power*, Trans. Amer. Math. Soc. 114 (1965), pp. 514–538.

[10] W. Richter, *Constructive transfinite number classes*, Bull. Amer. Math. Soc. 73 (1967), pp. 261–265.

[11] H. Rogers, Jr., *Theory of Recursive Functions and Effective Computability*, McGraw–Hill, New York 1967.

[12] G. Sacks, *Saturated Model Theory*, Benjamin, Reading, Massachusetts 1972.

HARVARD UNIVERSITY
and
MASSACHUSETTS INSTITUTE OF TECHNOLOGY

Accepté par la Rédaction le 20. 12. 1976

J. Barwise, H. J. Keisler and K. Kunen, eds., *The Kleene Symposium*
©North-Holland Publishing Company (1980) 201–222

Post's Problem, Absoluteness and Recursion in Finite Types

*Gerald E. Sacks**

Massachusetts Institute of Technology and Harvard University, Science Center, 1 Oxford Street, Cambridge, MA 02138, U.S.A.

Dedicated to Professor S. C. Kleene on the occasion of his 70th birthday

Abstract: The unresolved character of the power set operation stymies the solution of elementary problems arising in Kleene's theory of recursion in objects of finite type. E.g. Post's problem for ^{3}E has a positive solution if $V=L$ (NORMANN, 1975), and a negative if AD holds. Let $\mathbb{S}$ be the class of all sets $R\subseteq 2^{\omega}$ such that R is recursive in ^{3}E,b for some real b. A forcing construction shows $\mathbb{S}$ is not recursively enumerable in ^{3}E when there is a recursively regular well-ordering of 2^{ω} recursive in ^{3}E. It follows that the concepts of Σ^*, and weak Σ^*, definability differ. With the aid of the notions of indexicality and ordinal recursiveness, an absolute version of Post's problem for ^{3}E is devised, and studied when certain recursively enumerable projecta are equal.

The problem raised by POST (1944) concerning the Turing degrees of recursively enumerable sets of numbers has persisted in the study of generalizations of ordinary recursion theory. A seemingly temperate version of his problem in the context of Kleene's theory of recursion in objects of finite type (KLEENE, 1959) is as follows.

Post's Problem for ^{n}E: Does there exist a type n object A such that A is recursively enumerable in ^{n}E; A is not complete; and A is not recursive in ^{n}E,b for any object b of type less than n?

Kleene's theory of recursion in normal objects of finite type is reviewed several pages ahead. Until then make do with the following. The objects of type 0 are the natural numbers. An *object of type n* ($n>0$) is a collection of objects of type less than n. ^{n}E is the equality functional for objects of type less than n. Thus

$$ {}^nE(x,y)=\begin{cases}0 & \text{if } x=y,\\ 1 & \text{if } x\neq y,\end{cases} $$

*This paper is based on an address given at the Kleene Symposium in May 1978. Its preparation was supported in part by National Science Foundation Grant MCS 7807344. The author is indebted to S. Homer, D. Normann and T. Slaman.

where x and y are arbitrary objects of type less than n. Kleene's notion of reducibility for objects of finite type is written $\leqslant$, and read "recursive in". Thus $A \not\leqslant {}^n\mathrm{E}, B$ means A is not recursive in ${}^n\mathrm{E}, B$. Let A be type n and recursively enumerable in ${}^n\mathrm{E}$; A is *complete* if $B \leqslant A, {}^n\mathrm{E}$ for every type n B recursively enumerable in ${}^n\mathrm{E}$. An important result of Kleene, and a key witness to the correctness of his theory, says

$$X \leqslant {}^2\mathrm{E} \leftrightarrow X \text{ is hyperarithmetic},$$

where X is any type 1 object, e.g. real number.

An affirmative answer to Post's problem faces two impedimenta. Both have to do with the computations developed in the course of recursively enumerating a collection of objects of finite type. The meaning of "computation" will have to await the promised review of Kleene's theory, but for now there is no harm in thinking of a computation as a well-founded tree that serves as a proof that some object belongs to some recursively enumerable collection. The two obstacles are:

(I) a violation of parity between elements of recursively enumerable collections and computations;

(II) lack of a suitable well-ordering of computations.

The first obstacle goes back to KREISEL's early work (1961) on the foundations of metarecursion theory. A violation of parity is at the least a difference in type or complexity. There is a parity (and it is quite useful) between elements and computations in ordinary recursion theory. In that case a computation has the form of a finite tree, hence can be effectively encoded by a number, and consequently there is little difference between a complete, recursively enumerable set of numbers and the set of all computations.

The next proposition is a classic example of obstacle (I) at the type 2 level.

Proposition 1. *Let A be a set of natural numbers. If A is recursively enumerable, but not recursive, in ${}^2\mathrm{E}$, then A is complete* (SPECTOR, 1955).

Proof. Since A is not recursive in ${}^2\mathrm{E}$, the natural enumeration of A is cofinal with the natural enumeration of C, the complete, recursively enumerable in ${}^2\mathrm{E}$ subset of ω. Hence $m \in C$ iff

$$(\mathrm{E}n)[n \in A \,\&\, m \text{ is enumerated in } C \text{ before } n \text{ is enumerated in } A],$$

and the latter is recursive in $A, {}^2\mathrm{E}$.

Kreisel was the first to point out the violation of parity inherent in Proposition 1. A number is enumerated in A iff an appropriate computation exists. In this case a computation has the form of a countable tree encoded by a real recursive in ${}^2\mathrm{E}$. Kreisel's remedy is almost immediate. Let $\mathrm{REC}({}^2\mathrm{E})$ be the class of all reals recursive in ${}^2\mathrm{E}$. Enumerate subclasses of $\mathrm{REC}({}^2\mathrm{E})$ rather than subsets of ω. The associated computations will still be encoded by elements of $\mathrm{REC}({}^2\mathrm{E})$, and so obstacle (I) will have vanished. The remedy has a further prescription somewhat less immediate. The desire for parity requires some revision of the notion of completeness. Let $A \subseteq \mathrm{REC}({}^2\mathrm{E})$ be recursively enumerable in ${}^2\mathrm{E}$. A is said to be *complete* if B is recursive in $A, {}^2\mathrm{E}$ *on* $\mathrm{REC}({}^2\mathrm{E})$ for every recursively enumerable (in ${}^2\mathrm{E}$) $B \subseteq \mathrm{REC}({}^2\mathrm{E})$. To be more precise, there is a Kleene partial recursive functional $\{e\}$ such that for all $R \in \mathrm{REC}({}^2\mathrm{E})$:

$$R \in B \leftrightarrow \{e\}({}^2\mathrm{E}, A, R) = 0;$$

$$R \notin B \leftrightarrow \{e\}({}^2\mathrm{E}, A, R) = 1.$$

Thus $\{e\}$ correctly answers all membership questions about B from A, so long as those questions are restricted to $\mathrm{REC}({}^2\mathrm{E})$. To require $\{e\}$ to do more is to violate parity, because computations involving arguments outside $\mathrm{REC}({}^2\mathrm{E})$ have little hope of being encoded by elements of $\mathrm{REC}({}^2\mathrm{E})$. Note that $\{e\}$ is not required to be total.

The next theorem attests to the efficacy of Kreisel's remedy. It is a positive solution to Post's problem at the type 2 level analogous to the FRIEDBERG (1957)–MUCHNIK (1956) solution at the type 1 level.

Theorem 2. *There exist* A *and* B, *each a recursively enumerable* (*in* ${}^2\mathrm{E}$) *subclass of* $\mathrm{REC}({}^2\mathrm{E})$, *such that* $A \nleqslant B, {}^2\mathrm{E}$ *and* $B \nleqslant A, {}^2\mathrm{E}$ *on* $\mathrm{REC}({}^2\mathrm{E})$.

Proof. Implicit in Sacks (1966). Let A_1 and B_1 be hyperregular, ω_1^{CK}-recursively enumerable sets such that neither is ω_1^{CK}-recursive (i.e. metarecursive) in the other. Let $\{H_\delta \mid \delta < \omega_1^{\mathrm{CK}}\}$ be the natural enumeration of the hyperarithmetic subsets of ω obtained by iterating the Turing jump. Define

$$A = \{H_\delta \mid \delta \in A_1\} \quad \text{and} \quad B = \{H_\delta \mid \delta \in B_1\}.$$

Let $\{e\}$ be a Kleene partial recursive functional, and suppose $B \leqslant A, {}^2\mathrm{E}$ on $\mathrm{REC}({}^2\mathrm{E})$ via $\{e\}$. The hyperregularity of A implies that the computation of $\{e\}({}^2\mathrm{E}, A, H)$, where H is a hyperarithmetic real, is encodable by a hyperarithmetic real. The collection of all such computations is an ω_1^{CK}-recursively enumerable set. It follows that B_1 is ω_1^{CK}-recursive in A_1.

The next proposition states that impediment (I) occurs not only at the type 2 level, as noted by Spector, but also at all higher levels.

Proposition 3. *Assume* $n \geqslant 2$. *Let* A *be a type* $(n-1)$ *object. If* A *is recursively enumerable in* ${}^n\mathrm{E}$, *but not recursive in* ${}^n\mathrm{E}, a$ *for any object* a *of type less than* $n-1$, *then* A *is complete.*

Proof. Same as that of Proposition 1.

The Kreisel remedy is readily applied at all type levels beyond the first. Fix $n > 2$. An *individual* is defined to be any object of type less than $n-1$. By a *set* is meant any collection R of individuals such that $R \leqslant {}^{n}\mathrm{E}, a$ for some individual a. To remedy the violation of parity in Proposition 3, enumerate classes of sets rather than classes of individuals. The associated computations are encodable by sets, and so parity has been restored.

Impediment (II) is not manifest at the type 2 level. Computations arising from recursion in ${}^{2}\mathrm{E}$ have numerical codes, and consequently lend themselves to a suitable well-ordering. At the type 3 level the codes are reals, and there need not be any effective well-ordering of the reals in sight. The situation is especially grim with respect to Post's problem if AD, the axiom of determinateness, is assumed, as is seen in Proposition 4.

From now on it will be best to focus on recursion in ${}^{3}\mathrm{E}$. All that follows, save Proposition 4, applies equally well to ${}^{n}\mathrm{E}$ when $n > 3$, but is more easily grasped when $n = 3$.

In accord with the definition of set given above, a *set* R is a collection of reals such that $R \leqslant {}^{3}\mathrm{E}, a$ for some real a. A set R is said to be *indexical*[1] if there exists a nonempty set I of reals such that $I \leqslant {}^{3}\mathrm{E}, R$ and

$$a \in I \rightarrow R \leqslant {}^{3}\mathrm{E}, a$$

for all a. The notion of indexicality is less technical than a first glance suggests. Its intuitive meaning will be discussed after the review of Kleene recursion. For now it must suffice to say that the class of indexical sets is recursively enumerable in ${}^{3}\mathrm{E}$, but the class of all sets need not be.[2]

Proposition 4. *Assume AD. Let A be a recursively enumerable* (*in* ${}^{3}\mathrm{E}$) *class of sets. Then there exists a real b such that either* (i) *or* (ii) *holds.*

(i) $A \leqslant {}^{3}\mathrm{E}, b$ *on the class of all indexical sets.*

(ii) $C \leqslant {}^{3}\mathrm{E}, A, b$ *where C is the complete, recursively enumerable* (*in* ${}^{3}\mathrm{E}$) *sets of reals.*

Proof. Modeled on Simpson (1974, p.439). Let cA be the class of all indexical sets not in A. If $R \in cA$, then there is an e such that

$$I = \{e\}({}^{3}\mathrm{E}, R), \quad \varnothing \neq I \subseteq \omega \times 2^{\omega},$$

and

$$\langle e_0, a\rangle \in I \rightarrow R = \{e_0\}({}^{3}\mathrm{E}, a) \quad \text{for all } \langle e_0, a\rangle \in I.$$

[1]Cf. the definition of indexical given in Webster's Third International Dictionary.

[2]This appears to be the reason that Normann (1975) introduces ${}^{1}M$.

Gandy's selection theorem for recursively enumerable (in $^3\mathrm{E}, R$) classes of numbers makes it possible to pass effectively from an indexical set R to a unique e (denoted by e_R) that witnesses the indexicality of R. Define

$$I_R = \{e_R\}(^3\mathrm{E}, R) \quad \text{and} \quad i[cA] = \bigcup \{I_R | R \in cA\}.$$

According to Wadge, AD yields a continuous function f satisfying either (1) or (2).

(1) $\quad x \in i[cA] \leftrightarrow f(x) \in C \quad$ (for all real x).

(2) $\quad x \notin C \leftrightarrow f(x) \in i[cA].$

Let b be a real that encodes f, that is b encodes rational approximations of f on all rational arguments. Assume (1). Then

$$R \in cA \leftrightarrow f[I_R] \subseteq C$$

for every indexical set R, and so cA is recursively enumerable in $^3\mathrm{E}, b$. Hence (i) holds, since A is recursively enumerable in $^3\mathrm{E}$. Now assume (2). Then

$$x \notin C \leftrightarrow f(x) = \langle e_0, a \rangle \;\&\; R = \{e_0\}(^3\mathrm{E}, a) \\ \&\; R \in cA,$$

and (ii) holds.

Two remedies have been found for impediment (II), one by HARRINGTON (1973), and the other by NORMANN (1975). The first approach is to restrict attention to sets and computations that of necessity lend themselves to a suitable well-ordering. In the case of $^3\mathrm{E}$ this means narrowing down to the class of sets recursive in $^3\mathrm{E}$, namely REC(^{3}E). For $^n\mathrm{E}$ ($n > 3$) the restriction is to the subrecursive sets, an uncountable, but bounded subfamily of the indexical sets (cf. SACKS (1977) for details). The second approach is to assume a suitable well-ordering of 2^ω recursive in $^3\mathrm{E}$.

Harrington's Remedy: There exist recursively enumerable (in ^{3}E) classes $A, B \subseteq \mathrm{REC}(^3\mathrm{E})$ such that $A \not\leq B, {}^3\mathrm{E}$ and $B \not\leq A, {}^3\mathrm{E}$ on REC(^{3}E).

Normann's Remedy: Assume that the cardinality c of the continuum is regular and that some well-ordering of the real of height c is recursive in ^{3}E. Then there exist recursively enumerable (in ^{3}E) classes A and B of sets such that for every set R,

$$A \not\leq B, {}^3\mathrm{E}, R \quad \text{and} \quad B \not\leq A, {}^3\mathrm{E}, R$$

on the class of all sets.

The virtue of Harrington's remedy is its explicitness. No assumptions are made about ^{3}E. There is no reference, at least no direct reference, to an indeterminate continuum. All sets to be enumerated, as well as the associated computations, have numerical codes, and Post's problem is solved by an adaptation of the α-recursion theoretic methods that succeeded at the type 2 level. But a virtue pursued to the limit, particularly the virtue of explicitness, becomes a fault. Harrington's result says very little about recursively enumerable (in ^{3}E) classes of sets. It avoids the problem of the continuum, a problem that Normann confronts more directly. His assumption concerning the regularity of the continuum and its computability from ^{3}E is true in L and elsewhere. He in fact manages with a weaker regularity assumption made explicit just before Lemma 5 below. (This paper and its sequel deal with the singular case.) An unexpected consequence of Normann's regularity assumption is the non-recursive enumerability in ^{3}E of the class of all sets (Theorem 9), one of the reasons for introducing indexical sets.

The absolute approach to Post's problem for ^{3}E formulated below owes more to Normann's remedy than it does to Harrington's. It consists of staying as close as possible to the ordinals arising as lengths of computations generated by recursion in ^{3}E. These ordinals define sets that will take the place of the sets mentioned in Normann's remedy. Some of his sets are lost, sacrificed on the altar of absoluteness, but the loss will be seen to be less than one might think.

The time of the promised review of Kleene recursion in ^{3}E is at hand. Recall how the hyperarithmetic sets of numbers are obtained by iteration of the Turing jump through the recursive ordinals. The class of all sets is similarly obtained by iteration of the ^{3}E jump through the ordinals recursive in ^{3}E and an arbitrary real. Do not forget that "set" has been defined above to mean a collection of reals recursive in $^3\mathrm{E},a$ for some real a. The sets and ordinals in question are defined by a simultaneous recursion. The hierarchy developed is equivalent to an initial segment of $L(2^\omega)$. Thus the sets recursive in $^3\mathrm{E},a$ for some real a are obtained by starting with the continuum, iterating first order definability, and stopping at the first ordinal that cannot be justified from below in a certain predicative fashion.

Let $X\subseteq 2^\omega$. The ^{3}E jump of X, denoted by $j(X)$, is an encoding of certain definable collections of reals. Let $\{e\}^X_\pi$ be the eth function from 2^ω into ω definable over $\langle 2^\omega,\varepsilon,X\rangle$ by means of a first order formula whose only parameters are nonnegative integers, and whose atomic parts are of the form $y\in z$ or $y\in\chi$, where y and z are real variables and χ is a constant naming X.

$j(X)$ is the collection of all $\langle e,a,m\rangle$ such that $\{e\}^X_\pi(a)=m$.

$\mathcal{O}$ is the collection of notations for ordinals recursive in $^3\mathrm{E},a$ for some real a. Notations are reals $a,b,c,\ldots$; if $b\in\mathcal{O}$, then $|b|$ is the ordinal represented by b. H_σ is the set obtained by σ iterations of the ^{3}E jump j. The simultaneous recursive definition of $\mathcal{O}$ and H_σ has three clauses.

(1) $\langle 1,a\rangle \in \mathcal{O}$ and $|\langle 1,a\rangle|=0$ for all $a\in 2^\omega$. And $H_0=\varnothing$.

(2) Suppose $\langle e,a\rangle \in \mathcal{O}$ and $|\langle e,a\rangle|=\sigma$. Then $\langle 2^e,a\rangle \in \mathcal{O}$, $|\langle 2^e,a\rangle|=\sigma+1$, and $H_{\sigma+1}=j(H_\sigma)$.

(3) Suppose $\langle m,a\rangle \in \mathcal{O}$ and $|\langle m,a\rangle|=\sigma$. Define

$$W^{H_\sigma}_{e,a}=\{b\,|\,\{e\}^{H_\sigma}(a*b)=0\},$$

where $a*b$ is a real code for $\langle a,b\rangle$. Suppose $W^{H_\sigma}_{e,a}\subseteq \mathcal{O}$. Then $\langle 3^m\cdot 5^e,a\rangle \in \mathcal{O}$ and $|\langle 3^m\cdot 5^e,a\rangle|=\lambda$, where λ is the least limit ordinal greater than σ and greater than $|b|$ for all $b\in W^{H_\sigma}_{e,a}$. And H_λ is the set of $\langle b,c\rangle$ such that $b\in\mathcal{O}$, $|b|<\lambda$ and $c\in H_{|b|}$.

Let κ_1 be the supremum of $|b|$ for all $b\in\mathcal{O}$. The hierarchy $\{H_\sigma|\sigma<\kappa_1\}$ is equivalent to a hierarchy of all the computations that arise in the computing of partial functions from ^{3}E and real arguments.

$\{e\}(^3\mathrm{E},a)$ is defined and equal to m if $e=\langle e_0,e_1\rangle$, $\langle e_0,a\rangle\in\mathcal{O}$, $|\langle e_0,a\rangle|=\sigma$ and $\{e_1\}^{H_\sigma}_\pi(a)=m$.

Thus to compute $\{e\}(^3\mathrm{E},a)$ is to iterate the ^{3}E jump until $\langle e_0,a\rangle$ is seen to be a notation for an ordinal σ and then to pluck the value from H_σ in a first order fashion dictated by e_1.

A function f is recursive in ^{3}E,a (in symbols $f\leqslant {}^3\mathrm{E},a$) if there is an e such that $f(b)=\{e\}(^3\mathrm{E},a*b)$ for all real b.

A collection R of reals in recursive in ^{3}E,a (in symbols $R\leqslant {}^3\mathrm{E},a$) if the characteristic function of R is recursive in ^{3}E,a. (Such an R has been termed a "set" above.)

An ordinal γ is *constructive* in ^{3}E,a if γ is $|\langle e,a\rangle|$ for some $\langle e,a\rangle\in\mathcal{O}$. γ is *recursive* in ^{3}E,a if γ is the height of a prewell-ordering of 2^ω recursive in ^{3}E,a. If γ is constructive in ^{3}E,a, then γ is recursive in ^{3}E,a. If γ is recursive in ^{3}E,a, then γ is less than or equal to some ordinal constructive in ^{3}E,a.

Clause 3 of the definition of $\mathcal{O}$ and H implies a *bounding principle*: if $f:2^\omega\to\mathcal{O}$ is recursive in ^{3}E,a, then there exists an ordinal γ constructive in ^{3}E,a such that $|f(b)|<\gamma$ for all b. It follows from the bounding principle that R is recursive in ^{3}E,a iff there is a σ constructive in ^{3}E,a and an e such that

$$R=W^{H_\sigma}_{e,a}.$$

An *index* for a *set* R is $\langle d,e,a\rangle$, where $|\langle d,a\rangle|=\sigma$ and the above equation relating R and W is satisfied by σ, e and a. From now on assume that indices for sets are encoded by reals. Note that if b is an index for R, then $R\leqslant {}^3\mathrm{E},b$.

Let S be an arbitrary collection of reals. Recursion relative to ^{3}E,S is defined in the same hierarchic fashion as recursion in ^{3}E. The only difference is beginning with S rather than $\varnothing$. Thus the simultaneous recursive definition of $\mathcal{O}^S$ and H_σ^S has three clauses identical with those given for $\mathcal{O}$ and H_σ save that $H_0^S = S$. κ_1^S is of course the supremum of $|b|$ for all $b \in \mathcal{O}^S$, and may exceed κ_1. $\{e\}({}^3\mathrm{E}, S, a)$ is defined and equal to m if $e = \langle e_0, e_1 \rangle$, $\langle e_0, a \rangle \in \mathcal{O}^S$, $|\langle e_0, a \rangle| = \sigma$ and $\{e_1\}_\pi^{H_\sigma^S}(a) = m$.

$\mathcal{O}$ and $\{H_\sigma | \sigma < \kappa_1\}$ can be imbedded effectively in $\mathcal{O}^S$ and $\{H_\sigma^S | \sigma < \kappa_1^S\}$. In other words recursion in ${}^3\mathrm{E}$ is effectively included in recursion in ${}^3\mathrm{E}, S$. One can pass effectively from e to e^* so that

$$\{e\}({}^3\mathrm{E}, a) \simeq \{e^*\}({}^3\mathrm{E}, S, a)$$

for all S and a. ($\simeq$ is Kleene's symbol for equality of graphs of partial functions.) The imbedding can be reversed if S is a set (i.e. recursive in ${}^3\mathrm{E}, a$ for some a). If S is a set, then $\kappa_1^S = \kappa_1$, and the H_σ^S hierarchy can be imbedded in the H_σ hierarchy by means of an effective transfinite recursion that begins at the location of S in the H_σ hierarchy. In short the computations for recursion in ${}^3\mathrm{E}, S$ are essentially the same as those for recursion in ${}^3\mathrm{E}$ if S is a set.

Let A be a class of collections of reals. A is *recursively enumerable* in ${}^3\mathrm{E}$ if there is an e such that

$$S \in A \leftrightarrow \{e\}({}^3\mathrm{E}, S) \quad \text{is defined}$$

(for every collection S of reals). As in Normann's remedy given above, interest tends to center on recursively enumerable classes of sets. Let A be a recursively enumerable (in ${}^3\mathrm{E}$) class of sets for the sake of sharpening the intuitions behind Kreisel's remedy for impediment (I) to the solution of Post's problem. A set R is put into A if and only if there exists a computation Z of $\{e\}({}^3\mathrm{E}, R)$. The existence of Z is equivalent to the existence of some H_σ^R with some first order property specified by e. Since R is a set, there is an a such that $R \leqslant {}^3\mathrm{E}, a$. Then Z, if it exists, will be recursive in ${}^3\mathrm{E}, a$, and so will be a set located in the H_σ hierarchy not far from R. Thus recursively enumerable (in ${}^3\mathrm{E}$) classes of sets are enumerated via computations which are themselves sets.

To carry the parity between elements of recursively enumerable classes and computations one step further, let A be a class of sets recursively enumerable in ${}^3\mathrm{E}, R_0$, where R_0 is a set; that is there is an e such that

$$R \in A \leftrightarrow \{e\}({}^3\mathrm{E}, R_0, R) \quad \text{is defined}$$

(for every set R). As in the previous paragraph, R lands in A if and only if there exists a computation Z to that effect. As above, if such a Z exists, it will be recursive in ${}^3\mathrm{E}, R_0, R$, and so will be a set. Now the requirements of parity have been fully met: the elements of A, the parameter R_0 that defines the procedure for enumerating A, and the associated computations all live on the same level of complexity; they are all sets, that is sets located in the H_σ hierarchy for recursion in ${}^3\mathrm{E}$.

But there is an unexpected fly in the ointment. The class of all sets is not recursively enumerable in ${}^3\mathrm{E}$ when ${}^3\mathrm{E}$ satisfies Normann's regularity assumption.

Let $\prec$ be a well-ordering of 2^ω. $\prec$ is said to be *recursively regular* if there is no function $f \leqslant {}^3\mathrm{E}, a$ for some real a such that the domain of f is a proper initial segment of $\prec$ and the range of f is unbounded in $\prec$.

NRA: *Normann's Regularity Assumption.* There exists a recursively regular well-ordering $\prec$ of 2^ω such that $\prec$ is recursive in ${}^3\mathrm{E}$.[3]

Lemma 5. *Suppose Normann's regularity assumption is satisfied by $\prec$. Let $A \subseteq 2^\omega$ be recursively enumerable in ${}^3\mathrm{E}, a$ for some real a, and let B be a proper initial segment of $\prec$. Then $A \cap B$ is recursive in ${}^3\mathrm{E}, b$ for some real b.*

Proof. An immediate consequence of the main theorem of NORMANN (1977).

The proof below that the class of all sets is not recursively enumerable in ${}^3\mathrm{E}$ (when Normann's regularity assumption holds) is largely a proof of the existence of a non-indexical set. This notion of set, defined just before Proposition 4, is in need of some clarification. Let R be a set. Thus $R \leqslant {}^3\mathrm{E}, a$ for some real a, and R is first order definable over $\langle 2^\omega, \varepsilon, H_\sigma \rangle$, where $\sigma = |\langle e, a \rangle|$ for some e. Let σ_R be the least such σ. In short σ_R is the first level of the $\{H_\sigma | \sigma < \kappa_1\}$ hierarchy where the set R occurs.

Proposition 6. *Let R be a set. R is indexical iff σ_R is recursive in ${}^3\mathrm{E}, R$.*

Proof. First suppose R is indexical. Thus there is a nonempty $I \subseteq 2^\omega$ such that $I \leqslant {}^3\mathrm{E}, R$ and

$$(a)[a \in I \rightarrow R \leqslant {}^3\mathrm{E}, a].$$

It follows from Gandy selection and bounding that there exists an ordinal γ recursive in ${}^3\mathrm{E}, R$ such that

$$(a)[a \in I \rightarrow |a| < \gamma].$$

Consequently $\sigma_R < \gamma$, and so σ_R is recursive in ${}^3\mathrm{E}, R$.

Now suppose σ_R is recursive in ${}^3\mathrm{E}, R$. To see that R is indexical, let I be the set of all real a such that $|(a)_0| = \sigma_R$ and $|(a)_1|$ is the parameter needed in the first order definition of R over $\langle 2^\omega, H_{\sigma_R}, \varepsilon \rangle$.

The above proposition is intended to express the intuitive content of the concept of indexicality. Each set R has a natural location in the hierarchy

[3]Analogous to the assumption in α-recursion theory that α^* is a regular α-cardinal.

of all sets obtained by iterating the ^{3}E jump through all ordinals recursive in ^{3}E, a for some real a. To say R is indexical is to say R is located at some ordinal recursive in ^{3}E, R.

Proposition 7. (i) *Every H_σ is indexical.*

(ii) *The class of indexical sets is recursively enumerable in* ^{3}E.

Proof. (i) Let I be $\{a \mid |a|=\sigma\}$. Then $I \leqslant {}^3\mathrm{E}, H_\sigma$, and $H_\sigma \leqslant {}^3\mathrm{E}, a$ for all $a \in I$.

(ii) By Proposition 6, R is indexical iff there is an e such that

$$\langle e,0\rangle \in \mathcal{O}^R \ \&\ R \text{ is first order definable over } H_{|\langle e,0\rangle|_R}.$$

Observe that every recursively enumerable (in ^{3}E) class A of indexical sets has a natural enumeration tied to that of $\mathcal{O}$. The members of $\mathcal{O}$ are enumerated in κ_1 steps. At step σ, all $a \in \mathcal{O}$ such that $|a|=\sigma$ are enumerated, and also all $R \in A$ such that $\sigma_R=\sigma$. Some such enumeration of A seems to be an essential part of priority arguments in the study of recursively enumerable classes of sets. This is why Normann (1975, p.8) restricts his attention from M to ${}^1M$. If A were allowed to have some non-indexical member S, then the reason for putting S into A might not become evident until S was located at stage σ_S of the H_σ hierarchy. By then it would be too late to put S in A, because σ_S exceeds all ordinals recursive in ^{3}E, S, and the latter are the only ordinals allowed to serve as the height of a computation that puts S in a recursively enumerable class.

Proposition 8. *Assume there is a well-ordering W of 2^ω recursive in* ^{3}E. *Then* (i), (ii) *and* (iii) *are equivalent.*

(i) *R is an indexical set.*

(ii) *There exists a real a such that $R \leqslant {}^3\mathrm{E}, a$ and $a \leqslant {}^3\mathrm{E}, R$.*

(iii) *There exists a $b \in \mathcal{O}$ such that $\sigma_R = |b|$ and $b \leqslant {}^3\mathrm{E}, R$.*

Proof. Note that if S is a nonempty set, then the least member of S (least with respect to W) is recursive in ^{3}E, S. Assume (i) to show (ii). Thus there is a nonempty $I \leqslant {}^3\mathrm{E}, R$ such that $R \leqslant {}^3\mathrm{E}, b$ for every $b \in I$. Let a be the W-least member of I. Then a satisfies (ii).

Assume a satisfies (ii). Then σ_R is recursive in ^{3}E, a. The set of all b that satisfy $\sigma_R = |b|$ is recursive in ^{3}E, a. The least such b is recursive in ^{3}E, R.

(i) follows from (iii) by Proposition 6.

Theorem 9. *Assume* ^{3}E *satisfies Normann's regularity assumption. Then the class of all sets, namely*

$$\{R \mid R \subseteq 2^\omega \ \&\ (\mathrm{E}a)(a \in 2^\omega \ \&\ R \leqslant {}^3\mathrm{E}, a)\},$$

is not recursively enumerable in ^{3}E, c *for any* $c \in 2^\omega$.

Corollary 10. *Assume* $^3\mathrm{E}$ *satisfies* NRA. *Then there exists a non-indexical set.*

Proof. Theorem 9 and Proposition 7(ii).

Corollary 11. *Assume the cardinality of the continuum is regular. Then there exists a normal type* 3 *object* F *such that*

$$\{R \mid R \subseteq 2^\omega \ \&\ (\mathrm{E}a)(a \in 2^\omega \ \&\ R \leqslant F, a)\}$$

is not recursively enumerable in F, b *for any* $b \in 2^\omega$.

Proof. Let W be a well-ordering of the continuum of minimal height. Define $F = \langle {}^3\mathrm{E}, W\rangle$ and imitate the proof of Theorem 9.

Normann (1975, p. 9) asks if the concepts of Σ^* and weak Σ^* definability coincide. Let A be a class of sets. Call A *weakly recursively enumerable* in $^3\mathrm{E}$ if there is an e such that for every set R,

$$R \in A \leftrightarrow (a)\big[R \leqslant {}^3\mathrm{E}, a \rightarrow \{e\}({}^3\mathrm{E}, R, a) \quad \text{is defined}\big].$$

Clearly every class recursively enumerable in $^3\mathrm{E}$ is also weakly recursively enumerable in $^3\mathrm{E}$. For $^3\mathrm{E}$, (weak) Σ^* definability is the same as (weak) recursive enumerability.

Corollary 12. *Assume* $^3\mathrm{E}$ *satisfies Normann's regularity assumption. Then*

$$\{R \mid R \subseteq 2^\omega \ \&\ (\mathrm{E}a)(a \in 2^\omega \ \&\ R \leqslant {}^3\mathrm{E}, a)\}$$

is weakly Σ^* *definable, but not* Σ^* *definable, in* $^3\mathrm{E}$.

The Ramified Language $\mathfrak{L}(\kappa_1, \mathfrak{T})$. The proof of Theorem 9 is a forcing argument that ranges over the H_σ hierarchy of sets associated with recursion in $^3\mathrm{E}$. The result of adjoining a generic $T \subseteq 2^\omega$ to $\{H_\sigma \mid \sigma < \kappa_1\}$ will be studied. The resulting structure, $\{H_\sigma^T \mid \sigma < \kappa_1\}$, is of greatest interest when $\kappa_1^T = \kappa_1$. The language $\mathfrak{L}(\kappa_1, \mathfrak{T})$ contains just what is needed to analyze $\{H_\sigma^T \mid \sigma < \kappa_1\}$. It includes: number variables $(x_0, y_0, \ldots)$; real variables $(x, y, \ldots)$; ordinal variables $(\sigma, \tau, \ldots)$; ordinal constants $(\boldsymbol{\alpha}, \alpha < \kappa_1)$; real constants $(\mathbf{a}, a \in 2^\omega)$; quantifiers on numbers, reals and ordinals; and propositional connectives. It also includes predicates

$$t_2 \in H_{t_1}^{\mathfrak{T}}, \qquad |t_2|_{\mathfrak{T}} = t_1 \quad \text{and} \quad t_2 \in W\big(e, t_3, H_{t_1}^{\mathfrak{T}}\big),$$

where t_1 is an ordinal term, and t_2 and t_3 are real terms.

Each formula $\mathfrak{F}$ has one or more Gödel numbers of the form $\langle e, a_1, \ldots, a_m, b_1, \ldots, b_n\rangle$: e specifies the logical form of $\mathfrak{F}$; a_i specifies a real constant $\mathbf{a}_i$ occurring in $\mathfrak{F}$; and b_j is a notation from $\mathcal{O}$ that specifies an ordinal constant $|\mathbf{b}_j|$ occurring in $\mathfrak{F}$. There is no need to choose a unique Gödel number for $\mathfrak{F}$, since each is recursive in every other via ^{3}E. Thus the ^{3}E degree of the Gödel number of $\mathfrak{F}$ is unique. Let $\ulcorner\mathfrak{F}\urcorner$ denote a Gödel number of $\mathfrak{F}$. The collection of all Gödel numbers is recursively enumerable in ^{3}E.

$\mathfrak{F}$ is said to be a *bounded formula* if every ordinal quantifier in $\mathfrak{F}$ is bounded (by some ordinal less than κ_1).

Every formula $\mathfrak{F}$ has a *rank* based on the range of its ordinal quantifiers, its ordinal constants and its logical complexity. It is a routine matter to define a rank function whose range is well-founded with the following properties:

(r1) rank $\mathfrak{F}(\boldsymbol{\alpha})<\text{rank}(\mathsf{E}\sigma)\mathfrak{F}(\sigma)$ for all $\alpha<\kappa_1$.

(r2) rank $\mathfrak{F}(\boldsymbol{\beta})<\text{rank}(\mathsf{E}\sigma)_{\sigma<\tau}\mathfrak{F}(\sigma)$ for all $\beta<\tau$.

(r3) rank $\mathfrak{F}(\mathbf{b})<\text{rank}(x)\mathfrak{F}(x)$ for all $b\in 2^\omega$.

(r4) rank $\mathfrak{F}<\text{rank}(\mathfrak{F}\ \&\ \mathfrak{G})$.

(r5) $\text{rank}(\sim\mathfrak{F})<\text{rank}\,\mathfrak{F}$.

(r6) $|\langle\mathbf{m},\mathbf{a}\rangle|_{\mathfrak{T}}=\boldsymbol{\alpha}$ has lower rank than $|\langle\mathbf{n},\mathbf{b}\rangle|_{\mathfrak{T}}=\boldsymbol{\beta}$ if $\alpha<\beta$.

In addition the graph of the rank function can be taken to be recursively enumerable in ^{3}E, since the rank of $\mathfrak{F}$ can be effectively obtained from a Gödel number of $\mathfrak{F}$.

The Forcing Relation. Assume NRA holds. Thus $\prec$ is a well-ordering of 2^ω recursive in ^{3}E and recursively regular. Forcing conditions are denoted by $p,q,r,\ldots$, and specify a small part of the characteristic function of T. A typical forcing condition is a function $f:\{c\,|\,c\prec a\}\to 2$ such that f is recursive in $^3\text{E},b$ for some real b. A condition p is encoded by a pair $\langle e,b\rangle$ such that $p=\{e\}(^3\text{E},b)$. p is extended by q (in symbols $p\geqslant q$) if graph $p\subseteq$ graph q.

The forcing relation $p\Vdash\mathfrak{F}$ is defined by recursion on the rank of $\mathfrak{F}$. Each clause of the definition either parallels some clause of the definition of $\{H_\sigma^T\,|\,\sigma<\kappa_1\}$ or is a standard feature of forcing technology. Some of the clauses are as follows:

(f1) $p\Vdash\mathbf{a}\in\mathfrak{T}$ iff $p(a)=1$. ($\mathfrak{T}$ is $H_0^{\mathfrak{T}}$.)

(f2) $p\Vdash\sim\mathfrak{F}$ iff $(q)_{p\geqslant q}\sim[q\Vdash\mathfrak{F}]$.

(f3) $p\Vdash(\mathsf{E}\sigma)\mathfrak{F}(\sigma)$ iff $p\Vdash\mathfrak{F}(\boldsymbol{\alpha})$ for some $\alpha<\kappa_1$.

(f4) $p\Vdash(\mathsf{E}x)\mathfrak{F}(x)$ iff $p\Vdash\mathfrak{F}(\mathbf{a})$ for some $a\in 2^\omega$.

(f5) $p\Vdash|\langle\mathbf{2}^e,\mathbf{a}\rangle|_{\mathfrak{T}}=\boldsymbol{\alpha}+\mathbf{1}$ iff $p\Vdash|\langle\mathbf{e},\mathbf{a}\rangle|_{\mathfrak{T}}=\boldsymbol{\alpha}$.

(f6) Let $\lambda<\kappa_1$ be a limit ordinal.

$$p\Vdash|\langle 3^m\cdot 5^e,\mathbf{a}\rangle|_{\mathfrak{T}}=\boldsymbol{\lambda}\text{ iff (f6.1)–(f6.3) hold for some }\alpha<\lambda.$$

(f6.1) $p\Vdash|\langle\mathbf{m},\mathbf{a}\rangle|_{\mathfrak{T}}=\boldsymbol{\alpha}$. Let W be $W(\mathbf{e},\mathbf{a},H_{\boldsymbol{\alpha}}^{\mathfrak{T}})$.

(f6.2) $p\Vdash(x)(\mathsf{E}\sigma)_{\sigma<\boldsymbol{\lambda}}[x\in W\to|x|_{\mathfrak{T}}=\sigma]$.

(f6.3) $p \Vdash (\sigma)_{\sigma<\lambda}(\mathrm{E}x)(\mathrm{E}n)_{n<\omega}[x \in W \;\&\; |x|_{\mathfrak{T}} + \mathbf{n} = \sigma]$.

(f7) The above clauses indicate, it is hoped, how to force $\mathbf{b} \in W(\mathbf{e}, \mathbf{a}, H_{\boldsymbol{\alpha}}^{\mathfrak{T}})$. Recall that $W(\mathbf{e}, \mathbf{a}, H_{\boldsymbol{\alpha}}^{\mathfrak{T}})$ denotes a first order definable subset of $\langle \mathbf{2}^{\omega}, \varepsilon, H_{\boldsymbol{\alpha}}^{\mathfrak{T}} \rangle$ whose definition is built up from the atomic predicate $x \in H_{\boldsymbol{\alpha}}^{\mathfrak{T}}$, real variable quantifiers, and logical connectives. Thus clauses such as (f2) and (f4) can be used to reduce the forcing of reals into $W(\mathbf{e}, \mathbf{a}, H_{\boldsymbol{\alpha}}^{\mathfrak{T}})$ to the forcing of reals into $H_{\boldsymbol{\alpha}}^{\mathfrak{T}}$, and the latter situation reduces to events occurring below level $\boldsymbol{\alpha}$.

Proposition 13. *There exists a recursive* (*in* ^{3}E) *set of codes for the forcing conditions.*

Proof. Call a collection R of reals a *small* set if it is contained in $\{c | c \prec a\}$ for some real a, and is recursive in $^3\mathrm{E}, b$ for some real b. A forcing condition is in essence a small set. According to Lemma 1 of NORMANN (1977), every small set occurs in the H_σ hierarchy for ^{3}E prior to level $|\prec|$, the ordertype of $\prec$, an ordinal recursive in ^{3}E. Thus there is a recursive (in ^{3}E) set of codes for the conditions defined at level $|\prec|$.

From now on assume that the forcing conditions are codes as in Proposition 13. It will be convenient to allow $p, q, r, \ldots$ to denote both forcing conditions and their codes.

The ordinal rank of a bounded formula $\mathfrak{F}$ is the greatest ordinal constant that occurs in $\mathfrak{F}$, either as a strict upper bound on some ordinal variable or as a parameter in some predicate.

Sublemma 14. *If* $a \in \mathcal{O}$, *then the relation*

$$p \Vdash \mathfrak{F}, \text{ restricted to } \mathfrak{F}\text{'s of ordinal rank } \leqslant |a|,$$

is recursive in $^3\mathrm{E}, a$ (*uniformly in* a).

Proof. By effective transfinite induction on the rank of $\mathfrak{F}$. Suppose by way of illustration that $\mathfrak{F}$ is $(\sigma)_{\sigma<\boldsymbol{\alpha}}\mathfrak{G}(\sigma)$. Then $p \Vdash \mathfrak{F}$ means

$$(\beta)_{\beta<\alpha}(q)_{p>q}(\mathrm{E}r)_{q>r}[r \Vdash \mathfrak{G}(\boldsymbol{\beta})].$$

The ordinal rank of $\mathfrak{G}(\boldsymbol{\beta})$ is at most that of $\mathfrak{F}$. The rank of $\mathfrak{G}(\boldsymbol{\beta})$ is less than that of $\mathfrak{F}$. Proposition 13 justifies the induction step up to $\mathfrak{F}$ from sentences of lower rank.

The next lemma will establish bounds on the ordinals recursive in $^3\mathrm{E}, T, a$ when T is generic. It will be seen that

$$\text{if } p \Vdash [\,|\mathbf{a}|_{\mathfrak{T}} \leqslant \kappa_1\,], \quad \text{then } p \Vdash |\mathbf{a}|_{\mathfrak{T}} < \boldsymbol{\beta}$$

for some β recursive in $^3\mathrm{E},p,a$. The proof of bounding seems to require the notion of weak forcing. Say $p\Vdash^*\mathfrak{F}$ (read p weakly forces $\mathfrak{F}$) if there is no q such that $p\geqslant q$ and $q\Vdash\sim\mathfrak{F}$. Note that Sublemma 14 remains true when $\Vdash$ is replaced by $\Vdash^*$, thanks to Proposition 13.

Lemma 15. *Suppose* $p\Vdash^*(\mathrm{E}\sigma)[|\mathbf{a}|_{\mathfrak{T}}=\sigma]$. *Then there exists a* q *and a* γ *such that*

$$p\geqslant q,\quad q\Vdash|\mathbf{a}|_{\mathfrak{T}}=\gamma,\quad q\leqslant{}^3\mathrm{E},p,a$$

and γ *is recursive in* $^3\mathrm{E},p,a$.

Proof. q and γ, regarded as partial recursive functions of $^3\mathrm{E}$, p and a, are defined by effective transfinite induction on the rank of $\langle p,a\rangle$. A rank can be assigned to $\langle p,a\rangle$, when

$$(1)\qquad p\Vdash^*(\mathrm{E}\sigma)[\,|\mathbf{a}|_{\mathfrak{T}}=\sigma\,],$$

by appealing to the inductive definition of $\Vdash^*$. (1) becomes

$$(2)\qquad (q)_{p>q}(\mathrm{E}r)_{q>r}(\mathrm{E}\gamma)[\,r\Vdash|\mathbf{a}|_{\mathfrak{T}}=\gamma\,].$$

The pair $\langle r,\gamma\rangle$, if $r\Vdash|\mathbf{a}|_{\mathfrak{T}}=\gamma$, of course has rank γ. The rank of $\langle p,a\rangle$, if (1) holds, is the supremum of the ranks of the pairs $\langle r,\gamma\rangle$ needed to verify the truth of (2). Logical complexity also plays a part in the assignment of rank to $\langle p,a\rangle$. For example, suppose a is $\langle 2^e,b\rangle$ and (1) holds. Then $\langle p,a\rangle$ has higher rank than $\langle p,\langle e,b\rangle\rangle$ because it has greater logical complexity. The logical difference has to be considered, since $\langle p,\langle 2^e,b\rangle\rangle$ and $\langle p,\langle e,b\rangle\rangle$ may have the same ordinal rank. Thus a pair $\langle p,a\rangle$, if (1) holds, will have an ordinal rank less than or equal to κ_1, and a logical rank developed in accord with clauses (f5)–(f7) of the definition of $\Vdash$.[4]

To see how q and γ are defined, let a be $\langle 3^m,5^e,c\rangle$ and assume (1) holds. Then

$$p\Vdash^*(\mathrm{E}\sigma)\,]|\langle\mathbf{m},\mathbf{c}\rangle|_{\mathfrak{T}}=\sigma\,].$$

By induction there exist q_0 and γ_0, each computed from $^3\mathrm{E},p,a$, such that

$$p\geqslant q_0\quad\text{and}\quad q_0\Vdash|\langle\mathbf{m},\mathbf{c}\rangle|_{\mathfrak{T}}=\gamma_0.$$

Let W be $W(\mathbf{e},\mathbf{c},H_{\gamma_0})$. Then

$$q_0\Vdash^*(x)(\mathrm{E}\sigma)[\,x\in W\rightarrow|x|_{\mathfrak{T}}=\sigma\,],$$

and so for each $b\in 2^\omega$,

$$(q)_{q_0>q}(\mathrm{E}r)_{q>r}[\,r\Vdash(\mathrm{E}\sigma)(\mathbf{b}\in W\rightarrow|\mathbf{b}|_{\mathfrak{T}}=\sigma)\,].$$

Fix b and $q\leqslant q_0$ in order to define $r(q,b)$ and $\gamma(q,b)$. If there is an r such

[4]T. Slaman has noted that the wellfoundedness of the rank relation for $\langle p,a\rangle$ requires a proof based on the countable closure property of the forcing conditons. Cf. Sacks (1980).

that $q \geqslant r$ and $r \Vdash \mathbf{b} \notin W$, then $r(q,b)$ is the least such r (least with respect to $\prec$), and $\gamma(q,b)=0$. If there is no such r, then

$$q \Vdash^* (\mathrm{E}\sigma)\left[\,|\mathbf{b}|_{\mathfrak{T}} = \sigma\right].$$

By induction there exist $r(q,b)$ and $\gamma(q,b)$, each computed from $^3\mathrm{E}, q, b$ such that

$$q \geqslant r(q,b) \quad \text{and} \quad r(q,b) \Vdash |\mathbf{b}|_{\mathfrak{T}} = \gamma(\mathbf{q},\mathbf{b}).$$

By Sublemma 14, the functions $r(q,b)$ and $\gamma(q,b)$ are recursive in $^3\mathrm{E}, p, a, q, b$. It follows that $\gamma(q,b)$ is bounded, for all $q \leqslant q_0$ and all b, by some γ_1 recursive in $^3\mathrm{E}, p, a$. Hence

$$q_0 \Vdash^* |\mathbf{a}|_{\mathfrak{T}} \leqslant \gamma_1. \tag{3}$$

The desired q and γ can now be obtained from (3) by an application of Sublemma 14 similar to the one above.

Lemma 16. *Suppose* $p \Vdash (x)(\mathrm{E}\sigma)[x \in W(\mathbf{e},\mathbf{a},H_\gamma) \to |x|_{\mathfrak{T}} = \sigma]$. *Then there exists an* $\alpha < \kappa_1$ *such that* $p \Vdash (x)(\mathrm{E}\sigma)_{\sigma<\alpha}[\cdots]$.

Proof. Let W be $W(\mathbf{e},\mathbf{a},H_\gamma)$. Then for each b,

$$(q)_{p>q}(\mathrm{E}r)_{q>r}\left[\,r \Vdash (\mathrm{E}\sigma)(\mathbf{b} \in W \to |\mathbf{b}|_{\mathfrak{T}} = \sigma)\right].$$

Now proceed as in the last part of the proof of Lemma 15 to obtain the bound on σ.

Let $T \subseteq 2^\omega$ and $\mathfrak{F}$ be a sentence of $\mathfrak{L}(\kappa_1, \mathfrak{T})$. T is said to be *generic over* $\mathfrak{F}$ if there is a p such that T satisfies p, and either $p \Vdash \mathfrak{F}$ or $p \Vdash \sim\mathfrak{F}$. T is said to be *generic* if T is generic over $\mathfrak{F}$. A familiar argument that proceeds by induction on the rank of sentences shows: if T is generic, then $\mathfrak{F}$ is a true statement about T if and only if there is a p satisfied by T such that $p \Vdash \mathfrak{F}$. It follows from Lemma 16 that $\kappa_1^T = \kappa_1$ when T is generic. Unfortunately the existence of a generic T is far from certain. If $|\prec|$ (the ordinal height of $\prec$, the well-ordering of 2^ω assumed to be recursive in $^3\mathrm{E}$) is a regular cardinal, then a straightforward recursion on $|\prec|$ yields a generic T. It seems unlikely that the regularity assumption of Theorem 9, namely that $|\prec|$ is recursively regular, is strong enough to yield a generic T, but it is enough to develop a T generic over all bounded sentences. The next two lemmas are needed to control κ_1^T when T is boundedly generic.

Lemma 17. *Suppose* $p \Vdash (\sigma)[|\mathbf{b}|_{\mathfrak{T}} \neq \sigma]$. *Then there exist a* q, *a real* c *and a bounded sentence* $\mathfrak{F}$ *with the following properties.*

(i) $p \geqslant q$ *and* $q \Vdash \mathfrak{F}$.

(ii) $q \Vdash (\sigma)[|\mathbf{c}|_{\mathfrak{T}} \neq \sigma]$.

(iii) *For all* $T \subseteq 2^\omega$: *if* $\mathfrak{F}$ *is a true statement about* T *and* $b \in \mathcal{O}^T$, *then* $c \in \mathcal{O}^T$ *and* $|b|_T > |c|_T$.

Proof. The definition of q, c and $\mathfrak{F}$ has three cases, only the last of which is of interest.

Case I: b is not of the form $\langle 2^e, a\rangle$ or $\langle 3^m\cdot 5^e, a\rangle$. Let $q=p$, $c=b$ and $\mathfrak{F}$ be $1=1$.

Case II: b is of the form $\langle 2^e, a\rangle$. Let $q=p$, $c=\langle e,a\rangle$ and $\mathfrak{F}$ be $1=1$.

Case III: b is $\langle 3^m\cdot 5^e, a\rangle$.

Subcase III_1: $p\Vdash(\sigma)[|\langle \mathbf{m},\mathbf{a}\rangle|_{\mathfrak{T}}\neq\sigma]$. Let $q=p$, $c=\langle m,a\rangle$ and $\mathfrak{F}$ be $1=1$.

Subcase III_2: otherwise. Thus there is a $q_0\leqslant p$ and a $\gamma<\kappa_1$ such that

$$q_0\Vdash|\langle \mathbf{m},\mathbf{a}\rangle|_{\mathfrak{T}}=\gamma.$$

Suppose (for the sake of a contradiction) that

$$(1)\qquad q_0\Vdash(x)(\mathsf{E}\sigma)\left[x\in W\left(\mathbf{e},\mathbf{a},H_\gamma^{\mathfrak{T}}\right)\rightarrow|x|_{\mathfrak{T}}=\sigma\right].$$

Then by Lemma 16, $q_0\Vdash(x)(\mathsf{E}\sigma)_{\sigma<\alpha}[\cdots]$ for some $\alpha<\kappa_1$, and consequently $q_0\Vdash[|b|_{\mathfrak{T}}<\alpha]$, which contradicts the supposition that $p\Vdash(\sigma)[|b|_{\mathfrak{T}}\neq\sigma]$. Thus (1) is false, and so there exist $q\leqslant q_0$ and c such that

$$q\Vdash\left[\mathbf{c}\in W\left(\mathbf{e},\mathbf{a},H_\gamma^{\mathfrak{T}}\right)\ \&\ (\sigma)(|\mathbf{c}|_{\mathfrak{T}}\neq\sigma)\right].$$

Let $\mathfrak{F}$ be $|\langle \mathbf{m},\mathbf{a}\rangle|_{\mathfrak{T}}=\gamma\ \&\ \mathbf{c}\in W(\mathbf{e},\mathbf{a},H_\gamma^{\mathfrak{T}})$.

Lemma 18. *Suppose* $p\Vdash(\sigma)[|\mathbf{b}|_{\mathfrak{T}}\neq\sigma]$. *Then there exist a* q *and a bounded sentence* $\mathfrak{F}$ *such that*

(i) $p\geqslant q$ *and* $q\Vdash\mathfrak{F}$; *and*

(ii) *for all* $T\subseteq 2^\omega$, *if* $\mathfrak{F}$ *is a true statement about* T, *then* $b\notin\mathcal{O}^T$.

Proof. Moschovakis (1967) showed that every co-recursively enumerable (in ^{3}E) predicate $P(x)$ is expressible in the form $(\mathsf{E}y)R(x,y)$, where $R(x,y)$ is recursively enumerable (in ^{3}E). The intuition behind R is the same as that behind the desired $\mathfrak{F}$. To say $b\notin\mathcal{O}^T$ is equivalent to saying any attempt to recursively enumerate (in ^{3}E, T) b in $\mathcal{O}^T$ must fail because the associated computation tree is not well-founded. $\mathfrak{F}$ specifies an infinite descending path through that tree. The path is obtained by repeated application of Lemma 17.

Let $p_0=p$ and $b_0=b$. Assume p_i and b_i satisfy the initial supposition of Lemma 17; let $\langle p_{i+1}, b_{i+1}, \mathfrak{F}_{i+1}\rangle$ bear the same relation to $\langle p_i, b_i\rangle$ that $\langle q, c, \mathfrak{F}\rangle$ bears to $\langle p, b\rangle$ in Lemma 17.

Let q be $\bigcup\{q_i|i<\omega\}$, and let $\mathfrak{F}$ be a bounded sentence logically equivalent to the infinite conjunction of the $\mathfrak{F}_i$'s. (The ordinal rank of $\mathfrak{F}$ is the supremum of the ordinal ranks of the $\mathfrak{F}_i$'s, and $\mathfrak{F}$ has a real parameter d such that $(d)_i$ is a Gödel number of $\mathfrak{F}_i$ for all $i>0$.) Clearly $q\Vdash\mathfrak{F}$.

Fix $T\subseteq 2^\omega$. Suppose (for the sake of a contradiction) that $\mathfrak{F}$ is a true statement about T and that $b\in\mathcal{O}^T$. By property (iii) of Lemma 17, $b_i\in\mathcal{O}^T$ for all i, and $|b_i|_T$ $(i<\omega)$ is an infinite descending sequence of ordinals.

Lemma 19. *For each p there exists a T that satisfies p and is generic on all bounded* $\mathfrak{F}$'s.

Proof. Let $\mathfrak{F}_a$ be the bounded sentence whose Gödel number is a. A function P will be defined so that for each a, if a is the Gödel number of a bounded sentence, then either $P(a)\Vdash\mathfrak{F}_a$ or $P(a)\Vdash\sim\mathfrak{F}_a$. The characteristic function of T will be $\bigcup\{P(a)|a\in 2^\omega\}$.

P is defined by transfinite recursion on $\prec$, a well-ordering of 2^ω that is recursive in ${}^3\mathrm{E}$ and recursively regular. Fix a. Let $P_0(a)$ be the union of the given p and the $P(b)$'s for all $b\prec a$. Suppose $a=\ulcorner\mathfrak{F}_a\urcorner$ and $P_0(a)$ is a forcing condition. (If the supposition is false, let $P(a)$ be $P_0(a)$.) Let $P(a)$ be the least q such that $q\leqslant P_0(a)$ and either $q\Vdash\mathfrak{F}_a$ or $q\Vdash\sim\mathfrak{F}_a$. ("least" means least with respect to $\prec$ with the understanding that forcing conditions are coded by reals as in Proposition 13.) By Sublemma 14, $P(a)$ can be found effectively from $P_0(a)$ and a.

The definition of P is successful if $P_0(a)$ is a forcing condition for all a. This last is proved by induction. Fix a, and assume $P_0(b)$ is a forcing condition for all $b\prec a$. Let D be the collection of all $b\prec a$ such that b is the Gödel number of a bounded sentence. Observe that $P\restriction a$ is recursive in ${}^3\mathrm{E},D,a$ because D decides if $P(b)$ (for each $b\prec a$) is to be $P_0(b)$ or an extension of $P_0(b)$ that forces $\mathfrak{F}_b$ or $\sim\mathfrak{F}_b$. Thus the definition by recursion of $P\restriction a$ becomes an effective transfinite recursion relative to ${}^3\mathrm{E},D,a$. By Lemma 5, D is recursive in ${}^3\mathrm{E},c,a$ for some real c, and so $P_0(a)$ is recursive in ${}^3\mathrm{E},c,a$. $P_0(a)$ is a union of compatible forcing conditions, hence it is the characteristic function of a set of reals. It need only be shown that the domain of $P_0(a)$ is bounded above (with respect to $\prec$) by some real d. For each $b\prec a$, let $Q(b)$ be the least upper bound of the domain of $P(b)$. Clearly $Q\restriction a$ is recursive in ${}^3\mathrm{E},c,a$. The recursive regularity of $\prec$ implies that the range of $Q\restriction a$, and hence the domain of $P_0(a)$, is bounded above by some d.

Proof of Theorem 9. Fix e and c. For the sake of a contradiction, assume that for all $S\subseteq 2^\omega$: $\{e\}({}^3\mathrm{E},c,S)$ is defined if and only if $S\leqslant{}^3\mathrm{E},a$ for some a. Let b be $\langle e_0,c\rangle$. Then $S\leqslant{}^3\mathrm{E},a$ for some a if and only if $b\in\mathcal{O}^S$. By Lemma 18 there is a forcing condition p such that either (B) or (C) holds.

(B) $p\Vdash|\mathbf{b}|_{\mathfrak{T}}=\gamma$ for some $\gamma<\kappa_1$.

(C) $p\Vdash\mathfrak{F}$, where $\mathfrak{F}$ is a bounded sentence with the following property. For all $T\subseteq 2^\omega$, if $\mathfrak{F}$ is a true statement about T, then $b\notin\mathcal{O}^T$.

By Lemma 19 there exists a T_1 that satisfies p and is generic on all bounded sentences. Let S be any set recursive in ${}^3\mathrm{E},a$ for some a. Clearly $\varnothing\Vdash\mathfrak{T}\neq\mathbf{S}$, where $\varnothing$ is the null forcing condition. It follows that T_1 is not recursive in ${}^3\mathrm{E},a$ for any a. Hence $b\notin\mathcal{O}^{T_1}$, and so (C) must hold.

Let $\sigma_0 < \kappa_1$ be a limit ordinal greater than the ordinal rank of $\mathcal{F}$ (mentioned in (C)), and greater than the ordinal needed to compute $\prec$ recursively from ${}^3\mathrm{E}$. Repeat the construction of Lemma 19 from the vantage point of σ_0 to obtain a T_2 that satisfies p, is generic on all bounded sentences of ordinal rank less than σ_0, and is first order definable over H_{σ_0}. Thus $T_2 \leqslant {}^3\mathrm{E}, a$ for any a such that $a \in \mathcal{O}$ and $|a| = \sigma_0$, and so $b \in \mathcal{O}^{T_2}$. But (C) implies $b \notin \mathcal{O}^{T_2}$.

Indexicality is one of two notions needed for the absolute formulation of Post's problem. The other is ordinal recursiveness. Let $R \subseteq 2^\omega$. R is said to be *ordinal recursive* in ${}^3\mathrm{E}$ if there exist an integer e and H-sets $H^1, \ldots, H^m$ such that

$$R = \{e\}({}^3\mathrm{E}, H^1, \ldots, H^m).$$

By an H-set is meant any set in the hierarchy $\{H_\sigma | \sigma < \kappa_1\}$. A typical ordinal recursive set is a set first order definable over some H_σ by means of a formula without real parameters. Suppose S is recursive in ${}^3\mathrm{E}, b$. The real b plays a double role. First $b_0 \in \mathcal{O}^T$ and $H_{|b_0|_T}$ is the universe over which S is defined. Second b_1 is the real parameter needed in the definition of S. Thus every set recursive in ${}^3\mathrm{E}, b$ for some real b is ordinal recursive if and only if every real is ordinal recursive.

Proposition 20. *Suppose there exist a well-ordering W of 2^ω such that W is recursive in ${}^3\mathrm{E}$. Then every set recursive in ${}^3\mathrm{E}, b$ for some $b \in 2^\omega$ is ordinal recursive in ${}^3\mathrm{E}$.*

Proof. Let γ be the height of W. Let f be the unique function that maps $\{H_\sigma | \sigma < \gamma\}$ in an order-preserving fashion onto W. Thus $f(H_\sigma) = b$ if and only if the height of b in W is σ. Since γ is recursive in ${}^3\mathrm{E}$, the function f^{-1} must also be recursive in ${}^3\mathrm{E}$. Hence b is recursive in ${}^3\mathrm{E}, f^{-1}(b)$ for each b, and so each real is ordinal recursive in ${}^3\mathrm{E}$.

The next lemma clarifies the connection between H-sets and ordinal recursive sets. Two sets of reals are said to have the same ${}^3\mathrm{E}$-*degree* if each is recursive in the other and ${}^3\mathrm{E}$.

Lemma 21. (i) *Let H^1 and H^2 be H-sets. Then there exists an H-set H^3 such that $H^1 \times H^2$ has the same ${}^3\mathrm{E}$-degree as H^3.*

(ii) *R is ordinal recursive in ${}^3\mathrm{E}$ if and only if R is first order definable over some H_σ by means of a formula whose only parameters are integers.*

Proof. (ii) follows from (i). The proof of (i) is an effective transfinite recursion that defines functions f and g with the following properties:

$H_\sigma \times H_\tau$ is first order definable over $H_{f(\sigma,\tau)}$ via the formula (with integer parameters only) whose Gödel number is $g(\sigma,\tau)$. Fix $\langle\gamma,\delta\rangle$ and assume f and g are already defined for all $\langle\sigma,\tau\rangle$'s below $\langle\gamma,\delta\rangle$. Define $f(\gamma,\delta)$ to be the strict least upper bound of $f(\sigma,\tau)$ for all $\langle\sigma,\tau\rangle$ less than $\langle\gamma,\delta\rangle$. The definition of $g(\gamma,\delta)$ has two cases. First suppose there is a $\rho<f(\gamma,\delta)$ such that $H_\gamma \times H_\delta$ is first order definable over H_ρ. Then $g(\gamma,\delta)$ is the Gödel number of a formula that defines (over $H_{f(\gamma,\delta)}$) the least such ρ, call it ρ_0, and also defines $H_\gamma \times H_\delta$ over H_{ρ_0}. Now suppose no such ρ exists. Then $\langle H_\gamma, H_\delta\rangle$ is the least pair of H-sets whose product is not first order definable over H_ρ for any $\rho<f(\gamma,\delta)$. This last sentence is the desired definition of $H_\gamma \times H_\delta$ over $H_{f(\gamma,\delta)}$.

It follows from Theorem 9 and Proposition 20 that the class of all sets ordinal recursive in ^{3}E need not be recursively enumerable in ^{3}E. That shortcoming is eliminated by the next definition, which combines the notions of ordinal recursiveness and indexicality. R is said to be *effectively ordinal recursive* in ^{3}E if there exist H-sets $H^1,\ldots,H^m$ such that R is recursive in $^3\mathrm{E}, H^1,\ldots,H^m$, and each H^i is recursive in $^3\mathrm{E}, R$.

Lemma 22. *R is effectively ordinal recursive if and only if R has the same ^{3}E-degree as some H-set.*

Proof. A consequence of Lemma 21(i).

A set is said to be H^* if it has the same ^{3}E-degree as an H-set. The absolute formulation of Post's problem for ^{3}E resembles Normann's remedy above save that H^*-sets stand in place of arbitrary sets recursive in ^{3}E and a real.

Absolute Post Problem for ^{3}E: Do there exist recursively enumerable (in ^{3}E and some fixed H^*-set) classes A and B of H^*-sets such that for each H^*-set R,

$$A \not\leq {}^3\mathrm{E}, B, R \quad \text{and} \quad B \not\leq {}^3\mathrm{E}, A, R$$

on the class of all H^*-sets?

The class of all H^*-sets is recursively enumerable in ^{3}E. The natural enumeration of the H^*-sets parallels the development of the ordinals less than κ_1 by iteration of the ^{3}E jump. It follows there is a useful well-ordering of the objects allowed to occur as members of recursively enumerable classes even when there is no well-ordering of 2^ω recursive in ^{3}E. And if such a well-ordering of 2^ω does exist, then Proposition 20 implies that every real is an H^*-set, and consequently that the incomparability of A and B asked for in the absolute problem is the most possible. Thus parity is maintained between the objects put into recursively enumerable classes and the objects that index reduction procedures without any weakening of incomparability requirements. The axioms of set theory leave open whether or not every real is ordinal recursive in ^{3}E, hence a positive solution to the absolute Post problem for ^{3}E leaves open how much incomparability is achieved, but nonetheless renders that incomparability as definite as the ordinals.

A last word about the choice of H^*-sets as the objects to be enumerated. An H^*-set has all the advantages of indexicality and ordinal recursiveness. The principal argument in favor of indexicality was given just before Proposition 8, and the one for ordinal recursiveness in the previous paragraph. But there is still another point to consider. Why H^* rather than H? The H-sets are of course indexical and ordinal recursive, and in addition they form a recursive (in ^{3}E) class. Why go to H^*? One answer is suggested by Lemma 22, namely that the H^*-sets constitute the largest class of objects suitable for the absolute study of Post's problem relative to ^{3}E. A second answer, perhaps more practical than the first, is that the H^*-sets have useful closure properties exemplified by the statement and proof of Lemma 21(i).

In order to state a morally certain conjecture concerning the absolute Post problem, it is necessary to define two notions of recursively enumerable projectum relative to ^{3}E.

The *greater projectum* is denoted by ρ. It is the least $\sigma \leqslant \kappa_1$ such that there exists an H-set H^1 with the following property: every H-set is recursive in ${}^3\mathrm{E}, H^1, H_\tau$ for some $\tau < \sigma$.

The *lesser projectum* is denoted by η. It is the least $\sigma \leqslant \kappa_1$ such that there exists an H-set H^1 with the following property: there is a recursively enumerable (in ${}^3\mathrm{E}, H^1$) class of H-sets contained in $\{H_\tau | \tau < \sigma\}$, but not recursive in ${}^3\mathrm{E}, H$ for any H-set H.

Proposition 23. (i) $\eta \leqslant \rho$.

(ii) *Suppose there exists a well-ordering of* 2^ω *recursive in* ${}^3\mathrm{E}, b$ *for some real* b. *Then the least height attained by any such well-ordering is equal to* ρ.

(iii) *Suppose there exists a recursively regular well-ordering of* 2^ω *recursive in* ${}^3\mathrm{E}, b$ *for some real* b. *Then* $\eta = \rho$.

Proof. (iii) follows from Lemma 5.

Inexorable limitations of space and time compel this paper to end with a mixture of questions, conjectures and halftruths that will be studied elsewhere.

(1) Suppose $\eta = \rho$. At this writing it appears very likely that the absolute Post problem has an affirmative answer.[5] NORMANN's work (1975, 1977)

[5]Yes, it does. Cf. SACKS (1980).

corresponds to the case when ρ is less than κ_1 and is a recursively regular cardinal. A quite different state of affairs ensues when ρ is singular; that case is handled with the aid of κ_r^b, the greatest reflecting ordinal relative to $\langle {}^3\mathrm{E}, b\rangle$, pinned down by HARRINGTON (1973) in his Ph.D. thesis. The supposition $\eta=\rho$ is analogous to a fundamental truth of α-recursion theory: an α-recursively enumerable subset of an ordinal less than α^*, the Σ_1 projectum of α, is α-finite. A closely related supposition, termed adequacy, is made by FENSTAD (1979).

(2) Does there exist a generic extension of L in which the absolute Post problem has a negative answer? in which every subset of 2^ω recursive in ${}^3\mathrm{E}$ and a real is indexical? The more appealing conjecture, in both cases, is yes.

(3) All the results of this paper, save for Proposition 5, lift to ${}^n\mathrm{E}$ when $n>3$. Some changes in definitions and proof are necessary. They are designed to take into account the difference between the Gandy and Grilliot selection principles, a difference manifest only if $n>3$.

(4) Is there an absolute version of Theorem 9? One that does not assume some well-ordering of the reals is recursive in ${}^3\mathrm{E}$? It might be possible to show that the class of all sets ordinal recursive in ${}^3\mathrm{E}$ is not recursively enumerable in ${}^3\mathrm{E}, R$, for any ordinal recursive set R, by assuming that $\rho<\kappa_1$ and ρ is recursively regular.

References

FENSTAD, J. E.
[1979] General Recursion Theory (Springer-Verlag, Berlin), to appear.
FRIEDBERG, R.
[1957] Two recursively enumerable sets of incomparable degrees of insolvability, *Proc. Nat. Acad. Sci.*, **43**, 236–238.
HARRINGTON, L.
[1973] Contributions to recursion theory in higher types, Ph.D. Thesis, M.I.T.
KLEENE, S. C.
[1959] Recursive functionals and quantifiers of finite type, *Trans. Am. Math. Soc.*, **91**, 1–52.
[1963] Recursive functionals and quantifiers of finite type, II, *Trans. Am. Math. Soc.*, **108**, 106–142.
KREISEL, G.
[1961] Set theoretic problems suggested by the notion of potential totality, in: Infinitistic Methods (Pergamon Press, Oxford), pp. 103–140.
MOSCHOVAKIS, Y. N.
[1967] Hyperanalytic predicates, *Trans. Am. Math. Soc.*, **138**, 249–282.

Muchnik, A. A.

[1956] On the unsolvability of the problem of reducibility in the theory of algorithms, *Dokl. Akad. Nauk. USSR*, **108**, 194–197.

Normann, D.

[1975] Degrees of functionals, Preprint Series in Mathematics No. 22, Oslo.

[1977] A note on reflection, preprint, Oslo.

Post, E.

[1944] Recursively enumerable sets of positive integers and their decision problem, *Bull. Am. Math. Soc.*, **50**, 284–316.

Sacks, G. E.

[1966] Post's problem, admissible ordinals and regularity, *Trans. Am. Math. Soc.* **124**, 1–23.

[1977] RE sets higher up, in: Logic, Foundations of Mathematics and Computability Theory (D. Reidel Publ. Co., Dordrecht).

Simpson, S.

[1974] Post's problem for admissible sets, in Generalized Recursion Theory (North-Holland, Amsterdam).

Spector, C.

[1955] Recursive wellorderings, *J. Symbolic Logic*, **20**, 151–163.

SOUTHEAST ASIAN CONFERENCE ON LOGIC
C.-T. Chong and M.J. Wicks (editors)

ON THE NUMBER OF COUNTABLE MODELS

Gerald E. Sacks*
Harvard University
Massachusetts Institute of Technology
U. S. A.

1. Introduction

Let C be a $PC_{\omega_1,\omega}$ class of countable structures, and let n(C) be the number of isomorphism types of elements of C. Morley [5] proved :

(1) $$n(C) < 2^{\omega} \rightarrow n(C) \leq \omega_1 .$$

Each countable structure M has a rank r(M), a countable ordinal that measures the length of the canonical Scott sentence for M, as defined in Section 2. Let $n(C,\beta)$ be the number of isomorphism types of structures in C of rank β. Morley obtained (1) by showing

$$n(C) < 2^{\omega} \rightarrow (\beta)_{\beta<\omega_1}[n(C,\beta) \leq \omega].$$

H. Friedman's example F shows Morley's result cannot be improved. F is $PC_{\omega_1,\omega}$, $n(F) = \omega_1$, and $n(F,\beta) \leq \omega$ for every countable β.

Vaught has conjectured that $n(C) \leq \omega$ if C is $EC_{\omega,\omega}$ and $n(C) < 2^{\omega}$. A stronger conjecture allows C to be $EC_{\omega_1,\omega}$; that is, C is the class of all countable models of some complete theory in some countable fragment of $L_{\omega_1,\omega}$. The purpose of this paper is to point out that $PC_{\omega_1,\omega}$ and $EC_{\omega_1,\omega}$ behave quite differently with respect to Vaught-type conjectures. The difference is a consequence of compactness arguments. The Friedman example F has the property that

$$(\beta)_{\beta<\omega_1}[n(F,\beta) < 2],$$

yet $n(F) = \omega_1$. On the other hand, Theorem 4.1 below implies

(2) $$(\beta)_{\beta<\omega_1}[n(C,\beta) < 2] \rightarrow n(C) \leq \omega$$

if C is $EC_{\omega_1,\omega}$. Note that (2) does not assume $N(C) < 2^{\omega}$.

Throughout the paper T is a complete theory in some countable fragment of $L_{\omega_1,\omega}$. Section 2 reviews Scott sentences and the fact that the rank of a structure M is at most $\omega_1^M + 1$, and ends with a lemma on the existence of an atomic model for T. Section 3 describes a Σ_1 enumeration of the countable models of T when $n(T,\beta) < 2^{\omega}$ for all countable β. The results of Sections 2 and 3 are combined in Section 4 to obtain (2) and the following result.

Theorem 4.2. Suppose the number of countable models of T is less than 2^{ω}. Assume that the canonical Scott rank of each countable model M of T is at most ω_1^M. Then T has only countably many countable models.

The above theorem suggests that the main obstacle to proving Vaught's conjecture has to do with structures M whose Scott rank is the greatest possible, namely $\omega_1^M + 1$.

Section 5 raises the question of a bound on the Scott ranks of the countable models of T when $n(T) \leq \omega$. Some bounds are offered in special cases. For example, the proof of Theorem 4.2 shows that every countable model of T has rank at most ω_2^T, where ω_2^T is the least Σ_2 admissible ordinal relative to T.

2. Scott Rank

Scott [1] discovered that each countable structure M can be characterized up to isomorphism by a single sentence F of $L_{\omega_1,\omega}$; that is

$$M \models F \quad \text{and} \quad [N \models F] \rightarrow N \cong M$$

for every countable structure N. A canonical choice for F, denoted by F_M, is developed from M by an inductive procedure that terminates at a countable ordinal r(M), the Scott rank of M.

Let $L(0,M)$ be the $L_{\omega,\omega}$ language whose primitive symbols correspond to the relations, functions and distinguished elements of M. For each limit ordinal λ, $L(\lambda,M)$ is $\cup\{L(\delta,M) \mid \delta < \lambda\}$. For each δ, let $T(\delta,M)$ be the complete theory of M in the language $L(\delta,M)$. And let $L(\delta+1,M)$ be the least fragment of $L_{\omega_1,\omega}$ that includes $L(\delta,M)$ and satisfies the following closure condition :

> if $n < \omega$ and $p(x_1,\ldots,x_n)$ is a non-principal n-type of $T(\delta,M)$ realized in M, then $\wedge\{F(\vec{x}) \mid F(\vec{x}) \in p\}$ belongs to the fragment.

Scott showed there exists a countable δ such that $L(\delta+1,M) = L(\delta,M)$. Let r(M) be the least such δ. Then M is the unique atomic model of T(r(M),M), and F_M, the canonical Scott sentence for M, is a sentence that asserts : "I am an atomic model of T(r(M),M)."

Scott's proof that r(M) exists and is countable is in essence a counting argument. If $L(\delta+2,M) \neq L(\delta+1,M)$, then there exist two n-tuples of M that are equivalent with respect to all n-place formulas of $L(\delta,M)$, but inequivalent with respect to some formula of $L(\delta+1,M)$. Once an inequivalence occurs, it becomes permanent. Since M is countable, there can be only countably many initial occurrences of inequivalence.

The set of inequivalent n-tuples of M can be construed as the end result of a π_1^1 monotonic inductive definition, and consequently, as Nadel [2] observed,

(1) $$r(M) \leq \omega_1^M + 1.$$

(ω_1^M is the least ordinal not recursive in M. In (1) M ambiguously denotes a countable structure M and some real that encodes M.)

For the remainder of the paper, let T be a complete, consistent theory in some countable fragment K of $L_{\omega_1,\omega}$. In this setting, "completeness" means every sentence of K or its negation belongs to T, and "consistency" requires that no infinite disjunction belong to T unless one of its components also belongs.

Define $n(T,\beta)$ to be the number of (isomorphism types of) models of T of Scott rank β.

<u>Lemma 2.1</u>. Suppose $n(T,\beta) < 2^{\beta}$ for all countable β. Then T has an atomic model.

<u>Proof</u>. It suffices to show $S_n(T)$, the set of n-types of T, is countable for all $n < \omega$. In that event, an atomic model of T is constructed by omitting all non-principle n-types.

Fix n and suppose $S_n(T)$ is uncountable. Then $S_n(T)$ has a perfect kernel Z whose cardinality is 2^{ω}. Let T ambiguously denote a coding of T by a real. The coding of T by a real gives rise to a coding of Z by a real Z_0 and a coding of n-types by reals. Thus the relation $p \in Z_0$ is arithmetic. According to the Gandy basis theorem, there exists a $p \in Z_0$ such that $\omega_1^{p,Z_0,T} = \omega_1^{Z_0,T}$. A slight variant of Gandy's argument yields a real Z_1 that codes a perfect subset of Z such that

$$(3) \qquad (p)[p \in Z_1 \rightarrow \omega_1^{p,Z_1,T} = \omega_1^{Z_1,T}] .$$

For each $p \in Z_1$, a Henkin construction provides a model M_p of T arithmetic in p, T and realizing p. Then

$$\omega_1^{M_p} \leq \omega_1^{p,T} \leq \omega_1^{p,Z_1,T} = \omega_1^{Z_1,T} ,$$

and consequently $r(M_p) \leq \omega_1^{Z_1,T} + 1$ by (2) above. The set $\{M_p | p \in Z_1\}$

has cardinality 2^ω, and gives rise to 2^ω isomorphism types of models of T. It follows from (3) that $n(T,\beta) = 2^\omega$ for some $\beta \leq \omega_1^{Z_1,T} + 1$.

3. Σ_1 Enumeration of Models

Let T be a complete theory in some countable fragment of $L_{\omega_1,\omega}$. Assume $n(T,\beta) < 2^\omega$ for all countable β. The reasoning that carries Lemma 2.1 leads to a natural enumeration of the canonical Scott sentences of the countable models of T. A tree T of countable, complete, consistent theories extending T is defined by recursion. Each countable model of T is the atomic model of some theory on the tree T.

Level 0 of T : theory T.

Level λ : S occurs on limit level λ iff there exists a strictly expanding sequence $\{T_\delta | \delta < \lambda$ and T_δ occurs on level $\delta\}$ whose union is S.

Level $\delta+1$: Suppose S occurs on level δ, and S is a complete theory in some countable fragment K of $L_{\omega_1,\omega}$. If every model of S is atomic, then S has no successors on level $\delta+1$. Suppose otherwise. If $n(S,\beta) = 2^\omega$ for some countable β, then $n(T,\gamma) = 2^\omega$ for some countable $\gamma \geq \beta$. Hence $n(S,\beta) < 2^\omega$ for all countable β. As in the proof of Lemma 2.1, Z, the set of n-types of S ($n < \omega$) is countable. Let W be $\{X | X \subseteq Z$ and X is the set of non-principal n-types of S realized in some model of S$\}$. If some coding of K by a real is fixed, then W can be regarded as a boldface Σ_1^1 set of reals. Consequently W is either countable or of cardinality 2^ω. If the latter, then $n(S,\beta) = 2^\omega$ for some countable β as in the proof of Lemma 2.1. Thus W is countable.

For each $X \in W$, let K_X be the least fragment of $L_{\omega_1,\omega}$ that includes K and all formulas $\bigwedge\{F(\vec{x}) | F(\vec{x}) \in p\}$ for all $p \in X$.

S' is a successor of S on level $\delta+1$ iff S' is a complete, consistent theory in K_X for some $X \in W$. End of definition of $\mathcal{T}$.

Let M be a model of T. M defines a countable path P_M through $\mathcal{T}$ that corresponds to the Scott analysis of M.

$P_M(0) = T$.

If $P_M(\gamma)$ is defined for all $\gamma < \lambda$, then $P_M(\lambda)$ is $\cup\{P_M(\gamma) | \gamma<\lambda\}$.

Suppose $P_M(\delta)$ is defined and occurs on level δ of $\mathcal{T}$, and $M \models P_M(\delta)$. If M is the atomic model of $P_M(\delta)$, then $P_M(\delta+1)$ is undefined. Suppose otherwise. Then the definition of $\mathcal{T}$ implies $P_M(\delta)$ has a unique successor S' on level $\delta+1$ of $\mathcal{T}$ such that $M \models S'$.

It follows that $P_M(r(M)) = T(M,r(M))$, and $P_M(r(M)+1)$ is undefined. Thus each canonical Scott theory extending T occurs on some countable level of $\mathcal{T}$. Conversely, each theory on $\mathcal{T}$ has an atomic model by Lemma 2.1. Hence $\mathcal{T}$ is an enumeration of all the canonical Scott theories extending T. Note that there are no gaps in the enumeration. The enumeration can be said to be continuous. A $PC_{\omega_1,\omega}$ class of countable structures of cardinality less than 2^ω need not be continuous in this sense.

Lemma 3.1. Let $\mathcal{T}^\alpha$ be the restriction of $\mathcal{T}$ to levels below α. Suppose $L(\alpha,T)$ is Σ_1 admissible. Then $\mathcal{T}^\alpha$ is a $\Sigma_1(L(\alpha,T))$ subset of $L(\alpha,T)$.

Proof. (Sketch) $\mathcal{T}^\alpha$ is defined by a $\Sigma_1(L(\alpha,T))$ recursion. Recall the successor step in the definition of $\mathcal{T}$. It was noted that W is Σ^1_1 relative to some countings of Z and S. Hyperarithmetic theory [3], starting with any countings of Z and S, yields a counting of W hyperarithmetic in those of Z and S. Countings of Z and S can be added generically to $L(\alpha,T)$ without violating Σ_1 admissibility. Then a counting of W belongs to every such generic extension, and so does a counting of each S'a successor of S. But then S' belongs to $L(\alpha,T)$. (Of course S' need not be countable in $L(\alpha,T)$.) The passage from S to S' is given by a $\Sigma_1(L(\alpha,T))$ formula.

4. Main Results

Let T be a complete theory in some countable fragment of $L_{\omega_1,\omega}$.

Theorem 4.1. Suppose T has at most one model of Scott rank β for each countable ordinal β. Then T has only countably many models.

It was remarked in Section 1 that there exists an uncountable $PC_{\omega_1,\omega}$ class K of countable structures such that at most one member of K has Scott rank β for each countable ordinal β. Thus Theorem 4.1 distinguishes between $PC_{\omega_1,\omega}$ and $EC_{\omega_1,\omega}$ classes with respect to an aspect of Vaught's conjecture.

The proof of 4.1 relies on $\mathcal{T}$, the natural enumeration of Section 3, and some type-omitting argumentation in the style of Barwise [4]. Let α be the least γ such that $L(\gamma,T)$ is Σ_2 admissible. For the sake of a contradiction, assume T has a model of rank greater than α. Then $\mathcal{T}$ has a unique theory on level α; call it T_α. According to Lemma 3.1, $\mathcal{T}^\alpha$, the restriction of $\mathcal{T}$ to levels below α, is $\Sigma_1(\alpha,T)$. It follows that T_α is $\Delta_1(L(\alpha,T))$.

Two models of T of rank $\alpha+1$ are constructed. The first, M_1, is a model of T_α atomic over the realization of some non-principal n-type of T_α. The second, M_2, is a model of T_α not atomic over any n-tuple of its elements.

Construction of M_1 : Since T has a model of rank greater than α, T_α has a non-principal n-type p. T_α is a complete theory in some fragment K_α. Let K'_α be the least fragment containing K_α

and new individual constants $\underline{c}_1, \ldots, \underline{c}_n$. Then

$$T'_\alpha = T_\alpha \cup \{F(\vec{c}) \mid F(\vec{x}) \in p\}$$

is a complete theory in K'_α. Since $n(T_\alpha, \beta) < 2$ for all $\beta < \omega_1$, it follows that $n(T'_\alpha, \beta) \leq \omega$ for all $\beta < \omega_1$. By Lemma 2.1 T'_α has an atomic model $\langle M_1, c_1, \ldots, c_n \rangle$. M_1 is a model of T_α atomic over $\langle c_1, \ldots, c_n \rangle$. $r(M_1)$ cannot be less than α, because for each $\delta < \alpha$, $T_{\delta+1}$ includes a sentence asserting the realization of some type non-principal over T_δ. $r(M_1) \neq \alpha$, since M_1 realizes a type non-principal over T_α. Thus $r(M_1) > \alpha$. The language $L(\alpha+1, M_1)$ includes the formula $\wedge\{F(\vec{x}) \mid F(\vec{x}) \in p\}$ and so has enough expressive power to complete the Scott analysis of M_1. The fact that each n-type of T_α realized in M_1 is atomic over p implies that M_1 is an atomic model of its complete theory in $L(\alpha+1, M_1)$. Hence $r(M_1) = \alpha+1$.

Construction of M_2 : There exists a $\Sigma_2(L(\alpha,T))$ set A of axioms in the language of set theory with the following properties: If $V \models A$, then

(i) V is a proper end extension of $L(\alpha,T)$,

(ii) $L(\alpha,T) \prec_1 V$, and

(iii) in V there is a model M_2 of T_α.

The axioms of A that insure (i) and (iii) are $\Sigma_1(L(\alpha,T))$, since T_α is $\Delta_1(L(\alpha,T))$. The Σ_2 flavor of A is derived solely from (ii).

Let V be a model of A that omits α. Then $\omega_1^{M_2} \leq \alpha$, because otherwise α would occur in V. Hence $r(M_2) \leq \alpha+1$ by (2) of Section 2.

To see that $r(M_2) = \alpha+1$, suppose $r(M_2) = \alpha$. Then M_2 is the atomic model of T_α. Let f be a function that carries each n-tuple of M_2 $(n < \omega)$ to a formula of K_α that defines the atom realized by that n-tuple. f is $\Sigma_2(L(\alpha, M_2))$, since the notion of atom is $\pi_1(L(\alpha, M_2))$, a consequence of the Δ_1 definability of T_α over $L(\alpha, M_2)$.

The Σ_2 definition of f, when interpreted in V, yields f again thanks to property (ii) of A. It follows that the range of f is contained in $L(\delta,M)$ for some $\delta < \alpha$, because α is omitted from V. But then $r(M_2) < \alpha$, a possibility already ruled out in the discussion of M_1.

Similar reasoning also shows that M_2 is not atomic over any $\langle c_1,\ldots,c_n\rangle \in M_2^n$. The above argument goes through with $\langle M_2, c_1,\ldots,c_n\rangle$ in place of M_2. Note that the Scott analysis of $\langle M_2, c_1,\ldots,c_n\rangle$ is the same in V as it is in the real world at all levels below α thanks to property (ii).

<u>Theorem 4.2.</u> Suppose the number of countable models of T is less than 2^ω. Assume that the canonical Scott rank of each countable model M of T is at most ω_1^M. Then T has only countably many countable models.

<u>Proof.</u> By Morley [5] $n(\beta,T) \leq \omega$ for all $\beta < \omega_1$. Let α be a countable ordinal such that $L(\alpha,T)$ is a model of a fragment of ZFC strong enough to carry out the following manipulations. By Lemma 3.1, the tree $\mathcal{T}$ of Section 2, restricted to levels below α, is $\Sigma_1(L(\alpha,T))$. Let β be aleph-one in the sense of $L(\alpha,T)$. Assume that T has a model of rank at least β, and let T_β be a theory on level β of $\mathcal{T}$. The structure $L(\beta,T_\beta)$ is Σ_2 admissible, since β is a regular cardinal of $L(\alpha,T)$ and $T_\beta \in L(\alpha,T)$.

As in the construction of M_2 in the proof of Theorem 4.1, there is a $\Sigma_2(L(\beta,T))$ set A of axioms such that : if $V \models A$, then

(i) V is a proper end extension of $L(\beta,T)$,

(ii) $L(\beta,T) \prec_1 V$, and

(iii) in V there is a model M of T_β.

Let V be a model of A that omits β. The argument given in 4.2 that $r(M_2) = \alpha+1$ now shows $r(M) = \beta+1$. And $\omega_1^M \leq \beta$, since $\beta \notin V$. Hence

$r(M) = \omega_1^M + 1$. But by assumption $r(M) \leq \omega_1^M$.

5. Further Results and Question

Suppose $n(T)$, the number of countable models of T, is countable. Then there exists a countable ordinal β such that $r(M) < \beta$ for every M a model of T. Let the least such β be $r(T)$.

What can be said about $r(T)$? Clearly $r(T)$ is Δ_2^1 relative to T. Is there a sharper bound? The argument of Section 4, suitably modified, yield the following bounds.

I. If $(\beta)_{\beta<\omega_1}[n(\beta,T) < 2]$, then $r(T) \leq \omega_2^T + 1$. (ω_2^T is the least Σ_2 admissible ordinal relative to T.)

II. If $n(T) < 2^\omega$ and $r(M) \leq \omega_1^M$ for every M a model of T, then $r(T) \leq \omega_2^T + 1$.

III. If $n(T) < 2^\omega$ and $r(M) < \omega_1^M$ for every M a model of T, then $r(T) < \omega_1^T$.

Can the bounds in (I)-(III) be improved?

Ramas Salam [6] has extended (III) to the domain of Vaught's invariant model theory. Can the same be done for (I) and (II)?

If $n(T) < \omega$, then every model of T must complete its canonical Scott analysis on some finite level of T, and so $r(T) < \omega$. It also follows that each model of T has a presentation hyperarithmetic in T. T. Millar [7] has shown that "hyperarithmetic" cannot be improved to "arithmetic" in the previous sentence.

Suppose $(\beta)_{\beta<\omega_1}[n(\beta,T) < \omega]$. Does it follow that $n(T) \leq \omega$? The proof of Theorem 4.1 can be stretched to show

$$(\beta)_{\beta<\omega_1}[n(\beta,T) < 3] \rightarrow n(T) \leq \omega.$$

References

1. D. Scott, Invariant Borel sets, Fund. Math. 56(1964), 117-128.

2. M. Nadel, Scott sentences and amissible sets, Annals of Math. Log. 7 (1974), 267-294.

3. H. Rogers, Theory of Recursive Functions and Effective Computability, McGraw-Hill, 1967.

4. K. Barwise, Admissible Sets and Structures, Springer-Verlag, 1975.

5. M. Morley, The number of countable models, Jour. Symb. Log. 35(1970), 14-18.

6. R. Salam, On Vaught's conjecture in invariant model theory, in preparation.

7. T. Millar, Recursive theories with non-arithmetic models, Proceedings of the 1982 AMS Summer Institute in Recursion Theory, to appear.

* This paper is based on a talk given at the 1971 meeting of the International Congress of Logic, Methodology and Philosophy of Science in Bucharest. Its preparation was supported in part by NSF Grant MCS-8104266.

Proceedings of Symposia in Pure Mathematics
Volume **42**, 1985

Post's Problem in E-Recursion

GERALD E. SACKS[1]

Abstract. In the setting of E-recursion a positive solution to Post's problem is given for a class of structures that includes all E-closed, initial segments of L. The solution, based largely on fine structure and reflection properties, does not give rise to injuries when $L(\kappa)$ is Σ_1 inadmissible; i.e., each Friedberg-Muchnik requirement is attempted at most once.

1. Introduction. The partial E-recursive functions were discovered by Normann **[1]**, and independently by Moschovakis. They constitute a generalization of Kleene's theory of higher types recursion from normal objects of finite type to all sets. Thus they give a meaning to $\{e\}(x)$ for all x based on computations. The fundamentals of E-recursion are reviewed formally in §2. They are dealt with intuitively in the present section in order to provide a rough idea of the proof of the main result of the paper.

By a positive solution to Post's problem is usually meant the construction of two incomparable r.e. sets A and B. The meanings of "incomparable" and "r.e." depend on the choice of underlying structure $\mathscr{E}$, of which A and B are subsets, and the choice of recursion theory on $\mathscr{E}$. In this paper $\mathscr{E}$ is always E-closed; that is, $f(x) \in \mathscr{E}$ for all $x \in \mathscr{E}$ and every partial E-recursive f.

Let $E(x)$ be the least transitive E-closed set with x as an element. This paper was initiated by a striking result of Normann **[2]** concerning Post's problem for Kleene recursion in ${}^3E$. His result, restated in the language of E-recursion, is:

(1) If $E(2^\omega) \models [2^\omega$ is well ordered and of regular cardinality], then there exist two subsets of $E(2^\omega)$, each E-recursively enumerable on $E(2^\omega)$, such that neither is E-reducible to the other on $E(2^\omega)$.

1980 *Mathematics Subject Classification*. Primary 03D65; Secondary 03E45.

[1] The preparation of this paper was supported in part by National Science Foundation Grant MCS-81-04266. The author is grateful to Theodore Slaman for his generous assistance.

The proof of (1) bears only superficial resemblance to the positive solution of Post's problem given in classical recursion theory [**3**, **4**] or in α-recursion theory [**5**]. The results of the present paper originated in the desire to circumvent the assumption of "regular cardinality" in (1).

In order to clarify the consequent of (1), consider subsets A and B of some E-closed transitive set $\mathscr{E}$. A is E-r.e. on $\mathscr{E}$ if there are a partial E-rec. function $f(x, y)$ and a $p \in \mathscr{E}$ such that

$$(2) \qquad A = \{x | x \in \mathscr{E} \,\&\, f(x, p) \text{ converges}\}.$$

The schemes for computing E-recursive functions can be relativized to an arbitrary class C, thereby giving a meaning to $\{e\}^C(x)$, by adding the scheme

$$(3) \qquad g(x) = x \cap C.$$

A is E-reducible to B if there are a partial function f, E-rec. relative to B, and a $p \in \mathscr{E}$ such that $f(x, p)$ converges for all $x \in \mathscr{E}$, and $A = \{x | x \in \mathscr{E} \,\&\, f(x, p) = 0\}$.

Definition (2) is based on the idea of an r.e. set as the domain, rather than range, of a partial recursion function. Thus in (2) x lands in A iff some convergent computation c puts x into A. c is generated by starting with x and following the rules of computation. The idea of an r.e. set as the range of a partial recursive function does not, in general, yield an effective method of working back from an element y of an r.e. set B to a computation that puts y into B. In particular, there is a partial E-recursive function f whose range is not E-recursively enumerable. The trouble with f is that there is no partial E-recursive g such that for y in the range of f, $g(y)$ converges and $y = f(g(y))$.

One advantage of the idea of an r.e. set as the range of a partial recursive function is that it leads directly to the view of an r.e. set as an enumerable set when the underlying structure can be effectively well ordered. This dynamic view of an r.e. set has been essential in r.e. set constructions since the time of Post [**6**]. A new r.e. set is enumerated by simultaneously enumerating all r.e. sets and putting elements in the new set whenever dictated by information so far developed. Definition (2) leads indirectly to a sort of dynamic enumeration of A. Suppose $x \in A$. Then some computation c puts x into A. Since $\mathscr{E}$ is E-closed, both c and $|c|$, the ordinal length of c, belong to $\mathscr{E}$. Define

$$A^\sigma = \{x | x \in A \text{ by virtue of some computation } c \text{ such that } |c| \leqslant \sigma\}$$

for each $\sigma \in \mathscr{E}$. Then $A = \bigcup\{A^\alpha | \sigma < \kappa\}$, where κ is the least ordinal not in $\mathscr{E}$. Thus, stage σ of the dynamic enumeration of A consists of putting into A all those elements of A that belong to A by virtue of computations of length σ. If A is a subset of some $b \in \mathscr{E}$, then each $A^\sigma \in \mathscr{E}$, and $\lambda\sigma | A^\sigma$ is partial E-recursive on $\mathscr{E}$.

$\mathscr{E}$ is effectively well orderable if there exists a partial E-recursive-on-$\mathscr{E}$ map from κ, the least ordinal not in $\mathscr{E}$, onto $\mathscr{E}$. $\mathscr{E}$ is so orderable if $\mathscr{E}$ is $L(\kappa)$ or is $E(x)$ for some set x of ordinals. Note that the hypothesis of (1) implies $E(2^\omega)$ has the latter property. The assumption that $\mathscr{E}$ be effectively well orderable is essential to all existing positive solutions of Post's problem in the setting of E-recursion theory.

The central positive result of the paper is:

(4) If $L(\kappa)$ is E-closed, then there exist two subsets of $L(\kappa)$, each E-recursively enumerable on $L(\kappa)$, such that neither is E-reducible to the other on $L(\kappa)$.

A curious feature of the proof of (4) is the presence of a Friedberg-Muchnik scheme of requirements without injuries when $L(\kappa)$ is not Σ_1 admissible. At most, one positive attempt is made to satisfy each requirement, and no negative requirement is injured. Further results, such as Slaman's density theorem [7], do necessitate Friedberg-Muchnik injury arguments. The reasons that r.e. degree theory on E-closed structures is less troubled by injuries than its counterparts in classical or α-recursion theory have to do with reflection phenomena. There is no hope in any recursion theory worthy of the name of effectively enumerating all divergent computations. But something very close to such an enumeration can be realized in many E-closed structures.

To define the concept of reflection in E-recursion it is helpful to make a connection between $E(x)$ and $L(x)$. Say y is E-recursive in x if $y = \{e\}(x)$ for some e. Define

$$\kappa_0^x = \sup\{\gamma \mid \gamma \text{ is an ordinal } E\text{-recursive in } x\}$$

and

$$\kappa^x = \sup\{\kappa_0^{x,a} \mid a \in \mathrm{TC}(x)\}.$$

($\mathrm{TC}(x)$ is the transitive closure of x.) Then

$$E(x) = L(\kappa^x, \mathrm{TC}(\{x\})).$$

γ is said to be x-reflecting if

$$[L(\delta, \mathrm{TC}(\{x\})) \models \mathscr{F}] \rightarrow [L(\kappa_0^x, \mathrm{TC}(\{x\})) \models \mathscr{F}]$$

for every $\mathscr{F}$ a Σ_1^{ZF} sentence with x as parameter. Define

$$\kappa_r^x = \text{sup of all } x\text{-reflecting ordinals}.$$

Clearly, κ_r^x is the greatest x-reflecting ordinal. It can be shown that $\kappa_r^x \leqslant \kappa^x$. More important is the fact that:

(5) If $L(\kappa)$ is not Σ_1 admissible, then $\kappa_r^x < \kappa$ for all $x \in L(\kappa)$.

If x is a set of ordinals, then κ_r^x turns out to be large enough to decide whether or not $\{e\}(x)$ converges for all e. If $\{e\}(x)$ converges, the computation of $\{e\}(x)$ is a well-founded tree whose height is an ordinal recursive in x, hence less than κ_0^x. If $\{e\}(x)$ diverges, then any infinite, descending path through its computation tree is called a Moschovakis witness to the divergence of $\{e\}(x)$. The connection between reflection and divergence needed for the solution to Post's problem is:

(6) If x is a set of ordinals and $\{e\}(x)$ diverges, then a Moschovakis witness to the divergence is first order definable over $L(\kappa_r^x, x)$.

It follows from (5) and (6) that:

(7) If $L(\kappa)$ is not Σ_1 admissible, $x \in L(\kappa)$ and $\{e\}(x)$ diverges, then some Moschovakis witness to the divergence belongs to $L(\kappa)$.

It is (7) that makes possible a positive solution to Post's problem by a Friedberg-Muchnik scheme without injuries. A requirement is satisfied in §4 below by waiting for a certain computation to converge or diverge. If it converges, then the usual steps are taken to create and preserve an inequality. If it diverges, then (7) implies there is a stage of the construction by which the divergence is apparent. At that stage a witness to the divergence is chosen and its role as a witness is preserved forever.

If $L(\kappa) = E(\omega_1)$, then the requirements can be indexed by countable ordinals, and each can be satisfied in its turn as in the previous paragraph. For an arbitrary Σ_1 inadmissible $L(\kappa)$, an indexing of requirments has to be chosen short enough to allow all requirements to be met before time runs out. In the case of $E(\omega_1) = L(\kappa)$, any countable set of requirements can be satisfied by stage σ for some $\sigma < \kappa$. In general, the requirements are indexed by ordinals less than ρ, the greater r.e. projectum of $L(\kappa)$.

$$\rho = \mu_{\gamma \leqslant \kappa}(Ef)[f \text{ is a partial function, } E\text{-rec. on } L(\kappa), \text{ from } \gamma \text{ onto } L(\kappa)].$$

It is shown in §3 that:

(8) If $x < \rho$ and $p \in L(\kappa)$, then $\sup\{\kappa_r^{p,y} \mid y < x\} < \kappa$.

It follows from (8) that any set of requirements bounded below ρ can be satisfied by stage σ for some $\sigma < \rho$. (There is an additional complication having to do with the r.e. cofinality of ρ and a technique similar to Shore blocking, but it can wait until §§3 and 4.) The proof of (8) has two parts. Define the lesser r.e. projection

$$\eta = \mu\gamma_{\gamma \leqslant \kappa}(EA)[A \in 2^\gamma - L(\kappa) \,\&\, A \text{ is } E\text{-r.e. on } L(\kappa)].$$

A condensation argument given in §3 shows:

(9) If $L(\kappa)$ is E-closed, then $\eta = \rho$.

The proof of (8) consists of proving (8) with η in place of ρ and then invoking (9).

The problem of extending the solution of Post's problem from initial segments of L to other effectively well-orderable E-closed structures is discussed in §5. The only substantial difficulty resides in the proof of (9), a Gödel type condensation argument that appears to be useless when $L(\kappa)$ is replaced by $L[\kappa, A]$ for some $A \subseteq L(\kappa)$. An important special case is $E(2^\omega)$. Normann's result (1) can be redone as follows. In the first half of his proof he shows:

(10) If $E(2^\omega) \models [2^\omega$ is well ordered and of regular cardinality], then $\eta = \rho$ for $E(2^\omega)$.

Then the second half of his argument can be replaced by the proof given in §4. Thus Normann's assumption of "regularity" in (1) was needed only to show $\eta = \rho$. Slaman [8] has shown:

(11) There exists a model M of ZFC in which $E(2^\omega) \models [2^\omega$ is well ordered and of singular cardinality] and $\eta = \rho$ for $E(2^\omega)$.

It follows from the argument of §4 that in Slaman's model M, $E(2^\omega)$ admits a positive solution to Post's problem. Thus the study of Post's problem for $E(2^\omega)$, when 2^ω is well orderable in $E(2^\omega)$, appears to turn more on the equality of η and ρ than on the regularity of 2^ω. The existing technology of forcing over E-closed structures [**8**–**10**] gives no hint of how to create an $E(2^\omega)$ in which 2^ω is well ordered and for which $\eta \neq \rho$.

2. Review of *E*-recursion. The partial E-recursive functions are defined via Normann's schemes [**1**]. $\{e\}$ is the partial E-recursive function with index e. $x_1,\ldots,x_n$ are arbitrary sets.

(1) $$\{e\}(x_1,\ldots,x_n) = x_i \quad \text{if } e = \langle 1, n, i\rangle.$$

(2) $$\{e\}(x_1,\ldots,x_n) = x_i - x_j \quad \text{if } e = \langle 2, n, i, j\rangle.$$

(3) $$\{e\}(x_1,\ldots,x_n) = \{x_i, x_j\} \quad \text{if } e = \langle 3, n, i, j\rangle.$$

(4) $$\{e\}(x_1,\ldots,x_n) \simeq \bigcup\{\{c\}(y, x_2,\ldots,x_n) \mid y \in x_1\} \quad \text{if } e = \langle 4, n, c\rangle.$$

(5) $$\{e\}(x_1,\ldots,x_n) \simeq \{c\}(\{d_1\}(x_1,\ldots,x_n),\ldots,\{d_m\}(x_1,\ldots,x_n)) \quad \text{if } e = \langle 5, n, m, c, d_1,\ldots,d_m\rangle.$$

(6) $$\{e\}(c, x_1,\ldots,x_n, y_1,\ldots,y_m) \simeq \{c\}(x_1,\ldots,x_n) \quad \text{if } e = \langle 7, n, m\rangle.$$

$\{e\}(x)\downarrow$ means $\{e\}(x)$ is defined or converges, and $\{e\}(x)\uparrow$ means $\{e\}(x)$ is undefined or diverges. In scheme (4), $\{e\}(x_1,\ldots,x_n)\downarrow$ iff $\{c\}(y, x_2,\ldots,x_n)\downarrow$ for all $y \in x_1$. An important derived scheme, called *bounding*, is

(7) $$\{2^m \cdot 3^n\}(x) \simeq \{\{n\}(y) \mid y \in \{m\}(x)\}.$$

Implementation of the schemes generates the universal computation tree $>_U$. Each node of the tree is of the form $\langle e, x_1,\ldots,x_n\rangle$, or more simply $\langle e, x\rangle$, and is called a computation instruction. $a >_U b$ is read: b is a subcomputation instruction of a, and is defined by: There exists a finite sequence $b_0,\ldots,b_n$ such that $a = b_0$, $b_n = b$, and b_{i+1} is an immediate subcomputation instruction of b_i $(i < n)$. The inductive definition of immediate subcomputation instruction has a clause corresponding to each Normann scheme. It is reasonable to regard (7) as typical, because it combines (4) and (5). The corresponding clauses are:

(8) $\langle m, x\rangle$ is an immed. subcomp. instruc. of $\langle 2^m \cdot 3^n, x\rangle$.

(9) If $\{m\}(x)\downarrow$ and $y \in \{m\}(x)$, then $\langle n, y\rangle$ is an immed. subcomp. instruc. of $\langle 2^m \cdot 3^n, x\rangle$.

In addition:

(10) If c is not the index of a Normann scheme, then $\langle c, x\rangle$ is an immed. subcomp. instruc. of $\langle c, x\rangle$.

The tree-like object consisting of $\langle e, x\rangle$ and that part of $>_U$ below $\langle e, x\rangle$ is called the *computation* of $\{e\}(x)$. Note that $\{e\}(x)\downarrow$ iff the computation of $\{e\}(x)$ is well founded.

Let A be a class. A is *E-recursively enumerable* in x if, for some e,

$$A = \{y|\{e\}(x, y)\downarrow\}.$$

If $x = 0$, then A is E-recursively enumerable. B is said to be E-recursively enumerable in x on C if $B = A \cap C$ for some A E-recursively in x.

The relation $\{e\}(x)\downarrow$ is E-recursively enumerable, and consequently the relation "b is an immediate subcomputation instruction of c" is E-recursively enumerable. *Note well*: The subcomputation relation $>_U$ is not E-recursively enumerable.

If $\{e\}(x)\downarrow$, let $|\{e\}(x)|$ be the ordinal height of the computation of $\{e\}(x)$. $|\{e\}(x)|$ is a partial E-recursive function of e and x.

A set y is said to be E-recursive in a set x if $y = \{e\}(x)$ for some e; in symbols, $y \leqslant_E x$. Thus $|\langle e, x\rangle| \leqslant_E x$ when $\{e\}(x)\downarrow$. Let $\alpha, \beta, \gamma, \ldots$ denote ordinal numbers. Define

$$\kappa_0^x = \bigcup\{\gamma|\gamma \leqslant_E x\} \quad \text{and} \quad \kappa^x = \bigcup\{\kappa_0^{x,a}|a \in \mathrm{TC}(x)\}.$$

$\mathrm{TC}(x)$ is the transitive closure of x.

A set z is E-closed if

$$x \in z \,\&\, \{e\}(x)\downarrow \;\rightarrow\; \{e\}(x) \in z$$

for all $e < \omega$. Define $E(x)$, the E-closure of x, to be the least E-closed z such that $z \supseteq \mathrm{TC}(\{x\})$. Then

$$E(x) = L(\kappa^x, \mathrm{TC}(\{x\})).$$

δ is said to be x-reflecting if

$$[L(\delta, \mathrm{TC}(\{x\})) \vDash \mathscr{F}] \rightarrow [L(\kappa_0^x, \mathrm{TC}(\{x\})) \vDash \mathscr{F}]$$

for every Σ_1 sentence $\mathscr{F}$ of ZF with parameter x. Define κ_r^x to be the supremum of all x-reflecting ordinals. Thus κ_r^x is the greatest x-reflecting ordinal. Note that $\kappa_r^x \leqslant \kappa^x$.

κ_r originated in an application of forcing [11] to the theory of Kleene recursion in finite types, where it was necessary to know that

$$\kappa_r^{2^{2^\omega}} > \kappa_0^{2^{2^\omega}}.$$

Later Harrington [12] showed $\kappa_r^{2^\omega} > \kappa_0^{2^\omega}$, and more importantly, gave a characterization of $\kappa_r^{2^\omega}$ that inspired Lemma 2.1.

If $\{e\}(x)\uparrow$, then there exists an infinite descending path below $\langle e, x\rangle$ in $>_U$. Any such path is termed a Moschovakis witness to the divergence of $\{e\}(x)$. They were introduced by Moschovakis [13] to show $E(2^\omega)$ is not Σ_1 admissible. Observe that the relation "z is an M-witness to $\{e\}(x)\uparrow$" is E-recursively enumerable. Let $\mathscr{E}$ be E-closed. $\mathscr{E}$ is said to *admit M-witnesses* if for all $e < \omega$ and $x \in \mathscr{E}$, if $\{e\}(x)\uparrow$, then some M-witness to $\{e\}(x)\uparrow$ is a member of M. $E(2^\omega)$ admits M-witnesses, but $E(\omega)$ does not.

LEMMA 2.1 [9]. *Suppose some well-ordering of* $\mathrm{TC}(x)$ *is E-recursive in x. If* $\{e\}(x)\uparrow$, *then some Moschovakis witness to* $\{e\}(x)\uparrow$ *is first-order definable over* $L(\kappa_r^x, \mathrm{TC}(\{x\}))$.

Harrington [12] showed $\kappa_r^{2^\omega, a, b} \geqslant \kappa_r^{2^\omega, a}$ for all $a, b \in 2^\omega$. But Slaman [14] found sets x and y such that $\kappa_r^{x,y} < \kappa_r^x$. It follows from Lemma 2.1 that the function $\lambda x | \kappa_r^x$, restricted to $x \subseteq On$, is E-r.e.

LEMMA 2.2 [9]. *Suppose some well-ordering of* $\mathrm{TC}(x)$ *is E-recursive in x. Then* $\kappa_r^{x,y} \geqslant \kappa_r^x$ *for all y, and* $\kappa_r^z \leqslant \kappa_r^x$ *for all* $z \leqslant_E x$.

LEMMA 2.3 [9]. *Suppose x is a set of ordinals, and* $L(\kappa, x)$ *is E-closed and* Σ_1 *inadmissible. Then* $\kappa_r^{x,y} < \kappa$ *for all* $y \in L(\kappa, x)$.

THEOREM 2.4 [9]. *Let x be a set of ordinals. Then the following are equivalent.*

(i) $\kappa_r^{x,y} \in E(x)$ *for all* $y \in E(x)$.

(ii) $E(x)$ *admits Moschovakis witnesses.*

(iii) $E(x)$ *is* Σ_1 *inadmissible.*

THEOREM 2.5 [9]. *Suppose x is a set of ordinals and* $L(\kappa, x)$ *is E-closed. If* $L(\kappa, x)$ *is not* Σ_1 *admissible, then* $L(\kappa, x)$ *admits Moschovakis witnesses.*

Slaman [14] has devised an E-closed, Σ_1 inadmissible set that does not admit Moschovakis witnesses.

Let $\mathscr{E}$ be an E-closed structure. The notion of Turing degree generalizes from r.e. subsets of ω to r.e. subsets of $\mathscr{E}$ as follows. The principal step consists of relativizing the concept of E-recursiveness. Let A be a class. $\{e\}^A$ is the eth partial E-recursive function relative to A. $\{e\}^A$ is defined by Normann schemes (1)–(5), with $\{\ \}^A$ in place of $\{\ \}$ throughout, and one additional Normann scheme:

$$\{e\}^A(x_1, \ldots, x_m) = A \cap x_i \quad \text{if } e = \langle 6, n, i \rangle. \tag{11}$$

B is said to be E-reducible to A if for some e, $B(x) = \{e\}^A(x)$ for all x.

Note well the difference between (i) x is E-recursive in y, and (ii) x is E-reducible to y. In (i) y occurs as an argument or parameter, and in (ii) as an additional predicate. For example, suppose x and y are infinite subsets of ω. Then (i) means x is hyperarithmetic in y, and (ii) says x is Turing reducible to y.

Now let A and B be subsets of $\mathscr{E}$. A is E-recursive on $\mathscr{E}$ if there exist e and $p \in \mathscr{E}$ such that

$$\{e\}(x, p)\downarrow \text{ for all } x \in \mathscr{E} \quad \text{and} \quad A = \{x | x \in \mathscr{E} \,\&\, \{e\}(x, p) = 0\}.$$

A is E-recursively enumerable on $\mathscr{E}$ if there exist e and $p \in \mathscr{E}$ such that

$$A = \{x | x \in \mathscr{E} \,\&\, \{e\}(x, p)\downarrow\}.$$

Finally, A is E-reducible to B on $\mathscr{E}$ (in symbols, $A \leqslant_{\mathscr{E}} B$) if there exist e and $p \in \mathscr{E}$ such that

$$\{e\}^B(x, p)\downarrow \text{ for all } x \in \mathscr{E} \quad \text{and} \quad A = \{x | x \in \mathscr{E} \,\&\, \{e\}^B(x, p) = 0\}.$$

The notion of E-reducibility is not, in general, transitive. Call $A \subseteq \mathscr{E}$ subgeneric if for all e and $x \in \mathscr{E}$, if $\{e\}^A(x)\downarrow$, then the computation of $\{e\}^A(x)$ belongs to $\mathscr{E}$. For such an A, $\mathscr{E}$ may be said to be $\mathscr{E}$-closed relative to A, or in other words the structure $\langle \mathscr{E}, A\rangle$ is E-closed.

PROPOSITION 2.6. *Suppose $\mathscr{E}$ is E-closed and A, B, $C \subseteq \mathscr{E}$.*

(i) *If $B \leqslant_{\mathscr{E}} C$ and C is subgeneric, then B is subgeneric.*

(ii) *If $A \leqslant_{\mathscr{E}} B$, $B \leqslant_{\mathscr{E}} C$ and C is subgeneric, then $A \leqslant_{\mathscr{E}} C$.*

(iii) *If A is $\mathscr{E}$-r.e. and not subgeneric and B is E-recursively enumerable on $\mathscr{E}$, then $B \leqslant_{\mathscr{E}} A$.*

(iv) *If A, B and C are E-recursively enumerable on $\mathscr{E}$, $A \leqslant_{\mathscr{E}} B$ and $B \leqslant_{\mathscr{E}} C$, then $A \leqslant_{\mathscr{E}} C$.*

PROOF. (i) is immediate. (ii) follows from (i) and effective bounding. Similarly, (iv) follows from (iii). To prove (iii), observe that the failure of A to be subgeneric implies there is a function f, E-recursive in A on $\mathscr{E}$, such that

$$(\operatorname{dom} f) = d \in \mathscr{E} \quad \text{and} \quad \sup\{|f(x)| \mid z \in d\} = \kappa,$$

where κ is the least ordinal not in $\mathscr{E}$. Suppose B is $\mathscr{E} \cap \operatorname{dom}(\lambda x|\{e\}(x, p))$. Then, for all $x \in \mathscr{E}$,

$$x \in B \leftrightarrow (Ez)\big[z \in d \,\&\, |\{e\}(x, p)| \leqslant |f(z)|\big]. \quad \square$$

The proof of (iii) is essentially the same as Spector's proof [15] that every nonhyperarithmetic Π^1_1 subset of ω has the same hyperdegree as Kleene's $\mathcal{O}$.

Let A and B be E-recursively enumerable on $\mathscr{E}$. In the light of Proposition 2.6 it makes sense to say A and B have the same *degree* if each is E-reducible to the other on $\mathscr{E}$. The proof of Theorem 2.4, in particular the equivalence of 2.4(i) and (iii), yields a connection between E-recursion and recursion on an admissible set.

LEMMA 2.7. *Suppose x is a set of ordinals, $E(x)$ is Σ_1 admissible, and A, $B \subseteq E(x)$.*

(i) *A is E-recursively enumerable on $E(x)$ iff A is Σ_1^A.*

(ii) *If A is E-reducible to B on $E(x)$, then A is $E(x)$-recursive in B (in the sense of admissibility theory).*

Another notion of some use in the study of recursion on an E-closed set $\mathscr{E}$ is that of regularity. A set $A \subseteq \mathscr{E}$ is said to be *regular* if $(A \cap x) \in \mathscr{E}$ for all $x \in \mathscr{E}$.

LEMMA 2.8. *Suppose $\mathscr{E}$ is E-closed and A, $B \subseteq \mathscr{E}$ are E-recursively enumerable on $\mathscr{E}$.*

(i) *If A is not regular, then $B \leqslant_{\mathscr{E}} A$.*

(ii) *There exists a regular $C \subseteq \mathscr{E}$ such that C is E-recursively enumerable on $\mathscr{E}$ and has the same degree as A.*

PROOF. (i) Suppose $(A \cap x) \notin \mathscr{E}$. Then the lengths of computations needed to enumerate $A \cap x$ are unbounded in $\mathscr{E}$, and so A is not subgeneric. Now apply Proposition 2.6(iii).

(ii) By (i) it is safe to assume A is "complete". Thus it suffices to find a regular, r.e. C such that $A \leqslant_{\mathscr{E}} C$. The original construction of C is a transfinite recursion that applies some of the thinking behind the regular sets theorem of α-recursion theory. A much shorter proof was subsequently found by Slaman [7]. (2.7 is not needed in this paper, but is used in the proof of Slaman's density theorem [7].) □

Let A be a class. All of the definitions and results of this section through Theorem 2.4 relativitize to A. $\{e\}^A(x)$ has already been defined. The computation tree for $\{e\}^A$ differs from that of $\{e\}$ only by nodes corresponding to scheme (11). A set y is E-recursive, relative to A, in a set x if $y = \{e\}^A(x)$ for some e; in symbols, $y \leqslant_E^A x$. Define

$$\kappa_0^{A;x} = \bigcup\{\gamma | \gamma \leqslant_E^A x\} \quad \text{and} \quad \kappa^{A;x} = \bigcup\{\kappa_0^{A;x,a} | a \in \mathrm{TC}(x)\}.$$

A set z is E-closed relative to A if, for all e,

$$x \in z \,\&\, \{e\}^A(x)\downarrow \to \{e\}^A(x) \in Z.$$

Define $E^A(x)$, the E-closure of x relative to A, to be the least $z \supseteq \mathrm{TC}(\{x\})$ such that z is E-closed relative to A. Then

$$(12) \qquad E^A(x) = L[\kappa^{A;x}, A; \mathrm{TC}(\{x\})].$$

The right side of (12) is the set of all sets constructible via ordinals less than $\kappa^{A;x}$ with $\mathrm{TC}(\{x\})$ as the starting set and A as an additional predicate. (In addition to the usual Gödel operations, $A \cap x$ is allowed.)

δ is x-reflecting relative to a if

$$[L[\delta, A; \mathrm{TC}(\{x\})] \models \mathscr{F}] \to [L[\kappa_0^{A;x}, A; \mathrm{TC}(\{x\})] \models \mathscr{F}]$$

for every Σ_1 sentence $\mathscr{F}$ of ZF with parameter x and additional predicate $y \in A$. Let $\kappa_r^{A;x}$ be the greatest x-reflecting ordinal relative to A.

The treatment of κ_r and Moschovakis witnesses in 2.1–2.5 relativizes to A without any surprises. For example, Theorem 2.5 becomes

THEOREM 2.9. *Suppose x is a set of ordinals and $\langle L[\kappa, A; x], A\rangle$ is E-closed, but not Σ_1 admissible. Then $L[\kappa, A; x]$ admits Moschovakis witnesses.*

3. Fine structure. Assume $L(\kappa)$ is E-closed. This section concentrates on several structural properties of $L(\kappa)$ needed in the next section to establish a suitable indexing of Friedberg-Muchnik requirements. Define

$$\rho = \mu\gamma_{\gamma\leqslant\kappa}(Ef)\,[f \text{ is a partial } E\text{-rec. (on } \mathrm{L}(\kappa)) \text{ map of } \gamma \text{ onto } \mathrm{L}(\kappa)]$$

and

$$\eta = \mu\gamma_{\gamma\leqslant\kappa}(EA)[A \in 2^\gamma - L(\kappa) \,\&\, A \text{ is } E\text{-r.e. on } L(\kappa)].$$

LEMMA 3.1. $\eta = \rho$.

PROOF. First suppose that

$$(1) \qquad L(\kappa) \text{ is } \Sigma_1 \text{ admissible and equal to } E(\beta)$$

for some $\beta < \kappa$. By Proposition 2.7, E-r.e. coincides with $\Sigma_1(L(\kappa))$, and ρ and η are equivalent definitions of the $\Sigma_1(L(\kappa))$ projectum of $L(\kappa)$.

(2) Now suppose (1) is false.

To see $\eta \leqslant \rho$, let f be a partial E-rec. map of ρ onto $L(\kappa)$. If $\rho < \eta$, then $(\text{dom } f) \in L(\kappa)$, and consequently $(\text{range } f) \in L(\kappa)$.

To see $\eta = \rho$, fix $\gamma < \rho$ and let $A \subseteq \gamma$ be E-r.e. in some $p \in L(\kappa)$ via Gödel number e with the intention of showing $A \in L(\kappa)$. It is safe to assume p is an ordinal, since each member of $L(\kappa)$ is E-recursive in some ordinal less than κ. Let g be a universal, partial E-recursive function. Define

$$H = g[\gamma \cup \{p\}].$$

(H is said to be the partial E-recursive hull of $\gamma \cup \{p\}$.)

Suppose H is bounded below κ, i.e. $H \subseteq L(\delta)$ for some $\delta < \kappa$. Then for $y < \gamma$,

$$y \in A \leftrightarrow L(\delta) \models \{e\}(y, p)\downarrow,$$

and so $A \in L(\kappa)$.

Suppose H is unbounded in $L(\kappa)$. Observe that

$$z \in H \to \mathcal{O}(z) \in H, \tag{3}$$

where $\mathcal{O}$ is Gödel's order of constructibility function. If $z \in H$, then $z = g(x)$ for some $x \in \gamma \cup \{p\}$, and $g(x)$ is first-order definable over $L(x + |g(x)| + \omega)$. Since H is E-closed, $|g(x)| \in H$.

The unboundedness of H implies H admits M-witnesses. Fix $x \in H$. It follows from (2) and Lemma 2.3 that $\kappa_r^x < \kappa$. By (3) the supremum of all ordinals in H is κ. Choose $\delta \in H$ so that $\kappa_r^x \leqslant \delta$. Then $\kappa_r^x \leqslant_E \delta, x$, and so $\kappa_r^x \in H$. Consequently, H admits M-witnesses by Lemma 2.1.

It follows from (3) that every element of H is well ordered in H. Thus H is extensional, hence isomorphic, to a transitive set H_0 via the collapsing map t. (3) also implies $H_0 = L(\beta)$ for some $\beta \leqslant \kappa$. The partial E-recursive function g maps $\gamma \cup \{t(p)\}$ onto $L(\beta)$. Hence $\beta < \kappa$, since $\gamma < \kappa$.

Since H admits M-witnesses, the map t preserves divergence, as well as convergence, facts. Hence A is first-order definable over $L(\beta)$. □

PROPOSITION 3.2. *Let p be an ordinal and A a set of ordinals. Suppose the function $\lambda x|\kappa_r^{p,x}$ does not attain a maximum on A. Then the range of $\lambda x|\kappa_r^{p,x}$ is bounded on A by* $\sup\{\kappa_0^{p,y,z}|y, z \in A\}$.

PROOF. Suppose not. Then there exist $c, d \in A$ such that

$$\kappa_r^{p,c} > \kappa_r^{p,d} \geqslant \sup\{\kappa_0^{p,y,z}|y, z \in A\}. \tag{4}$$

It follows from Lemma 2.2 that $\kappa_r^{p,c,d} > \kappa_r^{p,d}$. The desired contradiction is: $\kappa_r^{p,c,d}$ is a $\langle p, d\rangle$-reflecting ordinal. Suppose some Σ_1 sentence $\mathcal{F}$ about $\langle p, d\rangle$ is true below $\kappa_r^{p,c,d}$. Then $\mathcal{F}$ is true below $\kappa_0^{p,c,d}$ by definition of κ_r. But then by (4), $\mathcal{F}$ is true below $\kappa_r^{p,d}$. □

Theorem 3.3. *Assume $L(\kappa)$ is E-closed and $p \in L(\kappa)$. Further assume $L(\kappa)$ is either Σ_1 inadmissible or not of the form $E(\beta)$ for any $\beta < \kappa$. If $\gamma < \rho$, then*

$$\sup\{\kappa_r^{p,\delta} | \delta < \gamma\} < \kappa.$$

Proof. As in the proof of 3.1, p can be assumed to be an ordinal. First observe that

$$\sup\{\kappa_0^{p,\delta} | \delta < \gamma\} < \kappa. \tag{5}$$

Define $f(e, \delta) \simeq |\{e\}(p, \delta)|$. f is a partial E-rec. map from $\omega \times \gamma$ into κ. Since ρ is either κ or an $L(\kappa)$-cardinal, it is safe to assume γ is closed under pairing in the following sense:

$$x, y < \gamma \rightarrow (E\delta)(\delta < \gamma \,\&\, x \leqslant_E \delta \,\&\, y \leqslant_E \delta). \tag{6}$$

Consequently, the domain of f can be construed as a subset of γ. It follows from Lemma 3.1 that (dom f) $\in L(\kappa)$, hence (range f) $\in L(\kappa)$, and so (5) is proved.

Suppose $\lambda\delta | \kappa_r^{p,\delta}$ does not attain a maximum on γ. Then Proposition 3.2 yields

$$\sup\{\kappa_r^{p,\delta} | \delta < \gamma\} \leqslant \sup\{\kappa_0^{p,x,y} | x, y \in \gamma\}. \tag{7}$$

The theorem now follows from (5) and (6).

If $\lambda\delta | \kappa_r^{p,\delta}$ does attain a maximum on γ, then that maximum is less than κ by Lemma 2.3. □

Cofinality considerations figure prominently in the solution to Post's problem. Let $\lambda \leqslant \kappa$. Define r.e. cf(λ), the r.e.-on-$L(\kappa)$ cofinality of λ, to be

$$\mu\gamma(EA)[\sup A = \lambda \,\&\, A \text{ is } E\text{-r.e. on } L(\kappa) \,\&\, \text{ordertype of } A \text{ is } \gamma].$$

Lemma 3.4. *r.e.* cf(κ) = *r.e.* cf(ρ).

Proof. Assume $L(\kappa)$ is either Σ_1 inadmissible or not of the form $E(\beta)$ for any $\beta < \kappa$. To show cf(κ) $\leqslant$ cf(ρ), let A be an unbounded subset of ρ and E-r.e. on $L(\kappa)$. Let p be the parameter needed to define a partial E-rec. f from ρ onto $L(\kappa)$. As in the proof of 3.1, p can be assumed to be an ordinal. Define

$$x^r = \sup\{\kappa_r^{p,y} | y \leqslant x\} \quad \text{and} \quad A^r = \{x^r | x \in A\}.$$

Theorem 3.3 implies $A^r \subseteq \kappa$. A^r is unbounded in κ, since $\kappa_r^{f(y)} \leqslant \kappa_r^{p,y}$ by Lemma 2.2 if $f(y)\downarrow$. $|A^r|$, the ordertype of A^r, is at most $|A|$. The predicate $v = \kappa_r^w$, restricted to $w \in \mathcal{O}n$, is E-rec. by Lemma 2.1. It follows that A^r is E-rec. on $L(\kappa)$.

To show cf(ρ) $\leqslant$ cf(κ) let B be an unbounded subset of κ, E-r.e. on $L(\kappa)$, and of ordertype cf(κ). B can be replaced by an E-rec.-on-$L(\kappa)$ set with the same ordertype as B. Assume (a) no such set exists. Define B^0 to be the set of all σ such that: σ is the length of a computation that put some x into B greater than any y put into B by any computation of length less than σ. Clearly B^0 is E-rec. on $L(\kappa)$. $|B^0| \leqslant |B|$ since B^0 is "strictly increasing". If B^0 is bounded below κ, then there exists a σ_0 such that computations of length at most σ_0 put an unbounded set of ordinals into B; but that contradicts assumption (a). Thus it is safe to assume B^0 is the set that replaces B.

B^0 divides κ into blocks. Let $\delta_0 < \delta_1 < \cdots < \delta_\tau < \cdots$ $(\tau < |B^0|)$ be a listing of B^0. Define block τ to be $[\delta_\tau, \delta_{\tau+1})$. The blocks are needed to describe the enumeration of A_0, a subset of dom f. A_0 will be unbounded in ρ, and its ordertype will be at most $|B^0|$.

A^0 is the range of a partial function g defined by a partial recursion on $|B^0|$.

$g(\tau) \simeq ux$ [x is enumerated in dom f via a computation of length in block τ and $(\beta)_{\beta<\tau}(g(\beta)$ defined $\rightarrow g(\beta) < x)$].

$|A_0| \leqslant |B^0|$ because g is strictly increasing.

To check that A_0 is unbounded in ρ, suppose otherwise. Thus $A_0 \subseteq (\text{dom } f) \cap z$ for some $z < \rho$. By Lemma 3.1, $((\text{dom } f) \cap z) \in L(\kappa)$, hence the computations needed to enumerate $(\text{dom } f) \cap z$ all lie below some δ_{τ_0}. It follows that g *is not* defined for any $\tau \geqslant \tau_0$. But $((\text{dom } f) - z) \notin L(\kappa)$, so some $x \in (\text{dom } f) - z$ must be enumerated by a computation of length greater than δ_{τ_0}, and consequently g *is* defined for some $\tau \geqslant \tau_0$.

x is recursively enumerated in A_0 as follows. First enumerate x in dom f. Suppose the length of computation used to put x into dom f belongs to block τ. Compute $g(\beta)$ for all $\beta \leqslant \tau$. Enumerate x in A_0 if $g(\tau) = x$.

The above instructions are effective largely because B^0 is E-recursive on $L(\kappa)$. □

PROPOSITION 3.5. *Suppose* $\gamma = r.e.$ cf(λ) *for some* $\lambda \leqslant \kappa$. *then there exists a* g: $\gamma \rightarrow \lambda$ *such that*:

(i) *the graph of* g *is E-r.e. on* $L(\kappa)$,

(ii) sup(range g) $= \lambda$,

(iii) g *is strictly increasing, and*

(iv) $\tau_1 < \tau_2 < \lambda \rightarrow |\langle \tau_1, g(\tau_1)\rangle| < |\langle \tau_2, g(\tau_2)\rangle|$.

PROOF. The proof is similar to the second half of the proof of 3.4. □

4. Solution to Post's problem.

THEOREM 4.1. *Suppose* $L(\kappa)$ *is E-closed. Then there exist two subsets of* $L(\kappa)$, *each E-recursively enumerable on* $L(\kappa)$, *such that neither is E-reducible to the other on* $L(\kappa)$.

Assume $L(\kappa)$ is either Σ_1 inadmissible or not equal to $E(\beta)$ for any $\beta < \kappa$. Otherwise Proposition 2.7 implies that the positive solution of Post's problem for $L(\kappa)$ provided by α-recursion theory [5] is also a solution in the sense of E-recursion theory. Thus $L(\kappa)$ has certain structural properties described in §§2 and 3. In particular, $\kappa_r^x < \kappa$ for every $x \in L(\kappa)$.

The subsets of $L(\kappa)$ to be constructed are A and B. All requirements are of the form $A \neq \{y\}^B$ or $B \neq \{y\}^A$, and are indexed by ordinals less than ρ. Let f be a partial function, E-rec. on $L(\kappa)$, from ρ onto $L(\kappa)$. Then

requirement $2x$ is $A \neq \{f(x)\}^B$ and requirement $2x+1$ is $B \neq \{f(x)\}^A$.

Each requirement belongs to a block labeled by an ordinal less than γ, the r.e.-on-$L(\kappa)$ cofinality of ρ. By Proposition 3.5, with ρ in place of λ, there exists a function g suitable for dividing ρ into blocks. Assume $g(0) = 0$. Define

$$(\text{req. } 2x) \in \text{block } 2\tau \quad \text{and} \quad (\text{req. } 2x+1) \in \text{block } 2\tau+1,$$

if $g(\tau) \leqslant x < g(\tau+1)$.

The construction has the form of an effective transfinite recursion on κ. At stage σ ($\sigma < \kappa$) all decisions are made by examining the set of all computations of length at most σ. (It will be convenient to assume that the "length" of a computation exceeds all ordinals mentioned in the computation.) Let

$$A^{<\sigma} = \text{set of all elements put in } A \text{ before stage } \sigma$$

and

$$A^{\sigma} = A^{<\sigma} \cup \{\text{elements put in } A \text{ at stage } \sigma\}.$$

Each requirement is acted on at most once.

Requirement $2x$ is *acted on* at stage σ if (i)–(iv) hold.

(i) req. $2x$ has not been acted on prior to stage σ.

(ii) $\langle \delta, g(\delta)\rangle$ has been computed for all $\delta \leqslant \tau$ prior to stage σ. (Req. $2x$) $\in$ block 2τ. $f(x)$ has been computed prior to stage σ.

(iii) Every requirement in block y, for all $y < 2\tau$, has been acted on prior to stage σ.

(iv) Let $\sigma(2\tau)$ be the first stage after all the activity of (ii) and (iii) has occurred. Then either (iv)(a) or (iv)(b) holds.

(iv)(a) $\{f(x)\}^{B<\sigma(2\tau)}(\langle\sigma(2\tau), f(x)\rangle)$ converges via a computation of length at most σ. (Let $\langle\sigma(2\tau), f(x)\rangle$ be an ordinal that encodes the ordered pair $(\sigma(2\tau), f(x))$ and exceeds both of its elements.)

(b) A Moschovakis witness to the divergence of $\{f(x)\}^{B^{<\sigma(2\tau)}}(\langle\sigma(2\tau), f(x)\rangle)$ is first-order definable over $L(\sigma)$.

Suppose req. $2x$ is acted on at stage σ. If (iv)(a) holds and

$$\{f(x)\}^{B^{<\sigma(2\tau)}}(\langle\sigma(2\tau), f(x)\rangle) = 1.$$

then $\langle\sigma(2\tau), f(x)\rangle$ is put in A at stage σ.

Requirement $2x+1$ is handled similarly. Simply replace $2x$ by $2x+1$, 2τ by $2\tau+1$, and exchange A and B in the above instructions.

Proposition 4.2. *Suppose every requirement in block z is acted on. Then there exists a σ such that every requirement in block z is acted on prior to stage σ.*

Proof. Suppose $z = 2\tau$ and (req. $2x$) $\in$ block 2τ. Let σ_{2x} be the unique σ such that req. $2x$ is acted on at stage σ. Let p be an ordinal that encodes $\langle B^{<\sigma(2\tau)}, \sigma(2\tau), q\rangle$, where q is the parameter needed to compute f. Then

$$\sigma_{2x} \leqslant p + x + \kappa_r^{p,x}$$

by Lemma 2.1. And $\sup\{\kappa_r^{p,x} | 2x \in \textit{block}\, 2\tau\} < \kappa$ by Theorem 3.3. □

PROPOSITION 4.3. *Assume* $z <$ *r.e.* cf(ρ). *Suppose every requirement in block* y, *for all* $y < z$, *is acted on. Then there exists a* σ *such that every requirement in block* y, *for all* $y < z$, *is acted on prior to stage* σ.

PROOF. If $u < v$, then no requirement in block v is acted on until every requirement in block u has been acted on. Consequently, if z is a successor ordinal, then the desired σ exists by Proposition 4.2. Assume z is a limit. For each $y < z$, let $\tau(y)$ be the least σ such that every requirement in block y is acted on prior to stage σ. $\tau(y) < \kappa$ by Proposition 4.2. Since

$$\tau(y) \leqslant \sigma(y+1) \leqslant \tau(y+1)$$

for all $y < z$, it remains only to show the supremum of

$$\mathscr{S} = \{\sigma(y) \mid y < z\}$$

is less than κ. $\mathscr{S}$ is E-r.e. on $L(\kappa)$, since the graph of $\lambda x | \kappa_r^x$ is E-r.e., as was noted after Lemma 2.1. The ordertype of $\mathscr{S}$ is at most z, since $\lambda y | \sigma(y)$ is nondecreasing. By Lemma 3.4, $z <$ r.e. cf(κ), hence sup $\mathscr{S} < \kappa$. □

Now the proof of Theorem 4.1 is easily completed. Fix $x < \rho$ in order to see req. $2x$ is acted on. Suppose (req. $2x$) $\in$ block 2τ. Assume for the sake of an induction that every requirement in block y, for all $y < 2\tau$, has been acted on. By Proposition 4.3, $\sigma(2\tau)$ is well defined. Suppose

$$\{f(x)\}^{B^{<\sigma(2\tau)}}(\langle\sigma(2\tau), f(x)\rangle) \tag{1}$$

converges. Then req. $2x$ is acted on at some stage $\sigma \geqslant \sigma(2\tau)$, clause (iv)(a) holds, and $A^\sigma(\langle\sigma(2\tau), f(x)\rangle)$ disagrees with (1).

The construction insures that

$$A^\sigma(\langle\sigma(2\tau), f(x)\rangle) = A(\langle\sigma(2\tau), f(x)\rangle);$$

and

$$B^{\sigma(2\tau)} = B^\sigma = B \cap \sigma. \tag{2}$$

Consequently $A \neq \{f(x)\}^B$.

Suppose (1) diverges. Then Theorem 2.5 implies req. $2x$ is acted on at some stage $\sigma \geqslant \sigma(2\tau)$ and clause (iv)(b) holds. Thus there is a Moschovakis witness w to the divergence of (1) first-order definable over $L(\sigma)$. It follows from (2) that w also witnesses the divergence of

$$\{f(x)\}^B(\langle\sigma(2\tau), f(x)\rangle).$$

Thus again $A \neq \{f(x)\}^B$.

To complete the induction observe that the above argument shows every requirement in block 2τ is acted on.

It remains only to show A and B are E-r.e. on $L(\kappa)$. Suppose z, a typical member of A, is put in A at stage σ. The decision is based on the set K_σ, the set of all computations of length at most σ. Since $K_\sigma \leqslant_E \sigma$ (uniformly in σ), it suffices to show $\sigma \leqslant_E z$.

Let z be $\langle \sigma(2\tau), f(x) \rangle$ and assume clause (iv)(a) of the definition of *acted on* holds. Then $\sigma \leqslant_E z$, $B^{<\sigma(2\tau)}$, and $B^{<\sigma(2\tau)} \leqslant_E K_{\sigma(2\tau)} \leqslant_E z$. □

5. Further results and open questions. As in §1, a positive solution to Post's problem consists of two "incomparable" r.e. sets. The method of §4 sheds a little light on Post's problem for structures other than initial segments of L. Assume $A \subseteq \kappa$. Let $L[\kappa, A]$ as usual be the set of all sets constructible via ordinals less than κ with $x \in A$ as an additional predicate. (Simply add $A \cap y$ to the nine Gödel operations that define constructibility.) The structure $\langle L[\kappa, A], A \rangle$ is said to be E-closed if $\{e\}^A(x) \in L[\kappa, A]$ when $x \in L[\kappa, A]$.

As described in §2, the definitions of κ_0 and κ_r relativize straightforwardly to A; $\{e\}$ is replaced by $\{e\}^A$ throughout. Similarly η, ρ and r.e. cofinality relativize to $\langle L[\kappa, A], A \rangle$. The proof of Theorem 4.1 shows:

(1) Assume $\langle L[\kappa, A], A \rangle$ is E-closed. If $\eta = \rho$ and $\kappa_r^{A;x} < \kappa$ for all $x \in L[\kappa, A]$, then $\langle L[\kappa, A], A \rangle$ admits a positive solution to Post's problem.

It seems unlikely that $\eta = \rho$ in general. An immediate corollary of a selection theorem of Griffor and Normann [**16**] states:

(2) If $L[\kappa, A] \models$ [there exists a greatest cardinal and it is regular], then $\eta = \rho$ for $\langle L[\kappa, A], A \rangle$.

The assumption in (1) on κ_r can be related to inadmissibility as in Theorem 2.9.

(3) If $\langle L[\kappa, A], A \rangle$ is E-closed, but not Σ_1 admissible, then $\kappa_r^{A;x} < \kappa$ for all $x \in L[\kappa, A]$.

The proof of Theorem 4.1 also applies to an arbitrary E-closed structure $\mathscr{E}$. Let κ be the least ordinal not in $\mathscr{E}$. $\mathscr{E}$ is said to be *effectively well orderable* if there exists a partial, E-recursive-on-$\mathscr{E}$ map from κ onto $\mathscr{E}$.

(4) Assume $\mathscr{E}$ is E-closed and effectively well orderable. If $\eta = \rho$ for $\mathscr{E}$, and $\mathscr{E}$ admits Moschovakis witnesses, then $\mathscr{E}$ admits a positive solution to Post's problem.

The proof of Theorem 4.1 yields a property stronger than subgenericity for those κ where blocking is unnecessary, that is, when r.e. $\text{cf}(\rho) = \rho$. The simplest example is:

(5) $E(\omega_1)$ admits a positive solution (A, B) to Post's problem such that for all $\delta < \omega_1$,

(6) $\kappa_0^{A;\omega_1,\delta} = \kappa_0^{B;\omega_1,\delta} = \kappa_0^{\omega_1,\delta}$.

A set $A \subseteq E(\omega_1)$ with property (6) is said to preserve the κ_0-spectrum of $E(\omega_1)$. It is not known if the κ_0-spectrum can be preserved when ρ fails to be r.e. regular.

A consequence of (4) and Theorem 2.5 is:

(7) Suppose x is a set of ordinals, and $L(\kappa, x)$ is E-closed but not Σ_1 admissible. If $\eta = \rho$ for $L(\kappa, x)$, then $L(\kappa, x)$ admits a positive solution to Post's problem.

At this writing virtually nothing is known about situations where Post's problem has a negative solution, or where $\eta \neq \rho$.

(8) Does there exist a set x of ordinals such that $L(\kappa, x)$ is E-closed and Σ_1 inadmissible, and does not admit a positive solution to Post's problem for E-recursion theory?

(9) Does there exist a set x of ordinals such that $L(\kappa, x)$ is E-closed and Σ_1 inadmissible, and $\eta \neq \rho$ for $L(\kappa, x)$?

According to (7) a negative answer for (8) implies a negative answer for (9). It might be easier, however, to deal with (9) directly. Both questions are good candidates for forcing arguments. Forcing over E-closed structures has been extensively studied in [**9**, **10** and **14**], but no technique so far developed appears to be of any use.

Normann's result on Post's problem for $E(2^\omega)$, quoted in §1, leads to the following terminal question:

Does there exist a model of ZFC in which

$$E(2^\omega) \models [2^\omega \text{ is well ordered}] \tag{10}$$

and $E(2^\omega)$ does not admit a positive solution to Post's problem?

References

1. D. Normann, *Set recursion*, Generalized Recursion Theory. II, North-Holland, Amsterdam 1978, pp. 303–320.

2. ______, *Degrees of functionals*, Preprint Series in Math., No. 22, Univ. of Oslo, Oslo, 1975.

3. R. Friedberg, *Two recursively enumerable sets of incomparable degrees of unsolvability*, Proc. Nat. Acad. Sci. U.S.A. **43** (1957), 236–238.

4. A Muchnik, *On the unsolvability of the problem of reducibility in the theory of algorithms*, Dokl. Akad. Nauk SSSR **108** (1956), 194–197.

5. G. Sacks and S. Simpson, *The α-finite injury method*, Ann. Math. Logic **4** (1972), 323–367.

6. E. Post, *Recursively enumerable sets of positive integers and their decision problem*, Bull. Amer. Math. Soc. **50** (1944), 284–316.

7. T. Slaman, *The E-recursively enumerable degrees are dense*, these PROCEEDINGS.

8, ______, *The extended plus-one hypotheses—a relative consistency result*, Nagoya Math. J. **92** (1983) (to appear).

9. G. Sacks, *The limits of E-recursive enumerability*, Nagoya Math. J. **92** (1983), 107–120.

10. G. Sacks and T. Slaman, *Inadmissible forcing*, Ann. Pure Appl. Logic (to appear).

11. G. Sacks, *The k-section of a type n object*, Amer. J. Math., **99** (1977), 901–917.

12. L. Harrington, *Contributions to recursion theory on higher types*, Ph.D. Thesis, M.I.T., Cambridge, Mass., 1973.

13. Y. Moschovakis, *Hyperanalytic predicates*, Trans. Amer. Math. Soc. **138** (1967), 249–282.

14. T. Slaman, *Forcing and reflection in E-recursion*, Ann. Pure and Appl. Logic (to appear).

15. C. Spector, *Recursive wellorderings*, J. Symbolic Logic **20** (1955), 151–163.

16. E. Griffor and D. Normann, *Effective cofinalities and admissibility in E-recursion*, preprint, University of Oslo, 1982.

17. R. Shore, *The recursively enumerable α-degrees are dense*, Ann. Math. Logic **9** (1976), 123–155.

DEPARTMENT OF MATHEMATICS, HARVARD UNIVERSITY, CAMBRIDGE, MASSACHUSETTS 02138

MASSACHUSETTS INSTITUTE OF TECHNOLOGY, CAMBRIDGE, MASSACHUSETTS 02139

Reprinted from Annals of Pure and Applied Logic 31 (1986) 87–120 with kind permission of Elsevier Science - NL, Sara Burgerhartstraat 25, 1055 KV Amsterdam, The Netherlands.

THE LIMITS OF E-RECURSIVE ENUMERABILITY

Gerald E. SACKS*

Harvard University, Massachusetts Institute of Technology, Cambridge, MA 02139, USA

Communicated by A. Nerode
Received 1 September 1984

A classical result of Kleene is: $2^\omega \cap \text{HYP}$ is recursively enumerable in ${}^2E$. Restated in E-recursion theory, it says: $2^\omega \cap E(\omega)$ is E-recursively enumerable in ω. Forcing, compactness and fine structure techniques decide various possible extensions of Kleene's result. Suppose $L(\kappa)$ is E-closed but not Σ_1 admissible. Let cf and gc denote cofinality and greatest cardinal in the sense of $L(\kappa)$. Then (i) $\leftrightarrow$ (ii). (i) $2^{\text{gc}(\kappa)} \cap L(\kappa)$ is E-recursively enumerable in some $a \in L(\kappa)$. (ii) $\omega < \text{cf}(\text{gc}(\kappa)) < \text{gc}(\kappa)$; or $\text{cf}(\text{gc}(\kappa)) = \omega$ and every amenable subset of $\text{gc}(\kappa)$ belongs to $L(\kappa)$. Thus (i) fails if $\text{gc}(\kappa)$ is regular (in $L(\kappa)$), or if $\text{cf}(\text{gc}(\kappa)) = \omega$ and κ is countable.

1. Introduction

This paper studies the question: is every naturally enumerable set recursively enumerable? At the lowest level, where 'set' means set of integers, Church's thesis seems to yield an affirmative answer. At many higher levels the main theorem (9.1) below appears to imply a negative answer. Examples of naturally enumerable sets abound in Gödel's L. Perhaps the simplest one beyond ω is $2^\omega \cap L(\omega_1^{\text{ck}})$, the set of all hyperarithmetic reals. Its natural enumeration is tied to the iteration of the Turing jump through the recursive ordinals. A classical result of Kleene [1] is:

(1) $2^\omega \cap L(\omega_1^{\text{ck}})$ is recursively enumerable in ${}^2E$.

The Kleene result, restated in the language of E-recursion, is:

(2) $2^\omega \cap E(\omega)$ is E-recursively enumerable in ω.

$E(\omega)$ is the E-closure of ω. The E-closure of a set x is the least transitive set with x as an element and closed under application of all partial E-recursive functions. The latter were discovered by Normann [2], and independently by Moschovakis. They extended Kleene's notion of recursion in normal objects of finite types to objects of all types, namely sets. Normann's definitions and fundamental results are reviewed in Section 2.

One generalization of (2) is:

(3) $2^{2^\omega} \cap E(2^\omega)$ is E-recursively enumerable in 2^ω, a for some $a \in 2^\omega$.

* The preparation of this paper was supported in part by National Science Foundation Grant MCS-81-04266. The author is grateful to Theodore Slaman for his generous assistance, and to the University of Chicago for its hospitality during the Spring of 1982.

It was shown in [3] that (3) is false if

(4) $E(2^\omega) \models [2^\omega$ is wellordered & and card(2^ω) is regular].

On the other hand (3) is a consequence of the axiom of determinateness.

Another generalization of (2) is:

(5) $2^{\mathrm{gc}(\kappa)} \cap L(\kappa)$ is E-recursively enumerable in some $a \in L(\kappa)$.

gc(κ) is the greatest cardinal in the sense of $L(\kappa)$, if there is one, and there is one if $L(\kappa)$ is not Σ_1 admissible. It is shown in Section 6 that (5) is false if $L(\kappa)$ is E-closed and not Σ_1 admissible, and gc(κ) is a regular cardinal in the sense of $L(\kappa)$. The assumption of Σ_1 inadmissibility is inspired by the fact that $E(2^\omega)$ is Σ_1 inadmissible, a result of Moschovakis [4]. Theorem 9.1 provides a complete resolution of the problem of enumerability in the inadmissible case.

Theorem 9.1. *Suppose $L(\kappa)$ is E-closed and not Σ_1 admissible. Then* (i) *and* (ii) *are equivalent.*

(i) $2^{\mathrm{gc}(\kappa)} \cap L(\kappa)$ *is E-recursively enumerable in some $a \in L(\kappa)$.*

(ii) $L(\kappa) \models [\omega < \mathrm{cofinality}(\mathrm{gc}(\kappa)) < \mathrm{gc}(\kappa)]$; *or* $L(\kappa) \models [\omega = \mathrm{cofinality}(\mathrm{gc}(\kappa))]$ *and every amenable subset of* gc(κ) *belongs to* $L(\kappa)$.

($x \subseteq \mathrm{gc}(\kappa)$ is *amenable* if $(x \cap \delta) \in L(\mathrm{gc}(\kappa))$ for all $\delta < \mathrm{gc}(\kappa)$.) The admissible case of the problem of enumerability is still under study. Some partial results are given in Section 9. One such is:

(5, Section 9) If z is a set of ordinals and $E(z)$ is Σ_1 admissible, then $E(z)$ is E-recursively enumerable in some $a \in E(z)$.

It is reasonable to regard (5, Section 9) as a more suitable generalization of (2) than (3) or (5). Some evidence in favor of this view is the selection theorem of [5], which implies $E(z)$ is Σ_1 admissible if z is a set of ordinals and $E(z) \models [\mathrm{cofinality}(\mathrm{gc}(\kappa)) = \omega]$.

The proof of Theorem 9.1 breaks into three cases defined by the cofinality of gc(κ) inside $L(\kappa)$.

(a) $L(\kappa) \models [\mathrm{gc}(\kappa)$ is regular]. In Section 6 it is shown that $2^{\mathrm{gc}(\kappa)} \cap L(\kappa)$ is not E-recursively enumerable in any $a \in L(\kappa)$. The argument is based on a forcing technique developed in Section 5, and geared to producing E-closed, generic extensions of E-closed structures. (In general set-generic extensions of E-closed, initial segments of L are not E-closed. For contrast recall that set-generic extensions of Σ_1 admissible sets are always Σ_1 admissible.) Section 4 is devoted to the verification of a fine structure fact needed to complete the forcing argument of Section 6 when $L(\kappa)$ is uncountable.

(b) $L(\kappa) \models [\omega < \mathrm{cofinality}(\mathrm{gc}(\kappa)) < \mathrm{gc}(\kappa)]$. In Section 8 it is shown that $2^{\mathrm{gc}(\kappa)} \cap L(\kappa)$ is E-recursively enumerable in some $a \in L(\kappa)$. The work of Silver [6] and S. Friedman [7, 8] on singular cardinals of uncountable cofinality is applied to show that Gödel's order of constructibility function O, restricted to $2^{\mathrm{gc}(\kappa)} \cap L(\kappa)$, is E-recursive in some $d \in L(\kappa)$.

(c) $L(\kappa) \models [\omega = \mathrm{cofinality}(\mathrm{gc}(\kappa))]$. In Section 7 it is shown that $2^{\mathrm{gc}(\kappa)} \cap L(\kappa)$ is not E-recursively enumerable in any $a \in L(\kappa)$, if there exists an amenable subset of $\mathrm{gc}(\kappa)$ outside $L(\kappa)$. The proof is inspired by Green's compactness and type-omitting theorems [9] for structures such as the Σ_1 admissible closure of Beth_ω.

The principal result of Section 3 is the existence of Moschovakis witnesses (to divergence) in $L(\kappa, x)$, if x is a set of ordinals and $L(\kappa, x)$ is E-closed but not Σ_1 admissible. The witnesses are needed for the forcing arguments of Sections 5 and 9.

Section 10 lists several open questions.

2. E-recursion

The partial E-recursive functions are defined by the Normann schemes [2]. $\{e\}$ denotes a partial E-recursive function with index e. $x_1, \dots, x_n$ are arbitrary sets.

(1) $\{e\}(x_1, \dots, x_n) = x_i$ if $e = \langle 1, n, i \rangle$.

(2) $\{e\}(x_1, \dots, x_n) = x_i - x_j$ if $e = \langle 2, n, i, j \rangle$.

(3) $\{e\}(x_1, \dots, x_n) = \{x_i, x_j\}$ if $e = \langle 3, n, i, j \rangle$.

(4) $\{e\}(x_1, \dots, x_n) \simeq \bigcup \{\{c\}(y, x_2, \dots, x_n) \mid y \in x_1\}$ if $e = \langle 4, n, c \rangle$. (In (4) $\{e\}(x_1, \dots, x_n)$ is not defined unless $\{c\}(y, x_2, \dots, x_n)$ is defined for all $y \in x_1$.)

(5) $\{e\}(x_1, \dots, x_n) \simeq \{c\}(\{d_1\}(x_1, \dots, x_n), \dots, \{d_m\}(x_1, \dots, x_n))$ if $e = \langle 5, n, m, c, d_1, \dots, d_m \rangle$.

(6) $\{e\}(c, x_1, \dots, x_n, y_1, \dots, y_m) \simeq \{c\}(x_1, \dots, x_n)$ if $e = \langle 6, n, m \rangle$.

In brief the Normann schemes are the rudimentary schemes for *partial* functions and an enumeration scheme. If $\{e\}(x)$ is defined (i.e. has a value), then $\{e\}(x)$ is said to converge, in symbols $\{e\}(x)\downarrow$. Otherwise $\{e\}(x)$ is said to diverge, in symbols $\{e\}(x)\uparrow$.

An important derived scheme, called *bounding*, is:

(7) $\{2^m \cdot 3^n\}(x) \simeq \{\{n\}(y) \mid y \in \{m\}(x)\}$.

Note that $\{2^m \cdot 3^n\}(x)\downarrow$ iff $\{m\}(x)\downarrow$ and $\{n\}(y)\downarrow$ for all $y \in \{m\}(x)$. A typical inductive definition or proof of this paper, if given in detail, would have six cases corresponding to the six Normann schemes. To simplify matters, it will be assumed from now on that there is only one scheme of genuine interest, the above bounding scheme. It has been formulated so as to combine the essential features of (4) and (5). (4) is central because it is the sole source of infinitely long computations.

The schemes give rise to the universal computation tree $>_{\mathrm{u}}$. Each node of the tree is of the form $\langle e, x_1, \dots, x_n \rangle$, or more simply, $\langle e, x \rangle$, and is called a computation instruction. $a >_{\mathrm{u}} b$ is read: b is a subcomputation instruction of a, and is defined by: there exists a finite sequence $b_0, \dots, b_n$ such that $a = b_0$, $b_n = b$, and b_{i+1} is an *immediate subcomputation instruction* of b_i $(i < n)$. The definition of immediate subcomputation instruction has a clause corresponding to each Normann scheme, but as promised above, only the bounding scheme will be considered.

(8) $\langle m, x\rangle$ is an immediate subcomputation instruction of $\langle 2^m \cdot 3^n, x\rangle$.

(9) If $\{m\}(x)\downarrow$ and $y \in \{m\}(x)$, then $\langle n, y\rangle$ is an immediate subcomputation instruction of $\langle 2^m \cdot 3^n, x\rangle$.

(10) If c is not the index of a Normann scheme, then $\langle c, x\rangle$ is an immediate subcomputation instruction of $\langle c, x\rangle$.

The tree-like object consisting of $\langle e, x\rangle$ and that part of $>_u$ below $\langle e, x\rangle$ is called the *computation* of $\{e\}(x)$. It is readily verified that $\{e\}(x)\downarrow$ iff the computation of $\{e\}(x)$ is wellfounded.

Let A be a class and x a set. A is said to be *E-recursively enumerable in x* if for some e,

$$A = \{y \mid \{e\}(x, y)\downarrow\}.$$

(A is E-recursively enumerable if $x = 0$.) B is E-recursively enumerable in x on C if $B = A \cap C$ for some A E-recursively enumerable in x.

The method of definition by effective transfinite recursion yields a partial E-recursion function f such that for all e and x, if $\{e\}(x)\downarrow$, then $f(e, x)\downarrow$ and equals the computation tree of $\{e\}(X)$. In addition there exists a partial E-recursive function $|\{e\}(x)|$ such that if $\{e\}(x)\downarrow$, then $|\{e\}(x)|$ is defined and equal to the ordinal height of the computation of $\{e\}(x)$.

Brief review of the method of definition by effective transfinite recursion: Let $<_R$ be a wellfounded relation and I a total recursive function from ω into ω. Suppose for all $e < \omega$, and all x in the field of $<_R$, that $\{I(e)\}(x)\downarrow$ whenever $\{e\}(y)\downarrow$ for all $y <_R x$. Then there exists a partial E-recursive function $\{c\}$ such that $\{c\}(x)\downarrow$ for all x in the field of $<_R$ and $\{I(c)\} \simeq \{c\}$. The existence of $\{c\}$ follows from scheme (6) by the usual fixed point argument.

Also by (6) the relation $\{e\}(x)\downarrow$ is E-recursively enumerable, and consequently the relation, b is an immediate subcomputation instruction of c, is E-recursively enumerable. Two tempting *false* moves at this point are: the immediate subcomputation relation is E-recursive; the subcomputation relation $>_u$ is E-recursively enumerable.

Assume $\{2^m \cdot 3^n\}(x)$ converges for the sake of considering the definition of $|\langle 2^m \cdot 3^n, x\rangle|$ in detail. $|2^m \cdot 3^n, x\rangle| = \sigma$ if

(11) $(E\gamma)_{\gamma<\sigma}[|\langle m, x\rangle| = \gamma]$,

(12) $(y)(E\tau)_{\tau<\sigma}[y \in \{m\}(x) \rightarrow |\langle n, y\rangle| = \tau]$, and

(13) $(\tau)_{\tau<\sigma}(Ey)[y \in \{m\}(x) \ \& \ |\langle n, y\rangle| \geq \tau]$.

A set y is said to be E-recursive in x if $y = \{e\}(x)$ for some e. Thus $|\langle e, x\rangle|$ is E-recursive in x when $\{e\}(x)\downarrow$. In symbols, $y \leq_E x$.

κ_0^x is the supremum of the set of all ordinals recursive in x.

κ^x is the supremum of the set of all ordinals γ such that γ is recursive in x, a, where $a \in \mathrm{TC}(x)$, the transitive closure of x.

Suppose $\{e\}(x)\downarrow$ and $|\{e\}(x)| = \delta$. It is not difficult to check that the computation tree for $\{e\}(x)$ is first-order definable over

$$L(\delta, \mathrm{TC}(\{x\})).$$

Consequently each set recursive in x belongs to $L(\kappa_0^x, \mathrm{TC}(\{x\}))$. A set Z is *E-closed* if

$$x \in Z \;\&\; \{e\}(x)\downarrow \;\rightarrow\; \{e\}(x) \in Z$$

for all $e < \omega$. Define $E(x)$, the *E-closure* of x, to be the least E-closed Z such that $\mathrm{TC}(\{x\}) \in Z$.

If y is a first-order definable subset of $L(\gamma)$, then y is E-recursive in γ and the parameter from $L(\gamma)$ occuring in the definition of y. It follows that

$$(14) \quad E(x) = L(\kappa^x, \mathrm{TC}(\{x\})).$$

The above equation not only exposes the absolute character of E-recursion; it also makes possible a straightforward definition of κ_r^x, the greatest x-reflecting ordinal. An ordinal δ is said to be *x-reflecting* if

$$L(\delta, \mathrm{TC}(\{x\})) \models \mathscr{F} \rightarrow L(\kappa_0^x, \mathrm{TC}(\{x\})) \models \mathscr{F}$$

for every Σ_1 sentence $\mathscr{F}$ of ZF with parameter x. Clearly the supremum of x-reflecting ordinals is x-reflecting, and so κ_r^x is welldefined.

κ_r had its origin in the application of forcing to Kleene recursion in finite types, e.g. [10] where it was observed that

$$\kappa_r^{2^{2^\omega}} > \kappa_0^{2^{2^\omega}}.$$

Then Harrington [11] showed $\kappa_r^{2^\omega} > \kappa_0^{2^\omega}$, and more importantly, gave a characterization of $\kappa_r^{2^\omega, a}$ $(a \in 2^\omega)$ that inspired the results of the next section.

The dynamic meaning of κ_r is as follows. Let $A \subseteq \mathrm{TC}(\{x\})$ be E-recursively enumerable in x. Suppose some b is seen to be an element of A via a computation of height less than κ_r^x. Then some computations of height less than κ_0^x, hence less than some δ E-recursive in x, must establish some elements of A. The shortest such computations, since they are first-order definable over $L(\delta, \mathrm{TC}(\{x\}))$ with x as the only parameter, must be E-recursive in x. Thus A has a nonempty subset E-recursive in x.

$\kappa_r^{u,v}$ denotes $\kappa_r^{\langle u,v\rangle}$, and similarly for $\kappa_0^{u,v}$.

Note that if $a \in \mathrm{TC}(x)$, then $E(\langle a, x\rangle)$ is $E(x)$ and $\kappa^{x,a} = \kappa^x$.

Proposition 2.1. (i) $\kappa_r^{x,a} \leq \kappa^x$ *for all* $a \in \mathrm{TC}(x)$.

(ii) *If* $E(x)$ *is not* Σ_1 *admissible, then* $\kappa_r^{x,a} < \kappa^x$ *for some* $a \in \mathrm{TC}(x)$.

Proof. (i) Suppose $\kappa_r^{x,a} > \kappa^x$. Then $E(\langle a, x\rangle) \in L(\kappa_r^{x,a}, \mathrm{TC}(\{x\}))$, and by reflection, $E(\langle x, a\rangle) \in L(\kappa_0^{x,a}, \mathrm{TC}(\{x\}))$, an impossibility since $\kappa^x \geq \kappa_0^{x,a}$.

(ii) Suppose $\kappa_r^{x,a} = \kappa^x$ for all $a \in \mathrm{TC}(x)$, and

$$E(x) \models (u)_{u \in c}(Ev)D(u, v, \boldsymbol{p}),$$

where D is a Δ_0 formula of ZF with parameter p. It is safe to assume $p = x$ and $c \subseteq \mathrm{TC}(x)$. For example, let $p = \{e\}(x, b)$ for some $b \in \mathrm{TC}(x)$. Then $D(u, v, p)$ is equivalent to the following Σ_1 formula:

$$(E\gamma)(Eb)_{b \in \mathrm{TC}(x)}[|\{e\}(x, b)| = \gamma \;\&\; D(u, v, \{e\}(x, b))].$$

By reflection, for each $a \in c \subseteq \mathrm{TC}(x)$, there is a $\delta \leqslant_E x$, a such that

$$L(\delta, \mathrm{TC}(\{x\})) \models (Ev)D(\boldsymbol{a}, v, \boldsymbol{x}).$$

by the remarks following Lemma 2.2 (Gandy selection), δ can be taken to be a partial E-recursive function of x and a, defined for all $a \in c$. The bounding scheme implies δ is bounded above by some $\gamma \leqslant_E x$, c as a ranges over c. Hence there is a $d \in E(x)$ such that

$$E(x) \models (u)_{u \in c}(Ev)_{v \in d}D(u, v, \boldsymbol{x}),$$

since $\gamma \in E(x)$. □

Lemma 2.2 (Gandy Selection). *There exists a partial E-recursive function $\phi(e, x)$ such that for all $e < \omega$ and all x:*

$$(En)_{n<\omega}[\{e\}(n, x)\downarrow] \leftrightarrow \phi(e, x)\downarrow,$$

$$\phi(e, x)\downarrow \rightarrow \{e\}(\phi(e, x), x)\downarrow.$$

The proof of Lemma 2.2 is the same as it is in the setting of Kleene's theory of recursion in normal objects of finite type (e.g. Fenstad [12]).

Suppose $P(x, y)$ is an E-recursively enumerable predicate and

$$(x)(Ey)[y \text{ is } E\text{-recursive in } x \;\&\; P(x, y)].$$

Then

$$(x)(En)[\{n\}(x)\downarrow \;\&\; P(x, \{n\}(x))].$$

Gandy selection allows n to be construed as an E-recursive function of x. Thus there is an E-recursive f such that $(x)P(x, f(x))$.

3. Existence of Moschovakis witnesses

The universal computation tree $>_{\mathrm{u}}$ was defined in Section 2, where it was noted that $\{e\}(x)\downarrow$ iff $>_{\mathrm{u}}$ below $\langle e, x\rangle$ is wellfounded. If $\{e\}(x)\uparrow$, then any infinite path in $>_{\mathrm{u}}$ below $\langle e, x\rangle$ is called a Moschovakis witness to the divergence of $\{e\}(x)$. They were introduced by Moschovakis [4], who used them to show, among other things, that $E(2^\omega)$ is not Σ_1 admissible. An E-closed structure $\mathscr{E}$ is said to *admit M-witnesses* if for all $e < \omega$ and $x \in \mathscr{E}$: if $\{e\}(x)\uparrow$, then there exists an M-witness to the divergence of $\{e\}(x)$ in M. $E(2^\omega)$ admits M-witnesses, but $E(\omega)$ does not. The principal result of the present section is the existence of M-witnesses in $E(x)$ if x is a set of ordinals and $E(x)$ is not Σ_1 admissible. It will be needed in Section 5 to show a certain kind of forcing extension of $L(\kappa)$ is E-closed if $L(\kappa)$ is E-closed and Σ_1 inadmissible.

Lemma 3.1. *Suppose y is E-recursive in x, and A is E-recursively enumerable in x. If $y - A$ is nonempty, then there exists a $b \in y - A$ such that $\kappa_r^{x,b} \leqslant \kappa_r^x$* (Kechris–Harrington basis theorem [11]).

Proof. First observe that $\kappa_r^{x,b} > \kappa_r^x$ implies $\kappa_0^{x,b} > \kappa_r^x$. Otherwise Σ_1 statements about x would reflect from $\kappa_r^{x,b}$ down to κ_r^x. Thus it suffices to find a $b \in y - A$ such that $\kappa_r^x \geqslant \kappa_0^{x,b}$. Now suppose no such b exists. Then

$$y \subseteq A \cup \{b \mid \kappa_r^x < \kappa_0^{x,b}\}.$$

The predicate $\kappa_r^x < \kappa_0^{x,b}$ is E-recursively enumerable in x, since it is equivalent to

$$(E\delta)[\delta\ E\text{-rec. in } x, b\ \&\ \delta \text{ is not } x\text{-reflecting}],$$

which is the same as

$$(Ee)[\{e\}(x, b)\downarrow\ \&\ |\{e\}(x, b)| \text{ is not } x\text{-reflecting}],$$

and the latter is E-recursively enumerable in x by Lemma 2.2. (A standard corollary of Gandy selection is: if $R(e, b)$ is E-recursively enumerable, then so is $(Ee)R(e, b)$.)

Hence for each $b \in y$ there is a convergent computation c (E-recursive in b, x) such that either c puts b in A or the height of c exceeds κ_r^x. c can be construed as an E-recursive (in x) function of b by Gandy selection. The bounding scheme, (7) of Section 2, provides an upper bound γ on the height of c (for all $b \in y$) such that γ is E-recursive in x. Since $\gamma < \kappa_0^x \leqslant \kappa_r^x$, it follows that c always puts b in A. But then $y - A$ is empty contrary to hypothesis. □

Proposition 3.2. *Suppose some wellordering of* TC(x) *is E-recursive in x. If $\langle c, y\rangle$ is an immediate subcomputation instruction of $\langle e, x\rangle$, then some wellordering of* TC(y) *is E-recursive in y, x.*

Proof. Suppose $e = 2^m \cdot 3^n$ and $c = n$. Then $\{m\}(x)\downarrow$ and $y \in \{m\}(x)$. The computation tree for $\{m\}(x)$ is first-order definable over

(1) $L(\delta, \mathrm{TC}(\{x\}))$,

where δ is any ordinal $\geqslant |\{m\}(x)|$. The only parameter occurring in the definition is x. Let w be a wellordering of TC(x) E-recursive in x. Fix δ so that δ is E-recursive in x, $\delta \geqslant |\{m\}(X)|$, and w belongs to (1). Then w gives rise to a wellordering v of (1) definable over (1). (1) is E-recursive in x, hence the restriction of v to TC(y) is E-recursive in y, x □

Lemma 3.3. *Suppose some wellordering of* TC(x) *is E-recursive in x. If* $\{e\}(x)\uparrow$, *then some Moschovakis witness to the divergence of* $\{e\}(x)$ *is first-order definable over*

$$L(\kappa_r^x, \mathrm{TC}(\{x\})).$$

Proof. Related to arguments of Moschovakis [4] and Harrington [11]. Both rely on the effective equivalence of x and x^ω, true if $x = 2^\omega$, but false in the present context. The witness $\lambda t \mid \langle e_t, x_t \rangle$ is defined by recursion on $t < \omega$. $\langle e_0, x_0 \rangle$ is $\langle e, x \rangle$. Assume $\langle e_t, x_t \rangle$ has been defined so that

(1) $\{e_t\}(x_t)\uparrow$,

(2) $x_t \in L(\kappa_r^x, \mathrm{TC}(\{x\}))$, and

(3) $\kappa_r^x \geq \kappa_r^{x_0,\ldots,x_t}$.

For simplicity assume $e_t = 2^m \cdot 3^n$.

Suppose $\{m\}(x_t)\uparrow$. Then $\langle e_{t+1}, x_{t+1} \rangle$ is $\langle m, x_t \rangle$.

Assumption (3) implies $\kappa_r^x \geq \kappa_0^{x_t}$. Consequently $\{m\}(x_t)\downarrow$ iff there is a convergent computation for $\{m\}(x_t)$ in $L(\kappa_r^x, \mathrm{TC}(\{x\}))$.

Suppose $\{m\}(x_t)\downarrow$. It follows from Proposition 3.2 that there exists a wellordering w of $\{m\}(x_t)$ such that w is E-recursive in $x_0, \ldots, x_t$. Then $\langle e_{t+1}, x_{t+1} \rangle$ is $\langle n, y \rangle$, where y is the w-least u such that

(4) $u \in \{m\}(x_t)$ & $|\{n\}(u)| \geq \kappa_r^x$.

It has to be shown that

(5) $\{n\}(y)\uparrow$ & $\kappa_r^{x_0,\ldots,x_t} \geq \kappa_r^{x_0,\ldots,x_t,y}$.

By Lemma 3.1 there is a $z \in \{m\}(x_t)$ with the properties ascribed to y by (5); let z_0 be the w-least such z.

It sufficies to show $y = z_0$. Clearly $y \leq_w z_0$, since

$$|\{n\}(z_0)| = \infty > \kappa_r^x.$$

Suppose $y <_w z_0$. Consider, as in the proof of 2.1, the following E-recursive (in $x_0, \ldots, x_t, z_0$) enumeration of $\{z \mid z <_w z_0\}$:

(I) If $\{n\}(z)\downarrow$, then enumerate z.

(II) If $\kappa_r^{x_0,\ldots,x_t} < \kappa_r^{x_0,\ldots,x_t,z}$, then enumerate z.

Step (II) is effective because the hypothesis of (II) implies

$$\kappa_r^{x_0,\ldots,x_t} < \kappa_0^{x_0,\ldots,x_t,z}.$$

By the bounding scheme there is a γ recursive in $x_0, \ldots, x_t, z_0$ that bounds the height of all computations occurring in (I) or (II). It follows that (II) never applies to any $z <_w z_0$; otherwise

$$\kappa_r^{x_0,\ldots,x_t} \leq \gamma < \kappa_0^{x_0,\ldots,x_t,z_0} \leq \kappa_r^{x_0,\ldots,x_t}.$$

is a consequence of (5) with z_0 in place of y.

Hence (I) applies to every $z <_w z_0$, and in particular to y, so $|\{n\}(y)| \leq \gamma$. But

$$|\{n\}(y)| \geq \kappa_r^x \geq \kappa_r^{x_0,\ldots,x_t} \geq \kappa_0^{x_0,\ldots,x_t,z_0} > \gamma$$

according to (4) with y in place of u, (3), and (5) with z_0 in place of y. □

The notion of left-most path can be used to clarify the statement and proof of Lemma 3.3. Here 'left-most' refers to the fact that x_{t+1} is the w-least u satisfying (4). Let $T_{\langle e,x\rangle}$ be that part of the universal computation tree $>_{\mathrm{u}}$ that lies below $\langle e, x\rangle$. Define $>^{\gamma}_{\mathrm{u}}$ by replacing the notion of convergence in the definition of $>_{\mathrm{u}}$ by convergence via computations of height less than γ. Let $T^{\gamma}_{\langle e,x\rangle}$ be that part of $>^{\gamma}_{\mathrm{u}}$ that lies below $\langle e, x\rangle$.

Restatement of Lemma 3.3. *Suppose some wellordering of* TC(x) *is E-recursive in x. If* $\{e\}(x)\downarrow$, *then* $T^{\kappa^x_{\mathrm{r}}}_{\langle e,x\rangle}$ *is not wellfounded and its left-most, infinite, descending path is first-order definable over* $L(\kappa^x_{\mathrm{r}}, \mathrm{TC}(\{x\}))$.

The restatement of 3.3 leads to yet another ordinal associated with an arbitrary set z. Define

$$\theta^z = \mu\gamma(e)[\{e\}(z)\uparrow \rightarrow T^{\gamma}_{\langle e,z\rangle} \text{ is not wellfounded}].$$

Some early results of Harrington [11] for recursion in ${}^3E$ are equivalent to

$$\kappa_{\mathrm{r}}^{2^\omega,a} = \theta^{2^\omega,a} \quad \text{and} \quad \kappa_{\mathrm{r}}^{2^\omega,a,b} \geqslant \kappa_{\mathrm{r}}^{2^\omega,a}$$

for all $a, b \in 2^\omega$. Slaman [13] has found sets x and y such that

$$\kappa^x_{\mathrm{r}} > \theta^x \quad \text{and} \quad \kappa^{x,y}_{\mathrm{r}} < \kappa^x_{\mathrm{r}}.$$

Lemma 3.4. *Suppose some wellordering of* TC(x) *is E-recursive in x. Then for all u and v*:

(i) $\kappa^u_{\mathrm{r}} \geqslant \theta^u$.
(ii) $\theta^{u,v} \geqslant \theta^u$.
(iii) $\kappa^x_{\mathrm{r}} = \theta^x$.
(iv) $\kappa^{x,u}_{\mathrm{r}} \geqslant \kappa^x_{\mathrm{r}}$.

Proof. (i) Suppose $\{e\}(u)\uparrow$. Proceed as in the proof of Lemma 3.3 with u in place of x. Let u_{t+1} be a y such that

$$\{u\}(y)\uparrow \;\&\; \kappa_{\mathrm{r}}^{u_0,\ldots,u_t} \geqslant \kappa_{\mathrm{r}}^{u_0,\ldots,u_t,y}.$$

The existence of such a y again follows from Lemma 3.1. Now the M-witness $\lambda t \mid \langle e_t, u_t\rangle$ need not be first-order definable over $L(\kappa^u_{\mathrm{r}}, \mathrm{TC}(\{u\}))$, but it is still an infinite, descending path through $T^{\kappa^u_{\mathrm{r}}}_{\langle e,u\rangle}$.

(ii) Let $*$ have the property that $\{e^*\}(u, v) \simeq \{e\}(u)$ for all e, u and v. Then

$$T^{\gamma}_{\langle e^*,\langle u,v\rangle\rangle} = T^{\gamma}_{\langle e,u\rangle}$$

for all γ.

(iii) There exists a recursive function h with the following properties:

(a) $\{e\}(x)\downarrow \leftrightarrow \{h(e)\}(x)\downarrow$.

(b) If $\{h(e)\}(x)\uparrow$, then $T_{\langle h(e),x\rangle}$ has a *unique* infinite, descending path.

Intuitively, if $\{e\}(x)\uparrow$, then $T_{\langle h(e),x\rangle}$ consists of the left-most infinite, descending path of $T_{\langle e,x\rangle}$ and everything in $T_{\langle e,x\rangle}$ to the left of that path. h is defined by not allowing the Normann schemes to continue beyond the left-most path. Suppose $e = 2^m \cdot 3^n$. The immediate subcomputation instructions of $\langle h(e), x\rangle$ are as follows.

(c) $\langle h(m), x\rangle$ is an immediate subcomputation instruction of $\langle h(e), x\rangle$.

(d) Suppose $\{h(m)\}(x)\downarrow$. Let $<_w$ be the wellordering of $\{h(m)\}(x)$, E-recursive in x, as developed in the proof of Proposition 3.2. If $\{h(n)\}(z)\downarrow$ for all $z <_w y$, then $\langle h(n), y\rangle$ is an immediate subcomputation instruction of $\langle h(e), x\rangle$.

Assume $\{e\}(x)\uparrow$. Then $T^{\theta^x}_{\langle h(e),x\rangle}$ has a unique infinite, descending path t that witnesses the divergence of $\{h(e)\}(x)$, and hence by (a), the divergence of $\{e\}(x)$. The wellfounded part of $T^{\theta^x}_{\langle h(e),x\rangle}$ to the left of t has rank at most θ^x. It follows that t is first-order definable over $L(\theta^x, x)$. Now assume $\theta^x < \kappa^x_r$. Then by reflection there is a witness to the divergence of $\{e\}(x)$ that is E-recursive in x. Thus $\{c \mid \{x\}(x)\downarrow\}$, the complete E-r.e. subset of ω, is E-rec. in x, a contradiction.

(iv) $\kappa^{x,u}_r \geq \theta^{x,u} \geq \theta^x = \kappa^x_r$ by (i), (ii) and (iii). □

Lemma 3.5. *Suppose x is a set of ordinals, and $L(\kappa, x)$ is E-closed but not Σ_1 admissible. Then $\kappa^{x,y}_r < \kappa$ for all $y \in L(\kappa, x)$.*

Proof. Suppose $\kappa^{x,y}_r \geq \kappa$ for some $y \in L(\kappa)$ with the intent of showing $L(\kappa, x)$ Σ_1 admissible. Let

$$L(\kappa, x) \models (u)_{u\in \boldsymbol{d}}(Ev)\mathcal{F}(u, v),$$

where $\mathcal{F}$ is a Δ_0 formula. For simplicity assume $\mathcal{F}$ has no parameters. For the moment assume

(1) $\kappa^{x,y,b}_r \geq \kappa^{x,y}_r$ for all $b \in d$.

Then for each $b \in d$, there is a $c \in L(\kappa^{x,y,b}_r, x)$ such that

$$L(\kappa, x) \models \mathcal{F}(\boldsymbol{b}, \boldsymbol{c});$$

by reflection there is such a c in $L(\gamma, \mathrm{TC}(\{x, y, b\}))$ for some $\gamma \leq_E x, y, b$. By Gandy selection γ can be taken to be a partial E-recursive function f of x, y and b defined for all $b \in d$. The bounding scheme implies $f(x, y, b) \leq \delta$ for some $\delta \leq_E x, y, d$, as b ranges over d. Hence

$$L(\kappa, x) \models (u)_{u\in \boldsymbol{d}}(Ev)_{v\in \boldsymbol{p}}\mathcal{F}(u, v),$$

where p is $L(\delta, \mathrm{TC}(\{x, y, b\}))$.

Now for the proof of (1). By Lemma 3.4(iv) it suffices to find a wellordering of $\mathrm{TC}(\{x, y\})$ E-recursive in x, y. Since x is a set of ordinals and $\kappa^{x,y}_r \geq \kappa$, there is a wellordering of $\mathrm{TC}(\{x, y\})$ in $L(\kappa^{x,y}_r, x)$. By reflection there is one E-recursive in x, y. □

Theorem 3.6. *Let x be a set of ordinals. Then* (i) $\leftrightarrow$ (ii) $\leftrightarrow$ (iii).

(i) $\kappa_r^{x,y} \in E(x)$ *for all* $y \in E(x)$.
(ii) $E(x)$ *admits Moschovakis witnesses.*
(iii) $E(x)$ *is* Σ_1 *inadmissible.*

Proof. (iii) → (i) by Lemma 3.5.

Assume (i) to prove (ii). Suppose $y \in E(x)$ and $\{e\}(y)\uparrow$. There exists a set w of ordinals such that $w \in E(x)$, $y \leqslant_E w$ and $\kappa_r^{x,w} < \kappa$. By Lemma 3.3 there is an M-witness to $\{e\}(y)\uparrow$ first-order definable over $L(\kappa_r^{x,w}, \mathrm{TC}(\{x, w\}))$.

Assume (ii) holds and (iii) fails. For each $e < \omega$ and $a \in \mathrm{TC}(x)$, let $f(e, a, x)$ be the least γ such that: $|\{e\}(a, x)| < \gamma$ or there is an M-witness to the divergence of $\{e\}(a, x)$ first-order definable over $L(\gamma, \mathrm{TC}(\{x\}))$. f is Σ_1 over $E(x)$, hence bounded by some $\gamma_0 \in E(x)$. But $E(x) \subseteq L(\gamma_0, \mathrm{TC}(\{x\})) \in E(x)$. □

Theorem 3.7. *Suppose x is a set of ordinals and $L(\kappa, x)$ is E-closed. If $L(\kappa, x)$ is not Σ_1 admissible, then $L(\kappa, x)$ admits Moschovakis witnesses.*

Proof. Suppose $y \in L(\kappa, x)$ and $\{e\}(y)\uparrow$. Proceed as in the proof of (iii) → (i) of Theorem 3.6. □

Slaman [13] has constructed an E-closed, Σ_1 inadmissible set that does not admit Moschovakis witnesses.

4. Amenable singularities

In this section a five structure fact is derived straightforwardly from the work of Jensen as presented in Devlin [14]. The fact is needed in Section 6 to circumvent the difficulty of finding a generic extension of $L(\kappa)$ when κ is uncountable. Let λ be a limit ordinal. A *singular sequence* through λ is a cofinal subset of λ whose ordertype is less than λ. A set $S \subseteq \lambda$ is said to be *amenable* if $(S \cap \delta) \in L(\lambda)$ for all $\delta < \lambda$.

Theorem 4.1. *Assume $V = L$. If λ is a singular limit ordinal, then there exists an amenable, singular sequence through λ.*

Proof. It is safe to assume $\omega\lambda = \lambda$. Thus $L(\lambda) = J_\lambda$. It can be shown that the $<_J$-least singularity through λ is amenable, however a less substantial analysis suffices to prove the theorem.

Let β be the least $\theta \geqslant \lambda$ such that some singular sequence through λ is definable over J_θ.

Case I: $\lambda \leqslant \rho_\beta^m$ *for all* $m \geqslant 0$. Let S be a singular sequence through λ, Σ_n definable over J_β. Then S is an amenable subset of $J_{\rho_\beta^n}$; and so for each

$\delta<\lambda$, $(S\cap\delta)\in J_\beta$. Either $\beta=\lambda$ or

$$J_\beta\models[\lambda \text{ is a cardinal}].$$

Either way $(S\cap\delta)\in J_\lambda$.

Case II: $\rho^n_\beta<\lambda\leqslant\rho^{n-1}_\beta$ *for some* $n>0$. Then there exists a partial $\Sigma_1(\langle J_{\rho^{n-1}_\beta}, A^{n-1}_\beta\rangle)$ map from $\omega\rho^n_\beta$ onto $J_{\rho^{n-1}_\beta}$. Consequently there is a partial $\Sigma_1(\langle J_{\rho^{n-1}_\beta}, A^{n-1}_\beta\rangle)$ map g from σ_0 into λ cofinally such that $\sigma_0\leqslant\omega\rho^n_\beta$. Assume σ_0 is as small as possible. Thus

$$\sigma_0=\Sigma_1(\langle J_{\rho^{n-1}_\beta}, A^{n-1}_\beta\rangle) \text{ cofinality of } \lambda.$$

For the moment assume

(1) $\sigma_0\leqslant\Sigma_1(\langle J_{\rho^{n-1}_\beta}, A^{n-1}_\beta\rangle)$ cofinality of $\omega\rho^{n-1}_\beta$.

It follows that

(2) $x<\sigma_0\rightarrow(g\restriction x)\in J_{\rho^{n-1}_\beta}$.

To verify (2) let h be $g\restriction x$. By (1) there is a $\gamma<\omega\rho^{n-1}_\beta$ such that $h\subseteq S_\gamma$ and the existential witnesses needed for the $\Sigma_1(\langle J_{\rho^{n-1}_\beta}, A^{n-1}_\beta\rangle)$ definition of h belong to S_γ. Hence h is Σ_1 definable over $\langle S_\gamma, A^{n-1}_\beta\cap S_\gamma\rangle$. But S_γ and $A^{n-1}_\beta\cap S_\gamma$ belong to $J_{\rho^{n-1}_\beta}$, so $h\in J_{\rho^{n-1}_\beta}$.

For $x<\sigma_0$, let $k(x)$ be sup range$(g\restriction x)$. Let S be the range of k. If $\delta<\lambda$, then (2) implies $(S\cap\delta)\in J_{\rho^{n-1}_\beta}$. Hence $(S\cap\delta)\in J_\beta$, and as in case I, $(S\cap\delta)\in J_\lambda$.

To prove (1) let σ_1 denote the right side of (1), and let t be a partial $\Sigma_1(\langle J_{\rho^{n-1}_\beta}, A^{n-1}_\beta\rangle)$ map from σ_1 into $\omega\rho^{n-1}_\beta$ cofinally. For each $y<\sigma_1$ let $g_0(y)$ be the supremum of all $g(x)$ such that: $x<\sigma_0$, $g(x)$ and $t(y)$ are defined, $g(x)\in S_{t(y)}$, and the existential witness needed for the $\Sigma_1(\langle J_{\rho^{n-1}_\beta}, A^{n-1}_\beta\rangle)$ definition of $g(x)$ belong to $S_{t(y)}$.

$g_0(y)<\lambda$. Otherwise there is a singular sequence S' through λ definable over $S_{t(y)}$. Let $t(y)=\omega\alpha+m$. Then S' is definable over J_α, contrary to the definition of β since $\alpha<\beta$.

g_0 is a partial $\Sigma_1(\langle J_{\rho^{n-1}_\beta}, A^{n-1}_\beta\rangle)$ map from σ_1 into λ cofinally. But then $\sigma_1\geqslant\sigma_0$. □

5. Countably closed forcing

Assume $L(\kappa)$ is E-closed and Σ_1 inadmissible. A well known condensation argument shows there exists a greatest cardinal in the sense of $L(\kappa)$; denote it by gc(κ). Let $G\subseteq$ gc(κ). The connections made in Section 2 (following the definition of κ^x) between convergent computations and the L hierarchy imply that $L(\kappa, G)$ equals

$$\{\{e\}(a, G)\mid a\in L(\kappa)\ \&\ |\{e\}(a, G)|<\kappa\}.$$

Consequently the forcing language $\mathscr{L}(\kappa, \mathscr{G})$, which will be used to study $L(\kappa, G)$

when G is generic, need only describe the computation of $\{e\}(a, G)$ when $a \in L(\kappa)$ and $|\{e\}(a, G)| < \kappa$.

The primitive terms of $\mathscr{L}(\kappa, \mathscr{G})$ are $\boldsymbol{a}$ ($a \in L(\kappa)$) and $\mathscr{G}$. A general term is $\{e\}(t_1, \ldots, t_n)$, where $t_1, \ldots, t_n$ are terms. Each term t is equivalent to one of normal form $\{e\}(\boldsymbol{a}, \mathscr{G})$. A unique *normal form* for t can be found effectively from t by elementary syntactical manipulations. Note that in general a term t does not name an element of $L(\kappa, G)$. Rather t names a computation. If that computation converges via an ordinal less than κ, then t names the resulting element of $L(\kappa, G)$. $\mathscr{L}(\kappa, \mathscr{G})$ has the following additional primitive symbols: $\boldsymbol{\sigma}$ ($\sigma < \kappa$), $\in$, $|\ |$, set and ordinal variables, quantifiers and propositional connectives.

A forcing condition is a characteristic function with domain $\delta < \mathrm{gc}(\kappa)$ that belongs to $L(\kappa)$. Forcing conditions are denoted by $p, q, r, \ldots$; $p \geqslant q$ means that domain $p \subseteq$ domain q and $q \restriction (\operatorname{dom} p)$ equals p. The set $\mathscr{P}$ of all forcing conditions belongs to $L(\kappa)$, since the usual Gödelian argument shows each bounded subset of $\mathrm{gc}(\kappa)$ in $L(\kappa)$ is constructed via an ordinal less than $\mathrm{gc}(\kappa)$. Thus $\mathscr{P}$ is E-recursive in $\mathrm{gc}(\kappa)$.

Three objects are defined simultaneously by recursion on $\sigma < \kappa$:

(i) the forcing relation $p \Vdash |\{e\}(\boldsymbol{a}, \mathscr{G})| = \boldsymbol{\sigma}$;

(ii) the set $\mathscr{T}(p, e, a, \mathscr{G}, \sigma)$ of terms;

(iii) the forcing relation $q \Vdash s \in \{e\}(\boldsymbol{a}, \mathscr{G})$.

(ii) will be defined only if $p \Vdash |\{e\}(\boldsymbol{a}, \mathscr{G})| = \boldsymbol{\sigma}$; in that event the elements of (ii) will suffice to name the elements of $\{e\}(a, G)$ for every generic G that satisfies p. (iii) will be defined only if (i) holds, $p \geqslant q$ and $s \in \mathscr{T}(p, e, a, \mathscr{G}, \sigma)$.

Each major clause of the recursion corresponds to a Normann scheme. Only the bounding scheme is of interest.

(1) $p \Vdash |\{2^m \cdot 3^n\}(\boldsymbol{a}, \mathscr{G})| = \boldsymbol{\sigma}$ iff

(1a) $(E\gamma)_{\gamma < \boldsymbol{\sigma}}[p \Vdash |\{m\}(\boldsymbol{a}, \mathscr{G})| = \boldsymbol{\gamma}]$,

(1b) $p \Vdash (s)(E\tau)_{\tau < \sigma}[s \in \mathscr{T}(p, m, a, \mathscr{G}, \gamma)\ \&$
$s \in \{m\}(\boldsymbol{a}, \mathscr{G}) \rightarrow |\{n\}(s)| = \tau]$, and

(1c) $p \Vdash (\tau)_{\tau < \boldsymbol{\sigma}}(Es)[s \in \mathscr{T}(p, m, a, \mathscr{G}, \gamma)\ \&\ s \in \{m\}(\boldsymbol{a}, \mathscr{G})\ \&\ |\{n\}(s)| \geqslant \tau]$.

(2) $\mathscr{T}(p, 2^m \cdot 3^n, a, \mathscr{G}, \sigma) = \{\{n\}(s) \mid s \in \mathscr{T}(p, m, a, \mathscr{G}, \gamma)\}$.

(3) $q \Vdash \{n\}(s) \in \{2^m \cdot 3^n\}(\boldsymbol{a}, \mathscr{G})$ iff $p \geqslant q, s \in \mathscr{T}(p, m, a, \mathscr{G}, \gamma)$ and
$q \Vdash [s \in \{m\}(a, \mathscr{G})\ \&\ (E\tau)_{\tau < \sigma}(|\{n\}(s)| = \tau)]$.

Quantifiers and propositional connectives are forced in the usual manner. p forces $(Ex)\mathscr{F}(x)$ if p forces $\mathscr{F}(i)$ for some appropriate instance i, p forces $\sim\mathscr{F}$ if no $q \leqslant p$ forces $\mathscr{F}$; and p forces $(x)\mathscr{F}$ if p forces $\sim(Ex)\sim\mathscr{F}$.

Some clarification of (1)–(3) is in order. Suppose p forces $\{2^m \cdot 3^n\}(\boldsymbol{a}, \mathscr{G})$ to converge via a computation of height σ. Then by definition of forcing, and in a manner consistent with the assignment of ordinals to computations (Section 2, (11)–(13)), p forces $m(\boldsymbol{a}, \mathscr{G})$ to converge via a computation of height γ for some

$\gamma < \sigma$. Since $\gamma < \sigma$, the set $\mathcal{T}(p, m, a, \mathcal{G}, \gamma)$ of terms, and the relation q forces $s \in \{m\}(\boldsymbol{a}, \mathcal{G})$ (where $q \leqslant p$ and $s \in \mathcal{T}(p, m, a, \mathcal{G}, \gamma)$), have already been defined. In addition the relation q forces $|\{n\}(S)| = \boldsymbol{\tau}$ has already been defined for all q and s, and all $\tau < \sigma$. (1b) can be rendered as

$$(1b^*)\quad p \Vdash (x)(E\tau)_{\tau<\sigma}[x \in \{m\}(\boldsymbol{a}, \mathcal{G}) \rightarrow |\{n\}(s)| = \tau],$$

if it is understood that x ranges over the terms in $\mathcal{T}(p, m, a, \mathcal{G}, \gamma)$. Since, as already noted, quantifiers and connectives are forced in the usual manner, no further analysis of (1b*) is needed. (1c) is dealt with similarly. With respect to (2), if the elements of $\mathcal{T}(p, m, a, \mathcal{G}, \gamma)$ suffice to name the elements of $\{m\}(a, G)$, then certainly the elements of $\mathcal{T}(p, 2^m \cdot 3^n, a, \mathcal{G}, \sigma)$ suffice for those of $\{2^m \cdot 3^n\}(a, G)$.

There are two points worth noting in definition (3). First q is required to extend p $(p \geqslant q)$ so that q will force $\{2^m \cdot 3^n\}(\boldsymbol{a}, \mathcal{G})$ to converge, and hence to be a set. Second q is required to force $\{n\}\{s\}$ to converge, and hence to be a set, and to converge via an ordinal $\tau < \sigma$ for the sake of the recursion.

The above approach to the definition of $\Vdash$ is termed 'from the top down'. Another approach, 'from the bottom up', can be found in Slaman [15], where the hierarchial view is maintained throughout.

If t is a term of $\mathcal{L}(\kappa, \mathcal{G})$, then $p \Vdash |\{e\}(T)| = \boldsymbol{\sigma}$ means $p \Vdash |\{e_t\}(\boldsymbol{a}_t, \mathcal{G})| = \boldsymbol{\sigma}$, where $\{e_t\}(\boldsymbol{a}_t, \mathcal{G})$ is the normal form of $\{e\}(t)$. Similarly

$$\mathcal{T}(p, e, t, \sigma) = \mathcal{T}(p, e_t, a_t, \mathcal{G}, \sigma).$$

Lemma 5.1. *The relations, $p \Vdash |\{e\}(t)| = \boldsymbol{\sigma}$ and $q \Vdash s \in \{e\}(t)$, and the set $\mathcal{T}(p, e, t, \sigma)$ $(p \geqslant q$ and $s \in \mathcal{T}(p, e, t, \sigma))$, are E-recursive in σ,* gc(κ) *(uniformly in σ).*

Proof. By effective transfinite recursion on $\check{\sigma}$. gc(κ) is needed because $\mathcal{P}$ is E-recursive in gc(κ). □

The *weak forcing* relation, $p \Vdash^* \mathcal{F}$, is given by $p \Vdash \sim\sim \mathcal{F}$. The notion of weak forcing plays a curious, but essential, role in the computation of a bound on $|\{e\}(t)|$ when p forces $|\{e\}(t)|$ to be an ordinal less than κ. The computation procedure succeeds only if a bound is sought on $|\{e\}(t)|$ when p weakly forces $(E\sigma)[|\{e\}(t)| = \sigma]$. (Note that an unbounded quantifier on σ ranges over the ordinals less than κ.)

The relation $>_{\mathrm{V}}$ is a forcing-theoretic counterpart of $>_{\mathrm{u}}$, the universal computation tree. $>_{\mathrm{V}}$ is termed the tree of possibilities, because it encompasses all values of $>_{\mathrm{u}}$ for all generic G's. A node of $>_{\mathrm{V}}$ is of the form $\langle p, e, t\rangle$, where p is a forcing condition and t is a term of $\mathcal{L}(\kappa, \mathcal{G})$. A preliminary definition of

$$\langle p, e, t\rangle >_{\mathrm{V}} \langle q, n, s\rangle$$

is $p \geqslant q$ and $q \Vdash^* \langle e, t\rangle >_{\mathrm{u}} \langle n, s\rangle$. To complete the definition of $>_{\mathrm{V}}$, it need only be indicated how the forcing of atomic sentences about $>_u$ can be reduced to the forcing of sentences from $\mathscr{L}(\kappa, \mathscr{G})$.

Typical details parallel the definition of $>_u$ given in Section 2, (8)–(10).

(8_V) $p \Vdash \langle 2^m \cdot 3^n, t\rangle >_u \langle m, t\rangle$ (all p and t).

(9_V) If $p \Vdash |\{m\}(t)| = \gamma$, $s \in \mathscr{T}(p, m, t, \gamma)$ and $p \Vdash s \in \{m\}(t)$, then

$$p \Vdash \langle 2^m \cdot 3^n, t\rangle >_u \langle n, s\rangle.$$

(10_V) If c is not the index of a Normann scheme, then $p \Vdash \langle c, t\rangle >_u \langle c, t\rangle$ (all p and t).

$>_V$ will turn out to give global properties similar to those of $>_u$. Assume that $\mathrm{gc}(\kappa)$ has uncountable cofinality in $L(\kappa)$. It will be seen that

(11) the immediate predecessor relation in $>_V$ is E-recursively enumerable in $\mathrm{gc}(\kappa)$ on $L(\kappa)$, and

(12) if $p \Vdash^* (E\sigma)[|\{e\}(t)| = \sigma]$, then that part of $>_V$ below $\langle p, e, t\rangle$ is wellfounded and E-recursive in $\langle p, e, t\rangle$, $\mathrm{gc}(\kappa)$.

It should be noted that (11) and (12) are not true for every set of forcing conditions that belongs to $L(\kappa)$. A counterexample can be obtained by taking $L(\kappa)$ to be $E(\omega_1)$ and Levy-collapsing ω_1 to ω. (11) and (12) are true in this paper because $\mathscr{P}$ is *countably closed*:

(13) $(n)[p_n \geqslant p_{n+1}] \to (Eq)(n)[p_n \geqslant q]$.

Of course $\mathscr{P}$ is countably closed in the present context because $\mathrm{gc}(\kappa)$ has uncountable cofinality in $L(\kappa)$. The part played by (13) is discussed further in [5].

It is (12) that gives rise to the desired bound on $|\{e\}(t)|$ when p weakly forces $(E\sigma)[|\{e\}(t)| = \sigma]$. The bound is roughly equal to the height of $>_V$ below $\langle p, e, t\rangle$.

Lemma 5.2. *Suppose* $p \Vdash^* (E\sigma)[|\{e\}(t)| = \sigma]$. *If* $>_V$ *is wellfounded below* $\langle p, e, t\rangle$, *then*

$$p \Vdash^* |\{e\}(t) \leqslant \gamma$$

for some γ *E-recursive in* p, t, $\mathrm{gc}(\kappa)$.

Proof. γ is computed by an effective transfinite recursion on that part of $>_V$ below $\langle p, e, t\rangle$. The only interesting case is $e = 2^m \cdot 3^n$. By recursion there is a δ-recursive in p, t, $\mathrm{gc}(\kappa)$ such that

$$p \Vdash^* |\{m\}(t)| \leqslant \boldsymbol{\delta},$$

since $\langle p, 2^m \cdot 3^n, t\rangle >_V \langle p, m, t\rangle$. Consider the set K defined by:

$$\langle p', \sigma\rangle \in K \leftrightarrow p \geqslant p' \;\&\; \sigma \leqslant \delta \;\&\; p' \Vdash |\{m\}(t)| = \boldsymbol{\sigma}.$$

K is the set of all possible outcomes for $|\{m\}(t)|$ consistent with p. By Lemma 5.1, K is E-recursive in p, t, $\mathrm{gc}(\kappa)$.

Fix $\langle p', \sigma \rangle \in K$ and $s \in \mathcal{T}(p', m, t, \sigma)$ in order to bound $|\{n\}(s)|$. Then

$$p' \Vdash^* (E\beta)[s \in \{m\}(t) \to |\{n\}(s)| = \beta].$$

Let J be $\{q \mid p' \geq q \,\&\, q \Vdash s \in \{m\}(t)\}$. By Lemma 5.1, J is E-recursive in $p', s, t, \sigma, \mathrm{gc}(\kappa)$. For each $q \in J$, there is by recursion a ρ such that

$$q \Vdash^* |\{n\}(s)| \leq \boldsymbol{\rho}$$

and ρ is E-recursive in $q, s, \mathrm{gc}(\kappa)$. Gandy selection (cf. remarks following Lemma 2.2) makes it possible to think of ρ as a partial E-recursive function of $q, s, \mathrm{gc}(\kappa)$.

As q varies over J, ρ is strictly bounded above by some ρ_J E-recursive in $J, s, \mathrm{gc}(\kappa)$. This last is a consequence of the bounding scheme. As s varies over $\mathcal{T}(p', m, t, \sigma)$, ρ_J is strictly bounded above by some $\rho_{\mathcal{T}}$ E-recursive in $p', t, \sigma, \mathrm{gc}(\kappa)$, since $\mathcal{T}(p', m, t, \sigma)$ is E-recursive in $p', t, \sigma, \mathrm{gc}(\kappa)$ by Lemma 5.1.

As $\langle p', \sigma \rangle$ varies over K, $\rho_{\mathcal{T}}$ is strictly bounded by some ρ_K E-recursive in $p, t, \mathrm{gc}(\kappa)$. Thus

$$p \Vdash^* (s)[s \in \{m\}(t) \to |\{n\}(s)| \leq \rho_K].$$

Let $\gamma = \sup^+(\delta, \rho_K)$. $\square$

Lemma 5.3. *If* $p \Vdash^* (E\sigma)[|\{n\}(t)| = \sigma]$, *then* $>_{\mathrm{V}}$ *is wellfounded below* $\langle p, e, t \rangle$.

Proof. The proof of Lemma 5.2 is largely a description of an E-recursive procedure for computing an ordinal γ from $p, e, t, \mathrm{gc}(\kappa)$. The procedure is welldefined whether or not $>_{\mathrm{V}}$ is wellordered below $\langle p, e, t \rangle$. Thus Lemma 5.2 can be rephased as follows. There exists a partial E-recursive (in $\mathrm{gc}(\kappa)$) function $\{g\}$ such that: if

$$p \Vdash^* (E\sigma)[|\{e\}(t)| = \sigma]$$

and $>_{\mathrm{V}}$ is wellfounded below $\langle p, e, t \rangle$, then $\{g\}(p, e, t)\downarrow$ and

$$p \Vdash^* |\{e\}(t)| \leq \{g\}(p, e, t).$$

The proof of the present lemma is in brief a comparison of $>_{\mathrm{u}}$ below $\langle g, \langle p, e, t \rangle \rangle$ with $>_{\mathrm{V}}$ below $\langle p, e, t \rangle$.

If $\{g\}$ converges on $\langle p, e, t \rangle$, then $>_{\mathrm{V}}$ is well founded below $\langle p, e, t \rangle$. This is because $\{g\}$ assigns an 'ordinal rank' to the nodes of $>_{\mathrm{V}}$ below $\langle p, e, t \rangle$. The value of $\{g\}$ on each $w \leq_{\mathrm{V}} \langle p, e, t \rangle$ is a *strict* upper bound on the values of $\{g\}$ on the predecessors of w in $>_{\mathrm{V}}$.

Now suppose $\{g\}$ diverges on $\langle p, e, t \rangle$. By Theorem 3.7 there is a Moschovakis witness z in $L(\kappa)$ to the divergence. Thus $z_0 = \langle g, \langle p, e, t \rangle \rangle$ and $z_n >_{\mathrm{u}} z_{n+1}$ for all n. There must be a z_m such that

$$z_m = \langle g, \langle q, n, s \rangle \rangle \quad \text{and} \quad \langle p, e, t \rangle >_{\mathrm{V}} \langle q, n, s \rangle.$$

The existence of z_m follows from the description of $\{g\}$ given in the proof of 5.2.

Since $\{g\}(p, e, t)\uparrow$, it must be that $\{g\}(q, n, s)\uparrow$ for some $\langle q, n, s\rangle$ an immediate predecessor of $\langle p, e, t\rangle$ in $>_{\mathrm{V}}$. (Note that z_m need not be an immediate predecessor of z_0 in $>_{\mathrm{u}}$, since the computation of $\{g\}$ at a given node w of $>_{\mathrm{V}}$ includes some rudimentary steps apart from the computation of $\{g\}$ at immediate precessors of w in $>_{\mathrm{V}}$.)

Similarly there is an $m' > m$ such that

$$z_{m'} = \langle g, \langle q', n', s'\rangle\rangle \quad \text{and} \quad \langle q, n, s\rangle >_{\mathrm{V}} \langle q', n', s'\rangle.$$

Thus z includes a built-in, infinite descending sequence z' in $>_{\mathrm{V}}$. If z is defined as in the proof of 3.3, then z' is seen to be defined simultaneously with z. Thus $z' \in L(\kappa)$. Let $z'_0 = \langle p, e, t\rangle$ and $z'_n = \langle p_n, e_n, t_n\rangle$. Then

$$\langle p_n, e_n, t_n\rangle >_{\mathrm{V}} \langle p_{n+1}, e_{n+1}, t_{n+1}\rangle$$

for all n. Since $\mathcal{P}$ is countably closed ((13) following 5.1), there is a q such that $p_n \geqslant q$ for all n. Then

$$q \Vdash^* (n)[\langle e_n, t_n\rangle >_{\mathrm{u}} \langle e_{n+1}, t_{n+1}\rangle].$$

Thus q weakly forces the existence of a Moschovakis witness to the divergence of $\{e\}(t)$, an impossibility since $q \leqslant p$ and p weakly forces $\{e\}(t)$ to converge. □

Lemma 5.4. *If $p \Vdash^* (E\sigma)[|\{e\}(t)| = \sigma]$, then there exists a γ E-recursive in p, t, $\mathrm{gc}(\kappa)$ such that*

$$p \Vdash^* |\{e\}(t)| \leqslant \boldsymbol{\gamma}.$$

Proof. Immediate from 5.2 and 5.3. □

The next lemma implies that $L(\kappa, G)$ is E-closed when G is sufficiently generic.

Lemma 5.5 (bounding). *Suppose $p \Vdash |\{m\}(t)| = \boldsymbol{\rho}$, and*

$$p \Vdash (s)(E\tau)[s \in \mathcal{T}(p, m, t, \rho) \;\&\; s \in \{m\}(t) \rightarrow |\{n\}(s)| = \tau].$$

Then $p \Vdash |\{2^m \cdot 3^n\}(t)| = \boldsymbol{\delta}$ for some δ E-recursive in p, t, $\mathrm{gc}(\kappa)$.

Proof. By Lemma 5.4, $\rho \leqslant \gamma_0$, where γ_0 is E-recursive in p, t, $\mathrm{gc}(\kappa)$. It follows from Lemma 5.1 that ρ, $\mathcal{T}(p, m, t, \rho)$ and the relation $s \in \{m\}(t)$ ($s \in \mathcal{T}(p, m, t, \rho)$) are E-recursive in p, t, $\mathrm{gc}(\kappa)$.

For each $s \in \mathcal{T}(p, m, t, \rho)$, the set of all q such that

(1) $\quad p \geqslant q \;\&\; q \Vdash s \in \{m\}(t)$

is E-recursive in p, t, $\mathrm{gc}(\kappa)$, s. For each such q,

$$q \Vdash^* |\{n\}(S)| \leqslant \boldsymbol{\gamma}$$

for some γ E-recursive in q, s, $\mathrm{gc}(\kappa)$ by Lemma 5.4. According to the remarks following Lemma 2.2 (Gandy selection), γ can be taken to be a partial E-recursive function of q, s, $\mathrm{gc}(\kappa)$. Two applications of the bounding scheme supply an upper bound δ on γ, for all $s \in \mathcal{T}(p, m, t, \rho)$ and all q satisfying (1), such that δ is E-recursive in p, t, $\mathrm{gc}(\kappa)$. □

Let $\mathcal{F}$ be a sentence of $\mathcal{L}(\kappa, \mathcal{G})$ and G a subset of $\mathrm{gc}(\kappa)$. G satisfies p, in symbols $G \in p$, if p is an initial segment of the characteristic function of G. G is said to be *generic on* $\mathcal{F}$ if there exists a p such that $G \in p$ and either $p \Vdash \mathcal{F}$ or $p \Vdash \sim\mathcal{F}$. G is said to be *sufficiently generic* if (i)–(iii) hold.

(i) G is generic on all sentences of the form

(2) $(E\sigma)[|\{e\}(t)| = \sigma]$.

(ii) If $p \Vdash |\{m\}(t)| = \boldsymbol{\rho}$, and $G \in p$, then G is generic on

(3) $(s)(E\tau)[s \in \mathcal{T}(p, m, t, \rho) \;\&\; s \in \{m\}(t) \rightarrow |\{n\}(s)| = \tau]$.

(iii) If G is generic on $\mathcal{F}$, then G is generic on every instance of every subformula of $\mathcal{F}$.

Explication of (iii): $|\{e\}(t)| = \boldsymbol{\gamma}$ is an instance of a subformula of (2). For each $s \in \mathcal{T}(p, m, t, \rho)$,

$$(E\tau)[s \in \{m\}(t) \rightarrow |\{n\}(s)| = \tau]$$

is an instance of a subformula of (3).

Theorem 5.6. *If G is sufficiently generic, then $L(\kappa, G)$ is E-closed.*

Proof. Suppose that $e < \omega$ and $a \in L(\kappa)$ are such that $\{e\}(a, G)\downarrow$ and $|\{e\}(a, G)| \leq \kappa$. As usual the only interesting possibility for e is $2^m \cdot 3^n$. Since G is sufficiently generic, there is a p and a $\sigma < \kappa$ such that $G \in p$,

$$p \Vdash |\{m\}(\boldsymbol{a}, \mathcal{G})| = \boldsymbol{\sigma},$$

and

$$p \Vdash (s)(E\tau)[s \in \mathcal{T}(p, m, s, \mathcal{G}, \sigma) \;\&\; s \in \{m\}(\boldsymbol{a}, \mathcal{G}) \rightarrow |\{n\}(s)| = \tau].$$

By Lemma 5.5 there is a δ E-recursive in p, t, $\mathrm{gc}(\kappa)$ such that

$$p \Vdash |\{2^m \cdot 3^n\}(\boldsymbol{a}, \mathcal{G})| = \boldsymbol{\delta}.$$

Then $|\{e\}(\boldsymbol{a}, G)| = \delta$. (Note that $\delta < \kappa$ because δ is E-recursive in p, t, $\mathrm{gc}(\kappa) \in K(\kappa)$, and $L(\kappa)$ is E-closed.) □

Theorem 5.6 is all that is needed for the applications of forcing made in this paper. The remainder of the present section is intended for purposes of clarification only.

Lemma 5.7. *The relation, $p \Vdash^* (E\sigma)[|\{e\}(t)| = \sigma]$, is E-recursively enumerable (in $\mathrm{gc}(\kappa)$) on $L(\kappa)$.*

Proof. $p \Vdash^* (E\sigma)[|\{e\}(t)| = \sigma]$ is equivalent to

(1) $\{g\}(p, e, t)\downarrow \;\&\; p \Vdash^* |\{e\}(t)| \leq \{g\}(p, e, t)$,

where $\{g\}$ is the partial E-recursive (in gc(κ)) function central to the proof of Lemma 5.3. By Lemma 5.1, (1) is E-recursively enumerable (in gc(κ)) on $L(\kappa)$.

□

Lemma 5.8. *Let $a \in L(\kappa)$. Suppose p does not weakly force $(E\sigma)[|\{e\}(\boldsymbol{a}, \mathscr{G})| = \sigma]$. Then there exists a $q \leq p$ such that q forces the existence of a Moschovakis witness (to $\{e\}(\boldsymbol{a}, \mathscr{G})\uparrow$) first-order definable over $L(\kappa_r^{a,p,\mathrm{gc}(\kappa)}, \mathscr{G})$.*

Proof. Recall $\{g\}$, the partial E-recursive (in gc(κ)) function occurring in the proof of Lemma 5.3. Observe that if $\{g\}(p, e, t)\downarrow$, then p weakly forces $|\{e\}(t)|$ to be at most $\{g\}(p, e, t)$. This can be verified by transfinite induction on $>_{\mathrm{V}}$, since $>_{\mathrm{V}}$ below $\langle p, e, t\rangle$ is wellfounded when $\{g\}(p, e, t)\downarrow$ (as was shown in the proof of 5.3).

Suppose $\{g\}(p, e, \langle \boldsymbol{a}, \mathscr{G}\rangle)\uparrow$. According to Lemma 3.3 the divergence of $\{g\}$ is witnessed by some z first-order definable over $L(\kappa_r^{p,a,\mathrm{gc}(\kappa)}, a)$. As in the proof of 5.3, z has a built-in, infinite descending sequence z' in $>_{\mathrm{V}}$. In fact the definition of z over $L(\kappa_r^{p,a,\mathrm{gc}(\kappa)}, a)$, as performed in the proof of 3.3, includes a definition of z' over $L(\kappa_r^{p,a,\mathrm{gc}(\kappa)}, a)$. $z_0' = \langle p, e, \langle \boldsymbol{a}, \mathscr{G}\rangle\rangle$, $z_n' = \langle p_n, e_n, t_n\rangle$ and

$$\langle p_n, e_n, t_n\rangle >_{\mathrm{V}} \langle p_{n+1}, e_{n+1}, t_{n+1}\rangle$$

for all n. Let $q \leq p_n$. Then q weakly forces $\lambda_n \mid \langle e_n, t_n\rangle$ to be a Moschovakis witness (to $\{e\}(\boldsymbol{a}, \mathscr{G})\uparrow$) first-order definable over $L(\kappa_r^{a,p,\mathrm{gc}(\kappa)}, \mathscr{G})$. □

Proposition 5.9. *There exists a c such that for all x: $\{x\}(x)\uparrow$ and there is no Moschovakis witness (to $\{c\}(x)\uparrow$) E-recursive in x.*

Proof. Let t be a recursive function such that for all e and x: $\{t(e)\}(x)\downarrow$ iff there exists an M-witness (to $\{e\}(x)\uparrow$) E-recursive in x. (The existence of t follows from Gandy selection, Lemma 2.2.)

Let c be a fixed point of t. Thus $\{c\}(x) \simeq \{t(c)\}(x)$. If $\{c\}(x)\downarrow$, then there exists an M-witness to $\{c\}(x)\uparrow$. Hence $\{c\}(x)\uparrow$, and so there is no M-witness (to $\{c\}(x)\uparrow$) E-recursive in x. □

Theorem 5.10 (T. Slaman [15]). *Assume $a \in \kappa$. If*

$$p \Vdash^* \kappa_r^{a,\mathscr{G}} > \boldsymbol{\delta},$$

then $\delta < \kappa_r^{a,p,\mathrm{gc}(\kappa)}$.

Proof. Let $\{c\}$ be as in Proposition 5.9. Then p cannot weakly force $\{c\}(\boldsymbol{a}, \mathscr{G})\downarrow$.

It follows from Lemma 5.8 that there is $q \leqslant p$ such that q forces the existence of a Moschovakis witness w (to $\{c\}(\boldsymbol{a}, \mathcal{G})\uparrow$) first-order definable over $L(\kappa_r^{a,p,\mathrm{gc}(\kappa)}, \mathcal{G})$. By 5.9, q weakly forces that no M-witness to $\{c\}(\boldsymbol{a}, \mathcal{G})\uparrow$ is E-recursive in $\langle \boldsymbol{a}, \mathcal{G}\rangle$. Hence q weakly forces w not be a member of $L(\kappa_r^{\boldsymbol{a},\mathcal{G}}, \mathcal{G})$. (By reflection, if there is an M-witness before κ_r, then there is one before κ_0, hence an E-recursive one.) Therefore q weakly forces $\kappa_r^{a,\mathcal{G}}$ to be at most $\kappa_r^{a,p,\mathrm{gc}(\kappa)}$. $\square$

All the arguments of the present section go smoothly for any notion of set forcing that is countably closed. A general treatment of countably closed set forcing over E-closed structures is given in [5], where a special case of countably closed, class forcing is also discussed. In the proof of Lemma 5.3 above, it was shown that if the attempt to compute a bound on $|\{e\}(t)|$ fails, then some q forces a Moschovakis witness (to $\{e\}(t)\uparrow$) into existence. The q in question lies below a infinite, descending sequence of forcing conditions, hence the need for countable closure.

In [16] it is proved that countable chain, set forcing over an E-closed structure preserves E-closure. This result, coupled with the fact that iterated countable chain forcing satisfies the countable chain condition, is applied by Slaman [17] to obtain a model of ZFC in which the continuum is a singular cardinal, yet the extended 2-section of ${}^4E$ is the extended 2-section of some normal, type 3 object.

The properties of countable chain forcing are established by a kind of selection argument. The tree of possibilities, $>_V$, is of no use. The results obtained are similar to, but stronger than, Lemma 5.4 and Theorem 5.10. Lemma 5.4 becomes: if there is a $q \leqslant p$ and a $\gamma < \kappa$ such that $q \Vdash |\{e\}(t)| = \gamma$, then some such q and γ are E-recursive in p, t, $\mathcal{P}$. ($\mathcal{P} = \langle P, \geqslant\rangle$ is the forcing relation.) Theorem 5.10 becomes: The null condition weakly forces $\kappa_r^{\boldsymbol{a},\mathcal{G}} \leqslant \kappa_r^{a,\mathcal{P}}$.

The bounds on $\kappa_r^{\boldsymbol{a},\mathcal{G}}$ for countably closed, and countable chain, forcing suggest it may be difficult to force κ_r to behave wildly.

6. Non-enumerability via forcing

Recall from Section 1 if $L(\kappa)$ is Σ_1 inadmissible, then there exists a greatest cardinal in the sense of $L(\kappa)$ denoted by $\mathrm{gc}(\kappa)$. In this section the forcing technique of Section 5 and the fine structure fact of Section 4 combine to show:

Theorem 6.1. *If $L(\kappa)$ is E-closed but not Σ_1 admissible, and*

$$L(\kappa) \models [\mathrm{gc}(\kappa) \textit{ is regular}],$$

then $2^{\mathrm{gc}(\kappa)} \cap L(\kappa)$ is not E-recursively enumerable in any $a \in L(\kappa)$.

Proof. Suppose in hope of a contradiction that

$$2^{\mathrm{gc}(\kappa)} \cap L(k) = \{x \mid \{e\}(a, x)\downarrow\}.$$

By Theorem 5.6, the null condition $\emptyset$ weakly forces $L(\kappa, \mathscr{G})$ to be E-closed. It follows from the absolute nature of the proof of Theorem 3.7 that $\emptyset$ weakly forces $L(\kappa, \mathscr{G})$ to admit Moschovakis witnesses. Thus either (1) or (2) holds.

(1) $\quad \emptyset \Vdash^* (E\sigma)[|\{e\}(\boldsymbol{a}, \mathscr{G})| = \alpha]$.

(2) There exist q, z, and $\gamma < \kappa$ such that q weakly forces: z is an M-witness to $\{e\}(\boldsymbol{a}, \mathscr{G})\uparrow$ first-order definable over $L(\gamma, \mathscr{G})$.

If (1) holds, then there will be a $G \subseteq \mathrm{gc}(\kappa)$ such that $\{e\}(a, G)\downarrow$ but $G \notin L(\kappa)$.

If (2) holds, then there will be a $G \subseteq \mathrm{gc}(\kappa)$ such that $G \in L(\kappa)$ but $\{e\}(a, G)\uparrow$.

Assume that (1) holds. By Lemma 5.4 there is a limit ordinal $\lambda < \kappa$ such that

(3) $\quad \phi \Vdash^* |\{e\}(\boldsymbol{a}, \mathscr{G})| < \lambda$.

$\mathscr{L}_\lambda(\mathscr{G})$, the set of all sentences of $\mathscr{L}(\kappa, \mathscr{G})$ of rank $< \lambda$, is defined as follows. (As usual only the bounding scheme is considered; $e = 2^m \cdot 3^n$.)

(4) (i) $(\mathrm{E}\sigma)_{\sigma<\lambda}[|\{e\}(\boldsymbol{a}, \mathscr{G})| = \sigma]$ belongs to $\mathscr{L}_\lambda(\mathscr{G})$.

(ii) If $\rho < \lambda$ and $p \Vdash [|\{2^m \cdot 3^n\}(t) = \boldsymbol{\rho}]$, then

$$(s)(E\tau)_{\tau<\boldsymbol{\rho}}[s \in \mathscr{T}(p, m, t, \rho) \;\&\; s \in \{m\}(t) \rightarrow |\{n\}(s)| = \tau],$$

$$(\tau)_{\tau<\boldsymbol{\rho}}(Es)[s \in \mathscr{T}(p, m, t, \rho) \;\&\; s \in \{m\}(t) \;\&\; |\{n\}(s)| \geq \tau],$$

and $(E\sigma)_{\sigma<\boldsymbol{\rho}}[|\{m\}(t)| = \sigma]$ belong to $\mathscr{L}_\lambda(\mathscr{G})$.

(iii) Every instance of every subformula of every sentence in $\mathscr{L}_\lambda(\mathscr{G})$ belongs to $\mathscr{L}_\lambda(\mathscr{G})$. (For explication of (iii) see the remarks just before Theorem 5.6.)

$\mathscr{L}_\lambda(\mathscr{G})$ has enough expressive power to give a complete account of the computation of $\{e\}(\boldsymbol{a}, \mathscr{G})$. By Lemma 5.1, $\mathscr{L}_\lambda(\mathscr{G})$ can be construed as an element of $L(\kappa)$. Let $f \in L(\kappa)$ be a map from $\mathrm{gc}(\kappa)$ onto $\mathscr{L}_\lambda(\mathscr{G})$.

There exists an $S \subseteq \mathrm{gc}(\kappa)$ such that $S \notin L(\kappa)$ but $(S \cap \delta) \in L(\kappa)$ for every $\delta < \mathrm{gc}(\kappa)$. If $\mathrm{gc}(\kappa)$ is not a regular cardinal of L, then S is provided by Theorem 4.1. If $\mathrm{gc}(\kappa)$ is a regular L-cardinal, then S exists because κ is not Σ_1 admissible, hence less than the next L-cardinal after $\mathrm{gc}(\kappa)$.

The desired G is generic on every sentence in $\mathscr{L}_\lambda(\mathscr{G})$, and encodes S. The characteristic function of G is a union of forcing conditions $p_\delta(\delta < \mathrm{gc}(\kappa))$ defined by recursion on δ.

(5) $\quad p_0 = \emptyset. \qquad p_\gamma = \bigcup \{p_\beta \mid \beta < \gamma\} \quad$ if γ is a limit.

$q_\delta =$ least $q \geq p_\delta$ such that $q \Vdash f_\delta$ or $q \Vdash {\sim} f_\delta$.

$$p_{\delta+1} = \begin{cases} q_\delta \;\langle 0 \rangle & \text{if } \delta \in S, \\ q_\delta \;\langle 1 \rangle & \text{if } \delta \notin S. \end{cases}$$

(6) $\quad p_\delta \in \mathscr{P}$, and

(7) $\quad \{p_\gamma \mid \gamma \leq \delta\} \leq_E S \cap \delta, \mathrm{gc}(\kappa), \lambda, f, \delta \qquad$ (uniformly in δ).

When δ is a limit, then (6) holds because $\mathrm{gc}(\kappa)$ is regular and S is amenable. When δ is a successor, then (7) follows from Lemma 5.1, since $(S \cap \delta) \in L(\kappa)$.

Thus G is well defined. By (3) and (4), $\{e\}(\boldsymbol{a}, G)\downarrow$. $G \notin L(\kappa)$ because $S \notin L(\kappa)$ and

(8) $\quad S \leqslant_E G, \mathrm{gc}(\kappa), \lambda, f.$

(8) is a consequence of

(9) $\quad \langle p_\delta, S \cap \delta \rangle \leqslant_E G, \mathrm{gc}(\kappa), \lambda, f, \delta \qquad$ (uniformly in δ);

and (9) is proved by recursion on δ. (The coding of S into G via (5) implies S and G have the same E-degree (modulo $\mathrm{gc}(\kappa), \lambda, f$).)

Now assume (2) holds. Thus q weakly forces: there exist c, d and $\delta < \kappa$ such that

(10) (i) $\{c\}(\boldsymbol{\gamma}, \mathscr{G})\downarrow$ and $\{d\}(\boldsymbol{\delta}, \mathscr{G})\downarrow$,
(ii) $\{c\}(\boldsymbol{\gamma}, \mathscr{G})$ is an M-witness to $\{e\}(\boldsymbol{a}, \mathscr{G})\uparrow$, and
(iii) $\{d\}(\boldsymbol{\delta}, \mathscr{G})$ is an ω-sequence of computations that establish the E-recursive enumerability facts need to verify (ii). (The relation, x is an M-witness to the divergence of y, is E-recursively enumerable.)

By Lemma 5.4 there is a $\lambda < \kappa$ and a $q_0 \leqslant q$ such that q_0 forces all computations mentioned in (10) to have height less than λ. Repeat the construction of G given in (5) with $p_0 = q_0$ and $S = 0$. Then G is generic on every sentence of $\mathscr{L}_\lambda(\mathscr{G})$, and so $\{e\}(a, G)\uparrow$ by (10). It follows from (7) that $G \leqslant_E \mathrm{gc}(\kappa), \lambda, f$. Hence $G \in L(\kappa)$. $\square$

7. Countable cofinality

It was noted in Section 1 that $L(\kappa)$ is enumerable when $\mathrm{cf}(\mathrm{gc}(\kappa)) = \omega$ and $L(\kappa) = E(a)$ for some $a \in L(\kappa)$. In this section compactness and type-omitting arguments of Green [9] are adapted to obtain the non-enumerability of $L(\kappa) \cap 2^{\mathrm{gc}(\kappa)}$ when $\mathrm{cf}(\mathrm{gc}(\kappa)) = \omega$, $L(\kappa)$ is Σ_1 inadmissible, and some amenable subset of $\mathrm{gc}(\kappa)$ exists outside $L(\kappa)$. Recall that $Z \subseteq \lambda$ is *amenable* if $(Z \cap \delta) \in L(\lambda)$ for every $\delta < \lambda$. (It would be interesting to know if the non-enumerability result of Section 6 could be obtained by a compactness, rather than forcing, argument.)

Theorem 7.1. *Suppose $L(\kappa)$ is E-closed, not Σ_1 admissible, and*

$$L(\kappa) \models \mathrm{cf}(\mathrm{gc}(\kappa)) = \omega.$$

Then (i) *and* (ii) *are equivalent.*

(i) *There exists an amenable subset of* $\mathrm{gc}(\kappa)$ *outside* $L(\kappa)$.

(ii) $L(\kappa) \cap 2^{\mathrm{gc}(\kappa)}$ *is not E-recursively enumerable in any* $a \in L(\kappa)$.

Assume that $L(\kappa)$ satisfies the three initial suppositions of Theorem 7.1. If $L(\kappa)$ is countable, then (i) holds, and so $L(\kappa)\cap 2^{\mathrm{gc}(\kappa)}$ is not E-recursively enumerable in any $a\in L(\kappa)$. If $L(\kappa)$ is uncountable, then the truth value of (i) is uncertain. If there exists a non-constructible real, then (i) holds. If (in V) the cardinality of $\mathrm{gc}(\kappa)$ has countable cofinality, then (i) holds. If $V=L$, then there exists an uncountable κ such that (i) fails (and the initial suppositions of 7.1 hold).

Proof of 7.1. Assume (i) fails. Then (ii) fails because $2^{\mathrm{gc}(\kappa)}\cap L(\kappa)$ equals the set of all amenable subsets of $\mathrm{gc}(\kappa)$.

Assume (i) holds. Suppose in hope of a contradiction that

$$L(\kappa)\cap 2^{\mathrm{gc}(\kappa)}=\{x\mid \{e\}(a,x)\downarrow\},$$

where $a\in L(\kappa)$. It is safe to assume $a\subseteq \mathrm{gc}(\kappa)$, since any member of $L(\kappa)$ can be encoded as a subset of $\mathrm{gc}(\kappa)$. Then $E(\{\mathrm{gc}(\kappa),a\})$ is Σ_1 admissible. This last follows from a selection theorem proved in [5, Corollary 5.2]. (It also follows from an earlier selection theorem of Moschovakis [18], studied in Kirousis [19], if it is noted that the Moschovakis argument, given for $a=0$, also works when a belongs to $L(\kappa)\cap 2^{\mathrm{gc}(\kappa)}$.) Thus

$$E(\{\mathrm{gc}(\kappa),\alpha\})=L(\kappa_1),$$

where κ_1 is a Σ_1 admissible ordinal less than κ. In addition κ_1^*, the $\Sigma_1^{\kappa_1}$ projection of κ_1, is $\mathrm{gc}(\kappa)$. By Lemma 7.4, there is an amenable $G\subseteq \mathrm{gc}(\kappa)$ such that $G\notin L(\kappa)$ and $L(\kappa_1,G)$ is Σ_1 admissible. Hence $\{e\}(a,G)\uparrow$, and $\kappa_r^G\leqslant\kappa^G\leqslant\kappa_1$. Lemma 3.3 implies there is an M-witness to the divergence of $\{e\}(a,G)$ first-order definable over $L(\kappa_1,G)$.

Let $b\in 2^{\mathrm{gc}(\kappa)}\cap L(\kappa)$ encode a map of $\mathrm{gc}(\kappa)$ onto κ_1. Then $E(\{\mathrm{gc}(\kappa),b\})$ is $L(\kappa_2)$, where κ_2 is Σ_1 admissible and $\kappa_1<\kappa_2<\kappa$. Also $\kappa_2^*=\mathrm{gc}(\kappa)$. There exists an infinitary sentence $\mathscr{F}\in\mathscr{L}(\kappa)$ that describes the construction over $L(\kappa_1,G)$ of an M-witness to the divergence of $\{e\}(a,G)$. The initial clauses of $\mathscr{F}$ are: $\mathscr{G}\subseteq\mathrm{gc}(\kappa)$; $z\in L(\kappa_1+1,\mathscr{G})$; and z is an M-witness to $\{e\}(a,\mathscr{G})\uparrow$. The remaining clauses of $\mathscr{F}$ insure that any model of $\mathscr{F}$ is an end extension of $L(\kappa_1+1,\mathscr{G})$. $\mathscr{F}$ has an amenable model, namely $L(\kappa_1+1,G)$. Hence Lemma 7.3 provides a model of $\mathscr{F}$ first-order definable over $L(\kappa_2)$. That model provides a $G_0\in L(\kappa_2+1)$ such that $\{e\}(a,G_0)\uparrow$, an impossibility since $G_0\in L(\kappa)\cap 2^{\mathrm{gc}(\kappa)}$. □

Lemmas 7.3 and 7.4 are modeled on results and ideas of Green [9]. She proved that $R(\omega+\omega)^+$, the least Σ_1 admissible set with $R(\omega+\omega)$ as an element, is Σ_1 compact in the sense of Barwise [20]. Let A denote $R(\omega+\omega)^+$, and call x A-finite if $x\in A$. Green's compactness theorem states: if $\mathscr{S}$ is a Σ_1^A set of sentences such that every A-finite subset of $\mathscr{S}$ has a model, then $\mathscr{S}$ has a model. The following crude sketch of her argument will be helpful in the proofs of 7.3 and 7.4.

Let $\mathscr{L}$ ambiguously denote the language of $\mathscr{S}$ and the set of all sentences in the language of $\mathscr{L}$. $\mathscr{L}$ is a Δ_1^A subset of A. A truth assignment v is any function whose range is contained in {true, false}. v is *acceptable* if v does not give any sentence of $\mathscr{S}$ the value false, and if v respects the propositional connectives. (In particular, if $v(\bigvee_i \mathscr{G}_i)$ is true, then $v(\mathscr{G}_i)$ is true for some i.) Assume that $\mathscr{L}$ and $\mathscr{S}$ have been amply provided with Henkin constants and axioms. Thus the problem of finding a model of $\mathscr{S}$ is in essence the problem of finding an acceptable truth assignment for $\mathscr{L}$.

Since A is the Σ_1 admissible closure of $R(\omega + \omega)$, there exists a Σ_1^A, one-one map from A into $R(\omega + \omega)$. For each n, define

$$\mathscr{L}^n = f^{-1}[f[\mathscr{L}] \cap R(\omega + n)].$$

$\mathscr{L}^n \in A$, hence $\mathscr{S} \cap \mathscr{L}^n$ has a model. Consequently there exists an acceptable truth assignment whose domain is $\mathscr{L}^n$. Note that if v is a truth assignment with domain $\mathscr{L}^n$, then $v = w \circ f$, where w is some truth assignment with domain $R(\omega + n)$. Let $\mathscr{T}$ be a tree with ω levels such that the nodes on the n-th level are all the truth assignments with domain $R(\omega + n)$. A path through $\mathscr{T}$ is a sequence $\{z_n \mid n < \omega\}$ of nodes such that for all n: z_n is on the n-th level and $z_n \subseteq z_{n+1}$. A model for $\mathscr{S}$ is equivalent to an acceptable path z, namely a z such that for all n, $z_n \circ f$ is an acceptable truth assignment.

The existence of an acceptable path follows from an appropriate syntactical notion of consistency. An A-finite set K of sentences is *consistent* if no contradiction can be derived from K via *proofs* in A generated by the usual axioms for $\mathscr{L}_{\infty,\omega}$ and some additional algebraic laws based on the presence in A of the power set operation restricted to elements of $R(\omega + \omega)$. For example, (1) is now an axiom.

$$(1) \qquad \bigvee_{g \in 2^I} \bigwedge_{i \in I} \mathscr{F}_{g(i)},$$

where $I \in R(\omega + \omega)$, $\{\mathscr{G}_i \mid i \in I\} \in A$, and $\mathscr{F}_{g(i)}$ is $\mathscr{G}_i$ if $g(i) = 0$, and $\sim\mathscr{G}_i$ if $g(i) = 1$. All of the additional algebraic laws are true in every $\mathscr{L}_{A,\omega}$ structure, hence the existence of a model for K establishes the consistency of K.

An acceptable path $\{z_n \mid n < \omega\}$ is defined by recursion on n. Suppose z_n has been chosen from level n. Identify z_n with the sentence

$$\bigwedge \{\mathscr{G} \mid \mathscr{G} \in \mathscr{L}^n \ \&\ (z_n \circ f)(\mathscr{G}) = \text{true}\}$$
$$\wedge \bigwedge \{-\mathscr{G} \mid \mathscr{G} \in \mathscr{L}^n \ \&\ (z_n \circ f)(\mathscr{G}) = \text{false}\}.$$

Suppose every A-finite subset of $\mathscr{S} \cup \{z_n\}$ is consistent. Let C be the set of all nodes on the $(n+1)$-th level of $\mathscr{T}$. For z_{n+1}, choose some $c \in C$ such that every A-finite subset of $\mathscr{S} \cup \{z_n\} \cup \{c\}$ is consistent. Suppose there is no such c. Then for each $c \in C$, there is an A-finite $\mathscr{S}_c \subseteq \mathscr{S}$ such that $\mathscr{S}_c \cup \{z_n\} \cup \{c\}$ is inconsistent. The relation of $\mathscr{S}_c$ to c is Σ_1^A, since the notion of proof alluded to above is Σ_1^A. Since A is Σ_1 admissible, there exists an A-finite $\mathscr{S}_0 \subseteq \mathscr{S}$ such that for all $c \in C$, $\mathscr{S}_0 \cup \{z_n\} \cup \{c\}$ is inconsistent. But then

$$\mathscr{S}_0 \cup \{z_n\} \cup \{\bigvee \{c \mid c \in C\}\}$$

is inconsistent. But axiom (1) implies $\bigvee \{c \mid c \in C\}$ is provable, hence $\mathscr{S}_0 \cup \{z_n\}$ is inconsistent.

Recall that α^*, the Σ_1^α projection of α, is

$$\mu\gamma(Ef)_{f \in \Sigma_1^\alpha}\left[f : \alpha \xrightarrow[\text{into}]{1\text{-}1} \gamma\right].$$

Also B is an amenable subset of λ if $(B \cap u) \in L(\lambda)$ for all $u \in L(\lambda)$. The next lemma is a straightforward adaptation of Green's compactness theorem.

Lemma 7.3. *Assume $L(\alpha)$ is Σ_1 admissible, $\alpha^* < \alpha$ and*

$$L(\alpha) \models \text{cofinality}(\alpha^*) = \omega.$$

Suppose $\mathscr{S}$ is a Σ_1^α set of sentences such that each α-finite subset of $\mathscr{S}$ has a model encoded by an amenable subset of α^*. Then $\mathscr{S}$ has a model first-order definable over $L(\alpha)$ and encoded by an amenable subset of α^*.

Proof. Same as that sketched above for Green's compactness theorem. A becomes $L(\alpha)$, and $R(\omega+\omega)$ becomes $L(\alpha^*)$. Let $\{\tau_n \mid n < \omega\}$ be a sequence of $L(\alpha)$-cardinals whose limit is α^*. $R(\omega+n)$ becomes $L(\tau_n)$. Axiom (1) is now interpreted to mean $I \in L(\alpha^*)$ and 2^I is the power set of I in $L(\alpha)$. An α-finite set K of sentences is consistent if no contradiction can be derived from K via proofs in $L(\alpha)$ generated by the usual axioms of $\mathscr{L}_{\infty,\omega}$ and additional algebraic laws such as (1). All of the additional laws are true in every structure encoded by an amenable subset x of α^* because $x \cap L(\tau_n)$ belongs to $L(\tau_{n+1})$. Hence the existence of an amenable model for an α-finite K establishes the consistency of K.

The nodes on the n-th level of $\mathscr{T}$ are all truth assignments in $L(\alpha)$ with domain τ_n. Every path through $\mathscr{T}$ is an amenable subset of α^*, hence every model of $\mathscr{S}$ provided by Green's argument is encoded by an amenable subset of α. The notion of consistency is π_1^α, hence there is a Δ_2^α acceptable path through $\mathscr{T}$, and consequently a model of $\mathscr{S}$ first-order definable over $L(\alpha)$. □

Lemma 7.4. *Assume $L(\kappa_1)$ is Σ_1 admissible, $\kappa_1^* < \kappa_1$ and*

$$L(\kappa_1) \models \text{cofinality}(\kappa_1^*) = \omega.$$

Suppose $\kappa_1 \leq \kappa$ and there exists an amenable subset of κ_1^* not in $L(\kappa)$. Then there exists an amenable $G \subseteq \kappa_1^*$ not in $L(\kappa)$ such that $L(\kappa_1, G)$ is Σ_1 admissible.

Proof. Similar to that of 7.3. Some additional steps are needed to insure

$G \notin L(\kappa)$ and $L(\kappa_1, G)$ is Σ_1 admissible. Let $\mathcal{S}$ be the following $\Sigma_1^{\kappa_1}$ set of axioms:

$$\mathcal{G} \subseteq \boldsymbol{\kappa}_1^*,$$
$$\mathcal{G} \neq \boldsymbol{d} \quad (d \in 2^{\kappa_1} \cap L(\kappa_1))$$
$$(x)[x \in \boldsymbol{a} \leftrightarrow \bigvee \{x = \boldsymbol{b} \mid b \in a\}] \quad (a \in L(\kappa_1)),$$
$$c > \boldsymbol{\gamma} \quad (\gamma < \kappa_1),$$
$$\bigvee \text{ is } \Sigma_1 \text{ admissible.}$$

Any model of $\mathcal{S}$ is a Σ_1 admissible structure that properly end extends $L(\kappa_1)$. Each κ_1-finite subset of $\mathcal{S}$ has a model of the form $L(\delta, G_0)$, where $\delta < \kappa_1$ and $G_0 \in L(\kappa_1)$, since new subsets of κ_1^* are constructed at arbitrarily high levels of $L(\kappa_1)$. $L(\kappa_1, G_0)$ can be encoded as an amenable subset of κ_1^* by application of f, a one-one $\Sigma_1^{\kappa_1}$ map of $L(\kappa_1)$ into κ_1^*. (Recall that every $\Sigma_1^{\kappa_1}$ subset of κ_1^* is amenable.)

By Lemma 7.3, $\mathcal{S}$ has a model M encoded by an amenable subset of κ_1^*. G is a distinguished element of M, hence G is amenable. $G \notin L(\kappa_1)$. $L(\kappa_1, G) \subseteq M$, but $L(\kappa_1, G) \neq M$.

If $\kappa_1 \notin M$, then $L(\kappa_1, G)$ is Σ_1 admissible, since a failure of admissibility could be used to define κ_1 over $L(c, G)$, where c is any non-standard ordinal of M. The additional steps needed in the proof of 7.3 to omit κ_1 from M are conventional. Suppose stage n of the construction of M concludes with the choice of node z_n on the tree $\mathcal{T}$ such that $\mathcal{S} \cup \{z_n\}$ is consistent. Let d be a Henkin constant such that $\mathcal{S} \cup \{z_n\}$ proves d is an ordinal. Suppose every κ_1-finite subset of

(2) $\quad \mathcal{S} \cup \{z_n\} \cup \{d > \boldsymbol{\gamma} \mid \gamma < \kappa_1\}$

is consistent. Let e be a *new* constant not occurring in $\mathcal{S} \cup \{z_n\}$. Then every κ_1-finite subset of

(3) $\quad \mathcal{S} \cup \{z_n\} \cup \{d > e > \boldsymbol{\gamma} \mid \gamma < \kappa_1\}$

is consistent. If not, $\mathcal{S} \cup \{z_n\}$ proves $d \leq e \vee e \leq \boldsymbol{\gamma}_0$ for some $\gamma_0 < \kappa_1$. But then $\mathcal{S} \cup \{z_n\}$ proves $d \leq \boldsymbol{\gamma}_0 + 1$, contrary to the consistency of (2). If (3) is added to $\mathcal{S} \cup \{z_n\}$ before the choice of z_{n+1}, then d will not denote κ_1 in M.

Since M is constructed in ω steps, it is necessary at stage n to carry out the above argument for all $d_i \in I$, where $I \in L(\kappa_1)$ and has $L(\kappa_1)$-cardinality less than κ_1^*. Suppose $\mathcal{S} \cup \{z_n\}$ proves d_i is an ordinal for each $d_i \in I$, and also proves I is linearly ordered by $<_I$.

Let I_1 be the set of all $d_i \in I$ such that d_i denotes κ_1 in some model of $\mathcal{S} \cup \{z_n\}$ coded by an amenable subset of κ_1^*. If $d_i \in I_i$, then $\{y \mid y \in I \,\&\, y <_I d_i\}$ is wellordered. Hence I_1 is wellordered by $<_I$. Let d be the $<_I$-least member of I_1. As above, (3) is consistent, and the addition of (3) to $\mathcal{S} \cup \{z_n\}$ prevents d from denoting κ_1 in M. The definition of d implies no member of I can denote κ_1 in M.

In order to insure that $G \notin L(\kappa)$, the axioms of $\mathcal{S}$ of the form $\mathcal{G} \neq \boldsymbol{d}$ $(d \in L(\kappa_1))$ are replaced by axioms of the form

(1) $\mathcal{G}$ is not defined by a path through $\boldsymbol{W}$,

where W is any narrow subtree of $\mathcal{T}$ in $L(\kappa_1)$. (For simplicity assume that incompatible nodes of $\mathcal{T}$ always make incompatible statements about $\mathcal{G}$.) A subtree W of $\mathcal{T}$ is called *narrow* if the set of nodes of W has $L(\kappa_1)$-cardinality less than κ_1^*, hence at most at τ_n. Let $|W|$ be the set of all paths through W. Then

(2) $(|W| \cap L(\kappa_1))$ is κ_1-finite,

since a path through W and in $L(\kappa_1)$ corresponds to an ω-sequence through τ_n and in $L(\kappa_1)$, if τ_n is the number of nodes in w. It follows from (2), and the existence of new subsets of κ_1^* at arbitrarily high levels of $L(\kappa_1)$, that each κ_1-finite subset of $\mathcal{S}$ (with the added narrow tree axioms (1)) has a κ_1-finite model, hence a model encodeable by an amenable subset of κ_1^*.

Let H be an amenable subset of κ_1^* not in $L(\kappa)$. The narrow tree axioms make it possible to code H into G during the construction of M. Observe that the amenability of H allows it to be construed as a function H_0 from ω into κ_1^*. $H_0(n)$ is the least ordinal that constructs $H \cap \tau_n$. Clearly $H_0(n) < \tau_{n+1}$. The coding of $H_0(n)$ into G is done immediately after the completion of n-th stage of the construction of M. Thus z_n has been chosen and every κ_1-finite subset of $\mathcal{S} \cup \{z_n\}$ is consistent. Let W_n be the set of all nodes on $\mathcal{T}$ that extend z_n and are consistent with every κ_1-finite subset of $\mathcal{S} \cup \{z_n\}$. W_n cannot be narrow; if it were, then

"$\mathcal{G}$ is not defined by a path through W_n"

would be an axiom, and $\mathcal{S} \cup \{z_n\}$ would be inconsistent.

Thus the number of nodes on W_n is κ_1^*, and so some level of W_n has at least τ_{n+1} nodes. Go to the least such level, call it level m, and let z_{n+1} be the $H_0(n)$-th node on level m.

Since the notion of consistency is π_1 over $L(\kappa_1)$, there is a path $\mathcal{T}$, definable over $L(\kappa_1, H_0)$, such that $\mathcal{T}$ defines a model of M of $\mathcal{S}$; M omits κ_1, and G, the distinguished element of M, encodes H_0 as described above. Consequently $L(\kappa_1, G)$ is Σ_1 admissible. H_0 and M are simultaneously definable (by induction on n) over $L(\kappa_1, G)$. The definition of M at stage $n+1$, that is the definition of Z_{n+1}, involved the m-th level of W_n. According to the simplicity assumption made before the definition of narrow tree, any two distinct nodes on level m of W_n make incompatible statements about G. Consequently z_{n+1}, hence $H_0(n)$, can be recovered from the consistency predicate and G.

Since H_0 is definable over $L(\kappa_1, G)$ and $H_0 \notin L(\kappa)$, it follows that $G \notin L(\kappa)$. $\square$

8. Uncountable cofinality

In this section the works of Silver [6] and Friedman [7, 8] are applied to show $2^{\mathrm{gc}(\kappa)} \cap L(\kappa)$ is E-recursively enumerable in some $d \in L(\kappa)$, if $L(\kappa)$ is E-closed,

Σ_1 admissible, and

(1) $L(\kappa) \models \omega < \mathrm{cf}(\mathrm{gc}(\kappa)) < \mathrm{gc}(\kappa)$.

Silver showed with the help of Fodor's theorem that $2^{\omega_{\omega_1}} = \omega_{\omega_1+1}$ if $2^\gamma = \gamma^+$ for every cardinal $\gamma < \omega_{\omega_1}$. Friedman combined Silver's combinatoric approach with fine structure methods to show that the ω_{ω_1}-degrees above $0'$ are wellordered if $V = L$. Let

$$O(x) = \mu\gamma[x \in L_{\gamma+1} - L_\gamma].$$

Friedman's results will be used to show there exists a $d \in L(\kappa)$ such that $O(x) \leqslant_E x, d$ (uniformly) for all $x \in 2^{\mathrm{gc}(\kappa)} \cap L(\kappa)$, if $L(\kappa)$ is E-closed and satisfies (1). The main result of this section then follows easily.

In order to define α-degree, let L_α be Σ_1 admissible, $A \subseteq \alpha$ and

$$F_A = \{x \mid x \subseteq A \;\&\; x \in L_\alpha\}.$$

Then for any amenable $B \subseteq \alpha$, A is said to be α-recursive in B ($A \leqslant_\alpha B$) if F_A and $F_{\alpha-A}$ are Σ_1 over $\langle L_\alpha, B\rangle$. A and B have the same α-degree if $A \leqslant_\alpha B$ and $B \leqslant_\alpha A$. The α-degree of A is denoted by $\mathbf{A}$. $\mathbf{A} \leqslant_\alpha \mathbf{B}$ if $A \leqslant_\alpha B$.

Let A' be the complete Σ_1 set for $\langle L_\alpha, A\rangle$. Define $A^{(n+1)}$ to be $(A^{(n)})'$ for all $n < \omega$.

Lemma 8.1. *Suppose $L(\kappa)$ is E-closed, there is a greatest cardinal in the sense of $L(\kappa)$, and*

$$L(\kappa) \models \omega < \mathrm{cf}(\mathrm{gc}(\kappa)) < \mathrm{gc}(\kappa).$$

Then there exist $d \in L(\kappa)$ and e such that:

(i) *for all $x \in 2^{\mathrm{gc}(\kappa)} \cap L(\kappa)$, $\{e\}(x, d)\downarrow$; and*

(ii) *for all $x \in 2^{\mathrm{gc}(\kappa)}$, if $\{e\}(x, d)\downarrow$, then $\{e\}(x, d) = O(x)$.*

Proof. Let α be $\mathrm{gc}(\kappa)$. Let d be the $<_L$-least singular sequence through α. Clearly $d \in L(\kappa)$. Friedman [8] shows:

(a) The α-degrees in $L(\kappa)$ and above d are wellordered by $\leqslant_\alpha$.

(b) Suppose $x, y \in 2^\alpha \cap L(\kappa)$, $d \leqslant_\alpha x$, and $d \leqslant_\alpha y$. If $x, y \in L_{\gamma+1} - L_\gamma$, then $y \leqslant_\alpha x^{(n)}$ for some $n < \omega$.

The proof of (a) follows from Prikry's proof of Silver's theorem: subsets of α can be compared by comparing their cut-off functions and applying Fodor's theorem. The proof of (b) combines infinite combinatorics with fine structure considerations such as the 'collapsibility' of master codes to master codes. Friedman [8] shows much more than (b). For example, if $x \in 2^\alpha \cap L(\kappa)$ and $d \leqslant_\alpha x$, then x has the same α-degree as some master code.

A fine structure fact that goes back to Boolos and Putnam [21] says:

(c) Suppose $x \in 2^\alpha \cap (L_{\gamma+1} - L_\gamma)$. Then there exists a $y \in (L_{\gamma+1} - L_\gamma)$ such that y is a wellordering of a α and $\mathrm{ord}(y)$, the ordertype of y, equals γ.

The next observation follows from (b) and (c).

(d) Suppose $x \in 2^\alpha \cap L(\kappa)$ and $d \leqslant_\alpha x$. Then there exist $n < \omega$ and $y \leqslant_\alpha x^{(n)}$ such that y is a wellordering of α and $\text{ord}(y) = O(x)$.

Assume $x \in 2^\alpha$. The computation of $\{e\}(x, d)$ proceeds as follows. Let x_d be $x \oplus d$, the disjoint, effective union of x and d. Thus $x, d \leqslant_\alpha x_d$, and $O(x) \leqslant O(x_d)$. Let

$$K_x = \{y \mid y \subseteq \alpha \;\&\; (En)(y \leqslant_\alpha x^{(n)})\}.$$

K_x is E-recursive in x, d. One last observation is needed to further the computation of $\{e\}(x, d)$.

(e) Suppose $y \in L(\kappa)$ and y is a linear ordering of α. Then y is a wellordering iff y has no infinite descending sequence in $L(\alpha)$.

To prove (e), let $\{m\}(y)$ be the natural computation of the ordinal height of y via transfinite recursion. Thus $\{m\}(y)\downarrow$ iff y is a wellordering. By Theorem 3.7, $L(\kappa)$ admits M-witnesses. Every M-witness to $\{m\}(y)\uparrow$ includes an infinite descending sequence through y. $(\alpha^\omega \cap L_\kappa) \subseteq L_\alpha$, since α has uncountable cofinality in L_κ. Now let

$$K_x^0 = \{y \mid y \in K_x \;\&\; y \text{ is a linear ordering of } \alpha \;\&\; y \text{ has no inf. desc. seq. in } L(\alpha)\}.$$

K_x^0 is E-recursive in x, d. Define

$$\{e_0\}(x, d) \simeq \sup\{\text{ord}\, y \mid y \in K_x^0\},$$
$$\{e\}(x, d) \simeq \mu\gamma[\gamma \leqslant \{e_0\}(x, d) \;\&\; x \in L_{\gamma+1} - L_\gamma].$$

Assume $x \in 2^\alpha \cap L(\kappa)$. By (e), $\{e_0\}(x, d)$ converges to an ordinal which is, by (d), sufficient to construct x.

Theorem 8.2. *Suppose $L(\kappa)$ is E-closed, not Σ_1 admissible and*

$$L(\kappa) \models \omega < \text{cf}(\text{gc}(\kappa) < \text{gc}(\kappa).$$

Then $2^{\text{gc}(\kappa)} \cap L(\kappa)$ is E-recursively enumerable in some $a \in L(\kappa)$.

Proof. $2^{\text{gc}(\kappa)} \cap L(\kappa)$ is enumerated as follows. Let e and d be as in Lemma 8.1. First try to compute $\{e\}(x, d)$. If $\{e\}(x, d)$ converges to γ, then $x \in L_{\gamma+1} - L_\gamma$. x is enumerated iff $\gamma < \kappa$. The inadmissibility of L_κ makes it possible to decide whether or not $\gamma < \kappa$. Let $D(x, y)$ be Δ_0^κ such that

$$L(\kappa) \models (x)_{x<b}(Ey)D(x, y),$$

but

$$L(\kappa) \models \sim(Ez)(x)_{x<b}(Ey)_{y\in z}D(x, y).$$

Then $\gamma \geqslant \kappa$ iff

$$L(\gamma+1) \models (Ez)(x)_{x<b}(Ey)_{y\in z}D(x, y).$$

(The z in question is simply $L(\kappa)$.) $\square$

9. Main theorem & extensions

In this section the main theorem (9.1) is assembled, further consequences of the reasoning behind 9.1 are drawn for the admissible case, and a further result of Griffor and Normann [22] is stated. Finally, there is a non-enumerability result for $E(x)$ when x is an arbitrary set of ordinals.

Theorem 9.1. *Suppose $L(\kappa)$ is E-closed and not Σ_1 admissible. Then* (i) *and* (ii) *are equivalent.*

(i) $2^{\mathrm{gc}(\kappa)} \cap L(\kappa)$ *is E-recursively enumerable in some* $a \in L(\kappa)$.

(ii) $L(\kappa) \models \omega < \mathrm{cf}(\mathrm{gc}(\kappa)) < \mathrm{gc}(\kappa)$; *or* $L(\kappa) \models [\omega = \mathrm{cf}(\mathrm{gc}(\kappa))]$ *and every amenable subset of* $\mathrm{gc}(\kappa)$ *belongs to* $L(\kappa)$.

Theorem 9.1 is an immediate consequence of Theorems 6.1, 7.1 and 8.2. A clearer result is available if countability is assumed.

Corollary 9.2. *Suppose $L(\kappa)$ is E-closed, countable, and not Σ_1 admissible. Then* (i) *is equivalent to* (ii).

(i) $2^{\mathrm{gc}(\kappa)} \cap L(\kappa)$ *is E-recursively enumerable in some* $a \in L(\kappa)$.

(ii) $L(\kappa) \models \omega < \mathrm{cf}(\mathrm{gc}(\kappa)) < \mathrm{gc}(\kappa)$.

Proof. If $L(\kappa)$ is countable, then a routine diagonal argument provides an amenable subset of $\mathrm{gc}(\kappa)$ not in $L(\kappa)$. □

The arguments behind Theorem 9.1 yield some information about the admissible case as well. For convenience let $\mathrm{gc}(\kappa)$ be κ when there is no greatest cardinal in the sense of $L(\kappa)$. $L(\kappa)$ is said to be an E-closure if $L(\kappa) = E(b)$ for some $b \in L(\kappa)$.

Assume in (1)–(4) that $L(\kappa)$ is admissible.

(1) If $L(\kappa)$ is an E-closure, then $2^{\mathrm{gc}(\kappa)} \cap L(\kappa)$ is E-recursively enumerable in some $a \in L(\kappa)$.

(2a) If $L(\kappa)$ is not an E-closure and $\mathrm{gc}(\kappa)$ is a regular cardinal in the sense of $L(\kappa)$, then $2^{\mathrm{gc}(\kappa)} \cap L(\kappa)$ is E-recursively enumerable in some $a \in L(\kappa)$ iff every amenable subset of $\mathrm{gc}(\kappa)$ belongs to $L(\kappa)$.

(2b) If $L(\kappa)$ is not an E-closure and $\mathrm{gc}(\kappa) = \kappa$, then $2^{\mathrm{gc}(\kappa)} L(\kappa)$ is not E-recursively enumerable in any $a \in L(\kappa)$.

(3) If $L(\kappa)$ is not an E-closure and

$$L(\kappa) \models \mathrm{cf}(\mathrm{gc}(\kappa)) = \omega,$$

then $2^{\mathrm{gc}(\kappa)} \cap L(\kappa)$ is E-recursively enumerable in some $a \in L(\kappa)$ iff every amenable subset of $\mathrm{gc}(\kappa)$ belongs to $L(\kappa)$.

(4) The methods of 9.1 shed little light when $L(\kappa)$ is Σ_1 admissible, not an E-closure and satisfies $\omega < \mathrm{cf}(\mathrm{gc}(\kappa)) < \mathrm{gc}(\kappa)$.

Proof of (1). Let $L(\kappa)=E(b)$. b can be taken to be a set of ordinals, since in $L(\kappa)$ every set can be coded by a set of ordinals. By Theorem 3.6 there is a $y \in L(\kappa)$ such that $\kappa_r^y \geqslant \kappa$. There is a $z \in L(\kappa)$ such that $z \subseteq \kappa$ and $y \leqslant_E z$. By reflection there is such a z E-recursive in y. Since $z \equiv_E y$, it follows that $\kappa_0^z = \kappa_0^y$ and $\kappa_r^z = \kappa_r^y$. Thus $\kappa_r^z \geq \kappa$, and so by Lemma 3.4(iv)

$$\kappa_r^{z,x} \geqslant \kappa_r^z \quad \text{for all } x.$$

Suppose $x \in 2^{\mathrm{gc}(\kappa)} \cap L(\kappa)$ with the intent of computing $O(x)$ from z, x. $O(x)$ is the unique δ such that

$$\delta < \kappa \leqslant \kappa_r^{z,x} \,\&\, x \in L_{\delta+1} - L_\delta.$$

By reflection δ is E-recursive in z, x.

It follows from Gandy selection that there is an e such that if $x \in 2^{\mathrm{gc}(\kappa)} \cap L(\kappa)$, then

$$\{e\}(z,x)\downarrow \,\&\, \{e\}(z,x) = O(x).$$

Enumerate $2^{\mathrm{gc}(\kappa)} \cap L(\kappa)$ as follows. First try to compute $\{e\}(z,x)$. If $\{e\}(z,x)\downarrow$, then enumerate x if $E(b)$ is not a subset of $L(\{e\}(z,x))$. □

The proof of (2) also establishes (5).

(5) If z is a set of ordinals and $E(z)$ is Σ_1 admissible, then $E(z)$ is E-recursively enumerable in some $a \in E(z)$.

Proof of (2). Suppose $\mathrm{gc}(\kappa) < \kappa$ and some amenable subset of $\mathrm{gc}(\kappa)$ lies outside $L(\kappa)$. The arguments of Section 6 show $2^{\mathrm{gc}(\kappa)} \cap L(\kappa)$ is not E-recursively enumerable in any $a \in L(\kappa)$. The only point to be checked is the existence of M-witnesses in $L(\kappa)$. Suppose $x \in L(\kappa)$. It suffices to show $\kappa > \theta^x$. $\kappa > \kappa^x$ since $L(\kappa)$ is not an E-closure. $\kappa^x \geqslant \theta^x$ by Lemma 3.4 and Proposition 2.1.

Suppose $\mathrm{gc}(\kappa)=\kappa$, $a \in L(\kappa)$ and $2^{\mathrm{gc}(\kappa)} \cap L(\kappa) = \{x \mid \{e\}(x,a)\downarrow\}$. First assume κ is a cardinal of L. Then $L(\kappa)$ is a Σ_1 substructure of L. But

$$(E\gamma)[\gamma \in \mathrm{On} \,\&\, \{e\}(\gamma,a)\uparrow] \qquad (\text{e.g. } \gamma=\kappa)$$

holds in L, hence in $L(\kappa)$.

Now assume κ is not a cardinal of L. Let ρ be a regular cardinal in the sense of $L(\kappa)$ such that $a \in L(\rho)$ and ρ exceeds all L-cardinals in $L(\kappa)$. Let ρ^+ be the next $L(\kappa)$-cardinal after ρ. Then

$$2^\rho \cap L(\rho^+) = \{x \mid x \subseteq \rho \,\&\, \{e\}(x,a)\downarrow\}.$$

The methods of Section 6 show $2^\rho \cap L(\rho^+)$ is not E-recursively enumerable in a. $L(\rho^+)$ admits M-witnesses because it is a Σ_1 substructure of $L(\kappa)$.

Proof of (3). The proof of Theorem 7.1 applies after one minute change. In 7.1 it was argued that $\kappa_2 < \kappa$ because $L(\kappa_2)$ is Σ_1 admissible and $L(\kappa)$ is not. Now it is argued that $\kappa_2 < \kappa$ because $L(\kappa_2)$ is an E-closure and $L(\kappa)$ is not.

Although (4) defines an open case, the reasoning behind Theorem 9.1, (1), (2) and (3) yields a complete solution to the enumerability problem for $E(\gamma)$, where γ is an ordinal. Let $\mathrm{gc}(E(\gamma))$ denote the greatest cardinal in the sense of $E(\gamma)$.

Theorem 9.3. *If γ is an ordinal, then* (i) $\leftrightarrow$ (ii).
(i) *$E(\gamma)$ is not E-recursively enumerable in any $a \in E(\gamma)$.*
(ii) *$E(\gamma)$ is not Σ_1 admissible and $E(\gamma) \models [\mathrm{gc}(E(\gamma))$ is regular$]$.*

Proof. Left to the reader save for one point. Suppose

$$E(\gamma) \models [\omega < \mathrm{cf}(\mathrm{gc}(E(\gamma))) < \mathrm{gc}(E(\gamma))].$$

The proof of Lemma 8.1 describes a procedure P for computing $O(x)$ for all $x \in 2^{\mathrm{gc}(E(\gamma))} \cap E(\gamma)$. P can be extended to all $y \in E(\gamma)$ by effective transfinite recursion on the set-theoretic rank of y. The extension succeeds because each $\delta \in E(\gamma)$ can be mapped one–one into $\mathrm{gc}(E(\gamma))$ by some f E-recursive in δ, p. p is a fixed parameter from $E(\gamma)$ that maps γ one-one onto $\mathrm{gc}(E(\gamma))$. □

Griffor and Normann [22] have extended the concluding argument of the proof of Theorem 9.3 with the aid of their notion of non-standard stage comparison. They show:

Theorem 9.4 (Griffor & Normann [22]). *If γ is an ordinal, $E(\gamma)$ is not Σ_1 admissible, and*

$$E(\gamma) \models [\omega < \mathrm{cf}(\mathrm{gc}(E(\gamma))) < \mathrm{gc}(E(\gamma))],$$

then $E(\gamma)$ is E-recursive in some $a \in L(\kappa)$.

The next result is a partial generalization of Theorem 9.3.

Theorem 9.5. *Suppose x is a set of ordinals and*

$$E(x) \models \sim[\omega < \mathrm{cf}(\mathrm{gc}(E(x))) < \mathrm{gc}(E(x))].$$

Then (i) $\leftrightarrow$ (ii).
(i) *$E(x) \models [\mathrm{gc}(E(x))$ is regular$]$ and $E(x)$ is not Σ_1 admissible.*
(ii) *$2^{\mathrm{gc}(E(x))} \cap E(x)$ is not E-recursively enumerable in any $a \in E(x)$.*

Proof. Suppose (ii) holds. By (5) above, $E(x)$ is not Σ_1 admissible. Then

$$E(x) \models [\mathrm{cf}(\mathrm{gc}(E(x))) \neq \omega]$$

according to the selection theorem proved in [5].

Suppose (i) holds. Then the proof of Theorem 6.1 is sufficient after a few changes. Suppose

$$2^{\mathrm{gc}(E(x))} \cap E(x) = \{y \mid \{e\}(a, y)\downarrow\}$$

for some $a \in E(x)$. It is safe to assume that a is a set of ordinals, and that $x \leq_E a$. By Theorem 3.7, $E(x)$ admits Moschovakis witnesses. Hence the forcing technology of Section 5 applies to $E(x)$ (in place of $L(\kappa)$). It follows that either (1) or (2) is true, as in the proof of Theorem 6.1. If (1) of 6.1 holds, then there is a $G \subseteq \mathrm{gc}(E(x))$ such that

$$G \notin E(x) \quad \text{and} \quad \{e\}(a, G)\downarrow.$$

G is constructed as in 6.1 from an $S \subseteq \mathrm{gc}(E(x))$ such that $S \notin E(x)$ and

(a) $\quad (S \cap \delta) \in E(x) \quad$ for all $\delta < \mathrm{gc}(E(x))$.

S is obtained as follows. The universal partial E-recursive function maps $\mathrm{TC}(\{x\})$ onto $E(x)$. Consequently there is a partial E-recursive (in some $b \in E(x)$) map f from $\mathrm{gc}(E(x))$ onto $E(X)$. Let S be $\mathrm{dom} f$. By bounding $S \notin E(x)$. Clearly S is E-recursively enumerable in b. Since $\mathrm{gc}(E(x))$ is regular in $E(x)$, the Griffor–Normann selection theorem [23] implies (a).

If (2) of 6.1 holds, then the argument of 6.1 proceeds with no substantial change. □

The next corollary was first stated and proved in [3] in the language of Kleene's theory of recursion in finite types.

Corollary 9.6. *Suppose*

$$E(2^\omega) \models [2^\omega \textit{ is wellordered } \& \ \mathrm{card}(2^\omega) \textit{ is regular}].$$

Then $2^{2^\omega} \cap E(2^\omega)$ *is not E-recursively enumerable in any* $a \in E(2^\omega)$.

10. Questions

Which of the following are true?

1. There exists a set x of ordinals such that $E(x)$ is not Σ_1 admissible,

$$E(x) \models [\omega < \mathrm{cf}(\mathrm{gc}(E(x))) < \mathrm{gc}(E(x))],$$

and $E(x)$ is not E-recursively enumerable in some $a \in E(x)$.

2. It is consistent with ZFC that

$$E(2^\omega) \models [2^\omega \text{ is wellordered}]$$

and $2^{2^\omega} \cap E(2^\omega)$ is E-recursively enumerable in some $a \subseteq \omega$.

3. If $L(\kappa)$ is E-closed and not Σ_1 admissible, and

$$L(\kappa) \models [\omega < \mathrm{cf}(\mathrm{gc}(\kappa)) < \mathrm{gc}(\kappa)],$$

then $L(\kappa)$ is E-recursive in some $a \in L(\kappa)$.

4. There exist κ, x and y such that x is a set of ordinals, $L(\kappa, x)$ is E-closed and not Σ_1 admissible, $y \in L(\kappa, x)$, and $\kappa_r^y = \kappa$.

Finally, is there a simple criterion for the E-recursive enumerability of $L(\kappa)$ when $L(\kappa)$ lies in case (4) of Section 9? A criterion applicable to this case has been found by Griffor & Normann [22, §3].

References

[1] S.C. Kleene, Recursive functionals and quantifiers of finite types I and II, Trans. Amer. Math. Soc. 91 and 108 (1959 and 1963) 1–52 and 106–142.
[2] D. Normann, Set recursion, in: J.E. Fenstad et al., eds., Generalized Recursion Theory II (North-Holland, Amsterdam, 1978) 303–320.
[3] G.E. Sacks, Post's problem, absoluteness and recursion in finite types, in: J. Barwise et al., eds., The Kleene Symposium (North-Holland, Amsterdam, 1980) 201–222.
[4] Y.N. Moschovakis, Hyperanalytic predicates, Trans. Amer. Math. Soc., 138 (1967) 249–282.
[5] G.E. Sacks and T.A. Slaman, Inadmissible forcing, Advances in Math., to appear.
[6] J. Silver, On the singular cardinals problem, in: Proc. Int. Cong. Math., Vancouver, 1974.
[7] S. Friedman, Negative solutions to Post's problem II, Ann. of Math. 113 (1981) 25–43.
[8] S. Friedman, Uncountable admissibles II: Compactness, Israel J. Math. 40 (1981) 129–149.
[9] J. Green, Σ_1 compactness for next admissible sets, J. Symbolic Logic 39 (1974) 105–116.
[10] G.E. Sacks, The k-section of a type n object, Amer. J. Math. 99 (1977) 901–917.
[11] L. Harrington, Contributions to Recursion Theory on Higher Types, Ph.D. Thesis, M.I.T., Cambridge, MA, 1973.
[12] J.E. Fenstad, General Recursion Theory (Springer, Berlin, 1980).
[13] T.A. Slaman, Reflection and Forcing in E-recursion theory, Ann. Pure Appl. Logic 29 (1985) 79–106.
[14] K. Devlin, Aspects of Constructibility, Lecture Notes in Math. 354 (Springer, Berlin, 1973).
[15] T.A. Slaman, Aspects of E-Recursion, Ph.D. Thesis, Harvard University, Cambridge, 1981.
[16] G.E. Sacks, Selection and forcing in E-recursion, in preparation.
[17] T.A. Slaman, The extended plus-one hypothesis: a relative consistency result, Nagoya Math. J. 92 (1983).
[18] Y.N. Moschovakis, On the Grilliot–Harrington–MacQueen Theorem, in: Logic Year 79–80, Lecture Notes in Math. 859 (Springer, Berlin, 1981) 246–267.
[19] L. Kirousis, On abstract recursion theory and recursion in the universe of sets, Ph.D. Thesis, UCLA, 1978.
[20] J. Barwise, Admissible Sets and Structures (Springer, Berlin, 1975).
[21] G. Boolos and H. Putnam, Degrees of unsolvability of constructible sets of integers, J. Symbolic Logic 33 (1968) 497–513.
[22] E.R. Griffor and D. Normann, The definability of $E(\alpha)$, Preprint, University of Oslo, 1982.
[23] E.R. Griffor and D. Normann, Effective cofinalities and admissibility in E-recursion, Preprint, University of Oslo, 1982.

Reprinted from ADVANCES IN MATHEMATICS Vol. 66, No. 1, October 1987

Inadmissible Forcing

G. E. SACKS* AND T. A. SLAMAN

Harvard University,
Massachusetts Institute of Technology,
Cambridge, Massachusetts 02139

A structure is E-closed if it is closed under all partial E-recursive functions from V into V, a set theoretic extension of Kleene's partial recursive functions of finite type in the normal case. Let $L(\kappa)$ be E-closed and $\sum_1$ inadmissible. Then $L(\kappa)$ has reflection properties useful in the study of generic extensions of $L(\kappa)$. Every set generic extension of $L(\kappa)$ via countably closed forcing conditions is E-closed. A class generic construction shows: if $L(\kappa)$ is countable, and inside $L(\kappa)$ the greatest cardinal $\text{gc}(\kappa)$, has uncountable cofinality, then there exists a $T \subseteq \text{gc}(\kappa)$ such that $L(\kappa, T) = E(T)$, the least E-closed set with T as a member. A partial converse is obtained via a selection theorem that implies $E(X)$ is $\sum_1$ admissible when X is a set of ordinals and the greatest cardinal in the sense of $E(X)$ has countable cofinality in $E(X)$.

1. INTRODUCTION

The notion of partial recursive function with numerical arguments was extended by Kleene [1, 2] so as to allow arguments of finite type. His extension was defined by schemes which give rise to infinitely long computations. In the so-called normal case the equality predicate is treated as if it were recursive. Normann [3] and Moschovakis independently extended the normal Kleene theory so as to allow all sets to occur as arguments. Thus $\{e\}(x)$, the eth partial recursive function with argument x, has a meaning for all $x \in V$.

Normann's definition is inductive. $\{e\}$ is defined by interpreting e as the Gödel number of a scheme. The schemes include rudimentary functions such as pairing and union, composition, effective bounding, and also enumeration $(f(e, x) \simeq \{e\}(x))$. The attempt to compute $\{e\}(x)$ via schemes produces a tree that is well founded iff $\{e\}(x)$ is defined. Computation trees are discussed in Section 2, but it is worth noting now that the unique source of infinitely long computations is the *effective bounding*

* The preparation of this paper was partially supported by NSF Grant MCS 7307344. The authors are thankful for the assistance of E. Griffor, D. Normann, and V. Wiley.

scheme: *if* $\{m\}(x)$ is defined, and if $\{n\}(y)$ is defined for all $y \in \{m\}(x)$, then $\{2^m \cdot 3^n\}(x)$ is defined and is equal to

$$\{\{n\}(y) \mid y \in \{m\}(x)\}.^1$$

If $\{e\}(x)$ is defined, then $\{e\}(x)$ said to convergence (in symbols $\{e\}(x)\downarrow$). Otherwise $\{e\}(x)$ diverges ($\{e\}(x)\uparrow$).

Let A be a transitive set. A is *E-closed* if

$$x \in A \ \& \ \{e\}(x)\downarrow \rightarrow \{e\}(x) \in A.$$

Every $\sum_1$ admissible set is E-closed but not conversely. The *E-closure* of a set x, denoted $E(x)$, is the least E-closed set that contains $\{x\} \cup \mathrm{TC}(x)$.[2] The classic example of an inadmissible, E-closed set is $E(2^\omega)$. The least example is

$$E(\omega) \cup E(E(\omega)) \cup E(E(E(\omega))) \cup \cdots.$$

Some of the results of [4] were obtained by showing certain generic extension of $E(2^\omega)$ are E-closed. The language of [4] was that of Kleene [1], so for clarity it must be pointed out that $E(2^\omega)$ corresponds to recursion in ${}^3E$ as follows: $z \in E(2^\omega)$ iff z is codable by a relation on 2^ω recursive in ${}^3E$, b for some $b \in 2^\omega$.

Set forcing over a Σ_1 admissible structure poses few problems. It is well known that if A is $\sum_1$ admissible and G is generic over A with respect to some set P (in A) of forcing conditions, then $A(G)$ is $\sum_1$ admissible. Class forcing ($P \subseteq A$) over A can be more intricate, as in Abramson [5] or Steel [6].

Set forcing over an E-closed, inadmissible structure poses a new problem. Let t be a term that refers to a generic object. What does it mean to force $\{e\}(t)$ to converge? The answer varies greatly from one notion of forcing to the next. It is a consequence of Gandy selection that if a Levy-style, generic collapse of ω_1 to ω is adjoined to $E(\omega_1)$, then the resulting extension is not E-closed. In [4] it was shown that every Cohen-generic extension of $E(\omega_1)$ is E-closed. Here Cohen-generic refers to forcing conditions that are countable initial segments of characteristic functions of subsets of ω_1. The causal difference between the above Levy and Cohen extensions was pointed out by Slaman: the set of countable Cohen conditions on ω_1 is *countably closed*, that is,

$$(n)[p_n \geqslant p_{n+1}] \rightarrow (Eq)(n)[p_n \geqslant q].$$

[1] For the sake of exposition the bounding scheme given here differs somewhat from that of Normann [3].

[2] $\mathrm{TC}(x)$ is the transitive closure of x.

The countable closure condition is not necessary for E-closed, generic extensions. The countable chain condition also suffices by a mode of argument [8] not employed in the present paper.

The set forcing construction of [4] was extended to an arbitrary E-closed, inadmissible $L(\kappa)$ in [9]. The machinery of [9] is reviewed and amplified in Sections 2 and 3, and then applied to a class forcing construction in Section 4 to prove

THEOREM 4.8. *Let $L(\kappa)$ be countable, E-closed and $\sum_1$ inadmissible. Suppose the greatest cardinal of $L(\kappa)$ has uncountable cofinality in $L(\kappa)$. Then there exists an $S \subseteq \omega_1^{L(\kappa)}$ such that $L(\kappa, S)$ is the E-closure of S.*

(According to a well-known fine structure fact, Proposition 2.5,

$$L(\kappa) \models [\text{There is a greatest cardinal.}],$$

where $L(\kappa)$ is $\sum_1$ inadmissible.)

Theorem 4.8 resembles the following result of admissibility theory [7]: If $L(\kappa)$ is countable and $\sum_1$ admissible, then there exists a $T \subseteq \omega$ such that $L(\kappa, T)$ is the least $\sum_1$ admissible set with T as a member. A partial converse to 4.8 is obtained in Section 5 via a selection theorem.

COROLLARY. 5.2. *Let X be a set of ordinals. If in $E(X)$ the greatest cardinal has cofinality ω, then $E(X)$ is $\sum_1$ admissible.*

If $\{e\}(x)$ does not converge, then its computation tree has at least one infinite descending path. We call any such path a Moschovakis witness to the divergence of $\{e\}$ at x. They were introduced in Moschovakis [10] to show $E(2^\omega)$ is not $\sum_1$ admissible, and have since proved essential to the study of recursion in higher type objects. A structure A is said to admit M. witnesses if for all $e < \omega$ and $x \in A$: if $\{e\}(x)$ diverges, then some member of A is an M. witness for $\{e\}(x)$.

LEMMA 2.4 [9]. *If $L(\kappa)$ is E-closed but not $\sum_1$ admissible, then $L(\kappa)$ admits Moschovakis witnesses.*

The above reflection property is precisely what is needed to make countably closed set forcing succeed over an E-closed structure.

THEOREM 3.5. *Suppose A is E-closed and admits Moschovakis witnesses. Let $P \in A$ be countably closed in A. If G is P-generic, then $A(G)$ is E-closed.*

The paper ends with some open questions and conjectures.

2. Existence of Moschovakis Witnesses

In this section it is shown that many inadmissible, E-closed structures admit Moschovakis witnesses. The proof extracts a reflecting ordinal from the failure of $\sum_1$ admissibility, and then draws on a characterization of reflection devised by Harrington [11] for recursion in objects of finite type.

y is said to be *E-recursive in* x if there exists an e such that $y = \{e\}(x)$. Whether or not an arbitrary z belongs to y can be decided by first computing $\{e\}(x)$ and then checking whether or not z belongs to $\{e\}(x)$. The second step is legitimate according to the schemes of E-recursion .

B is said to be *E-recursively enumerable in* x if there exists an e such that

$$B = \{z \mid \{e\}(z, x)\downarrow\}.$$

From now on the prefix E will often be dropped.

The enumeration scheme yields the usual Kleene recursion theorems and thereby the method of defining partial recursive functions by effective transfinite recursion.

Let $<_R$ be a well-founded relation and I a recursive function from ω into ω. Suppose for all $e < \omega$ and all x in the field of $<_R$,

$$\{I(e)\}(x)\downarrow$$

whenever $\{e\}(y)\downarrow$ for all $y <_R x$. Then there exists a partial E-recursive function $\{c\}$ such that $\{c\}(x)\downarrow$ for all x in the field of $<_R$ and $\{I(c)\} \simeq \{c\}$.

The *universal computation tree* $>_U$ is defined by reference to the schemes of E-recursion. Each element of the field of $>_U$ is of the form $\langle e, x\rangle$ where $e < \omega$ and $x \in V$. $a >_U b$ is read: b is a *subcomputation* of a, and is defined by: there exists a finite sequence $z_0, \ldots, z_n$ such that $a = z_0$, $z_n = b$, and z_{i+1} is an *immediate subcomputation* of z_i $(i < n)$. The definition of immediate subcomputation has only two interesting clauses. They correspond to the effective bounding scheme of Section 1.

(1) $\langle m, x\rangle$ is an immediate subcomputation of $\langle 2^m \cdot 3^n, x\rangle$.

(2) If $\{m\}(x)$ is defined, and $y \in \{m\}(x)$, then $\langle n, y\rangle$ is an immediate subcomputation of $\langle 2^m \cdot 3^n, x\rangle$.

The remaining clauses, save one, correspond to the finitary schemes and so give rise to finitary branching in $>_U$. The final clause states: if c is not the Godel number of a scheme, then $\langle c, x\rangle$ is an immediate subcomputation of $\langle c, x\rangle$.

It is readily verified that the relation "b is an immediate subcomputation of a" is recursively enumerable, and that $\{e\}(x)\downarrow$ iff $>_U$ below $\langle e, x\rangle$ is well founded. The method of definition by effective transfinite recursion yields a partial recursive function f such that for all e and x, if $\{e\}(x)\downarrow$, then $f(e, x)$ is the set theoretic object consisting of $>_U$ below $\langle e, x\rangle$. Thus an E-closed set includes not only the end values of computations but also the computations themselves. (If $\{e\}(x)\downarrow$, then it seems appropriate to call $>_U$ below $\langle e, x\rangle$ the computation for $\{e\}(x)$.)

Similarly there exists a partial recursive function $|\{e\}(x)|$ such that if $\{e\}(x)\downarrow$, then $|\{e\}(x)|$ is the ordinal height of $>_U$ below $\langle e, x\rangle$. For example, $|\{2^m \cdot 3^n\}(a)| = \sigma$ if

(1) $(E\gamma)_{\gamma<\sigma}[|\{m\}(a)| = \gamma]$,

(2) $(x)(E\tau)_{\tau<\sigma}[x \in \{m\}(a) \rightarrow |\{n\}(x)| = \tau]$, and

(3) $(\tau)_{\tau<\sigma}(Ex)[(x \in \{m\}(a) \ \& \ |\{n\}(x)| \geqslant \tau) \vee |\{m\}(a)| = \tau]$.

An ordinal δ is said to be recursive in x if there exists a c such that $\delta = \{c\}(x)$. Thus $|\{e\}(x)|$ is recursive in x if $\{e\}(x)\downarrow$.

An essential result in E-recursion theory is the Gandy Selection Theorem [19]: There exists a partial recursive function $\phi(e, x)$ such that for all $e < \omega$,

$$(En)_{n<\omega}[\{e\}(n, x)\downarrow] \leftrightarrow \phi(e, x)\downarrow \ \& \ \{e\}(\phi(e, x), x)\downarrow.$$

Suppose $P(x, y)$ is a recursively enumerable predicate and

$$(x)(Ey)[y \text{ is recursive in } x \ \& \ P(x, y)].$$

Then by Gandy selection there exists a recursive function f such that $(x)\, P(x, f(x))$.

An E-closed set A admits *Moschovakis witnesses* if for all $e < \omega$ and $x \in A$: if $\{e\}(x)\uparrow$, then there is an infinite descending path through $>_U$ below $\langle e, x\rangle$ in A.

Three more definitions are needed for the main lemma (2.3) on the existence of Moschovakis witnesses in E-closed sets.

κ_0^x is the supremum of all ordinals recursive in x.

κ^x is the supremum of all ordinals γ such that γ is recursive in a, x, where $a \in \mathrm{TC}(x)$.

If $\{e\}(x)\downarrow$ and $|\{e\}(x)| = \delta$, then the computation tree for $\{e\}(x)$ is first-order definable over

$$L(\delta, \{x\} \cup \mathrm{TC}(x)).$$

Consequently each set recursive in x belongs to $L(\kappa_0^x, \{x\} \cup \mathrm{TC}(x))$, and

$$E(x) = L(\kappa^x, \{x\} \cup \mathrm{TC}(x)).$$

An ordinal γ is said to be *x-reflecting* if for every $\sum_1$ formula ϕ,

$$L(\gamma, \{x\} \cup \mathrm{TC}(x)) \models \phi(x)$$

iff

$$L(\kappa_0^x, \{x\} \cup \mathrm{TC}(x)) \models \phi(x).$$

κ_r^x is the greatest x-reflecting ordinal. To appreciate the definition of κ_r^x, let $D(u, v)$ be a Δ_0 formula (with free variables u and v), and suppose

$$L(\kappa_r^x, \{x\} \cup \mathrm{TC}(x)) \models (Ev)\, D(x, v).$$

Then

$$L(\delta, \{x\} \cup \mathrm{TC}(x)) \models (Ev)\, D(x, v),$$

for some δ recursive in x. Suppose further that there is a well ordering $>$ of $\mathrm{TC}(x)$ recursive in x. Then the presence of an existential witness for D below κ_r^x implies there is a witness recursive in x, namely the least one relative to the well ordering of $L(\delta, \{x\} \cup \mathrm{TC}(x))$ extending $>$.

LEMMA 2.1. *Suppose A is recursive in x and $\mathscr{B}$ is co-recursively enumerable in x. If $A \cap \mathscr{B}$ is nonempty, then there exists a $c \in A \cap \mathscr{B}$ such that $\kappa_r^x \geqslant \kappa_r^{x,c}$ (Harrington–Kechris basis theorem [11]).*

Proof. First, observe that $\kappa_r^{x,c} > \kappa_r^x$ implies $\kappa_0^{x,c} > \kappa_r^x$. Otherwise $\sum_1$ statements about x would reflect from $\kappa_r^{x,c}$ down to κ_r^x. Thus it suffices to find a $c \in A \cap B$ such that $\kappa_r^x \geqslant \kappa_0^{x,c}$. Now suppose no such c exists. Consider the following recursive (in x) enumeration of A:

(I) If $c \in A - \mathscr{B}$, then enumerate c.

(II) If $c \in A$ and $\kappa_r^x < \kappa_0^{x,c}$, then enumerate c.

(Note that κ_r^x can be effectively defined from x and any ordinal greater than κ_r^x. Thus (II) applies when there is an ordinal recursive in x, c and greater than κ_r^x.)

By effective bounding there is a γ recursive in x that bounds the height of all computations occurring in the above enumeration of A. Since $\gamma < \kappa_0^x \leqslant \kappa_r^x$, it follows that (II) never applies. Hence, every $c \in A$ is enumerated according (I) and so $A \cap \mathscr{B}$ is empty. ∎

LEMMA 2.1a. *Suppose there is a well ordering of $\mathrm{TC}(x)$ which is recursive in $\langle x, z\rangle$. Then there is a well ordering of $\{y \mid y$ is an immediate subcomputation of $\langle e, x\rangle\}$ which is (uniformly) recursive in $\langle x, z\rangle$. Moreover, if y is an immediate subcomputation of x then there is a well ordering of $\mathrm{TC}(y)$ which is (uniformly) recursive in $\langle x, y, z\rangle$.*

Proof. Assume $e=2^m\cdot 3^n$. If the only immediate subcomputation of $\langle e, x\rangle$ is $\langle m, x\rangle$ then the lemma is immediate. Suppose then that $\{m\}(x)\downarrow$. The value of $\{m\}$ at x and the entire computation of $\{m\}(x)$ from x is contained in $L(|\{m\}(x)|, \{x\}\cup \mathrm{TC}(x))$, which is recursive in x. The $\langle x, z\rangle$-recursive well-ordering of $\mathrm{TC}(x)$ induces a well ordering W of $L(|\{m\}(x)|, \{x\}\cup \mathrm{TC}(x))$. W restricted to $\{m\}(x)$ induces a well ordering of the immediate subcomputations of $\langle e, x\rangle$.

If $\langle e', y\rangle$ is an immediate subcomputation of x and $\langle e', y\rangle$ is not equal to $\langle m, x\rangle$ then $e'=n$ and $y\in\{m\}(x)$. So $\mathrm{TC}(y)$ is a subset of $L(|\{m\}(x)|, \{x\}\cup \mathrm{TC}(x))$ since this set is transitive and contains $\{m\}(x)$. W restricted to the elements of $\mathrm{TC}(y)$ is a well ordering of $\mathrm{TC}(y)$ and is recursive in $\langle x, y, z\rangle$. ∎

LEMMA 2.2. *Suppose some well ordering of* $\mathrm{TC}(x)$ *is recursive in* x. *If* $\{e\}(x)\uparrow$, *then a Moschovakis witness for* $\{e\}(x)$ *belongs to* $L(\kappa_r^x+1, \{x\}\cup \mathrm{TC}(x))$.

Proof. Related to arguments of Moschovakis [10] and Harrington [11]. The witness $\lambda t|\langle e_t, x_t\rangle$ is defined by recursion on t. $\langle e_0, x_0\rangle$ is $\langle e, x\rangle$. Assume that $\langle e_t, x_t\rangle$ has been defined so that

(a) $\{e_t\}(x_t)\uparrow$,

(b) $x_t\in L(\kappa_r^x, \{x\}\cup \mathrm{TC}(x))$,

(c) $\kappa_r^x\geqslant \kappa_r^{x_0,\dots,x_t}$,

(d) there is a well ordering of $\mathrm{TC}(x_t)$ uniformly recursive in $\langle x_0,\dots, x_t\rangle$.

For simplicity assume $e_t=2^m\cdot 3^n$. By the inductive assumption (d) and Lemma 2.1a let $<_w$ be recursive in $\langle x_0,\dots, x_t\rangle$ and well order $\{z|z$ is an immediate subcomputation of $\langle e_t, x_t\rangle\}$.

If $\{m\}(x_t)\uparrow$ let $\langle e_{t+1}, x_{t+1}\rangle = \langle m, x_t\rangle$. Assumption (c) on x_t implies that $\kappa_r^x\geqslant\kappa_0^{x_t}$. Consequently, $\{m\}(x_t)\downarrow$ if and only if $|\{m\}(x_t)|<\kappa_r^x$. Thus one can look down from κ_r^x and see whether or not $\{m\}(x_t)\uparrow$. The inductive hypothesis are immediate in this case.

Suppose $\{m\}(x_t)\downarrow$. Then define $\langle e_{t+1}, x_{t+1}\rangle$ to be $\langle n, y\rangle$ where $\langle n, y\rangle$ is the $<_w$ least pair $\langle n, u\rangle$ so that

$$u\in\{m\}(x_t)\ \&\ |\{n\}(n)|\geqslant\kappa_r^x.$$

Since $|\{m\}(x_t)|<\kappa_r^{x_0,\dots,x_t}\leqslant\kappa_r^x$, $y\in L(\kappa_r^x+1, \{x\}\cup \mathrm{TC}(x))$. It has to be shown that

$$\{n\}(y)\uparrow\ \&\ \kappa_r^{x_0,\dots,x_t}\geqslant\kappa_r^{x_0,\dots,x_t,y}.$$

By Lemma 2.1 there is a $z \in \{m\}(x_l)$ such that

$$\{n\}(z)\uparrow \;\&\; \kappa_r^{x_0,\ldots,x_l} \geqslant \kappa_r^{x_0,\ldots,x_l,z}.$$

Let $\langle n, z_0\rangle$ be the $<_w$ least pair so that z_0 is such a z. It suffices to show that $y = z_0$. Clearly, $\langle n, y\rangle \leqslant_w \langle n, z_0\rangle$ since $|\{n\}(z_0)| = \infty > \kappa_r^x$. Suppose $\langle n, y\rangle <_w \langle n, z_0\rangle$. Consider (as in the proof of Lemma 2.1) the following recursive (in $x_0,\ldots, x_l, z_0$) enumeration of $W \upharpoonright \langle n, z_0\rangle = \langle \{n, z\rangle \mid \langle n, z\rangle <_w \langle n, z_0\rangle\}$. By induction, $W \upharpoonright \langle n, z_0\rangle$ is recursive in $x_0,\ldots, x_l, z_0$.

(I) If $\{n\}(z)\downarrow$, then enumerate z.

(II) If $\kappa_r^{x_0,\ldots,x_l} < \kappa_r^{x_0,\ldots,x_l,z}$, then enumerate z.

(Note that $\kappa_r^{x_0,\ldots,x_l} < \kappa_r^{x_0,\ldots,x_l,z}$ implies that $\kappa_r^{x_0,\ldots,x_l} < \kappa_0^{x_0,\ldots,x_l,z}$.)

By effective bounding there is a γ recursive in $x_0,\ldots, x_l, z_0$ that bounds the height of all computations occuring in the above enumeration. It follows that (II) never applies, since otherwise

$$\kappa_r^{x_0,\ldots,x_l} \leqslant \gamma < \kappa_0^{x_0,\ldots,x_l,z_0} \leqslant \kappa_r^{x_0,\ldots,x_l}.$$

Hence (I) applies to every z so that $\langle n, z\rangle <_w \langle n, z_0\rangle$. In particular, if $y \neq z_0$ then $|\{n\}(y)| \leqslant \gamma$. But

$$|\{n\}(y)| \geqslant \kappa_r^x \geqslant \kappa_r^{x_0,\ldots,x_l,z_0} > \gamma$$

according to assumption (c). Thus $z_0 = y$.

Finally to complete the induction, Lemma 2.2 uniformly provides a well ordering of TC(y) which is recursive in $x_0,\ldots, x_l, y$. ∎

LEMMA 2.3. *Suppose $L(\kappa)$ is E-closed but not $\sum_1$ admissible. Then $\kappa_r^x < \kappa$ for all $x \in L(\kappa)$.*

Proof. Let $\kappa_r^x \geqslant \kappa$ for some fixed x in $L(\kappa)$ with the object of showing that $L(\kappa)$ is $\sum_1$ admissible.

Since any element of $L(\kappa)$ can be well ordered in $L(\kappa)$,

$$L(\kappa) \models (EW)[W \text{ is a well ordering of } \mathrm{TC}(x)].$$

By assumption $\kappa_r^x \geqslant \kappa$ so there is a γ which is recursive in x such that

$$L(\gamma) \models (EW)[W \text{ is a well ordering of } \mathrm{TC}(x)].$$

Thus the L-least well-ordering of W_x of TC(x) is recursive in x.

To prove that $L(\kappa)$ satisfies $\sum_1$-bounding, suppose

$$L(\kappa) \models (u)_{u \in \mathbf{d}}(Ev)\,\mathscr{F}(u, v),$$

where $\mathscr{F}$ is Δ_0 with parameter p. Let γ be less than κ so that there is a well ordering $W_{p,d}$ of $\mathrm{TC}(\langle p, d\rangle)$ which is recursive in p, d, γ. Then there is a well ordering of $\mathrm{TC}(\langle p, d, \gamma\rangle)$ which is recursive in p, d, γ. To simplify the discussion, the parameters p, d, γ together are referred to as p^*.

For each b in d

$$\kappa_r^{x, p^*, b} \geqslant \kappa_r^x.$$

This last follows from (1)–(3).

(1) Let θ^z be the least ordinal, if there is one, needed to construct (from z) Moschovakis witnesses for $\{e\}(z)$ for all e such that $\{e\}(z)\uparrow$. Then $\theta^z \geqslant \kappa_r^z$. If not, then, by reflection, θ^z would be recursive in z; this would make the complete recursively enumerable in z subset of ω recursive, and absurdity.

(2) $\theta^{\langle x, p^*\rangle, b} \leqslant \kappa_r^{\langle x, p^*\rangle, b}$ for each b in d by Lemma 2.2. (The parameter γ was introduced to provide the well ordering needed to apply Lemma 2.2.)

(3) For any u and v, $\theta^u \leqslant \theta^{u,v}$, since a Moschovakis witness for a divergent u computation $\{e\}(u)$ can be construed as a Moschovakis witness for a divergent $\langle u, v\rangle$-computation $\{e^*\}(u, v)$.

Thus $\kappa_r^{\langle x, p^*\rangle, b} \geqslant \theta^{\langle x, p^*\rangle, b} \geqslant \theta^x \geqslant \kappa_r^x \geqslant \kappa$. Hence for each $b \in d$, there is a c such that

$$L(\kappa) \models \mathscr{F}(\mathbf{b}, \mathbf{c})$$

and c is recursive (by reflection) in x, p^*, b. The set of all e such that

$$\{e\}(x, p^*, b)\downarrow \quad \text{and} \quad L(\kappa) \models \mathscr{F}(\mathbf{b}, \{e\}(x, p^*, b))$$

is recursively enumerable in x, p^*, b. By Gandy selection (Sect. 2) one such e can be chosen by an effective method which is uniform in x, p^*, b. It follows there is a partial recursive function f such that for each $b \in d$,

$$f(x, p^*, b)\downarrow \quad \text{and} \quad L(\kappa) \models \mathscr{F}(b, f(x, p^*, b)).$$

By effective bounding $\{f(x, p^*, b) \mid b \in d\}$ is recursive in x, p^*, d and so must be an element of $L(\kappa)$ since $L(\kappa)$ is E-closed. ∎

LEMMA 2.5 [9]. *If $L(\kappa)$ is E-closed but not $\sum_1$ admissible, then $L(\kappa)$ admits Moschovakis witnesses.*

Proof. For $z \in L(\kappa)$ let γ be the least ordinal so that there is a well ordering of $\mathrm{TC}(z)$ which is recursive in z, γ. By Lemma 2.4, $\kappa_r^{z,\gamma} < \kappa$. By

Lemma 2.3, if $\{e\}(z, \gamma)$ is divergent, then there is a Moschovakis witness to this effect in $L(\kappa_r^{z,\gamma}+1, \mathrm{TC}(z \cup \gamma) \cup \{z, y\})$. In particular, the relevant Moschovakis witnesses for z are elements of $L(\kappa_r^{z,\gamma}+1, \mathrm{TC}(z \cup \gamma) \cup \{z, \gamma\})$ which is a subset of $L(\kappa)$.

PROPOSITION 2.5 (well known). *Suppose $L(\kappa)$ is E-closed but not Σ_1 admissible. Then*

$$L(\kappa) \models [\textit{There exists a greatest cardinal.}]$$

Proof. Let γ be a cardinal in the sense of $L(\kappa)$. A standard downward Lowenheim–Skolem argument shows that $L(\gamma)$ is a Σ_1 substructure of $L(\kappa)$. If f is Σ_1 over $L(\kappa)$ and its domain and defining parameters belong to $L(\gamma)$, then its range belongs to $L(\gamma)$. ∎

The greatest cardinal of $L(\kappa)$, when it exists, is denoted by $\mathrm{gc}(\kappa)$.

3. Generic Extensions of E-Closed Sets

Let A be an E-closed set, and G a subset of some element of A. Define $A(G)$ to be the set of all z such that for some e and a:

$$z = \{e\}(a, G), \qquad a \in A \quad \text{and} \quad |\{e\}(a, G)| \in A.$$

Suppose $\mathscr{P}$ is a partial ordering of forcing conditions such that $\mathscr{P} \in A$ and is countable closed in A. In this section, it will be shown that $A(G)$ is E-closed, if G is $\mathscr{P}$-generic over A, and A admits Moschovakis witnesses.

The forcing language $\mathscr{L}_A$ has the power to describe the computation of $\{e\}(a, G)$ when a and $|\{e\}(a, G)|$ belong to A. The primitive terms of $\mathscr{L}_A$ are a $(a \in A)$ and $\mathscr{G}$. A general term is $\{e\}(t_1,..., t_n)$, where $t_1,..., t_n$ are terms. Each term t is equivalent to one of so-called normal form $\{e\}(a, \mathscr{G})$. A unique normal form for t can be found effectively from t by elementary syntatical manipulations. $\mathscr{L}_A$ also includes among its primitive symbols: σ $(\sigma \in A)$, $\in$, $|\ |$, variables, quantifiers and propositional connectives.

Let $\mathscr{P} = \langle P, \geqslant \rangle \in A$ be a partial ordering of forcing conditions $p, q, r,...$; $p \geqslant q$ means p is extended by q (i.e., p says less than q about G).

The forcing relation $p \Vdash |\{e\}(a, \mathscr{G})| = \sigma$ and the set $\mathscr{T}(p, e, a, \mathscr{G}, \sigma)$ of terms are defined simultaneously by recursion on $\sigma \in A$. Among the elements of $\mathscr{T}(p, e, a, \mathscr{G}, \sigma)$ are terms that name the elements of $\{e\}(a, \mathscr{G})$ when $\mathscr{G}$ is generic, $p \in G$ and p forces $|\{e\}(a, \mathscr{G})|$ to be σ. Each major clause of the recursion corresponds to one of the Normann schemes, as in the definition of $<_U$, the universal computation tree of Section 2. $\mathcal{T}(p, e, a, \mathcal{G}, \sigma)$ shall be used to represent a set of terms and also as a term itself. The effective bounding scheme is handled as follows.

Define $p \Vdash |\{2^m \cdot 3^n\}(a, \mathcal{G})| = \sigma$ by:

$$(E\gamma)_{\gamma<\sigma}[p \Vdash |\{m\}(a, \mathcal{G})| = \gamma];$$

$$p \Vdash (u)(E\tau)_{\tau<\sigma}[u \in \mathcal{T}(p, m, a, \mathcal{G}, \gamma)$$
$$\& \, u \in \{m\}(a, \mathcal{G}) \rightarrow |\{n\}(u)| = \tau];$$

$$p \Vdash (\tau)_{\tau<\sigma}(Eu)[(u \in \mathcal{T}(p, m, a, \mathcal{G}, \gamma)$$
$$\& \, u \in \{m\}(a, \mathcal{G}) \,\&\, |\{n\}(u)| \geqslant \tau) \vee |\{m\}(a, \mathcal{G})| = \tau].$$

Define $\mathcal{T}(p, 2^m \cdot 3^n, a, \mathcal{G}, \sigma)$ to be

$$\{\{n\}(s) \mid s \in \mathcal{T}(p, m, a, \mathcal{G}, \gamma)\}.$$

The quantifiers and propositional connectives are forced in standard fashion. Thus p forces $(Ex)\,\mathcal{F}(x)$ if p forces $\mathcal{F}(i)$ for some appropriate i, p forces $(x)\,\mathcal{F}(x)$ if p forces $\sim(Ex)\sim\mathcal{F}(x)$, and p forces $\sim\mathcal{F}$ if $\mathcal{F}$ is not forced by any $q \leqslant p$. One detail above is the definition of

$$q \Vdash s \in \{m\}(a, \mathcal{G})$$

when $p \leqslant q$ and $s \in \mathcal{T}(p, m, a, \mathcal{G}, \gamma)$. It is filled in by adding one more step to the recursion. Define

$$q \Vdash \{n\}(s) \in \{2^m \cdot 3^n\}(a, \mathcal{G})$$

to be $q \Vdash \{s \in \{m\}(a, \mathcal{G}) \;\&\; (E\tau)_{\tau<\sigma}(|\{n\}(s)| = \tau)]$.

The *weak forcing relation*, $p \Vdash {}^*F$, is given by $p \Vdash \sim\sim\mathcal{F}$.

$p \Vdash |\{e\}(t)| = \sigma$ means $p \Vdash |\{e_t\}(a_t, \mathcal{G})| = \sigma$, where $\{e_t\}(a_t, \mathcal{G})$ is the normal form of $\{e\}(t)$. Similarly

$$\mathcal{T}(p, e, t, \sigma) = \mathcal{T}(p, e_t, a_t, \mathcal{G}, \sigma).$$

LEMMA 3.1. *$p \Vdash |\{e\}(t)| = \sigma$, $\mathcal{T}(p, e, t, \sigma)$ and $q \Vdash s \in \{e\}(t)$ (where $p \geqslant q$ and $s \in \mathcal{T}(p, e, t, \sigma)$) are recursive in σ, $\mathcal{P}$ (uniformly in σ).*

Proof. Be effective transfinite recursion on σ. ∎

The relation $>_V$ is needed for the development of an effective bound on $|\{e\}(t)|$ when p weakly forces $|\{e\}(t)|$ into A. An approximate version of

$$\langle p, e, t_1 \rangle >_V \langle q, n, t_2 \rangle$$

is simply $p \leqslant q$ and $q \Vdash^* \langle e, t_1 \rangle >_U \langle n, t_2 \rangle$. U is the universal computation tree of Section 2. A typical detail of the precise definition of $>_V$ is

$$q \Vdash \langle 2^m \cdot 3^n, t_1 \rangle >_U \langle n, t_2 \rangle$$

if there is a $\gamma \in A$ such that $q \Vdash |\{m\}(t_1)| = \gamma$,

$$t_2 \in \mathscr{T}(q, m, t_1, \gamma) \qquad \text{and} \qquad q \Vdash t_2 \in \{m\}(t_1).$$

$>_V$ is the forcing counterpart of $>_U$. It includes all possible values of $>_U$ for all generic G's. It will be seen that $p \Vdash$ "b is an immediate subcomputation of a" is a recursively enumerable relation on appropriate triples $\langle p, a, b \rangle$;

$$p \Vdash^* (E\sigma)[|\{e\}(t)| = \sigma],$$

as a relation on p, e, and t, is recursively enumerable in $\mathscr{P}$; and that $>_V$ is well founded below $\langle p, e, t \rangle$ when p weakly forces $|\{e\}(t)|$ into A. These three facts are interwoven in an essential way with the existence of an effective bound on $|\{e\}(t)|$.

LEMMA 3.2. *Suppose $>_V$ is well founded below $\langle p, e, t \rangle$. If*

$$p \Vdash^* (E\sigma)[|\{e\}(t)| = \sigma],$$

then there exists a γ recursive in p, t, $\mathscr{P}$ such that

$$p \Vdash^* |\{e\}(t)| \leqslant \gamma.$$

Proof. γ is computed by an effective transfinite recursion on the height of $\langle p, e, t \rangle$ in $>_V$. The case of maximum interest occurs when $e = 2^m \cdot 3^n$. By recursion there is a δ recursive in p, t, $\mathscr{P}$ such that

$$p \Vdash^* |\{m\}(t)| \leqslant \delta.$$

By Lemma 3.1 the set K, defined by

$$\langle p_0, \sigma \rangle \in K \leftrightarrow p \geqslant p_0 \ \& \ p_0 \Vdash |\{m\}(t)| = \sigma,$$

is recursive in p, t, $\mathscr{P}$. Assume $\langle p_0, \sigma \rangle \in K$ and $s \in \mathscr{T}(p_0, m, t, \sigma)$. Then

$$p_0 \Vdash^* (E\beta)[s \in \{m\}(t) \rightarrow |\{n\}(s)| = \beta].$$

Suppose $p_0 \geqslant q$. Then there is an r such that $q \geqslant r$ and either

$$r \Vdash s \in \{m\}(t) \qquad \text{or} \qquad r \Vdash s \notin \{m\}(t).$$

By Lemma 3.1 the set of all $r \leqslant q$ that force $s \in \{m\}(t)$ is recursive in $q, s, t, \sigma, \mathscr{P}$. For each such r there is by recursion a ρ recursive in $r, s, \mathscr{P}$ such that

$$r \Vdash^* |\{n\}(s)| \leqslant \rho.$$

The effective bounding priciple and Gandy selection yield a strict upper bound γ on ρ (for *all* relevant r and s) such that γ is recursive in $p, t, \mathscr{P}$. It is safe to assume $\delta < \gamma$. ∎

LEMMA 3.3. *Assume A admits Moschovakis witnesses and $\mathscr{P}$ is countably closed in A. If*

$$p \Vdash^* (E\sigma)[|\{e\}(t)| = \sigma],$$

then $>_V$ is well founded below $\langle p,e,t \rangle$.

Proof. Let $\{g\}(p, e, t)$ $(=\gamma)$ be the partial recursive function defined in the proof of Lemma 3.2. If $\{g\}$ converges on $\langle p, e, t \rangle$ then $>_V$ is well founded below $\langle p, e, t \rangle$. This is so because the value of $\{g\}$ on any $w \leqslant_V \langle p, e, t \rangle$ is a strict upper bound for the values of $\{g\}$ on the predecessors of w in $>_V$. Thus g assigns rank.

Suppose $\{g\}(p, e, t)\uparrow$. Let $z \in A$ be a Moschovakis witness for $\langle g, \langle p, e, t \rangle\rangle$. Thus $z_0 = \langle g, \langle p, e, t \rangle\rangle$, and $z_n >_U z_{n+1}$ for all n. There must be a z_m such that

$$z_m = \langle g, \langle q, n, s \rangle\rangle \qquad \text{and} \qquad \langle p, e, t \rangle >_V \langle q, n, s \rangle.$$

The existence of such *a* z_m is a consequence of the details of the definition of $\{g\}$ by effective transfinite recursion on $>_V$. The immediate predecessors of $\langle g, \langle p, e, t \rangle\rangle$ in $>_U$ need not have the form of z_m, but every Moschovakis witness, as it winds its way down below $\langle g, \langle p, e, t \rangle\rangle$, must encounter elements of the form z_m infinitely often. Thus each Moschovakis witness to the divergence of $\{g\}(p, e, t)$ has a built-in, infinite, descending sequence in $>_V$ below $\langle p, e, t \rangle$. It follows there is a sequence $\langle p_n, e_n, t_n \rangle$ $(n \geqslant 0)$ in A such that

$$\langle p_0, e_0, t_0 \rangle = \langle p, e, t \rangle$$

and

$$\langle p_n, e_n, t_n \rangle >_V \langle p_{n+1}, e_{n+1}, t_{n+1} \rangle.$$

Since $\mathscr{P}$ is countably closed in A, there is a q such that for all n, $p_n \geqslant q$. But then

$$q \Vdash^* (n)[\langle e_n, t_n \rangle >_U \langle e_{n+1}, t_{n+1} \rangle].$$

Thus q weakly forces the existence of an Moschovakis witness to the divergence of $\{e\}(t)$, and so p ($\geqslant q$) cannot weakly force $\{e\}(t)$ to convergence. ∎

LEMMA 3.4. *Assume A admits Moschovakis witnesses, and $\mathscr{P}$ is countably closed in A. If*

$$p \Vdash {}^{*}(E\sigma)[|\{e\}(t)| = \sigma],$$

then there exists a γ recursive in p, t, $\mathscr{P}$ such that

$$p \Vdash {}^{*}|\{e\}(t)| \leqslant \gamma.$$

Proof. Combine Lemmas 3.2 and 3.3. ∎

Let G be a path through $\mathscr{P}$. G is $\mathscr{P}$-generic with respect to a sentence $\mathscr{F}$ if there is a $p \in G$ such that either $p \Vdash \mathscr{F}$ or $p \Vdash \sim \mathscr{F}$. G is *$\mathscr{P}$-generic over A* if:

(i) G is generic with respect to every sentence of the form

$$|\{e\}(t)| = \sigma \qquad \text{or} \qquad (E\sigma)[|\{e\}(t)| = \sigma].$$

(ii) Suppose there is a $p \in G$ such that $p \Vdash |\{m\}(t)| = \sigma$. Then G is generic with respect to

$$\begin{aligned}&(u)(E\tau)[u \in \mathscr{T}(p, m, t, \sigma)\\ &\qquad \& \; u \in \{m\}(t) \rightarrow |\{n\}(u)| = \tau],\end{aligned}$$

and to every sentence of the form $s \in \{m\}(t)$ for all $s \in \mathscr{T}(p, m, t, \sigma)$.

THEOREM 3.5. *Suppose A is E-closed and admits Moschovakis witnesses. Let $\mathscr{P} \in A$ be countably closed in A. If G is $\mathscr{P}$-generic over A, then $A(G)$ is E-closed.*

Proof. Let $a \in A$. Suppose $|\{e\}(t)| \leqslant \kappa$, where κ is the least ordinal not in A, with the intent of showing $|\{e\}(t)| < \kappa$. As usual assume $e = 2^m \cdot 3^n$. Then there are $p \in G$ and $\sigma < \kappa$ such that

$$p \Vdash |\{m\}(t)| = \sigma$$

and

$$\begin{aligned}p \Vdash \; &(u)(E\tau)[u \in \mathscr{T}(p, m, t, \sigma)\\ &\qquad \& \; u \in \{m\}(t) \rightarrow |\{n\}(u)| = \tau].\end{aligned}$$

For each $s \in \mathscr{T}(p, m, t, \sigma)$ and $q \leqslant p$, there is an $r \leqslant q$ such that either

(i) $r \Vdash s \in \{m\}(t)$ or

(ii) $r \Vdash s \notin \{m\}(t)$.

The set of all $r \leqslant q$ such that (1) holds is recursive in $q, s, t\ \sigma, \mathscr{P}$ by Lemma 3.1. If r satisfies (i) then

$$r \Vdash {}^*|\{n\}(s)| \leqslant \rho$$

for some ρ recursive in $r, s, \mathscr{P}$ by Lemma 3.4. The effective bounding principle, as in proof of Lemma 3.2, yields a bound γ on ρ (for *all* relevant r and s) such that γ is recursive in $p, t, \mathscr{P}$. Since A is E-closed, $\gamma \in A$. Thus p forces $|\{e\}(t)|$ to be at worst $\max(\sigma, \gamma)$. ∎

The proof of Theorem 3.5 shows: if p forces $|\{e\}(a, \mathscr{G})| \leqslant \kappa$, then p forces $|\{e\}(a, \mathscr{G})|$ to be less than $\kappa_0^{p, a, \mathscr{P}}$. Further information can be obtained if A is of the form $L(\kappa)$. For example, for each $a \in L(\kappa)$ and p, there must be a $q \geqslant p$ such that q forces $K_r^{a, \mathscr{G}} \leqslant \kappa_r^{p, a, \mathscr{P}}$. Such matters are discussed in [9] and [13].

The present section concludes with an application of Theorem 3.5 needed for the main result of the next section. Recall Proposition 2.5. The next result states that under suitable hypotheses, the greatest cardinal of an E-closed structure can be collapsed to ω_1.

Corrolary 3.6. *Assume $L(\kappa)$ is countable, E-closed and $\sum_1$ inadmissible. Suppose* $\mathrm{gc}(\kappa)$, *the greatest cardinal in the sense of $L(\kappa)$, has uncountable cofinality in $L(\kappa)$. Then there exists an onto*

$$f\colon\ \omega_1^{L(\kappa)} \to \mathrm{gc}(\kappa)$$

such that $L(\kappa, f)$ is E-closed, its greatest cardinal is $\omega_1^{L(\kappa)}$, and $\omega_1^{L(\kappa)} = \omega_1^{L(\kappa, f)}$.

Proof. Let $\mathscr{P}$ be the set of all $p \in L(\kappa)$ of the form

$$p\colon\ \delta \to \mathrm{gc}(\kappa) \qquad (\delta < \omega_1^{L(\kappa)}).$$

$\mathscr{P}$ is an element of $L(\kappa)$ by a standard condensation argument, and $\mathscr{P}$ is countably closed in $L(\kappa)$, since $\mathrm{cf}(\mathrm{gc}(\kappa)) > \omega$ in $L(\kappa)$. By Lemma 2.4; $L(\kappa)$ admits Moschovakis witnesses. Let

$$f\colon \omega_1^{L(\kappa)} \to \mathrm{gc}(\kappa)$$

be $\mathscr{P}$-generic. Then f is an onto map, and by Theorem 3.5, $L(\kappa, f)$ is E-closed. $\omega_1^{L(\kappa)} = \omega_1^{L(\kappa, f)}$ because $\mathscr{P}$ is countably closed. ∎

4. E-Pointed Perfect Forcing

Let $L(\kappa)$ be countable, E-closed and $\sum_1$ inadmissible. By Proposition 2.9, $L(\kappa)$ has a greatest cardinal, namely $\mathrm{gc}(\kappa)$. Assume $\mathrm{gc}(\kappa)$ has uncountable cofinality in $L(\kappa)$. In this section it will be shown that there exists a $T \subseteq \mathrm{gc}(\kappa)$ such that $L(\kappa, T)$ is the E-closure of T. T will be generic with respect to uncountable, E-pointed, perfect forcing conditions. The ideas underlying the proof of Theorem 3.5 will guide the proof that $L(\kappa, T)$ is E-closed. Some modifications are necessary because the collection of forcing conditions is now a "class" rather than a "set." Pointed, perfect forcing was applied in [7]. Iterated perfect forcing was developed by Baumgartner and Laver [14] for countable conditions, and by Kanamori [15] for uncountable conditions.

By Corollary 3.6 there is an f such that

$$L(\kappa, f) \models [\text{There is only one uncountable cardinal.}]$$

and $L(\kappa, f)$ is E-closed. The forcing conditions will belong to $L(\kappa, f)$, and the desired T will be generic over $L(\kappa, f)$, and the desired T will be generic over $L(\kappa, f)$. Let ω_1 denote the greatest cardinal of $L(\kappa, f)$. Since f can be construed as a subset of ω_1, it follows that each bounded subset of ω_1 (in $L(\kappa, f)$) is constructible from a countable (in $L(\kappa, f)$) initial segment of f via an ordinal less than ω_1. Thus Seq, defined by

$$\sigma \in \mathrm{Seq} \leftrightarrow \sigma\colon \alpha \to \{0, 1\} \;\&\; \alpha < \omega_1 \;\&\; \sigma \in L(\kappa, f),$$

is a member of $L(\kappa, f)$.

Suppose $p \subseteq \mathrm{Seq}$. If $\sigma \in p$, then σ is said to *split* in p if $\sigma * \langle 0 \rangle$ and $\sigma * \langle 1 \rangle$ belong to p. (If $\mathrm{domain}(\sigma) = \alpha$, then $\mathrm{domain}(\sigma * \langle 0 \rangle) = \alpha + 1$ and $(\sigma * \langle 0 \rangle)(\alpha) = 0$.) p is *perfect* if:

(a) $(\sigma)_{\sigma \in p}(\alpha)_{\alpha \in \mathrm{dom}(\sigma)}[(\sigma \restriction \alpha) \in p]$.

(b) $(\sigma)_{\sigma \in p}(Et)_{t \in p}[\sigma \subseteq t \;\&\; t \text{ splits in } p]$.

(c) $(\alpha)_{\alpha < \omega_1}(\sigma)[(\mathrm{domain}(\sigma) = \alpha \;\&\; \alpha \text{ is a limit } \&\; (\beta)_{\beta < \alpha}((\sigma \restriction \beta) \in p)) \to \sigma \in p]$.

(d) $(f)_{f \in p^{\omega}}[(m)(n)(n < m \to f(n) \subseteq f(m)) \;\&\; (n)(f(n) \text{ splits in } p) \to \bigcup_n f(n) \text{ splits in } p]$.

Clause (d) says that p splits on a *closed* unbounded subset of each branch through p. A branch is a function $g\colon \omega_1 \to 2$ such that every countable initial segment of g belongs to p. Let $[p]$ be the collection of all branches through p.

p is *E-pointed* if

$$(E\delta)_{\delta < \omega_1}(T)[T \in [p] \to p \leqslant_E T, \delta, f].$$

In short P can be computed from any branch through p.

Let P be the class of all E-pointed, perfect p's in $L(\kappa, f)$. Note that $P \notin L(\kappa, f)$. Let $p, q, r, \ldots$, denote elements of P. $p \geqslant q$ (read p is extended by q) if $p \supseteq q$. Define $\mathscr{P}$ to be $\langle P, \geqslant \rangle$.

PROPOSITION 4.1. *$\mathscr{P}$ is countably closed in $L(\kappa, f)$.*

Proof. Let p_n $(n < \omega)$ be an infinite descending sequence in $\mathscr{P}$, and let q be $\bigcap\{p_n \mid n < \omega\}$. q satisfies clause (d) of the definition of perfect, because the filter of closed unbounded subsets of ω_1 is countably closed. To see that q is E-pointed, observe that each countable sequence of countable ordinals is E-recursive in some countable ordinal. Thus the δ_n's needed to compute the p_n's from a common branch T can be combined into a single δ that serves to compute q from T.

Suppose $X, Y \subseteq \omega_1$. X and Y are said to have the same *degree* if there exist $\alpha, \beta < \omega_1$ such that

$$X \leqslant_E Y, \quad \alpha, f \qquad \text{and} \qquad Y \leqslant_E X, \beta, f.$$

PROPOSITION 4.2. *The degrees of the branches through p are just the degrees greater than or equal to that of p.*

Proof. Suppose $p \leqslant_E X, \alpha, f$ for some $\sigma < \omega_1$. A $T \in [p]$ is defined by recursion on ω_1. If $T \restriction \gamma$ does not split in p, then $T(\gamma) = 0$ if $T \restriction \gamma * \langle 0 \rangle \in p$, and $= 1$ otherwise. Assume $T \restriction \gamma$ splits in p. Let β be the order type of

$$\{\delta \mid \delta < \gamma \;\&\; T \restriction \delta \text{ splits in } p\}.$$

Then

$$T(\gamma) = \begin{cases} 0 & \text{if } \beta \in X \\ 1 & \text{otherwise.} \end{cases}$$

T winds its way through P in the same manner that X does through 2^{ω_1}. Consequently

$$T \leqslant_E p, X \qquad \text{and} \qquad X \leqslant_E p, T.$$

The initial assumption on X implies the degree of T is at most that of X. The E-pointedness of p implies the degree of X is at most that of T. ∎

In the light of Proposition 4.2, an E-pointed p represents a cone of degrees whose vertex is the degree of p. If $T \in [p]$, then p "forces" the degree of T to be at least that of p without "forcing" any bound on the degree of T.

p is said to be *E-pointed via δ* $(\delta < \omega_1)$ if

$$(T)[T \in [p] \rightarrow p \leqslant_E T, \delta, f].$$

PROPOSITION 4.3. *Suppose* $Y \subseteq \omega_1$ *and* p *is E-pointed via* δ. *Then there exists a* q *such that* $p \geqslant q$, $q \leqslant_E p, Y$, *and* $Y \leqslant_E q, f, \delta$.

Proof. $\mathscr{s} \in q$ is defined by recursion on the domain of $\mathscr{s}$ with the intent of coding Y into every branch of $[q]$.

Suppose $\mathscr{s} \in q$ and $\mathscr{s}$ does not split in p. Then $\mathscr{s}*\langle 0\rangle \in q$ if $\mathscr{s}*\langle 0\rangle \in p$. Otherwise $\mathscr{s}*\langle 1\rangle \in q$.

Suppose $\mathscr{s}$ splits in p. Let τ be the order type of

$$\{\gamma \mid \gamma < \text{domain}(\mathscr{s}) \;\&\; \mathscr{s} \upharpoonright \gamma \text{ splits in } p\}.$$

If $\tau = \beta + 1$, then

$$\mathscr{s}*\langle 0\rangle \in q \qquad \text{if } \beta \in Y,$$

$$\mathscr{s}*\langle 1\rangle \in q \qquad \text{if } \beta \in Y.$$

If τ is a limit, then $\mathscr{s}$ splits in q.

By construction $p \leqslant q$. To see q is E-pointed, fix $T \in [q]$. Then $T \in [p]$, and so $p \leqslant_E T, f, \delta$. By construction $Y \leqslant_E T, p$. Thus $Y \leqslant_E T, f, \delta$. $q \leqslant_E p, Y$, so $q \leqslant_E T, f, \delta$.

The left-most branch of $[q]$ is recursive in q, hence $Y \leqslant_E q, f, \delta$. ∎

The language $\mathscr{L}(\kappa, f, \mathscr{T})$ has the power to analyze the computation of $\{e\}(x)$ when $x \in \mathscr{L}(\kappa, f, T)$ and $|\{e\}(x)| < \kappa$. The primitive terms are: $\mathscr{T}, f, a$ for each $a \in L(\kappa)$, and σ for each ordinal $\sigma < \kappa$. If $t_1, \ldots, t_n$ are terms, then $\{e\}(t_1, \ldots, t_n)$ is a term. The language also includes: $\in$, $|\;|$, variables, quantifiers and propositional connectives. The relations

$$L(\kappa, f, T) \models |\{e\}(t)| = \sigma,$$

$s \in \mathscr{T}(e, t, \sigma)$ and

$$L(\kappa, f, T) \models s \in \{e\}(t)$$

are defined simultaneously by recursion on σ. $\mathscr{T}(e, t, \sigma)$ is a set of terms that suffice to name the elements of $\{e\}(t)$ when $|\{e\}(t)| = \sigma$. For example, suppose $e = 2^m \cdot 3^n$.

Then $L(\kappa, f, T) \models |\{2^m \cdot 3^n\}(t)| = \sigma$ if there is a $\gamma < \sigma$ such that the following are true in $L(\kappa, f, T)$:

$$|\{m\}(t)| = \gamma,$$

$$(u)(E\tau)_{\tau<\sigma}[u \in \mathscr{T}(m, t, \gamma) \;\&\; u \in \{m\}(t) \rightarrow |\{n\}(u)| = \tau],$$

$$(\tau)_{\tau<\sigma}(Eu)[(u \in \mathscr{T}(m, t, \gamma) \;\&\; u \in \{m\}(t) \;\&\; |\{n\}(u)| \geqslant \tau) \vee |\{m\}(t)| = \tau].$$

$$\mathscr{T}(2^m \cdot 3^n, t, \sigma) = \{\{n\}(s) \mid (E\gamma)_{\gamma<\sigma}(s \in \mathscr{T}(m, t, \gamma))\};$$

and

$$L(\kappa, f, T) \models \{n\}(s) \in \{2^m \cdot 3^n\}(t)$$

if $s \in \mathcal{T}(m, t, \sigma)$ and $L(\kappa, f, T) \models s \in \{m\}(t)$.

A sentence $\mathcal{F}$ is said to be *ranked* (and of rank at most σ) if it is of the form $|\{e\}(t)| = \sigma$, or $s \in \{e\}(t)$, where $s \in \mathcal{T}(e, t, \sigma)$. The *forcing relation* $p \Vdash \mathcal{F}$ is *defined by*

$$(T)[T \in [p] \rightarrow L(\kappa, f, T) \models \mathcal{F}]$$

when $\mathcal{F}$ is ranked, and in a standard fashion when $\mathcal{F}$ is unranked (e.g., has an unbounded ordinal quantifier).

The weak forcing relation $p \Vdash {}^*\mathcal{F}$ is given by $(q)_{p \geqslant q}(Er)_{q \geqslant r}[r \Vdash \mathcal{F}]$.

The reason for defining forcing for ranked sentences in terms of truth is to avoid quantification over a class of forcing conditions. The next lemma shows that the definition of $\Vdash$ is workable.

LEMMA 4.4. *Suppose $\mathcal{F}$ has rank at most σ. Let W be a well ordering of ω_1 of height σ. Then for each p, there is a q such that $p \leqslant q$, either $q \Vdash \mathcal{F}$ or $q \Vdash \sim\mathcal{F}$, and $q \leqslant_E p$, W, $\ulcorner \mathcal{F} \urcorner$ (uniformly).*

Proof. By Proposition 4.3 there is an r such that $p \geqslant r$, $r \leqslant_E W$, $\ulcorner \mathcal{F} \urcorner$ and $\langle W, \ulcorner \mathcal{F} \urcorner \rangle \leqslant_E r, f, \delta_0$ for some $\delta_0 < \omega_1$. q will be constructed by local forcing. Each branch through q will be generic in the sense of $\Vdash^r_\sigma$. The forcing conditions associated with $\Vdash^r_\sigma$ are the elements of r. The only sentences eligible for forcing are those of rank at most σ from the language $\mathcal{L}(\kappa, f, \mathcal{T})$. The definition of $\Vdash^r_\sigma$ is in essence the same as that of $\Vdash$, given at the beginning of Section 3. The only changes are the restrictions to r and σ.

Let $W^0 = \{\mathcal{F}_\sigma \mid \delta < \omega_1\}$ be a well ordering of all sentences of $\mathcal{L}(\kappa, f, \mathcal{T})$ that arise in the unraveling of the definition of $\Vdash \mathcal{F}$. Note that $W^0 \leqslant_E W$, $\ulcorner \mathcal{F} \urcorner$. A map h: Seq $\rightarrow r$ is defined by recursion on the length α of $t \in$ Seq. The range of h is the desired q.

$\alpha = 0$: $h(t)$ is the shortest $s \in r$ such that either $s \Vdash^r_\sigma \mathcal{F}$ or $s \Vdash^r_\sigma \sim\mathcal{F}$. (Thus $h(0)$ decides whether q forces $\mathcal{F}$ or $\sim\mathcal{F}$.)

$\alpha = \beta + 1$, where β is not a limit: let $\langle s_0, s_1 \rangle$ be the least pair of incomparable extensions of $h(t \restriction (\beta + 1))$ in r such that for each $i < 2$, either

$$s_i \Vdash^r_\sigma \mathcal{F}_\beta \quad \text{or} \quad s_i \Vdash^r_\sigma \sim\mathcal{F}_\beta.$$

Define $h((t \restriction (\beta + 1))^* \langle i \rangle) = s_i$.

$$\alpha = \lambda: \qquad h(t) = \bigcup \{h(t \restriction \gamma) \mid \gamma < \lambda\},$$

$$\alpha = \lambda + 1: \qquad h(t \restriction \lambda)^* \langle i \rangle = h(t \restriction \lambda)^* \langle i \rangle.$$

(Note that $h(t \upharpoonright \lambda)$ must split in r, because $h(t \upharpoonright (\beta+1]))$ splits when β is not a limit.)

With the aid of Lemma 3.1, $q \leqslant_E r, W, \ulcorner \mathscr{F} \urcorner$, hence $q \leqslant_E p, W, \ulcorner \mathscr{F} \urcorner$ (uniformly). q is perfect because it is a homeomorphic image of Seq. To see q is E-pointed, fix $T \in [q]$. Then $T \in [p]$, and so $p \leqslant_E, T, f, \delta_1$ for some $\delta_1 < \omega_1$. Thus $q \leqslant_E T, f, \delta_1, r, \delta_0$. But $T \in [r]$, so $r \leqslant_E T, f, \delta_2$ for some $\delta_2 < \omega_1$. Consequently $q \leqslant_E T, f, \delta_0, \delta_1, \delta_2$. ∎

As in Section 3 the relation $>_V$ is invoked to obtain an effective bound on $|\{e\}(t)|$ when p weakly forces $|\{e\}(t)| < \kappa$. The proof of the next lemma is similar to that of Lemma 3.4 save for a fusion construction needed to overcome the failure of $\mathscr{P}$ to be an element of $L(\kappa)$.

LEMMA 4.5. *Assume* $\ulcorner t \urcorner \subseteq \omega_1$. *Suppose* $>_V$ *is well founded below* $\langle p, e, t\rangle$. *If*

$$p \Vdash^* (E\sigma)[|\{e\}(t)| = \sigma],$$

then there exists $\langle q, \gamma, W(\gamma)\rangle$ *recursive in* $p, \ulcorner t \urcorner, f$ *such that* $p \geqslant q$,

$$q \Vdash |\{e\}(t)| = \gamma,$$

and $W(\gamma)$ *is a well ordering of* ω_1 *of height* γ.

Proof. $\langle q, \gamma, W(\gamma)\rangle$ is computed by an effective transfinite recursion on the height of $\langle p, e, t\rangle$ in $>_V$. As always assume $e = 2^m \cdot 3^n$. By recursion there exists $\langle p_0, \gamma_0, W(\gamma_0)\rangle$ recursive in $p, \ulcorner t \urcorner, f$ such that $p \geqslant p_0$,

$$p_0 \Vdash |\{m\}(t)| = \gamma_0,$$

and $W(\gamma_0)$ is a well ordering of ω_1 of height γ_0. By Proposition 4.3 there is a p_1 such that $p_0 \geqslant p_1$, $p_1 \leqslant_E p_0, W(\gamma_0), \ulcorner t \urcorner$, and $\langle W(\gamma_0), \ulcorner t \urcorner\rangle \leqslant_E p_1, f, \delta_0$ for some $\delta_0 < \omega_1$.

Let $W^0(\gamma_0) = \{t_\alpha \mid \alpha < \omega_1\}$ be a well ordering of all terms that belong to $\mathscr{T}(m, t, \gamma_0)$. A straightforward definition of $W^0(\gamma_0)$ yields: $W^0(\gamma_0) \leqslant_E W(\gamma_0), \ulcorner t \urcorner$; and $\ulcorner t_\alpha \urcorner \subseteq \omega$, for all $\alpha < \omega_1$.

A contracting sequence $\{q_\alpha \mid \alpha < \omega_1\}$ of forcing conditions is defined by effective transfinite recursion on α.

$$q_0 = p_1, \qquad q_\lambda = \bigcap \{q_\alpha \mid \alpha < \lambda\},$$

$$q_{\alpha+1} = \bigcup \{q^{\jmath}_{\alpha+1} \mid \jmath \in q_\alpha \cap 2^{\alpha+1}\}.$$

The definition of $q^{\jmath}_{\alpha+1}$ has two parts. First, by Lemma 4.4, there is an r_α such that:

$$q_\alpha \geqslant r_\alpha, \qquad r_\alpha \cap 2^{\alpha+1} = \{\jmath\};$$

$$r_\alpha \leqslant_E q_\alpha, \jmath, W(\gamma_0), \ulcorner t \urcorner, \ulcorner t_\alpha \urcorner;$$

and

(i) $r_\alpha \Vdash t_\alpha \in \{m\}(t)$ or

(ii) $r_\alpha \Vdash t_\alpha \notin \{m\}(t)$.

If (ii) holds, then $q^{s}_{\alpha+1} = r_\alpha$. If (i) holds, then

$$\langle p, e, t\rangle >_V \langle r_\alpha, n, t_\alpha\rangle;$$

and so by recursion on $>_V$ there is $\langle q^{s}_{\alpha+1}, \gamma^{s}_{\alpha+1}, W(\gamma_{\alpha+1})\rangle$ recursive in r_α, $\ulcorner t_\alpha \urcorner$, f (uniformly by Gandy selection) such that $r_\alpha \geqslant q^{s}_{\alpha+1}$,

$$q^{s}_{\alpha+1} \Vdash |\{n\}(t_\alpha)| = \gamma^{s}_{\alpha+1},$$

and $W(\gamma_{\alpha+1})$ is a well ordering of height $\gamma_{\alpha+1}$. Define

$$q_\infty = \bigcap\{q_\alpha \mid \alpha < \omega_1\}$$

and

$$\begin{aligned}\gamma_\infty = {} & \text{strict upper bound of} \\ & \{\gamma_0, \gamma^{s}_{\alpha+1} \mid \alpha < \omega_1 \;\&\; s \in q_\alpha \cap 2^{\alpha+1}\}.\end{aligned}$$

Then $q_\infty \leqslant p_1$, $W(\gamma_0)$, $\ulcorner t \urcorner$, $W^0(\gamma_0) \leqslant_E p$, $\ulcorner t \urcorner$, f. To see q_∞ is E-pointed, fix $T \in [q_\infty]$. Then $T \in p_1$, hence $p_1 \leqslant_E$, T, f, δ_1 for some countable δ_1 not depending on T. Thus $q_\infty \leqslant T, f, \delta_1, \delta_0$. To check that $[q_\infty]$ is perfect, let

$$C_\alpha = \{\beta \mid T \restriction \beta \text{ splits in } q_\alpha\}$$

for all $\alpha \leqslant \omega_1$. C_α $(\alpha \leqslant \omega_1)$ is closed and unbounded, hence $C_{\omega_1} = \bigcap\{C_\alpha \mid \alpha < \omega_1\}$ is closed. To verify that C_{ω_1} is unbounded, fix $\delta_0 < \omega_1$. Choose $\delta_{n+1} \in C_{\delta_n}$ so that $\delta_{n+1} > \delta_n$. Let $\delta_\omega = \bigcup\{\delta_n \mid n < \omega\}$. Then

$$\delta_\omega \in \bigcap\{C_\beta \mid \beta \in \delta_\omega\} = C_{\delta_\omega},$$

and so $T \restriction \delta_\omega$ splits in C_{δ_ω}. For all $\alpha \geqslant \delta_\omega$,

$$q_\alpha \cap 2^{\delta_\omega + 1} = q_{\delta_\omega} \cap 2^{\delta_\omega + 1}.$$

Hence $\delta_\omega \in C$.

Since $q_\infty \Vdash |\{e\}(t)| \leqslant \gamma_\infty$, it follows from a local forcing construction as in Lemma 4.4 that there exists a q such that $q_\infty \geqslant q$,

$$q \Vdash (E\beta)_{\beta \leqslant \gamma_0}[|\{e\}(t)| = \beta],$$

and $q \leqslant_E q_\infty$, $W(\gamma_\infty)$, $\ulcorner t \urcorner$. The desired γ is the definite value of $|\{e\}(t)|$ forced by q. $W(\gamma_\infty)$ can be obtained effectively from the $W(\gamma_{\alpha+1})$'s. ∎

A closer examination of Lemmas 3.2 and 4.5 will clarify the proof of Lemma 4.6. In both 3.2 and 4.5 a partial recursive function $\{g\}$ was defined by effective transfinite recursion. It was shown by induction that if $<_V$ is well founded below $\langle p, e, t\rangle$ then $\{g\}(p, e, t)\downarrow$ and its value includes an ordinal γ and a condition $q \leqslant p$ such that q forces $|\{e\}(t)|$ to be γ.

There is a second viewpoint. The partial recursive function $\{g\}$ is defined formally via a fixed point construction without reference to $<_V$. If $\{g\}(p, e, t)\downarrow$, then $<_U$ below $\langle g, \langle p, e, t\rangle\rangle$ is a well-founded relation closely associated with $<_V$ below $\langle p, e, t\rangle$. The arguments of 3.2 and 4.5 applied to $<_U$ below $\langle g, \langle p, e, t\rangle\rangle$ show that the value of $\{g\}(p, e, t)$ includes a γ and a q as above.

The second approach yields no new information concerning set forcing, because in that case $<_U$ below $\langle g, e, t\rangle\rangle$ is essentially the same as $<_V$ below $\langle p, e, t\rangle$. It is in the context of class forcing that further insight is gained. If $\mathscr{P}$ is a class, then the immediate predecessors of $\langle p, e, t\rangle$ in $<_V$ form a class rather than a set, and this is so even if $<_V$ below $\langle p, e, t\rangle$ is well founded with height less than κ. The fusion construction of Lemma 4.5 supplies a q, and a set version of $<_V$ below $\langle q, e, t\rangle$ extracted from $<_U$ below $\langle g, \langle p, e, t\rangle\rangle$. The set version of $<_V$ below $\langle q, e, t\rangle$ is dense in the full class version of $<_V$ below $\langle q, e, t\rangle$.

According to the second viewpoint the two cases that occur in the analysis of $\langle p, e, t\rangle$ are $\{g\}(p, e, t)$ converges or diverges, rather than $<_V$ is well founded or ill founded below $\langle p, e, t\rangle$.

LEMMA 4.6. *If* $\ulcorner t \urcorner \subseteq \omega_1$ *and*

$$p \Vdash^* (E\sigma)[|\{e\}(t)| = \sigma],$$

then $\{g\}(p, e, t)\downarrow$.

Proof. Virtually identical with that of Lemma 3.3. If $\{g\}(p, e, t)\uparrow$, then there is a Moschovakis witness to the divergence of $\{g\}(p, e, t)$ in $L(\kappa, f)$ by Lemma 2.4. The witness yields an infinite descending sequence in $<_V$ below $\langle p, e, t\rangle$. Since $\mathscr{P}$ is countably closed by Lemma 4.1, the infinite descending sequence contracts to a q below p that forces $\{e\}(t)$ to diverge. ∎

LEMMA 4.7. *If* $\ulcorner t \urcorner \subseteq \omega_1$ *and*

$$p \Vdash^* (E\sigma)[|\{e\}(t)| = \sigma],$$

then there exist $\langle q, \gamma, W(\gamma)\rangle$ *recursive in* $p, \ulcorner t \urcorner f$ *such that* $p \geqslant q$,

$$q \Vdash |\{e\}(t)| = \gamma,$$

and $W(\gamma)$ *is a wellordering of* ω_1 *of height* γ.

Proof. By Lemmas 4.5 and 4.6, and the intervening remarks. ∎

Let $T \subseteq \omega_1$. T is $\mathscr{P}$-generic on a sentence $\mathscr{F}$ of $\mathscr{L}(\kappa, f, \mathscr{T})$ if there is a p such that $T \in [p]$, and either $p \Vdash \mathscr{F}$ or $p \Vdash \sim \mathscr{F}$. T is $\mathscr{P}$-generic if:

(i) T is $\mathscr{P}$-generic on every $\mathscr{F}$ that is ranked or of the form $(E\sigma)[|\{e\}(t)| = \sigma]$ and

(ii) if $T \in [p]$ and $p \Vdash |\{m\}(t)| = \sigma$, then there is a $q \leqslant p$ such that $T \in [p]$ and q forces the following sentence or its negation,

$$(u)(E\tau)[u \in \mathscr{T}(m, t, \sigma) \,\&\, u \in \{m\}(t) \rightarrow |\{n\}(u)| = \tau].$$

THEOREM 4.8. *Let* $L(\kappa)$ *be countable, E-closed and* $\sum_1$ *inadmissible. Suppose the greatest cardinal of* $L(\kappa)$ *has uncountable cofinality in* $L(\kappa)$. *Then there exists an* $S \subseteq \omega_1^{L(\kappa)}$ *such that* $L(\kappa, S)$ *is the E-closure of S.*

Proof. Let T be $\mathscr{P}$-generic as above. Then S is $\langle f, T\rangle$ encoded as a subset of $\omega_1^{L(\kappa)}$. By Proposition 4.3 and the genericity of T, each $Y \in 2^{\omega_1} \cap L(\kappa, f)$ is recursive in T, f, δ for some countable δ. Hence $E(S) \supseteq L(\kappa, f, T)$. To check $E(S) \subseteq L(\kappa, f, T)$, suppose $\ulcorner t \urcorner \leqslant \omega_1$ and

$$L(\kappa, f, T) \models |\{2^m \cdot 3^n\}(t)| \leqslant \kappa.$$

Then there is a p such that $T \in [p]$, $p \Vdash (E\sigma)[|\{m\}(t)| = \sigma]$, and

$$p \Vdash (u)(E\tau)[u \in \mathscr{T}(m, t, \sigma) \,\&\, u \in \{m\}(t) \\ \rightarrow |\{n\}(u)| = \tau].$$

By Lemma 4.7 there is $\langle p_0, \gamma_0, W(\gamma_0)\rangle$ recursive in $p, \ulcorner t \urcorner, f$ such that $p \geqslant p_0$,

$$p_0 \Vdash |\{m\}(t)| = \gamma_0,$$

and $W(\gamma_0)$ is a wellordering of ω_1 of height γ_0. Now the fusion argument of Lemma 4.5 can be repeated to obtained q_∞ and γ_∞ such that $p_0 \geqslant q_\infty$,

$$q_\infty \Vdash^* |\{2^m \cdot 3^n\}(t)| \leqslant \gamma_\infty,$$

and $\gamma_\infty \leqslant_E p, \ulcorner t \urcorner, f$. Thus q_∞ weakly forces $|\{2^m \cdot 3^n\}(t)|$ to be less than κ. The set of such q_∞'s is dense in p, and so p weakly forces $(E\sigma)[|\{2^m \cdot 3^n\}(t)| = \sigma]$. (Note that p does not in general force a universal bound on $|\{2^m \cdot 3^n\}(t)|$ for all generic elements of $[p]$. ∎

5. Selection

In this section a selection theorem is proved to obtain a partial converse to Theorem 4.8. The theorem is inspired by a result of Y. Moschovakis [18] (also cf. Kirousis [20]). Let $E(R(\alpha))$ be the E-closure of the set of all sets of rank less than α. Moschovakis showed: if α has countable cofinality in $E(R(\alpha))$, then $E(R(\alpha))$ is Σ_1 admissible. Corollary 5.2 states: let X be a set of ordinals; if in $E(X)$ the greatest cardinal has cofinality ω, then $E(X)$ is Σ_1 admissible. Corollary 5.3 is the intended partial converse to Theorem 4.8. Let X be a set of ordinals. Some fundamental facts concerning the structure of $E(X)$ will prove helpful below.

(R0) Let $a, b, c, \ldots, \in \sup X$. For all z, $z \in E(X)$ iff $z \leqslant_E X, a$ for some a.

(R1) Each element of $E(X)$ can be coded by a subset of $\sup X$. More precisely, if $z \in E(X)$, then there exists a $z^* \subseteq \sup X$ such that z is recursive in z^*, and $z^* \in E(X)$. z^* is defined by a recursion on the length of computation of z from x.

(R2) In $E(X)$ there is a greatest cardinal.

Call it $\mathrm{gc}(E(X))$. It is safe to assume $\mathrm{gc}(E(X))$ is $\sup X$. More precisely, there exists an $X_0 \subseteq \sup X$ such that $E(X_0) = E(X)$ and $\mathrm{gc}(E(X_0)) = \sup X_0$.

Let $A \subseteq \sup X$. Recall from Section 2 that A is said to be recursively enumerable in X if there exists an e such that for all $a \in \sup X$,

$$a \in A \leftrightarrow \{e\}(a, X)\downarrow.$$

Each a corresponds to a node $\langle e, \langle a, X\rangle\rangle$ on the universal computation tree $>_U$ defined in Section 2. a belongs to A iff $>_U$ is well founded beneath $\langle e, \langle a, X\rangle\rangle$. For simplicity the node $\langle e, \langle a, X\rangle\rangle$ will be written a. Thus $a \in A$ iff $a\downarrow$. Fact (R1) makes it possible to think of each node below $\langle e, \langle a, X\rangle\rangle$ as being of the form $\langle e_0, \langle b, X\rangle\rangle$ for some $b \in \sup X$, or more simply b. Define

$$\min A = \min\{|a| \mid a \in A\}.$$

$|a|$ is the height of $>_U$ beneath a. $|a|$ is defined by recursion to be the strict least upper bound ($\sup^+$) of all $|b|$ such that b is an immediate subcomputation of a (as defined in Sect. 2).

Theorem 5.1. *Let X be a set of ordinals. Suppose in $E(X)$ there is an ascending sequence $\{\kappa_j \mid j < \omega\}$ of cardinals (in the sense of $E(X)$) such that*

$$\sup X = \mathrm{gc}(E(X)) = \sup\{\kappa_j \mid j < \omega\}.$$

If $A \subseteq \sup X$ is nonempty and recursively enumerable in X, then min A is recursive in X, $\{\kappa_j \mid j < \omega\}$ (uniformly).

Proof. The predicate, b is an immediate subcomputation of a, is recursively enumerable. The following modification, b is an immediate computation of a via β, is recursive. The idea is to let β bound the height of the computation needed to show b is an immediate subcomputation of a. For example:

(1_β) $\langle m, u\rangle$ is an immediate subcomputation of $\langle 2^m \cdot 3^n, u\rangle$ via β.

(2_β) If $\{m\}(u)\downarrow$, $|\{m\}(u)| \leqslant \beta$, and $v \in \{m\}(u)$, then $\langle n, v\rangle$ is an immediate subcomputation of $\langle 2^m \cdot 3^n, u\rangle$ via β.

The theorem is proved by an effective transfinite recursion on min A, henceforth called the main recursion. There are countably many cases woven together at the finish by the Gandy selection principle.

Case $2^i \cdot 3^j$. Suppose $i < j$ and there is an $a \in A \cap \kappa_i$ such that

$$\begin{aligned}\min A &= \min(A \cap \kappa_i) = |a| \\ &= \sup{}^+\{|b| \mid b < \kappa_j \text{ \& } b \text{ is immediate subcomputation of } a\}.\end{aligned}$$

In this case $\min(A \cap \kappa_i)$ is computed by a recursion of length κ_{j+1}. Fix $\alpha < \kappa_{j+1}$ and assume $\beta(\gamma)$ has been computed for all $\gamma < \alpha$. Define

$$\beta^-(\alpha) = \sup\{\beta(\gamma) \mid \gamma < \alpha\}.$$

The definition of $\beta(\alpha)$ has two subcases. In the first, $\beta^-(\alpha) + 1 \geqslant \min(A \cap \kappa_i)$ and $\beta(\alpha)$ is defined to be $\min(A \cap \kappa_i)$. In the second,

$$\beta^-(\alpha) + 1 < \min(A \cap \kappa_i),$$

and $\beta(\alpha)$ is computed as follows. Let Z_α be the set of all b such that

> $b < \kappa_j$, $b\downarrow$, $\beta^-(\alpha) < |b|$ and b is an immediate subcomputation of some $a \in \kappa_i$ via $\beta^-(\alpha)$.

Z_α is nonempty thanks to the a mentioned in the supposition of Case $2^i \cdot 3^j$. Z_α is recursively enumerable in $\beta^-(\alpha)$, X, κ_i, hence the main recursion can be applied to compute min Z_α. Define

$$\beta(\alpha) = \min Z_\alpha.$$

It follows that $\{\beta(\alpha) \mid \alpha < \kappa_{j+1}\}$ is recursive in X and the κ_n sequence. Note that for all sufficiently large α, $\beta(\alpha)$ is $\min(A \cap \kappa_i)$. If not, then for all sufficiently large α, the second subcase holds. For each such α there is a $b \in \kappa_j$ such that

$$\beta(\alpha) = |b| \qquad \text{and} \qquad \beta(\alpha) > \beta(\gamma) \quad \text{for all } \gamma < \alpha.$$

Each such b is associated with at most one α, but there are κ_{j+1} α's.

Case 5^i. Suppose $\min A = \min(A \cap \kappa_i)$. Let $\gamma(j, a)$ be the supremum$^+$ of all $|b|$ such that $b < \kappa_j$, and b is an immediate subcomputation of a. $\gamma(j, a)$ is a partial function, and is defined iff every $b < \kappa_j$ (and an immediate subcomputation of a) converges. Suppose further that

$$a \in A \cap \kappa_i \;\&\; |a| = \min A \rightarrow \gamma(j, a) < \min A$$

for all $j < \omega$. In this case $\min(A \cap \kappa_i)$ is computed by a recursion of length κ_{i+1}. Fix $\alpha < \kappa_{i+1}$ and assume $\beta(\gamma)$ has been computed for all $\gamma < \alpha$. Define

$$\beta^-(\alpha) = \sup\{\beta(\gamma) \mid \gamma < \alpha\}.$$

The definition of $\beta(\alpha)$ has two subcases. In the first, $\beta^-(\alpha) + 1 \geqslant \min(A \cap \kappa_i)$ and $\beta(\alpha)$ is $\min(A \cap \kappa_i)$. In the second,

$$\beta^-(\alpha) + 1 < \min(A \cap \kappa_i),$$

and $\beta(\alpha)$ is computed as follows. Let $\gamma(j, a, \beta)$ be the supremum$^+$ of all $|b|$ such that $b < \kappa_j$ and b is an immediate subcomputation of a via b. $\gamma(j, a, \beta)$ is partial recursive in j, a, β, and is defined iff every $b < \kappa_j$ (and an immediate subcomputation of a via β) converges. Let Y_α be the set of $\langle j, a \rangle$ such that

$$j < \omega \;\&\; a \in \kappa_i \;\&\; \beta^-(\alpha) < \gamma(j, a, \beta^-(\alpha)).$$

Y_α is nonempty by virtue of the suppositions that define Case 5^i. Y_α is recursively enumerable in $\beta^-(\alpha)$, X, κ_i, hence the main recursion can be applied to compute

$$\beta(\alpha) = \min\{\gamma(j, a, \beta^-(\alpha)) \mid \langle j, a \rangle \in Y_\alpha\}.$$

It follows that $\{\beta(\alpha) \mid \alpha < \kappa_{i+1}\}$ is recursive in X and the κ_n sequence. For all sufficiently large α, $\beta(\alpha)$ is $\min(A \cap \kappa_i)$. If not, then the second subcase obtains for all sufficiently large α. For each such α there is a $\langle j, a \rangle \in \omega \times \kappa_i$ such that

$$\beta(\alpha) = \gamma(j, a, \beta^-(\alpha)) \qquad \text{and} \qquad \beta(\alpha) > \beta(\gamma) \quad (\gamma < \alpha).$$

Each such $\langle j, a \rangle$ is associated with at most two α's, but there are κ_{i+1} α's. For a given $\langle j, a \rangle$ the first α might occur at β_0 when the only immediate subcomputation of a via β_0 is $\langle m, a, X\rangle$, and the second at β_1 when $\beta_1 > |\langle m, a, X\rangle|$.

An effective procedure has been defined for each case q above. It converges iff the suppositions of the case are true . If it converges, then it converges to min A. Let B be the set of all q such that the procedure for case q converges. Recall the Gandy selection principle from the proof of Lemma 2.3. B is a nonempty subset of ω recursively enumerable in X, $\{\kappa_i \mid i<\omega\}$. By Gandy selection, an element of B can be computed from X, $\{\kappa_i \mid i<\omega\}$. ∎

The proof of Theorem 5.1 establishes more than is stated. $E(X)$ can be replaced by an arbitrary E-closed structure $\mathscr{E}$ with an additional predicate. The only structural fact needed for the proof is somewhat weaker than the existence of a greatest cardinal of cofinality ω in $\mathscr{E}$. As above no assumption need be made about the power set operation inside $\mathscr{E}$ (cf. [13] for details).

Corollary 5.2. *Let X be a set of ordinals. If in $E(X)$ the greatest cardinal has cofinality ω, then $E(X)$ is $\sum_1$ admissible.*

Proof. Let $D(a, z)$ be Δ_0. Suppose

$$(b)_{b\in X}(Ez)_{z\in E(X)}D(b, z)$$

in the hope of bounding z. Then

$$(b)_{b\in X}(Ea)_{a\in X}(Ee)[\{e\}(X, a)\downarrow \ \& \ D(b, \{e\}(X, a))].$$

The set

$$\{|\{e\}(X, a)| \mid \{e\}(X, a)\downarrow \ \& \ D(b, \{e\}(X, a))\}$$

is recursively enumerable in X, b. Thus its min is recursive in X, b, $\{\kappa_i \mid i<\omega\}$ by Theorem 5.1. Consequently z is a computable function of b, hence bounded in $E(X)$. ∎

Corollary 5.3. *Suppose $L(\kappa)$ is E-closed and not $\sum_1$ admissible. If $\delta<\kappa$, $S\subseteq\delta$ and $E(S)=L(\kappa, S)$, then in $L(\kappa, S)$ the greatest cardinal has uncountable cofinality.*

Proof. Apply Corollary 5.2.

Corollary 5.4. *Suppose $L(\kappa)$ is E-closed and not $\sum_1$ admissible. Let $\{\omega_n^{L(\kappa)} \mid n\leqslant\omega\}$ be the set of all infinite cardinals in the sense of $L(\kappa)$. Then there do not exist $\delta<\kappa$ and $S\subseteq\delta$ such that $E(S)=L(\kappa, S)$.*

Proof. Suppose δ and S exist. Let gc(S) be the greatest cardinal in the sense of $E(S)$. By remark (R2) at the beginning of Section 5, it is safe to assume $S \subseteq \mathrm{gc}(S)$. It follows from Corollary 5.3 that $\mathrm{gc}(S) = \omega_n^{L(\kappa)}$ for some positive $n < \omega$, and that $\omega_n^{L(\kappa)}$ is regular in $E(S)$. There is a $t \in E(S)$ such that

$$t\colon\ \omega_\omega^{L(\kappa)} \xrightarrow[1-1]{\text{onto}} \omega_n^{L(\kappa)}.$$

t will be used to define a violation of the E-closedness of $E(S)$.

Suppose $Z \subseteq \omega_\omega^{L(\kappa)}$ and $Z \in L(\kappa)$. Z can be effectively coded by a countable subset of $\omega_\omega^{L(\kappa)}$ as follows. Let $z(n)$ be the least triple $\langle \gamma, p, n \rangle$ such that $Z \cap \omega_n$ is first order definable over $L(\gamma)$ via formula n with parameter p. A standard downward Skolem–Lowenheim argument shows $\langle \gamma, p, n \rangle \in \omega_{n+1}^{L(\kappa)}$. Thus $\{z(n) \mid n < \omega\}$ encodes Z, and the processes of encoding and decoding are effective.

Let Y be a countable subset of $\omega_\omega^{L(\kappa)}$ in $L(\kappa)$. Then $t(Y)$ is a bounded subset of $\omega_n^{L(\kappa)}$ in $E(S)$, since $\omega_n^{L(\kappa)}$ is uncountable and regular in $E(S)$. By a well-known theorem of Hajnal, $t(Y) \in L(\beta, S)$ for some $\beta < \omega_n^{L(\kappa)}$.

Thus every subset of $\omega_\omega^{L(\kappa)}$ in $L(\kappa)$ can be coded by a countable subset of $\omega_\omega^{L(\kappa)}$ in $E(S)$, which is constructible from S below $\omega_n^{L(\kappa)}$. This process, reversed and slightly modified, violates the E-closedness of $E(S)$. Let K be the set of all bounded subsets of $\omega_n^{L(\kappa)}$ in $E(S)$, and let $w \in K$. Define $h(w)$ as follows. If $t^{-1}[w]$ encodes a subset of $w_\omega^{L(\kappa)}$ that is the graph of a well-founded relation, then $h(w)$ is the ordinal height of that relation. Otherwise $h(w) = 0$.

h is recursive in S, t, but maps $K \in E(S)$ onto κ. ∎

6. Further Questions and Results

(Q1) Is the converse of Theorem 4.8 true? To make the question definite, let $L(\kappa)$ be countable and $\sum_1$ inadmissible, and assume there exist $\delta < \kappa$ and $S \subseteq \delta$ such that $L(\kappa, S) = E(S)$. Does it follow that gc(κ), the greatest cardinal of $L(\kappa)$, has uncountable cofinality in $L(\kappa)$? Corollary 5.3 requires the greatest cardinal of $L(\kappa, S)$ to have uncountable cofinality in $L(\kappa, S)$. It seems possible for gc(κ) to have countable cofinality in $L(\kappa)$, and to collapse in $L(\kappa, S)$ to a cardinal of uncountable cofinality.

(Q2) Suppose κ, δ, and S are as in (Q1). Can the degree of S be made minimal? In the present setting "minimal" means:: for all $R \in E(S)$, if $S \notin E(R)$, then $L(\kappa) \nsubseteq E(R)$. An affirmative answer is morally certain via the methods of [7]. There it was shown that if α is countable, greater than ω and $\sum_1$ admissible, then there exists an $S \subseteq \omega$ such that

(i) $E(S) = L(\alpha, S)$ and

(ii) for all $R \in E(S)$, if $S \notin E(R)$, then $L(\alpha) \nsubseteq E(R)$.

(Q3) Suppose $L(\kappa)$ is uncountable, E-closed and $\sum_1$ inadmissible. Assume gc(κ) has uncountable cofinality in $L(\kappa)$. Do there exist $\delta < \kappa$ and $S \subseteq \delta$ such that $L(\kappa, S) = E(S)$? S. Friedman [16, 17] has found a virtually complete answer to this question when $V = L$ and cardinality of $L(\kappa)$ is δ.

The analysis of $E(X)$ made in the various sections of the present paper depends strongly on the assumption that X is a set of ordinals. Slaman [13] has shown: Let $L(\kappa)$ be countable and E-closed. Then there exists a countable $X \subseteq 2^\omega$ such that

(a) $L(\kappa, X) = E(X)$;

(b) $L(\kappa, X)$ does not admit Moschovakis witnesses; and

(c) if $L(\kappa)$ is $\sum_1$ admissible but not the E-closure of any $Z \in L(\kappa)$, then $L(\kappa, X)$ is $\sum_1$ admissible but the notions of E-recursive enumerability and boldface $\sum_1$ do not agree on $L(\kappa, X)$.

References

1. S. C. Kleene, Recursive functionals and quantifiers of finite type I, *Trans. Amer. Math. Soc.* **91** (1959), 1–52.
2. S. C. Kleene, Recursive functionals and quantifiers of finite type II, *Trans. Amer. Math. Soc.* **108** (1963), 106–142.
3. D. Normann, "Set Recursion, in Generalized Recursion Theory, II," pp. 303–320, North-Holland, Amsterdam, 1978.
4. G. E. Sacks, Post's problem, absoluteness and recursion in finite types, *in* "The Kleene Symposium," pp. 201–222, North-Holland, Amsterdam, 1980.
5. F. G. Abramson, Sacks forcing does not always produce a minimal upper bound, *Advan. in Math.* **31** (1979), 110–130.
6. J. Steel, Forcing and tagged trees, *Ann. Math. Logic* **15** (1978).
7. G. E. Sacks, Countable admissible ordinals and hyperdegrees, *Advan. in Math.* **20** (1976), 213–262.
8. G. E. Sacks, Selection and forcing, in preparation.
9. G. E. Sacks, On the limits of recursive enumerability, in preparation.
10. Y. N. Moschovakis, Hyperanalytic predicates, *Trans. Amer. Math. Soc.* **138** (1967), 249–282.
11. L. Harrington, "Contributions to Recursion Theory in Higher Types," PhD thesis, M.I.T., Cambridge, Mass., 1973.
12. J. Moldstad, "Computations in Higher Types," Lecture Notes in Math. Vol. 574, Springer-Verlag, Berlin, 1977.
13. T. A. Slaman, "Aspects of E-Recursion," PhD thesis, Harvard, Univ. Cambridge, Mass., 1981.

14. J. Baumgartner and R. Laver, Iterated perfect-set forcing, *Ann. Math. Logic* **17** (1979), 271–288.
15. A. Kanamori, Perfect-set forcing for uncountable cardinals, *Ann. Math. Logic* **19** (1980), 97–114.
16. S. Friedman, Uncountable admissibles. I. Forcing, *Trans. Amer. Math. Soc.*, in press.
17. S. Friedman, Uncountable admissibles. II. Compactness, *Israel J. Math.*, in press.
18. Y. N. Moschovakis, On the Grilliot–Harrington–MacQueen theorem, *in* "Logic Year 1979–80," Lecture Notes in Math., Vol. 859, pp. 246–267, Springer-Verlag, Berlin, 1981.
19. J. E. Fenstad, "General Recursion Theory," Springer-Verlag, Berlin, 1980.
20. L. Kirousis, A selection theorem (abstract), *J. Symbolic Logic* **46** (1981), 38.

Reprinted from Annals of Pure and Applied Logic 81 (1996) 171–185 with kind permission of Elsevier Science - NL, Sara Burgerhartstraat 25, 1055 KV Amsterdam, The Netherlands.

Effective forcing versus proper forcing

Gerald E. Sacks [1]

Harvard University, Massachusetts Institute of Technology, Cambridge, MA 01238, USA

Received 1 July 1995

Communicated by A. Nerode

Dedicated to the Memory of S.C. Kleene

Abstract

$\mathscr{P}$, a notion of forcing over $E(\omega_1)$, the E-closure of $L(\omega_1)$, is said to be effective if every sideways $\mathscr{P}$-generic extension preserves E-closure. There are set notions of forcing in $E(\omega_1)$ that do not preserve E-closure. The main theorem below asserts that $\mathscr{P}$ is effective if and only if it is locally proper, a weak variant of Shelah's notion of proper.

1. Introduction

Effecting forcing alludes to forcing over an E-closed structure. The prime example is $E(\omega_1)$, the E-closure of ω_1, the least E-closed structure with ω_1 as an element. $E(\omega_1)$ equals $L(\kappa)$ for some ordinal κ less than the first Σ_1 admissible beyond ω_1. E-closed means:

$$\{e\}(x)\downarrow \longrightarrow \{e\}(x) \in L(\kappa)$$

for all $x \in L(\kappa)$ and $e < \omega$. $\{e\}(x)$ is Normann's extension of Kleene's concept of partial recursive function from objects of finite type to all sets x. In this paper $L(\kappa)$ is always $E(\omega_1)$ except in Section 2, where the elements of E-recursion are sketched. Let $\mathscr{P} \in L(\kappa)$ be a notion of set forcing. $\mathscr{P}$ is said to be *effective* if every $\mathscr{P}$-generic G preserves E-closure, that is, $L(\kappa, G)$ is E-closed. (G can be assumed to be a subset of ω_1, since every member of $L(\kappa)$ can be injected into ω_1 in $L(\kappa)$.) Locally proper is a weak variant of Shelah's proper forcing. The main result below is the equivalence of effective and locally proper forcing.

To force over E-closed structures is to force computations to converge or diverge. Forcing computations differs radically from forcing sets. Every set notion of forcing

[1] The author is grateful to A. Kanamori and T. Slaman for valuable mathematical suggestions and other acts of friendship. Early support was provided by the NSF (DMS).

preserves Σ_1 admissibility. Not so for E-closure. If G collapses ω_1 to ω, then $E(\kappa, G)$ is not E-closed, because $E(\kappa, G) = E(\kappa, A)$ for some $A \subseteq \omega$ and $E(B)$ is Σ_1 admissible for all $B \subseteq \omega$. The connection made in Section 3 between proper forcing and preserving E-closure is a consequence of how computations are forced to diverge. A Moschovakis divergence witness, defined in Section 2, is forced to exist by a construction that ranges unboundedly over a countable co-re substructure of $L(\kappa)$. (The latter is a weakening of the notion of Σ_1 substructure. Recall that one version of proper forcing uses countable elementary substructures and forcing conditions generic over those substructures.)

Sections 4 and 5 contain imperfect results along the following lines. There exists a forcing relation that is locally proper but not proper, and another that preserves ω_1 but not the E-closure of ω_1.

$E(\omega_1)$ **Proviso.** Almost everything below makes sense for $L(\kappa)$, an arbitrary E-closed, initial segment of L. On the other hand, this paper focuses on $E(\omega_1)$. From now on $L(\kappa)$, with no qualification, is arbitrary, and when $E(\omega_1)$ is meant, some qualification such as $(= E(\omega_1))$ will be inserted.

$I \in L(\kappa)$ is a *co-re substructure* of $L(\kappa)$ if I is E-closed, and if for all $e < \omega$ and $m, a \in I$:

$$\begin{aligned} &\exists x[x \in m \wedge \{e\}(x,a)\uparrow] \\ (1.1)\qquad &\quad\longrightarrow \exists x[x \in (m \cap I) \wedge \{e\}(x,a)\uparrow]. \end{aligned}$$

I is said to be *pointed* if $\exists b[b \in I \wedge \kappa_r^b \notin I]$. κ_r, the greatest reflecting ordinal, is defined in Section 2.

$\mathscr{P} \in L(\kappa)$ $(= E(\omega_1))$, a set forcing relation, is *locally proper* if for each $p \in P$ and $a \in L(\kappa)$ there is a countable, pointed co-re substructure $I \in L(\kappa)$ such that

(1.2) (i) $p, a \in I$ and

(ii) $\exists q[q \in P \wedge p \geqslant q \wedge q$ generic over $I]$.

($p \geqslant q$ is "p is extended by q".) The last part of (ii) means: if D is a dense set of conditions, suitably definable over I, then $q \subseteq \bigcup(D \cap I)$, i.e.,

$$\forall s[q \geqslant s \rightarrow \exists r(r \in D \cap I \wedge s \cap r \neq \emptyset)].$$

D is of the form $\{r | r \Vdash \mathscr{F} \text{ or } r \Vdash \neg\mathscr{F}\}$, where $\mathscr{F}$ is

(1.3) (i) $\exists\gamma[\gamma \in I \wedge |\{m\}(\underline{a})| = \underline{\gamma}]$, or

(ii) $\forall b \exists \gamma[\underline{b} \in \underline{c} \rightarrow \underline{\gamma} \in I \wedge |\{n\}(b)| \leqslant \underline{\gamma}]$ $(a, c \in I)$.

The genericity of q will be seen to imply that p cannot force $\{e\}(a)$ to converge outside I.

For now the definition of proper forcing in $E(\omega_1)$ is left open-ended. The following clauses certainly belong. There exists a *closed*, unbounded class C of countable, co-re substructures that covers $E(\omega_1)$, and such that for each $I \in C$,

(1.4) $\forall p \exists q[p \in I \rightarrow q$ generic over $I]$.

Familiar notions of set forcing in $L(\kappa)(= E(\omega_I))$ are proper in the above sense. These include *ccc*, countably closed and perfect. It seems reasonable to require that C be E-recursive on $L(\kappa)$.

2. *E*-recursion revue

$\{e\}(x)$, the eth partial E-recursive function, is defined by the Normann schemes [4]. A leisurely, but not overly detailed, account of E-recursion can be found in [8]. From now on the "E" will be dropped as often as clarity allows, e.g., "recursive" in place of "E-recursive". If e is not the Gödel number of a scheme, then $\{e\}(x)$ diverges ($\uparrow$). One of several rudimentary schemes is

$$\{2\}(x, y) = \{x, y\}.$$

There are also schemes for bounding, composition and enumeration. The last is

$$\{7\}(e,x, y) \simeq \{e\}(x, y).$$

The $\simeq$ symbol denotes equality for partial functions: their graphs are equal. A computation instruction is an $(n+1)$-tuple of the form $\langle e, x_0, \ldots, x_{n-1}\rangle$. For simplicity let n be 1. $\langle e,x\rangle$ is an instruction to try and compute $\{e\}(x)$. In order to shrink several cases of definitions and proofs to one case, scheme T is introduced.

$$\begin{aligned} &\{2^m \cdot 3^n\}(x) \downarrow \text{(converges)} \quad \text{if} \\ (2.1) \qquad &\{m\}(x) \downarrow \quad \text{and} \\ &\forall y[y \in \{m\}(x) \longrightarrow \{n\}(y) \downarrow]. \end{aligned}$$

Scheme T (for typical) combines the most salient features of bounding and composition. From now on scheme T will be the only scheme considered. The concept of computation is derived from that of immediate subcomputation instruction.

(2.2) $\langle m,x\rangle$ is an immed. subcomp. instr. of $\langle 2^m \cdot 3^n, x\rangle$.

(2.3) If $\{m\}(x) \downarrow$, then $\langle n, y\rangle$ is an immed. subcomp. instr. of $\langle 2^m \cdot 3^n, x\rangle$ for all $y \in \{m\}(x)$.

(2.4) If d is not the Gödel number of a Normann scheme, then $\langle d,z\rangle$ is an immed. subcomp. instr. of $\langle d,z\rangle$.

The relation, b is a subcomputation instruction of a, in symbols $a >_U b$, is the transitive closure of the relation, b is an immed. subcomp. instr. of a. The latter is *re*, but the former is not. $T_{\langle e,x\rangle}$ is the computation tree generated by $\langle e,x\rangle$; its top node is $\langle e,x\rangle$; underneath are all subcomputation instructions of $\langle e,x\rangle$. By transfinite induction,

(2.5) $T_{\langle e,x\rangle}$ is wellfounded $\longleftrightarrow \{e\}(x)\downarrow$.

A class is E-recursively enumerable (or more simply, *re*) if it is the domain of a partial E-recursive function. The following predicates are *re*:

(2.6) $|a| < |b|$.

(2.7) $|a| \leqslant |b| < \infty$.

∞ is the length of every divergent computation. If b is a convergent computation instruction, then $|b|$, the length of b, is the ordinal height of T_b.

Gandy selection says there is a partial recursive function s such that for all $e, n < \omega$:

(2.8) $\exists n\{e\}(n)\downarrow \longleftrightarrow s(e)\downarrow \wedge \{e\}(s(e))\downarrow$.

The proof of (2.8) is a fixed point argument based on the *re*-ness of (2.6) and (2.7); it assigns the least possible value to $|\{e\}(s(e))|$. Expressions (2.6) and (2.7) were initiated by Moschovakis, who aptly dubbed them *stage comparison*.

Gandy selection is the most general selection principle available in E-recursion theory. Note that selection fails for subsets of ω_1 *re* in ω_1; otherwise $E(\omega_1)$ would be Σ_1 admissible. Perhaps this is why the study of forcing over $E(\omega_1)$ is so convoluted. In the absence of strong selection E-recursion makes do with reflection. γ is an *a-reflecting* ordinal if

$$L(\gamma) \models \mathscr{F} \quad \text{implies} \quad L(\kappa_0^a) \models \mathscr{F}$$

for every Σ_1 formula $\mathscr{F}$ of *ZF* with a (and ω_1) as its only non-integer parameters. κ_0^a is the supremum of all ordinals recursive in a (and ω_1). δ is recursive in a (and ω_1) if $\delta = \{e\}(a, \omega_1)$. Since $L(\kappa) = E(\omega_1)$,

$$\kappa = \sup\{\kappa_0^a | a < \omega_1\}.$$

Let

$$\kappa_r^a = \sup\{\gamma | \gamma \quad \text{is } a\text{-reflecting}\}.$$

Clearly $\kappa_0^a \leqslant \kappa_r^a$. Slightly less clear is $\kappa_r^a < \kappa$; it follows from the Σ_1 inadmissibility of $L(\kappa)$. Much less clear is $\kappa_0^a < \kappa_r^a$. An inequality of the last sort was proved in [5] in order to make a forcing argument work for the E-closure of 2^{2^ω}. Later Harrington [2] showed

(2.9) $\kappa_0^{2^\omega} < \kappa_r^{2^\omega}$,

and it is his approach that leads to $\kappa_0^a < \kappa_r^a$ in the setting of $E(\omega_1)$.

Recall the $E(\omega_1)$ Proviso of Section 1. In accord with it, from now on ω_1 is a hidden parameter in computations of the form $\{e\}(x)$ for $x \in E(\omega_I)$. For example, ω_1 is a recursive ordinal.

Kechris's basis theorem states:

(2.10)
$$\begin{aligned} &\exists b[b \in m \wedge \{e\}(b,x)\uparrow] \\ &\longrightarrow \exists b[b \in m \wedge \{e\}(b,x)\uparrow \wedge \kappa_r^{m,x,b} \leqslant \kappa_r^{m,x}]. \end{aligned}$$

Harrington's approach to κ_r in (2.9), together with (2.10), leads to:

(2.11) If $\{e\}(a)\uparrow$, then a Moschovakis witness to $\{e\}(a)\uparrow$ is first-order definable over $L(\kappa_r^a, \omega_1)$.

The witness of (2.11) is a function on ω defined by an ω-recursion whose limit is κ_r^a and whose nth stage invokes (2.10). A variation on this ω-recursion is used to prove that locally proper forcing preserves E-closure in Section 3.

Let I be a *co-re* substructure as defined in Section 1. I is E-closed. Suppose $\{e\}(a)\uparrow$ for some $a \in I$. A divergence witness for $\{e\}(a)$ is definable over I as hinted in the previous paragraph. The simplest example of a *co-re* substructure (of $L(\kappa) = E(\omega_1)$) is:

$$I_0 = \{x | \kappa_r^x \leqslant \kappa_r \wedge x \in L(\kappa_r)\} \qquad (\kappa_r = \kappa_r^{\emptyset}).$$

I_0 is countable, and $I_0 \cap \omega_1 = k_0 \cap \omega_1$ for some $k_0 < \omega_1$. Slaman [9] introduced the κ_r spectrum:

$$\kappa_{r,0} = \kappa_r;$$

(2.12) $$\kappa_{r,\delta+1} = \kappa_r^{\kappa_{r,\delta}};$$

$$\kappa_{r,\lambda} = \sup\{\kappa_{r,\delta} | \delta < \lambda\}.$$

$\kappa_{r,\delta}$ is defined for all $\delta < recf(\kappa)$, the *re* cofinality of κ (same as the $\Sigma_1^{L(\kappa)}$ cofinality of κ). Since $L(\kappa) = E(\omega_1)$, the only possible values for $recf(\kappa)$ are ω and $\omega_1^{L(\kappa)}$. The κ_r spectrum is so called, because for all $B \subseteq \omega_1$ in $L(\kappa)$,

(2.13) $\exists\delta[\delta \text{ not a limit } \wedge \kappa_r^B = \kappa_{r,\delta}]$.

Associated with each $\kappa_{r,\delta}$ is a canonical *co-re* substructure

(2.14) $I_\delta = \{x | \kappa_r^x \leqslant \kappa_{r,\delta} \wedge x \in L(\kappa_{r,\delta})\}$.

There are of course other *co-re* substructures. For example, by forcing there is a *co-re* substructure $J \in L(\kappa)$ such that

$$\forall x[x \in J \longrightarrow \kappa_r^x \leqslant \kappa_r] \wedge J \nsubseteq L(\kappa_r).$$

Forcing supplies an x such that $\kappa_r^x \leqslant \kappa_r$ but $x \in L(\kappa) - L(\kappa_r)$. It turns out that for every *co-re* substructure I, the supremum of the ordinals in I is a member of the κ_r spectrum.

An arbitrary set forcing relation $\mathscr{P} \in L(\kappa)$ can be sketched as follows. A generic G is a subset of ω_1. Forcing conditions are coded by countable ordinals. A typical element of $L(\kappa, G)$, for any $G \subseteq \omega_1$, is of the form $\{e\}(a, G)$, where $a \in L(\kappa)$ and $|\{e\}(a, G)| < \kappa$. Among the formulas of the language $\mathscr{L}(\kappa, \mathscr{G})$ are:

(i) $|\{e\}(\underline{a}, \mathscr{G})| = \underline{\sigma}$ $(\sigma < \kappa)$;
(ii) $\exists\sigma[|\{e\}(\underline{a}, \mathscr{G})| = \sigma]$;
(iii) $x \in \{e\}(\underline{a}, \mathscr{G})$.

(i) and (iii) are ranked formulas. In (ii) the range of the existential quantifier is κ. The meaning of (iii) will be clearer after $\mathscr{T}(p,e,a,\sigma)$, a set of terms that name the elements of $\{e\}(a,G)$, is defined. A simultaneous recursion on σ defines

(2.15a) $p \Vdash |\{e\}(a)| = \underline{\sigma}$ ("$\mathscr{G}$" is omitted for notational simplicity);

(2.15b) $\mathscr{T}(p,e,a,\sigma)$;

(2.15c) $q \Vdash s \in \{e\}(a)$.

When $\sigma = 0$, $\{e\}(\underline{a})$ is $\mathscr{G}$. For all p, $p \Vdash |\mathscr{G}| = \underline{0}$. $\mathscr{T}(p,e,a,0) = \{\underline{\delta} | \delta < \omega_1\}$. And $q \Vdash \underline{\delta} \in \mathscr{G}$ iff q says so. A complete account of the recursion when $\sigma > 0$ would include a clause for each Normann scheme. One clause corresponding to scheme T (2.1) will suffice to show how (2.15) is built up. Thus

$$p \Vdash |\{2^m \cdot 3^n\}(\underline{a})| = \underline{\sigma} \quad \text{iff}$$

(2.16a) $\exists \gamma_{<\underline{\sigma}}[p \Vdash |\{m\}(a)| = \gamma]$,

(2.16b) $p \Vdash \forall x \exists \tau_{<\underline{\sigma}}[x \in \{m\}(a) \to |\{n\}(x)| = \tau]$,

(2.16c) $p \Vdash \forall \tau_{<\underline{\sigma}}[|\{m\}(\underline{a})| = \tau \vee \exists x[x \in \{m\}(a) \wedge |\{n\}(x)| \geqslant \tau]]$.

In (2.16b) and (2.16c), x ranges over $\mathscr{T}(p,m,a,\gamma)$.

$$\mathscr{T}(p,2^m \cdot 3^n,a,\sigma) = \{\{n\}(t) | t \in \mathscr{T}(p,m,a,\gamma)\}.$$

$q \Vdash s \in \{2^m \cdot 3^n\}(\underline{a})$ iff for some σ,

$q \Vdash |\{2^m \cdot 3^n\}(a)| = \underline{\sigma}$ and $s \in \mathscr{T}(q,2^m \cdot 3^n,a,\sigma)$.

From now on $\mathscr{P}$ will be a suppressed parameter, as is ω_1. Thus (2.15a) and (2.15c) are relations E-recursive on $L(\kappa)$, and (2.15b) is a function E-recursive *on* $L(\kappa)$.

A set D of forcing conditions is said to be *dense* if

$$\forall q \exists p[q \geqslant p \in D].$$

Call G *generic* if for every dense D definable over $L(\kappa)$ there is a $p \in D$ such that $G \in p$ (i.e., G satisfies p). In Section 3 it is shown that only the Σ_1 D's matter. $\mathscr{P}$ is said to be *effective* if for every $\mathscr{P}$-generic G, $L(\kappa,G)$ is E-closed. In short, there are no p and a such that

$$p \Vdash |\{e\}(a)| = \kappa.$$

The primary tool for studying $\mathscr{P}$ is $>_V$, the tree of possibilities. Each node of $>_V$ is of the form $\langle p,e,s\rangle$, where s is a term of $\mathscr{L}(\kappa,\mathscr{G})$.

(2.17) $\langle p,e,s\rangle >_V \langle q,f,t\rangle$ iff $p \geqslant q \wedge q \Vdash^* \langle e,s\rangle >_U \langle f,t\rangle$.

$>_V$ anticipates all the possibilities for $>_U$. $\Vdash^*$ is the weak forcing relation:

$$p \Vdash^* \mathscr{F} \quad \text{iff} \quad \neg \exists q[p \geqslant q \wedge q \Vdash \neg \mathscr{F}].$$

To clarify (2.17) unpack it in the case of scheme (T).

$$\langle p, 2^m \cdot 3^n, s\rangle >_V \langle q, n, t\rangle \quad \text{iff}$$
$$p \geqslant q,$$
$$q \Vdash |\{m\}(s)| = \underline{\sigma} \qquad (\text{some } \sigma < \kappa),$$
$$q \Vdash t \in \{m\}(s).$$

$>_V$ has the *wellfoundedness* property if $>_V$ below $\langle p, e, t\rangle$ is wellfounded whenever

$$p \Vdash \exists\sigma[|\{e\}(t)| = \sigma].$$

A countably closed P has wellfoundedness, but a *ccc* P need not. This is why the original proofs that countably closed and *ccc* preserve E-closure are so different [6,7]. It might be possible to characterize all $\mathscr{P}$'s with wellfoundedness along lines proposed by Baumgartner.

3. Effective equals locally proper

Keep in mind that $L(\kappa)$ is $E(\omega_1)$, and that ω_1 and $\mathscr{P}$ are suppressed parameters.

Theorem 3.1 *Let $\mathscr{P} \in L(\kappa)(= E(\omega_1))$ be a set notion of forcing. Then $\mathscr{P}$ is effective iff $\mathscr{P}$ is locally proper.*

Proof. Assume $\mathscr{P}$ is effective. Fix p, a, b, c and G so that G is generic and $G \in p$. Choose q so that $p \geqslant q$ and for some θ

$$q \Vdash \kappa_r^{\mathscr{G},p,a,b,c} = \underline{\theta}.$$

The reasoning behind (2.13) also shows θ is $\kappa_{r,\delta}$, where $\delta = 0$ or δ is a successor. The natural choice for the I of (1.2) is I_δ. I_δ is countable and pointed; the point is $\kappa_{r,\delta-1}$ if $\delta > 0$. From now on the point is yet another suppressed parameter. It is safe to assume p, a, b, c are effectively equivalent to countable ordinals, i.e., have the same E-degree as a countable ordinal modulo the suppressed parameters. So $p, a \in I_\delta$ and

$$\kappa_r^{p,a} \leqslant \kappa_r^{G,p,a,b,c} = \kappa_{r,\delta}.$$

Consider an arbitrary generic $G_0 \in q$. Let D be a dense set of forcing conditions. Then $G_0 \in r$ for some $r \in D$. Suppose

$$(3.1) \qquad r \Vdash |\{m\}(\underline{a})| = \underline{\gamma};$$

the set of all such r is computable from a, γ. Let r_0 be the least such r with $G_0 \in r$. Then

$$r_0 \leqslant_E G_0, \gamma, a.$$

But $\kappa_r^{G_0,a} \leqslant \kappa_{r,\delta}$ because $G_0 \in q$. So $r_0 \in I$. Note that the set of all r such that

$$(3.2) \qquad \exists\gamma[\gamma \in I \wedge r \Vdash |\{m\}(a)| = \underline{\gamma}]$$

is *re*, because γ, if there is one in I, has to be computable from r, a by reflection.

Suppose r forces $\neg(3.2)$. The set of all such r is co-*re*, since, by reflection, r does not force $\neg(3.2)$ iff

$$\exists s[s \leqslant_E r \wedge r \geqslant s \wedge s \Vdash (3.2)].$$

By Kechris's basis theorem (2.10), there is an r_0 such that $G_0 \in r_0$, $r_0 \Vdash \neg(3.2)$, and $\kappa_r^{G_0,a,r_0} \leqslant \kappa_{r,\delta}$. Again $r_0 \in I$.

Now suppose r forces

$$(3.3) \qquad \forall b \exists \gamma[\underline{b} \in \underline{c} \rightarrow \gamma \in I \wedge |\{n\}(\underline{b})| \leqslant \underline{\gamma}].$$

In (3.3) γ can be computed from b, c, G_0 since

$$\kappa_r^{G_0,b,c} \leqslant \kappa_{r,\delta}.$$

But then γ is bounded above by some γ_0 computable from c, G_0. Hence the least r, call it r_0, that forces (3.3) is computable from c, G_0. So $r_0 \in I$ since $\kappa_r^{G_0,c} \leqslant \kappa_{r,\delta}$. Note that the set of all r that force (3.3) is *re*.

Finally suppose r forces $\neg(3.3)$. The set of all such r is co-*re*, since, by reflection, r does not force $\neg(3.3)$ iff

$$\exists s[s \leqslant_E r \wedge r \geqslant s \wedge s \Vdash (3.3)].$$

Now assume $\mathscr{P}$ is locally proper. Fix p, a. Then there is a co-*re* substructure I such that $p, a \in I$. In addition there is a $q \leqslant p$ such that

$$q \subseteq \cup(D \cap I),$$

where D is any dense set of conditions as just before (1.3). Also I is pointed; there is a $z \in I$ such that $\kappa_r^z \notin I$. z can be taken to be a relation on ordinals, so $\kappa_r^{z,x} \notin I$ for all $x \in I$. From now on z will be suppressed as in ω_1. Let $\sup I$ be the supremum of all ordinals in I. Suppose

$$(3.4) \qquad p \Vdash |\{e\}(\underline{a})| \geqslant \sup I.$$

The plan is to show that q forces a divergence witness for $\{e\}(a)$, a witness first-order definable over $L(\sup I, G)$ for all generic $G \in q$. The tree of possibilities below $\langle p, e, a\rangle$ will be developed in considerable detail. Each level of the tree will contain countably many r's (from I) that cover q. Each such r will be associated with an f and $b \in I$ such that

$$(3.5) \qquad r \Vdash |\{f\}(\underline{b})| \geqslant \sup I.$$

Then each generic $G \in q$ will belong to some r_n (with associated f_n and b_n as in (3.5)) on level n of the tree below $\langle p, e, a\rangle$, and the sequence $\lambda n|\langle f_n, b_n\rangle$ will define a divergence witness for $\{e\}(a)$.

The r's for level n are found by recursion on n. At level 0, the p, e and a of (3.4) serve as the needed r, f and b. Fix $n \geqslant 0$, and let r_n, f_n and b_n be associated with level n:

$$r_n, b_n \in I;$$
$$(3.6) \qquad r_n \cap q \neq \emptyset; \quad \text{and}$$
$$r_n \Vdash |\{f_n\}(\underline{b}_n)| \geqslant \sup I.$$

Assume f_n is an instance of Scheme T: $f_n = 2^u \cdot 3^v$. Let D be the dense set of conditions that force

$$(3.7) \qquad |\{f_n\}(\underline{b}_n)| \geqslant \sup I$$

or force $\neg$ (3.7). Then $r_n \cap q$ is covered by $D \cap I$. The r's in $D \cap I$ that force (3.7) are in accord with the plan. Consider an $r \in D \cap I$ that touches $r_n \cap q$ and forces $\neg$ (3.7). Then

$$r \Vdash \{u\}(\underline{b}_n) = \underline{c}$$

for some $c \in I$. The plan requires that $r \cap r_n \cap q$ be covered by conditions $s \in I$ such that

$$(3.8) \qquad s \Vdash \exists b[b \in \underline{c} \wedge |\{v\}(b)| \geqslant \sup I].$$

Let E be the set of all s such that forces "$\exists b \ldots$" of (3.8) or forces its negation. $E \cap I$ covers $r \cap r_n \cap q$. Suppose (for a contradiction) that there is an $s \in E \cap I$ such that $r \cap r_n \cap q \cap s \notin \emptyset$ and s forces the negation of "$\exists b \ldots$" of (3.8). Thus

$$(3.9) \qquad s \Vdash \forall b \exists \gamma [b \in \underline{c} \longrightarrow \gamma \in I \wedge |\{v\}(b)| < \gamma].$$

Recall the proof that effective implies locally proper. The γ of (3.9) can be computed from s, b, c by reflection. So

$$s \Vdash \forall b[b \in \underline{c} \longrightarrow |\{n\}(b)| < \underline{\gamma}_0]$$

for some $\gamma_0 \in I$; hence

$$r \cap r_n \cap q \cap s \Vdash |\{f_n\}(\underline{b}_n)| < \sup I,$$

contrary to (3.6).

One last detail. The "s" of (3.8) is not of any use unless $s \in I$ implies some $b \in I$ satisfies (3.8). Such a b is available because

$$(3.10) \qquad \{b | b \in c \wedge s \Vdash |\{v\}(b)| \geqslant \sup I\}$$

is co-*re* and I is a co-*re* substructure of $L(\kappa)$. To check that (3.10) is co-*re*, observe that

$$s \nVdash |\{v\}(b)| \geqslant \sup I$$

is equivalent to

(3.11) $\exists t \exists \gamma [s \geqslant t \wedge t \Vdash |\{v\}(b)| \leqslant \gamma < \sup I]$.

Furthermore, t and γ are computable from s, b by reflection; so (3.11) is *re*.

4. Preserving cardinals but not E-closure

This section is devoted to an example of forcing that preserves ω_1 but not E-closure. Once again $L(\kappa)$ equals $E(\omega_1)$. Assume the co-*re* cofinality of ω_1 is ω_1.

Theorem 4.1 *There exist $B \subseteq \omega_1$ and $\mathscr{P} \in L(\kappa, B)$ such that:*

$L(\kappa, B)$ is E-closed;
$\mathscr{P}$ preserves ω_1;
but $\mathscr{P}$ does not preserve E-closure.

Proof. If $L(\kappa, B)$ is E-closed, then $\omega_1^{L(\kappa)}$ is preserved in $L(\kappa, B)$. Otherwise $L(\kappa, B) = E(A)$ for some $A \subseteq \omega$, and then κ is Σ_1 admissible. B, as constructed below, is somewhat, but not quite, generic via conditions that are countable initial segments of ω_1. Forcing with such conditions is countably closed. It is not difficult to verify that countably closed forcing is locally proper, and so preserves E-closure by Theorem 3.1. B is designed to exclude the κ_r spectrum in the following sense. Each $\kappa_{r,\delta}$ has a canonical code

$$j_\delta = \mu\gamma(\kappa_{r,\delta} \leqslant_E \gamma, \omega_1).$$

Note that $\gamma < \omega_1$ since $L(\kappa) = E(\omega_1)$. j_δ will be kept out of B for all δ.

The forcing conditions in $\mathscr{P}$ are all countable, closed initial segments of ω_1 that are subsets of B. Suppose $\mathscr{P}$ preserved E-closure. Let H be $\mathscr{P}$-generic over $L(\kappa, B)$. Then $L(\kappa, B, H)$ is E-closed. By Proposition 4.3, the canonical code for $\kappa_r^{B,H}$ belongs to H. But then it belongs to B despite its exclusion.

First a standard argument to show $\mathscr{P}$ preserves ω_1. Let t be a term in the forcing language for H over $L(\kappa, B)$ such that

$$p \Vdash t : \omega \to \omega_1$$

for some $p \in P$. Let $q_0 = p$. For each n, choose the least $\langle q_{n+1}, \delta_n \rangle$ such that $q_n \geqslant q_{n+1}$ and $q_{n+1} \Vdash t(n) = \underline{\delta}_n$. $\lambda_n | \langle q_n, \delta_n \rangle$ is computable from p, t. Let

$$\gamma_0 = \sup \text{ dom } \bigcup_n q_n.$$

Repeat (with q_0 replaced by the least condition that extends q_n for all n) to obtain γ_1. Repeat further to generate γ_β $(\beta < \omega_1)$. For successor β, proceed as above. For limit β, γ_β is the sup of everything previous. Thus

$$\{\gamma_\beta | \beta < \omega_1\}$$

is a closed subset of ω_1 in $L(\kappa, B)$. Part of the design of B below ensures that B is stationary with respect to all closed subsets of ω_1 in $L(\kappa, B)$. So some $\gamma_\beta \in B$. The definition of γ_β implies there is a sequence $\lambda n|r_n$ such that sup dom $\bigcup_n r_n = \gamma_\beta$. Hence $\bigcup_n r_n$ is a condition, and

$$\bigcup_n r_n \Vdash \text{range}(t) \subseteq \gamma_\beta.$$

Now the definition of B. $B(= \bigcup\{p_\delta | \delta < \omega_1\})$ is constructed in ω_1 stages. The p_δ's are as follows:

(b1) $\text{dom}(p_\delta) = j_\delta$. $p_{\delta+1}(j_\delta) = 0$. (Thus, $j_\delta \notin B$.)
If δ is a limit, then $p_\delta = \bigcup\{p_\gamma | \gamma < \delta\}$. If $\alpha < \beta$, then $p_\alpha > p_\beta$.

(b2) $\{\langle p_\delta, \delta\rangle | \delta \leqslant \beta\}$ is definable over $L(\kappa_{r,\delta})$ uniformly in δ.

(b3) Suppose δ is not a limit and even. Let p_δ^- be $p_{\delta-1}$ if $\delta \neq 0$, and null otherwise. Let

$$\lambda_\beta|\langle e_\beta, \gamma_\beta\rangle \quad (\beta < \omega_1 \wedge \beta \text{ not a limit} \wedge \beta \text{ even})$$

be an enumeration of $\omega \times \omega_1$ E-recursive in ω_1.

$$e_\beta \text{ denotes } \{e_\beta\}(\mathscr{B}, \beta, \omega_1).$$

$$p_\delta^- \text{ preserves } \kappa_{r,\delta}, \text{ i.e., } \kappa_r^{p_\delta^-} \leqslant \kappa_{r,\delta}.$$

Of course $p_\delta^- \geqslant p_\delta$. In addition, either

$$p_\delta \Vdash |e_\delta| < \kappa_{r,\delta}, \qquad \text{or}$$

p_δ forces a Moschovakis witness to $e_\delta \uparrow$ to be first-order definable over $L(\kappa_{r,\delta}, \mathscr{B})$.

(b4) Suppose δ is not a limit and odd. Let C be any closed subset of ω_1 E-recursive in $\kappa_{r,\delta-1}, \omega_1$. Then $p_\delta(\gamma) = 1$ for some $\gamma \in C$. (Thus $B \cap C \neq \emptyset$.)

The object of (b3) is to make $L(\kappa, B)$ E-closed. In order to see how (b3) is satisfied, recall the proof that countably closed forcing preserves E-closure as presented in [8] or [5]. That argument shows p_δ^- can be extended in one of two ways. The first produces a q and a θ such that

$$q, \theta \leqslant_E p_\delta^-, \delta \wedge q \Vdash |e_\delta| = \theta.$$

Then dom $q < j_\delta$; otherwise $j_\delta \leqslant_E q, \omega_1$ $(\alpha < \beta < \omega_1 \rightarrow \alpha \leqslant_E \beta, \omega_1)$ and then p_δ^- does not preserve $\kappa_{r,\delta}$. The second produces a q that forces some $\lambda n/w(n)$ to witness the divergence of e_β. q is of the form $\bigcup_n q_n$. Each q_n forces w_n to be the nth leg of a divergence witness and preserves $\kappa_{r,\delta}$, hence has domain $< j_\delta$. The q_n's and w_n's are defined in a first-order fashion over $L(\kappa_{r,\delta})$. Countable closure is needed to insure that $\bigcup_n q_n$ is a forcing condition.

The object of (b4) is to make B touch every closed subset of ω_1 in $L(\kappa)$. (b4) is satisfied with the help of two points. First, each subset of ω_1 in $L(\kappa)$ is E-recursive in δ, ω_1 for all sufficiently large countable δ. And $j_\delta \equiv_E \kappa_{r,\delta}, \omega_1$ for all δ. Second, Lemma 4.2 implies: if C is a closed subset of ω_1 E-recursive in $\kappa_{r,\delta-1}, \omega_1$, then

$$j_\delta = \sup(C \cap j_\delta).$$

Lemma 4.2 *Suppose C is a closed subset of ω_1 in $E(\omega_1)$. Let j be the canonical code for some element of the κ_r spectrum. If*

$$C \leqslant_E b, \omega_1 \wedge b < j,$$

then

$$j = \sup(C \cap j).$$

Proof. Let f enumerate C in increasing order. Then $f \leqslant_E C, \omega_1$. It suffices to show

$$b < \gamma < j \longrightarrow f(\gamma) < j.$$

Assume $b < \gamma < j$. Then $j \nleqslant_E \gamma, \omega_1$. So $j \nleqslant_E \gamma, b, \omega_1$, since $b \leqslant_E \gamma, \omega_1$. Hence $j \nleqslant_E f(\gamma), b, \omega_1$ and so $j \nleqslant_E f(\gamma), \omega_1$. Finally $f(\gamma) < j$.

Proposition 4.3 *Suppose $L(\kappa, B, H)$ is E-closed. Assume $\omega_1^{L(\kappa)} = \omega_1^{L(\kappa,B,H)}$ and H is a closed unbounded subset of $\omega_1^{L(\kappa)}$. Then the canonical code for $\kappa_r^{B,H}$ belongs to H.*

Proof. Let f enumerate H in strictly increasing order. Then

$$f(\delta) \leqslant_E B, H, \delta \qquad \text{(uniformly in } \delta\text{)}.$$

Let j be the canonical code for $\kappa_r^{B,H}$. Observe that

$$\forall \delta(\delta < j \longrightarrow f(\delta) < j).$$

Otherwise, for some $\delta < j$: $f(\delta) \geqslant j$, and so $j \leqslant_E f(\delta)$; but then $j \leqslant_E B, H, \delta$. Hence $f(j) = j$ and $j \in H$.

5. Locally proper but not proper

The main outcome of this section is the existence of a set forcing $\mathcal{P} \in L(\kappa)$ such that $\mathcal{P}$ is locally proper but does not preserve Σ_1 stationary subsets of $\omega_1^{L(\kappa)}$. $S \subseteq \omega_1^{L(\kappa)}$ is said to be Σ_1 stationary if S touches every $X \subseteq \omega_1^{L(\kappa)}$ such that X is closed, unbounded and $\Sigma_1^{L(\kappa)}$. Thus it seems reasonable to assert $\mathcal{P}$ is not proper despite the fact that "proper" has not yet been defined. Marcus [3] offers a definition of "proper" and shows that proper forcing preserves Σ_1 stationary sets (Lemma 5.2 below).

The re-cofinality of $\gamma \leqslant \kappa$ is the least λ such that for some $W \subseteq L(\kappa)$:

W is E-re on $L(\kappa)$;

$\sup W = \gamma$;

ordertype of $W = \lambda$.

Since $L(\kappa) = E(\omega_1^{L(\kappa)})$, $recf(\kappa)$ is either ω or $\omega_1^{L(\kappa)}$. In this section, as in Section 4, $recf(\kappa) = \omega_1^{L(\kappa)}$. Then $recf(\omega_1^{L(\kappa)}) = \omega_1^{L(\kappa)}$, as are the $\Sigma_1^{L(\kappa)}$ – cofinalities of $\omega_1^{L(\kappa)}$ and κ.

From now on, ω_1 is $\omega_1^{L(\kappa)}$. In essence, $\mathscr{P}$ is the notion of forcing introduced by Baumgartner, Harrington and Kleinberg. $\mathscr{P}$ will be seen to preserve E-closure and so is locally proper by Theorem 3.1. $\diamond_E$ is an adaptation of Jensen's $\diamond$ to $E(\omega_1)$.

$\diamond_E$: There exists a sequence $\{S_\alpha | \alpha < \omega_1\}$ in $L(\kappa)$ such that

(i) $S_\alpha \subseteq \alpha$ for all $\alpha < \omega_1$;

(ii) for each $\Sigma_1^{L(\kappa)}$-definable $S \subseteq \omega_1$, the set

$$\{\alpha | S \cap \alpha = S_\alpha\}$$

is Σ_1-stationary;

(iii) $\{S_\alpha | \alpha < \omega_1\}$ is lightface $\Sigma_1^{L(\omega_1)}$-definable.

The proof of $\diamond_E$ is a slight alteration of the proof of $\diamond$ in L given by Devlin [1, p. 125]. S_α is defined by $\Sigma_1^{L(\omega_1)}$ recursion on ω_1 so (iii) is immediate. At limit α's, whenever possible, an $L(\kappa)$-least pair $\langle S_\alpha, C_\alpha\rangle$ is chosen so that C_α is a closed, unbounded subset of α, and does not touch $\{\beta | S_\alpha \cap \beta = S_\beta\}$.

Suppose clause (ii) of $\diamond_E$ fails for S. Then there is a closed, unbounded $C \subseteq \omega_1$, $\Sigma_1^{L(\kappa)}$-definable, and not touching $\{\alpha | S \cap \alpha = S_\alpha\}$. Let $\langle S, C\rangle$ be $L(\kappa)$-least. Let H be the $\Sigma_1^{L(\kappa)}$ hull of $\{\omega_1, S, C\}$. Let π map H isomorphically onto its transitive collapse H_0. Since the $\Sigma_1^{L(\kappa)}$-cofinality of κ is ω_1, it must be that H, hence H_0, belongs to $L(\kappa)$. Let $\pi(\omega_1) = \alpha$. Then $\pi(S) = S_\alpha$, $\pi(C) = C_\alpha$, $S \cap \alpha = S_\alpha$, $C \cap \alpha = C_\alpha$ and $\alpha \in C$.

Theorem 5.1 *There exists a set notion of forcing $\mathscr{P} \in E(\omega_1)$ such that $\mathscr{P}$ is locally proper but not proper.*

Proof. Let $\{S_\alpha | \alpha < \omega_1\}$ be a $\diamond_E$ sequence for $E(\omega_1)$ as above. Define Σ_1 stationary sets

$$St_0 = \{\alpha | \alpha = S_\alpha\}, \quad \text{and}$$
$$St_1 = \{\alpha | \alpha - \{0\} = S_\alpha\}.$$

Then St_0 and its complement are stationary and definable over $L(\omega_1)$; one of them, say St_0, contains an unbounded subset of $\{j_{\delta+1} | \alpha < \omega_1\}$, the set of canonical codes for $\{\kappa_{r,\delta+1} | \delta < \omega_1\}$. The forcing relation $\mathscr{P}$ is designed to kill the stationarity of the complement of St_0. A forcing condition p is the characteristic function of a countable closed subset of St_0. A generic G is the limit of such p's, hence a closed subset of St_0.

Fix $p \in P$ and $b \in \omega_1$ to see $\mathscr{P}$ preserves E-closure. It suffices to find $q \leqslant p$ and $\theta < \kappa$ such that

$$q \Vdash \kappa_r^{\mathscr{G}, \underline{b}} \leqslant \underline{\theta}.$$

Choose $c \in St_0$ and $d < c$ so that $p, b \leqslant_E d$. c is the canonical code for some $\kappa_{r,\delta+1}$ that will serve as θ. There is an e_0 such that for all $G \subseteq \omega_1$:

(i) $\{e_0\}(G, b)\uparrow$;

(ii) $\kappa_r^{G,b}$ is the least γ such that a witness to the divergence of $\{e_0\}(G, b)$ is first-order definable over $L(\gamma, TC(\{G, b\}))$ (cf. [8, p. 271]).

$\mathscr{P}$ is not countably closed; nonetheless the plan is to build q as if $\mathscr{P}$ were. Thus q is the limit of a sequence q_n ($n < \omega$), where each q_n forces the existence of the nth leg of a Moschovakis witness w to the divergence of $\{e_0\}(\mathscr{G},\underline{b})$. Then q forces the existence of w. Some assembly is required to make (dom q) $\in St_0$; otherwise q is not a condition. Choose $(\lambda n|c_n) \in L(\kappa)$ so that $\bigcup_n c_n = c$. The q_n's can be chosen so that:

(i) $p_0 \geqslant q$ and $q_n \geqslant q_{n+1}$;

(ii) $c_n \in \text{dom } q_n$;

(iii) $q_n \leqslant_E \omega_1, c_m$ for some m.

Consequently dom $q = c \in St_0$. Observe that c is a limit point of St_0; otherwise $c \leqslant_E x, \omega_1$ for some $x < c$, since $St_0 \leqslant_E \omega_1$.

Lemma 5.2 (Marcus [3]) *Suppose $\mathscr{P} \in E(\omega_1)$ $(= L(\kappa))$ is proper, G is $\mathscr{P}$-generic and $S \in E(\omega_1)$ touches every closed unbounded $\Sigma_1^{L(\kappa)}$ subset of ω_1. Then S touches every closed unbounded subset of ω_1 in $L(\kappa, G)$.*

Proof. $\mathscr{P}$ is locally proper, so $L(\kappa, G)$ is E-closed by Theorem 3.1. The κ_r spectrum of $L(\kappa, G)$ is a subclass of the κ_r spectrum of $L(\kappa)$; also the canonical codes for the κ_r spectrum relative to G are a subclass of those for $L(\kappa)$. To be precise, consider $\kappa^G_{r,\delta}$. There is a γ such that $\kappa_{r,\gamma} = \kappa^G_{r,\delta}$. The canonical codes are:

$$j_\gamma = \mu\beta[\beta < \omega_1 \wedge \kappa_{r,\gamma} \leqslant_E \beta, \omega_1];$$
$$j^G_\delta = \mu\beta[\beta < \omega_1 \wedge \kappa^G_{r,\delta} \leqslant_E \beta, \omega_1, G].$$

Then $j_\gamma = j^G_\delta$. A related fact is $I_\gamma \subseteq I^G_\delta$.

Let D be a closed subset of ω_1 in $L(\kappa, G)$. Suppose $D \leqslant_E b, \omega_1, G$. By Lemma 4.2 relativized to G, if j is a canonical code for some member of the κ^G_r spectrum, then

$$(5.1) \qquad b < j \longrightarrow j \in D.$$

$\mathscr{P}$ is proper, so there exists C, a *closed*, rec-on-$L(\kappa)$ class of co-*re* substructures associated with $\mathscr{P}$. Recall that $j(\kappa_{r,\gamma})$ is the canonical code for $\kappa_{r,\gamma}$. For any p, let

$$C^p = \{j(\sup I) | (\exists I)[p \in I \wedge I \in C]\}.$$

C^p is a closed unbounded $\Sigma_1^{L(\kappa)}$ subset of ω_1, and so has an unbounded intersection with S. Then there is a $q \leqslant p$ and an $I \in C$ such that

$$b < j(\sup I) \in C^p \cap S, \quad \text{and}$$
$$q \quad \text{is subgeneric over} \quad I.$$

Since $\mathscr{P}$ is proper, q forces $\sup I$ to be an element of the $\kappa^{\mathscr{G}}_r$ spectrum. Since G is $\mathscr{P}$-generic, G belongs to such a q, and so $j(\sup I) \in D$ by (1).

6. Further results and questions

A trove of further information on Effective Forcing can be found in Marcus's Ph.D. Thesis [3]. She studies E-closed $L(\kappa)$'s of the form $E(\gamma)$, where γ is a regular cardinal of $E(\gamma)$. She makes explicit the definition of proper left implicit in the proof of Lemma 5.2. She also raises many questions.

An ambitious reader might wish to verify directly that ccc forcing is locally proper by exploiting the tree-of-possibilities approach of Section 3.

The example of Section 4 could stand some improvement. Surely there is a $\mathscr{P} \in L(\kappa)$ such that $\mathscr{P}$ preserves ω_1 but not E-closure. (Beware false proofs of this last conjecture.)

Slaman has asked for a forcing relation that is not *re* (cf. [8, 270) but is effective. Most likely there is one by the arguments of Sections 4 and 5.

Farewell to higher recursion theory (but not to recursion theory; there is no way to say goodbye to recursion theory).

References

[1] K. Devlin, Constructibility (Springer, Berlin, 1986).
[2] L. Harrington, Contributions to recursion theory in higher types, Ph.D. Thesis, MIT, Cambridge, MA, 1973.
[3] S. Marcus, Forcing in E-recursion, Ph.D. Thesis, MIT, Cambridge, MA, 1993.
[4] D. Normann, Set recursion, in: Generalized Recursion Theory II (North-Holland, Amsterdam, 1978) 303–320.
[5] G.E. Sacks, The k-section of a type n object, Amer. J. Math. 99 (1977) 901–917.
[6] G.E. Sacks, Post's problem, absoluteness and recursion in finite types, in: The Kleene Symposium (North-Holland, Amsterdam, 1980) 201–222.
[7] G.E. Sacks, On the limits of recursive enumerability, Ann. Pure and App. Logic 31 (1986) 87–120.
[8] G.E. Sacks, Higher Recursion Theory (Springer, Berlin, 1990).
[9] T. Slaman, Aspects of E-recursion, Ph.D. Thesis, Harvard Univ., Cambridge, MA, 1981.